Graphdiyne

Graphdiyne

Fundamentals and Applications in Renewable Energy and Electronics

Edited by

Yuliang Li

WILEY-VCH

Editor

Prof. Yuliang Li
Institute of Chemistry
Chinese Academy of Sciences
Zhongguancun North First Street 2
100190 Beijing
P.R. China

Cover Design: Wiley
Cover Image: © LAGUNA DESIGN/Getty Images

Library of Congress Card No.:
applied for

British Library Cataloguing-in-Publication Data
A catalogue record for this book is available from the British Library.

Bibliographic information published by the Deutsche Nationalbibliothek
The Deutsche Nationalbibliothek lists this publication in the Deutsche Nationalbibliografie; detailed bibliographic data are available on the Internet at <http://dnb.d-nb.de>.

Print ISBN: 978-3-527-34787-2
ePDF ISBN: 978-3-527-82849-4
ePub ISBN: 978-3-527-82848-7
oBook ISBN: 978-3-527-82847-0

Typesetting Straive, Chennai, India
Printing and Binding

Printed on acid-free paper

10 9 8 7 6 5 4 3 2 1

Contents

Preface

In 2010, a new carbon allotrope was born, which brought a new member to the family of carbon materials. Its unique chemical and electronic structures provide unlimited space for scientific innovation of scientists. It shows infinite charm and great potential to promote the development of fundamental and applied science. Chinese scientists named the material "Shimoque." New materials have become one of the keys to break through the bottleneck of science and technology. Inspired by this, many researchers are committed to discover or develop new materials in nontraditional architectures. The hybridization of carbon allotrope with sp-hybridization is a very interesting research topic, because the acetylenic bond of sp can cause significant changes in the properties of carbon. The sp and sp^2 hybrid structures, and its intrinsic properties and performances illustrate that graphdiyne exhibits transformative properties and performances in the fields of catalysis, energy, optoelectronic and intelligent information, and so on.

Graphdiyne (GDY) shows the characteristics of sp- and sp^2-hybridized carbon atoms, which are fundamentally different from the sp^3 and sp^2 hybridization of traditional carbon materials. It is rich in chemical bonds, highly conjugated, superlarge π structures, and has infinitely distributed cavities on the surface. GDY shows also high chemical activity and the functions of chemical reaction, chemical and physical doping, and chemical modification. Due to these exclusive structural features, GDY is expected to be a perfect and peculiar new carbon allotrope. As an important material, if you want to expand its application space, the material must be able to do "chemistry." GDY represents a great advantage in chemical modification. Several methods have been developed to obtain GDY-based materials such as invoke strains, B and N-doping, halogen doping, as-prepared nanostructures, controllable growth of aggregate structure with different dimensions, and hydrogenation, bromination, and fluorination have been developed for regulating band gap of GDY. As a new kind of carbon material, GDY's chemical and physical properties are of great concern to scientists. Therefore, many studies have focused on its basic properties for understanding the physical and chemical properties of GDY. These fundamental studies provide very important informations for GDY's further basic and applied research. Researchers can truly understand the structural and natural advantages of GDY and its development trend. Early fundamental and applied research gave researchers great confidence, that is, GDY demonstrated excellent and potential performance in catalysts, lithium ion batteries, sodium ion batteries, zinc water batteries, and fuel cells photoelectric conversion, optical devices and electrochemical intelligent and information devices, gas separation, and water purification. The infinite π bond on the GDY surface leads to high surface activity, which can interact with many units of organisms and can be used efficiently in life science–related research. So many biological and life scientists are also actively engaged in the field of research, aiming at toxicity, drug delivery, and therapy.

On the 10th anniversary of GDY discovery, at the invitation of Wiley publishers, we are very happy to write this book, which records the course and achievements of GDY research since 2010, from preliminary research to maturity, from fundamental research to application. The results demonstrate that GDY has strong potential for fundamental and applied researches. Preparation determines the future! In this book, we try to describe GDY-related theoretical calculations and simulations, chemical and physical models, synthetic methodologies, controlled growth of aggregated state structures, structural characterization, fundamental physical and chemical properties, and GDY applications in many fields. After nearly 10 years of persistent work, scientific researches in different regions have shown that GDY has strong advantages in the fields of energy, catalysis and photoelectricity, electrochemical intelligent devices, and so on. We believe that in the next 10 years, GDY-based materials should move rapidly toward the route of interdiscipline with different disciplines, and it is possible to show a powerful role in multidisciplinary crossing, and to become a model of cross-fusion in important fields such as chemistry, physics, information science, material science, and environmental science. GDY demonstrated potential and exciting results, which prompted us to better complete this book and convey graphdiyne's theoretical and practical progress to students, teachers, and practitioners who wish to participate in the exciting development of the subject.

December 2020

Yuliang Li
Beijing

1

Introduction

Yongjun Li and Yuliang Li

Institute of Chemistry, Chinese Academy of Sciences, Zhongguancun North First Street 2, 100190, Beijing, PR China

1.1 The Development of Carbon Materials

The pursuit of new materials with nontraditional architecture is one of the hot spots in current research [1–3]. Carbon-based nanomaterials (such as fullerenes, carbon nanotubes, and graphene) have attracted much attention due to their special structures and chemical and physical properties [4–7]. Carbon materials have experienced a long history of development. The contact of carbon materials with humans can be traced back to the earliest appearance of humans on the earth.

The first known existence of carbon was charcoal and soot. Diamond is a famous allotype of carbon, which was discovered by humans as early as 4000 BCE. Graphite is the most widely used allotrope of carbon and was found in the sixteenth century. Although carbon is one of the oldest elements, it is surprising that it constantly shows great vitality for the discovery of new allotropes (Figure 1.1), such as fullerene (1985), carbon nanotubes (1991), graphene (2004), and graphdiyne (GDY) (2010). The application of carbon materials can even be considered to promote the progress of human society and the development of other materials. In fact, the unique valence bond hybrid forms of carbon molecules, namely sp, sp^2, and sp^3, ensure that carbon allotropes can be constructed in various possible forms and exhibit different intrinsic properties. Diamond is composed of sp^3-hybridized carbon, while graphite, fullerene, carbon nanotubes, and graphene are composed of sp^2-hybridized carbon. sp^2 hybrid carbon can enhance the conjugation of materials and exhibit good electrical conductivity, while sp^3 hybrid carbon has three-dimensional (3D) spatial configuration in carbon materials, which can further improve the rigidity of related materials. The sp hybrid carbon has a linear structure, which can improve the porosity and provide enough active or storage sites for other atoms. At present, the reasonable design of carbon materials and the full use of the advantages of the three hybrid carbon materials are of great significance in many research fields.

An interesting family of carbon allotropes is represented by the so-called graphynes (GYs) and GDYs. In general, these allotropes are flat one-atom-thin carbon

Graphdiyne: Fundamentals and Applications in Renewable Energy and Electronics,
First Edition. Edited by Yuliang Li.

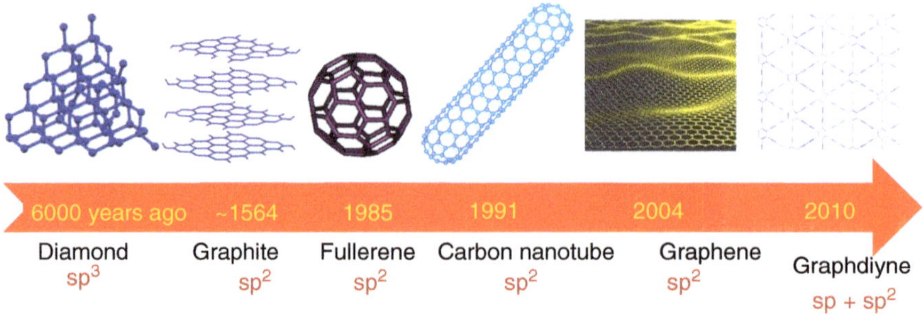

Figure 1.1 The development of carbon materials.

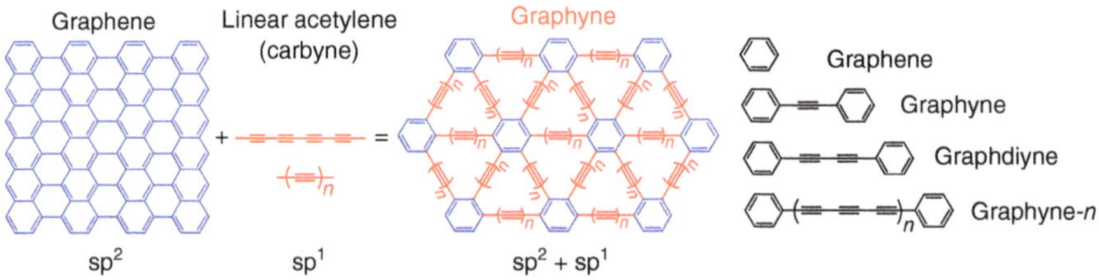

Figure 1.2 Schematic structures of graphene, carbyne, and graphynes, which comprise exclusively sp-atoms, sp^2-atoms, and both types: ($sp + sp^2$) atoms, respectively.

networks (such as graphene), which can be constructed by replacing some =C=C= bonds in graphene by uniformly distributed acetylenic bonds —C≡C— (graphynes) or diacetylenic bonds —C≡C—C≡C— (GDYs). In both cases, the resulting network consists of two nonequivalent types of carbon atoms: threefold coordinated sp^2-hybridized atom and twofold coordinated sp-hybridized atom. In this context, these flat carbon networks can be regarded as the "intermediate" ($sp^2 + sp$) systems between two famous carbon allotropes: graphene (containing only sp^2-like atoms) and carbyne (containing only sp-like atoms) [8], see Figure 1.2. We can simply classify these materials according to the number of "—C≡C—" bonds connecting two adjacent sp^2-hybridized carbon atoms. As shown in Figure 1.2, they are called graphyne, GDY, and graphyne-*n* [9].

The history of systematic study of ($sp + sp^2$) allotrope family began in 1987, when Baughman et al. [10] first proposed the structural model of graphynes and discussed some macrocyclic subunits suitable for creating these networks. Ten years later (1997), the structure of GDY was predicted and many small diacetylene molecules had been synthesized, which had become the "hot" spot of synthetic chemistry for a period of time. These studies started in the mid-1990s and continued into the new millennium. The chemists began some computational simulation and theoretical studies during this period, and the related materials with different sizes and dimensions [11–16], as well as some of their B–N and B–C–N analogues [15, 17–19] were also experimented by theoretical simulation. On the other hand, the experimental efforts in the synthesis of subunits of these systems were closely related to organic chemistry, that is, new synthetic routes in annulene chemistry

[20]. However, research in these areas has not advanced much because of the serious lack of innovation in synthetic methods, leading to the study of the synthesized GDY in the synthesis and properties of some small-molecule diacetylene. It was in 2010 that the synthesized bottleneck of GDY was broken, and this was a great success. A new allotrope of carbon was born, which opened up a new field for research in carbon materials.

1.2 Models and Nomenclature

In 1968, Balaban et al. first proposed a rich and diverse planar carbon network (consisting of only sp^2-bonded atoms with a threefold coordination) [21]. The search line was actively extended, and then a large number of related two-dimensional periodic carbon networks were constructed from non-C6 carbon polygons. For example, so-called pentaheptites [22, 23] (formed by periodically distributed pentagons C5 and heptagons C7) or haeckelites [24] (including pentagons C5, hexagons C6, and heptagons C7, see Figure 1.3), as well as some other related types of carbon networks, sometimes referred to as graphene allotropes [9, 27–35], were proposed and successfully investigated. Here, the so-called two-dimensional supracrystals [25, 36] can also be mentioned, Figure 1.3. These hypothetical low-stable polycyclic networks are composed of strained cycles such as C3, C4, and C12; therefore, their synthesis seems very suspicious. The recently studied 2D "square carbon" [25] also belongs to this category.

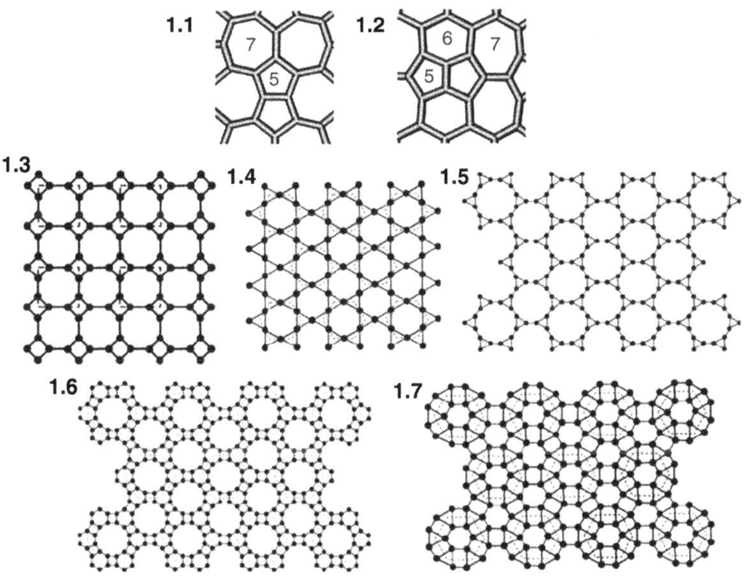

Figure 1.3 Structural motifs of some 2D carbon networks: **1.1**: pentaheptites [22, 23], **1.2**: haeckelites [24], and **1.3–1.7**: some hypothetical so-called 2D carbon supracrystals – polycyclic networks (based on Kepler's nets) composed of strained cycles such as C3, C4, and C12 [25]. Source: Ivanovskii [26]. © 2013, Elsevier.

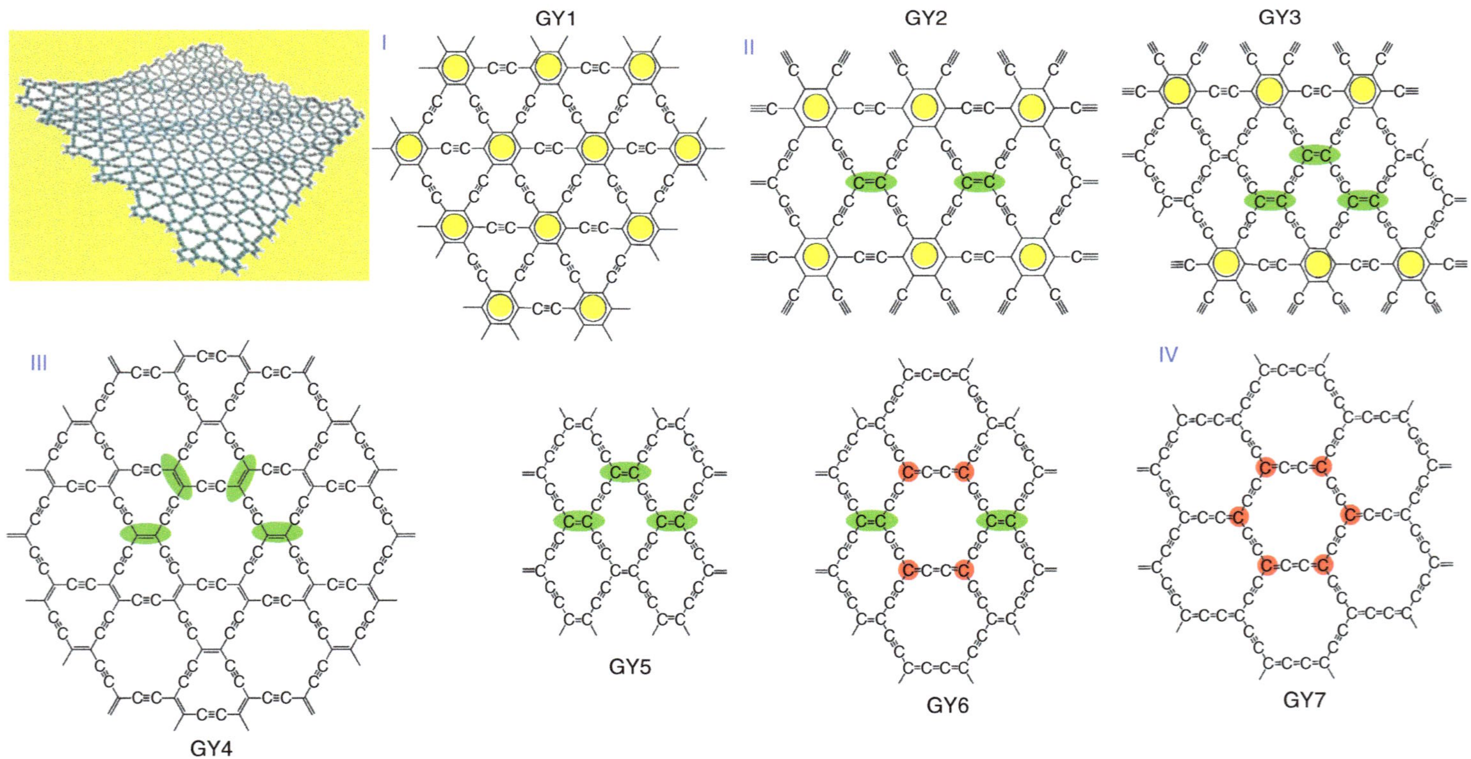

Figure 1.4 Ideal atomic models [10] of possible graphynes (GY1–GY7). The threefold coordinated sp^2 atoms (forming hexagons, pairs, or as isolated atoms) are marked. On the left: thermal ripples of GY1 network according to molecular dynamics (MD) simulation at $T = 300\,K$ [37]. Source: Ivanovskii [26]. © 2013, Elsevier.

Graphynes are a series of stable two-dimensional crystalline carbon allotropes composed of sp- and sp^2-hybridized carbon atoms. Their structural models were first proposed by R. H. Baughman et al. [10]. They have a two-dimensional structure similar to graphite and contain acetylenic linkages (sp components), referred to as graphyne. Accordingly, sp- and sp^2-hybridized carbon atoms can be connected to each other according to certain hybrid rules, producing a variety of 2D structures [26]. Some of such GYs are depicted in Figure 1.4. These (and related) networks fall into four categories: I–IV, see Figure 1.4. Therefore, the structure of group I (GY1) includes hexagons C6, which are connected to each other by —C≡C— linkages. The two networks (GY2, GY3) of the second family consist of hexagonal C6 and a pair of sp^2 atoms (C=C bonds), which are interconnected by —C≡C— linkages. The three networks of group III (GY4–GY6) have no hexagonal C6 and only contain paired sp^2 atoms (C=C bond). They are connected by —C≡C—bonds (GY4, GY5), or by paired sp^2 atoms and isolated sp^2 atoms (GY6). Finally, the network of group IV (GY7) consists of isolated sp^2 atoms, which are connected to each other by —C≡C— linkages. This network (so-called supergraphene) can be seen as a graphene-like structure, in which all C=C bonds are replaced by acetylenic linkages —C≡C—. Therefore, GY7 has the same hexagonal *p6m* symmetry as graphene.

Today, there is still no standard classification of such graphyne systems. In the first work, Baughman et al. [10] designated the GY networks to be considered in the simplified nomenclature, which defines the number of carbon atoms in different rings forming a given network. According to this method, graphynes can be named as a, b, and g-graphyne, where a and b represent the number of carbon atoms in the smallest ring of the graphynes (a ring) and number of carbon atoms in the adjacent smallest ring of the graphynes (b ring), respectively. Among them, rings a and b are connected by $C(sp^2)C(sp)C(sp)C(sp^2)$. The index g is the number

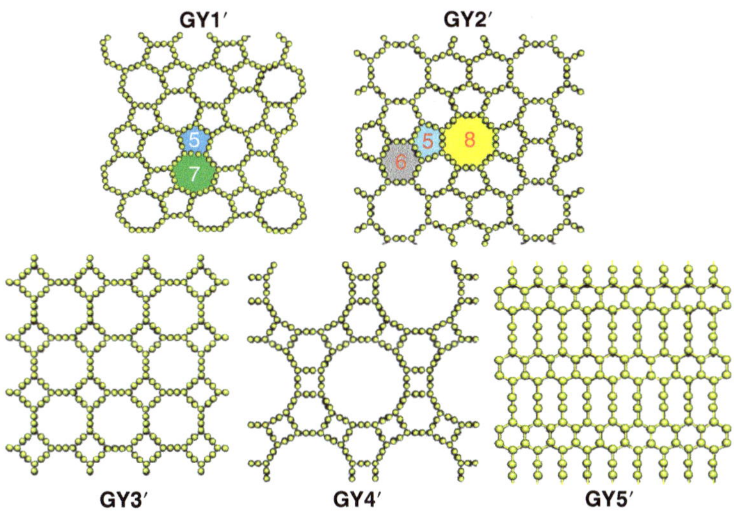

Figure 1.5 Some possible atomic motifs of graphyne-like structures, which are termed in the text as GY1′–GY5′. Source: Ivanovskii [26]. © 2013, Elsevier.

of carbon atoms in the third ring of graphynes, which is connected to ring b by $C(sp^2)C(sp)C(sp)C(sp^2)$. For example, GY2 network is called 6,6,12-graphyne, GY4 is named 12,12,12-graphyne, and supergraphene (GY7) is called 18,18,18-graphyne. In addition, for convenience, several kinds of graphynes are commonly named after the Greek alphabet [36], which can be called as the customary nomenclature: α-graphyne (GY7) [38], β-graphyne (GY4) [39], and γ-graphyne (GY1) [12].

Coming back to possible types of graphynes, the structures of GY1′ and GY2′ (Figure 1.5) can be easily constructed from pentaheptite or haeckelite networks by simple replacement of all C=C bonds by acetylenic linkages —C≡C—; the structures of GY3 and GY4′ are graphyne-like analogues of some 2D carbon supracrystals depicted in Figure 1.3. Besides, various graphene/graphyne "hybrids" can be supposed. A simple example is GY5′, which includes "stripes" of hexagons C6 bonded

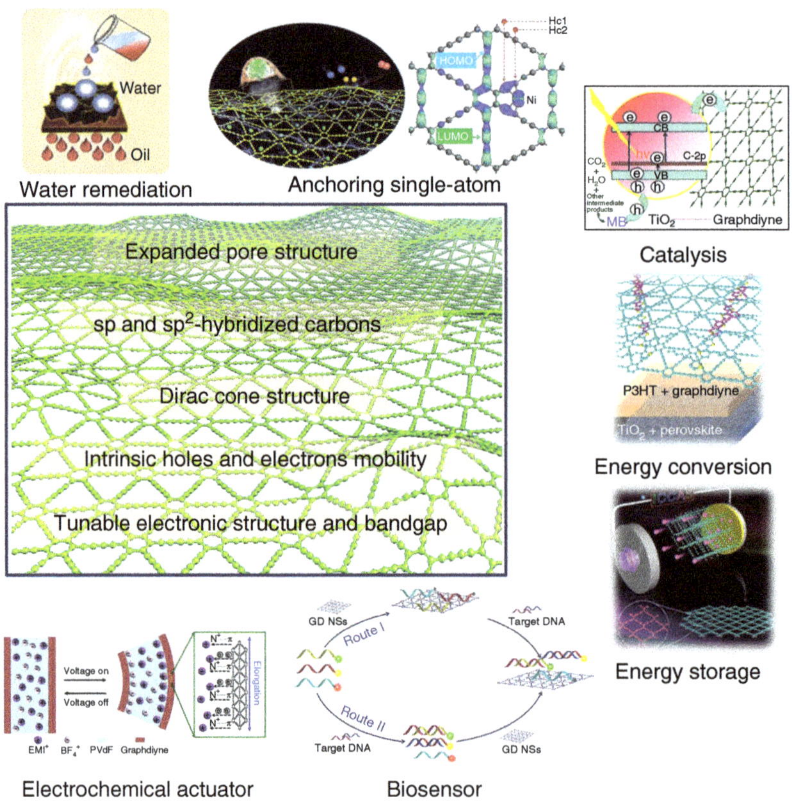

Figure 1.6 Graphdiyne with special properties and the related potential applications. EMI; 1-ethyl-3methylimidazolium, PVdF; poly(vinylidene fluoride), BF_4; tetrafluoroborate, GD NS; Graphdiyne nanosheets, P3HT; poly(3-hexylthiophene-2,5-diyl), HOMO; highest occupied molecular orbital, LUMO; lowest unoccupied molecular orbital. Source: (a) Gao et al. [40]. © 2016, John Wiey & Sons, (b) Xue et al. [41]. ©2018, Springer Nature/CC License 4.0, (c) Wang et al. [42]. ©2012, John Wiley & Sons, (d) Xiao et al. [43]. ©2015, John Wiley & Sons, (e) Jia et al. [44]. ©2017, Elsevier, (f) Parvin et al. [45]. John Wiley & Sons, (g) Lu et al. [46]. ©2018, Springer Nature/CC License 4.0.

by acetylenic linkages —C≡C—, etc. On the other hand, in all of the described graphynes, the sp^2 atoms are bonded by "single" —C≡C— linkages. Therefore, one more way of construction of graphyne-like networks is to increase the length of linear carbine-like atomic chains between sp^2 atoms, i.e. to replace (—C≡C—) by (—C≡C—C≡C—) or (—C≡C—C≡C—C≡C—) chains, etc. which connect either hexagons C6, or pairs of sp^2 atoms, or individual sp^2 atoms.

1.3 Brief Introduction of Graphdiyne

The chemical study of carbon-rich molecules has and will continue to produce significant structures in size, topology, and spatial direction. Nevertheless, the achievements of early chemists were indeed remarkable in the current synthetic and analytical techniques that modern chemists take for granted. Advanced synthesis methods for alkyne chemistry have been developed through Sonogashira cross-coupling reactions or oxidative acetylenic coupling reactions catalyzed by Cu-[47, 48] or Pd/Cu [49]. The on-surface chemistry also provides a new way for the development of GDY.

GDY has butadiyne linkage between two adjacent aromatic rings. The development of GDY prepared by in situ Glaser coupling reaction of hexaethynylbenzene (HEB) monomers on a copper (Cu) substrate by Professor Yuliang Li's group [50] in 2010, is widely recognized as a great breakthrough regarding the structure of carbon materials. One of the most important features of the chemical structure of GDY is the presence of quantitative sp carbon, which gives it some characteristics that other carbon materials do not have [51, 52].

Theoretical analysis shows that GDY has a direct natural bandgap (0.46 eV) [53] and a Dirac cone structure, which can be attributed to the inhomogeneous π-bonding between the sp and sp^2-hybridized carbon (Figure 1.6). GDY has excellent electrical properties, such as high carrier mobility and small carrier effective mass, which make it promising for nanoelectronics [54]. Both the intrinsic holes and electrons mobility of GDY at room temperature can reach up to 10^5 cm^2 V^{-1} s^{-1}[55]. As the number of GDY layers increases, the band gap of GDY decreases and the direct band gap remain unchanged. The mechanical properties of GYs are considered as a function of the number and arrangement of acetylenic linkages [9, 56]. The expanded pores surrounded by the butadiyne linkers and benzene rings in the structure provide additional space for the storage and diffusion of metal atoms such as lithium and sodium. Moreover, the uniformly distributed in-plane pores of GDY can also promote the vertical transfer of ions [57].

Another unique feature of GDY -based materials is that they can be prepared by chemical methods, which is conducive to adjusting and optimizing their morphology and some fundamental chemical properties, including the conductivity, size and distribution of the pores, and affinity to certain metal atoms. In addition, the position and number of heteroatoms introduced in GDY can be well controlled by this preparation method [58]. GDY has been synthesized under different experimental conditions in the forms of films, nanowires, nanotube arrays, nanowalls, 3D foams, nanosheets and ordered stripe arrays, etc.

The above structural features and performance advantages make it possible and convenient to adjust and optimize the electrochemical properties of GDY, leading to the wide application of GDY in efficient separation, energy storage, photoelectric and energy conversion (Figure 1.6). The abundant distribution of alkyne bonds makes the charge distribution on the GDY surface extremely uneven, which endows it with more active sites, leading to higher intrinsic activity, which can effectively promote the catalytic reaction process. Therefore, GDY should be a valuable complement to popular sp^2-hybridized carbon materials for constructing new concepts and highly active metal-free catalysts and understanding their catalytic mechanisms.

In Chapters 2–4, we will introduce the fundamental characteristics of GDY in terms of experiment and theory, namely electrical, mechanical, and optical properties [51, 52, 58, 59]. More importantly, we will focus on the application of GDY in catalysis [41, 42] (Chapter 5), energy conversion and storage [43, 44] (Chapters 6, 7), electronic devices [46] (Chapter 8), detectors, biomedicine and treatment [45], and water purification [40] (Chapter 9).

References

1 Coleman, J.N., Lotya, M., O'Neill, A. et al. (2011). Two-dimensional nanosheets produced by liquid exfoliation of layered materials. *Science* 331: 568–571.

2 Nicolosi, V., Chhowalla, M., Kanatzidis, M.G. et al. (2013). Liquid exfoliation of layered materials. *Science* 340: 1226419-1–1226419-18.

3 Wang, Q.H., Kalantar-Zadeh, K., Kis, A. et al. (2012). Electronics and optoelectronics of two-dimensional transition metal dichalcogenides. *Nature Nanotechnology* 7: 699–712.

4 Diederich, F. and Thilgen, C. (1996). Covalent fullerene chemistry. *Science* 271: 317–323.

5 Franklin, A.D. (2013). Electronics: the road to carbon nanotube transistors. *Nature* 498: 443–444.

6 Geim, A.K. and Novoselov, K.S. (2007). The rise of graphene. *Nature Materials* 6: 183–191.

7 Inagaki, M. and Kang, F. (2014). Graphene derivatives: graphane, fluorographene, graphene oxide, graphyne and graphdiyne. *Journal of Materials Chemistry A* 2: 13193–13206.

8 Chalifoux, W.A. and Tykwinski, R.R. (2009). Synthesis of extended polyynes: toward carbyne. *Comptes Rendus Chimie* 12: 341–358.

9 Cranford, S.W., Brommer, D.B., and Buehler, M.J. (2012). Extended graphynes: simple scaling laws for stiffness, strength and fracture. *Nanoscale* 4: 7797–7809.

10 Baughman, R.H., Eckhardt, H., and Kertesz, M. (1987). Structure-property predictions for new planar forms of carbon - layered phases containing sp^2 and sp atoms. *Journal of Chemical Physics* 87: 6687–6699.
11 Narita, N., Nagai, S., Suzuki, S. et al. (2000). Electronic structure of three-dimensional graphyne. *Physical Review B* 62: 11146–11151.
12 Coluci, V.R., Braga, S.F., Legoas, S.B. et al. (2003). Families of carbon nanotubes: graphyne-based nanotubes. *Physical Review B* 68: 035430.
13 Coluci, V.R., Galvao, D.S., and Baughman, R.H. (2004). Theoretical investigation of electromechanical effects for graphyne carbon nanotubes. *Journal of Chemical Physics* 121: 3228–3237.
14 Coluci, V.R., Braga, S.F., Legoas, S.B. et al. (2004). New families of carbon nanotubes based on graphyne motifs. *Nanotechnology* 15: S142–S149, Pii s0957-4484(04)69902-8.
15 Enyashin, A.N., Makurin, Y.N., and Ivanovskii, A.L. (2004). Quantum chemical study of the electronic structure of new nanotubular systems: alpha-graphyne-like carbon, boron-nitrogen and boron-carbon-nitrogen nanotubes. *Carbon* 42: 2081–2089.
16 Baughman, R.H., Galvao, D.S., Cui, C.X. et al. (1993). Fullereneynes - a new family of porous fullerenes. *Chemical Physics Letters* 204: 8–14.
17 Enyashin, A.N., Ivanovskaya, V.V., Makurin, Y.N. et al. (2004). Electronic structure of new graphyne-like boron nitride nanotubes. *Doklady Physical Chemistry* 395: 62–66.
18 Enyashin, A.N., Sofronov, A.A., Makurin, Y.N. et al. (2004). Structural and electronic properties of new alpha-graphyne-based carbon fullerenes. *Journal of Molecular Structure: THEOCHEM* 684: 29–33.
19 Ivanovskaya, V.V., Enyashin, A.N., and Ivanovskii, A.L. (2006). Nonempirical calculations of the electronic properties of new boron nitride graphyne-like nanotubes. *Russian Journal of Physical Chemistry* 80: 372–379.
20 Spitler, E.L., Johnson, C.A. II,, and Haley, M.M. (2006). Renaissance of annulene chemistry. *Chemical Reviews* 106: 5344–5386.
21 Balaban, A.T., Rentia, C.C., and Ciupitu, E. (1968). Chemical graphs. 6. Estimation of relative stability of several planar and tridimensional lattices for elementary carbon. *Revue Roumaine de Chimie* 13: 17.
22 Crespi, V.H., Benedict, L.X., Cohen, M.L. et al. (1996). Prediction of a pure-carbon planar covalent metal. *Physical Review B: Condensed Matter and Materials Physics* 53: 3.
23 Deza, M., Fowler, P.W., Shtogrin, M. et al. (2000). Pentaheptite modifications of the graphite sheet. *Journal of Chemical Information and Computer Sciences* 40: 1325–1332.
24 Terrones, H., Terrones, M., Hernandez, E. et al. (2000). New metallic allotropes of planar and tubular carbon. *Physical Review Letters* 84: 1716–1719.
25 Pujari, B.S., Tokarev, A., and Saraf, D.A. (2012). Theoretical investigation of planar square carbon allotrope and its hydrogenation. *Journal of Physics: Condensed Matter: An Institute of Physics Journal* 24: 175501.

26 Ivanovskii, A.L. (2013). Graphynes and graphdyines. *Progress in Solid State Chemistry* 41: 1–19.

27 Terrones, H., Terrones, M., and Morán-López, J.L. (2001). Curved nanomaterials. *Current Science* 81: 19.

28 Lambin, P. and Biró, L.P. (2003). Structural properties of Haeckelite nanotubes. *New Journal of Physics* 5: 141.

29 Biró, L.P., Márk, G.I., Horváth, Z.E. et al. (2004). Carbon nanoarchitectures containing non-hexagonal rings: "necklaces of pearls". *Carbon* 42: 2561–2566.

30 Mpourmpakis, G., Froudakis, G.E., and Tylianakis, E. (2006). Haeckelites: a promising anode material for lithium batteries application. An ab initio and molecular dynamics theoretical study. *Applied Physics Letters* 89: 233125.

31 Lusk, M.T. and Carr, L.D. (2009). Creation of graphene allotropes using patterned defects. *Carbon* 47: 2226–2232.

32 Enyashin, A.N. and Ivanovskii, A.L. (2011). Graphene allotropes. *Physica Status Solidi B* 248: 1879–1883.

33 Xin, S., Guo, Y.G., and Wan, L.J. (2012). Nanocarbon networks for advanced rechargeable lithium batteries. *Accounts of Chemical Research* 45: 1759–1769.

34 Ivanovskii, A.L. (2005). New layered carbon allotropes and related nanostructures: computer simulation of their atomic structure, chemical bonding, and electronic properties. *Russian Journal of Inorganic Chemistry* 50: 1408–1422.

35 Pokropivnyi, V.V. and Ivanovskii, A.L. (2008). New nanoforms of carbon and boron nitride. *Uspekhi Khimii* 77: 39.

36 Srinivasu, K. and Ghosh, S.K. (2012). Graphyne and graphdiyne: promising materials for nanoelectronics and energy storage applications. *Journal of Physical Chemistry C* 116: 5951–5956.

37 Cranford, S.W. and Buehler, M.J. (2011). Mechanical properties of graphyne. *Carbon* 49: 4111–4121.

38 Malko, D., Neiss, C., Vines, F. et al. (2012). Competition for graphene: graphynes with direction-dependent Dirac cones. *Physical Review Letters* 108: 086804.

39 Kim, B.G. and Choi, H.J. (2012). Graphyne: hexagonal network of carbon with versatile Dirac cones. *Physical Review B* 86: 115435.

40 Gao, X., Zhou, J., Du, R. et al. (2016). Robust superhydrophobic foam: a graphdiyne-based hierarchical architecture for oil/water separation. *Advanced Materials* 28: 168–173.

41 Xue, Y., Huang, B., Yi, Y. et al. (2018). Anchoring zero valence single atoms of nickel and iron on graphdiyne for hydrogen evolution. *Nature Communications* 9: 1460.

42 Wang, S., Yi, L., Halpert, J.E. et al. (2012). A novel and highly efficient photocatalyst based on P25–graphdiyne nanocomposite. *Small* 8: 265–271.

43 Xiao, J.Y., Shi, J.J., Liu, H.B. et al. (2015). Efficient $CH_3NH_3PbI_3$ perovskite solar cells based on graphdiyne (GD)-modified P3HT hole-transporting material. *Advanced Energy Materials* 5: 1401943.

44 Jia, Z., Zuo, Z., Yi, Y. et al. (2017). Low temperature, atmospheric pressure for synthesis of a new carbon Ene-yne and application in Li storage. *Nano Energy* 33: 343–349.

45 Parvin, N., Jin, Q., Wei, Y. et al. (2017). Few-layer graphdiyne nanosheets applied for multiplexed real-time dna detection. *Advanced Materials* 29: 1606755.

46 Lu, C., Yang, Y., Wang, J. et al. (2018). High-performance graphdiyne-based electrochemical actuators. *Nature Communications* 9: 752.

47 Haley, M.M. (2008). Synthesis and properties of annulenic subunits of graphyne and graphdiyne nanoarchitectures. *Pure and Applied Chemistry* 80: 519–532.

48 Hay, A.S. (1962). Oxidative coupling of acetylenes. 2. *Journal of Organic Chemistry* 27: 3320–3321.

49 Fairlamb, I.J.S., Bauerlein, P.S., Marrison, L.R. et al. (2003). Pd-catalysed cross coupling of terminal alkynes to diynes in the absence of a stoichiometric additive. *Chemical Communications*: 632–633.

50 Li, G.X., Li, Y.L., Liu, H.B. et al. (2010). Architecture of graphdiyne nanoscale films. *Chemical Communications* 46: 3256–3258.

51 Gao, X., Liu, H., Wang, D. et al. (2019). Graphdiyne: synthesis, properties, and applications. *Chemical Society Reviews* 48: 908–936.

52 Li, Y., Xu, L., Liu, H. et al. (2014). Graphdiyne and graphyne: from theoretical predictions to practical construction. *Chemical Society Reviews* 43: 2572–2586.

53 Hybertsen, M.S. and Louie, S.G. (1986). Electron correlation in semiconductors and insulators - band-gaps and quasi-particle energies. *Physical Review B* 34: 5390–5413.

54 Narita, N., Nagai, S., Suzuki, S. et al. (1998). Optimized geometries and electronic structures of graphyne and its family. *Physical Review B* 58: 11009–11014.

55 Chen, J.M., Xi, J.Y., Wang, D. et al. (2013). Carrier mobility in graphyne should be even larger than that in graphene: a theoretical prediction. *Journal of Physical Chemistry Letters* 4: 1443–1448.

56 Zhang, Y.Y., Pei, Q.X., and Wang, C.M. (2012). Mechanical properties of graphynes under tension: a molecular dynamics study. *Applied Physics Letters* 101: 081909.

57 Du, Y., Zhou, W., Gao, J. et al. (2020). Fundament and application of graphdiyne in electrochemical energy. *Accounts of Chemical Research* 53: 459–469.

58 Huang, C., Li, Y., Wang, N. et al. (2018). Progress in research into 2D graphdiyne-based materials. *Chemical Reviews* 118: 7744–7803.

59 Jia, Z.Y., Li, Y.J., Zuo, Z.C. et al. (2017). Synthesis and properties of 2D carbon-graphdiyne. *Accounts of Chemical Research* 50: 2470–2478.

2

Basic Structure and Band Gap Engineering: Theoretical Study of GDYs

Feng He

Institute of Chemistry, Chinese Academy of Sciences, Zhongguancun North First Street 2, Beijing 100190, PR China

2.1 Structures

2.1.1 Theoretical Prediction and Classification

In 1987, Baughman et al. [1] predicted a new carbon material, consisting of sp^2- and sp-hybrid carbon atoms, which was called graphyne (GY). It has 2D planar structure and symmetry similar to layered graphite, which can be constructed by replacing one-third of —C=C— bonds in graphite by —C≡C— linkages. Compared to other carbon phases containing sp-atoms as the major structural elements, graphyne has much higher density and much lower formation energy. Although the formation energy of graphyne is higher than that of graphite, it is predicted to have high kinetic stability in the process of graphitization. Combining the retention of stabilized aromatic rings in the network backbone, graphyne shows more attractive thermal stabilities and synthesis possibilities than other carbon phases. Using the semiempirical self-consistent-field method, the lattice parameters of energy-minimized graphyne unit cell (containing 24 carbon atoms) were calculated to be $a = b = 6.86$ Å, $c = 6.72$ Å, $\gamma = 120°$.

The original structure of graphdiyne (GDY) was proposed as a large molecular one. This is unlike our concept of material design today where we aim at finding an advanced material with a two-dimensional (2D), layered, superlarge cavity structure. As a result, we have named it "shimoque" in Chinese since the date of discovery. In fact, it is quite different from the "graphdiyne" in English, because in our proposed material, we pay more attention to its layered, two-dimensional atomic arrangement and its electronic structure with uneven charge distribution, band structure, and so on. The difference is the number (n) of acetylenic linkages (—C≡C—) between two nearest-neighboring aromatic rings, for example, $n = 1$ in graphyne and $n = 2$ in GDY. GDY is the most stable carbon allotrope containing diacetylenic linkages and has large pores (about 2.5 Å) in the planar sheets, which can accommodate various large metal ions and favor the through-sheet transport of small ions. In addition,

Graphdiyne: Fundamentals and Applications in Renewable Energy and Electronics,
First Edition. Edited by Yuliang Li.

GDY could provide the intrasheet intercalation, which is not available for graphite with this type of dopant storage.

Narita et al. [2] systematically studied the atomic geometries of graphyne and GDY using a full-potential linear combination of atomic orbitals (LCAOs) method in the local density approximation. These structures consist of hexagons connected by linear carbon chains with different lengths. The binding energies are 7.95 eV atom^{-1} for graphyne and 7.78 eV atom^{-1} for GDY, respectively, and the corresponding lattice parameters are 6.86 and 9.44 Å, respectively. As the lengths of acetylenic chains in these graphyne allotropes expand, the binding energies are decreased, while the lattice parameters and lattice space are uniformly increased. For example, when adding each acetylenic connection unit, the lattice space is regularly increased by about 0.266 nm. More importantly, the extension of acetylenic chain does not lead to the obvious structural change. In addition, it is estimated theoretically that the formation energies of graphyne (i.e. 12.4 kcal mol^{-1} carbon) and GDY (i.e. 18.3 kcal mol^{-1} carbon) are much lower than those of any other carbon allotropes containing acetylenic linkages, thereby having the thermodynamic stability.

Using the density functional theory tight-binding (DFT-TB) calculations, Buehler and coworkers [3] systematically studied the relative stabilities and structural properties of each 2D planar graphynes network. Graphene and carbyne (the linear acetylenic carbon) consist of pure sp^2- and pure sp-hybridization, respectively, while the graphyne structures are a mixed configuration of sp^2- and sp-hybridization (Figure 2.1a). The differential energy per carbon atom (δE) can be used to assess the relative stability of each graphyne, which is defined as the difference in the total energies between the graphyne allotropes ($E_{\text{n-yne}}$) and pristine graphene (E_{graphene}):

$$\delta E = E_{n\text{–yne}} - E_{\text{graphene}} \tag{2.1}$$

The energy of graphynes can be predicted theoretically through the number of acetylenic groups, n (Figure 2.1b) or the hybridization, h (Figure 2.1c). The geometries of graphene ($n = 0$) and a few graph-n-ynes (namely $n = 1, 2, 3$, and 4) are shown in Figure 2.1d–h. With the increase of the number of acetylenic connection units [—C≡C—] in the network, the graphyne structures tend to be less stable since it approaches pure sp^1-hybridization (e.g. carbyne) with an ultimate $\delta E_{\text{carbyne}}$ of 1.17 eV atom^{-1}. However, it is noted that the graphynes have much lower ranges of energy differences (<1.17 eV atom^{-1}) than other possible carbon allotropes containing acetylenic linkages [2, 4], thereby being relatively more stable.

The graphyne family can be divided into α-graphyne, β-graphyne, γ-graphyne, 6,6,12-graphyne, etc. [5]. All these graphynes have two-dimensional layered planar structures, which are the combination of aromatic rings, large hexagons, or truncated triangular pores. Due to the multiple carbon bonds such as single bond, aromatic bond, and triple bond, the structural variability of graphynes is much stronger than that of graphene, thereby easily forming the curved nanowires or nanotubes with high stability [6]. The cohesion energies of various sp–sp^2 hybridized graphyne allotropes (e.g. α-, β-, and γ-graphyne) were studied using quantum Monte Carlo (QMC) calculations with full description of electron–electron correlation [7]. It is found that the cohesive energies of different types of graphynes

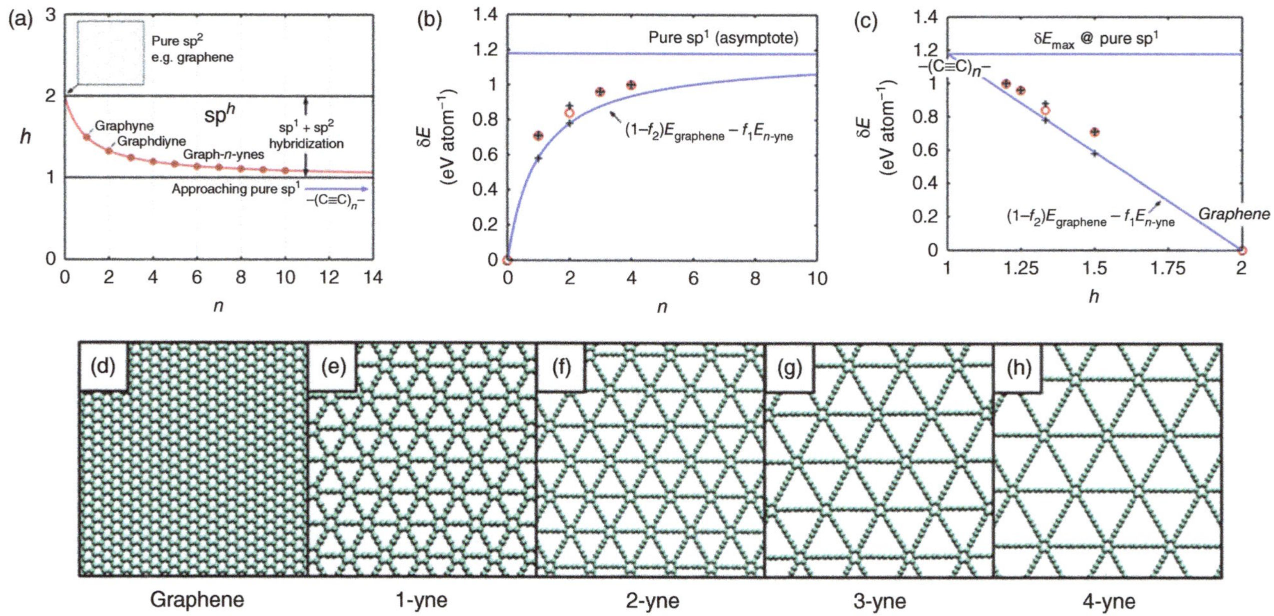

Figure 2.1 (a) A mixed-carbon configuration, which is classified as sp^h ($1 < h < 2$), according to the fraction of sp^1 and sp^2 hybridization. The energy of graphynes vs. (b) the number of acetylenic groups, *n*, or (c) the hybridization, *h*. For comparison, previous first-principles results [2, 4] are plotted with black crosses. Constructed full atomistic models of (d) graphene, (e) graphyne, (f) graphdiyne, (g) graphtriyne, and (h) graphtetrayne with approximately 100 Å × 100 Å in dimensions. Source: Cranford et al. [3]. © 2012, Royal Society of Chemistry.

decrease systematically as the ratio of sp-carbon atoms increases. Among them, γ-graphyne is the most energetically stable graphyne structure with a cohesive energy of 6.766(6) eV atom^{-1}. However, the cohesive energy is still smaller than that of graphene by 0.698(12) eV atom^{-1}, which could explain the experimental difficulty in synthesizing graphynes.

In 2010 year, Li's group successfully synthesized large-area γ-GDY films for the first time on the surface of copper foil via the cross-coupling reaction using hexaethynylbenzene [8]. The γ-GDY is the most widely studied graphyne structure and attracts wide attention. As a new 2D nanomaterial, GDY not only has much variable and controllable carbon atoms combination (i.e. sp- and sp^2 hybridization), but also can grow in situ on any substrate surface under low temperature and mild conditions. Arising from the abundant chemical bonds, large specific surface area, wide interplanar distance, high carrier mobility, good electrical conductivity, excellent chemical stability, and unique porous structures, GDY possesses excellent electrical, mechanical, optical, magnetic, and thermal properties. Therefore, it is successfully applied in many research fields such as photoelectric catalysis [9–12], energy storage [13–15], photoelectric conversion [16–19], solar–thermal conversion [20], oil–water separation [21], biological detection [22, 23], photoelectric detection [24], and electrochemical drives [25].

2.1.2 Geometric Structures of GDYs

The geometric structures of GDYs (such as α-, β-, and γ-graphyne) can be obtained by inserting the acetylenic linkages between two individual sp^2 carbon atoms (Figure 2.2a), two pairs of sp^2 carbon atoms (Figure 2.2b), and the vertices of two aromatic rings (Figure 2.2c), respectively [26]. In these structures, α-, β-, and γ-graphyne contain 100%, 66.67%, and 33.33% of acetylenic groups, respectively. As known, the carbon coordination number in graphene is 3. However, different hybridization combinations in graphynes lead to the noninteger average carbon coordination numbers, which are 2.25, 2.33, and 2.50 for α-, β-, and γ-graphyne, respectively. The stabilities of these graphyne allotropes structures can be evaluated by calculating the cohesive energy (E_C) of each carbon atom in graphyne sheet, which is defined as the energy required to form separated neutral carbon atoms in the ground state from the condensed phase at 0 K:

$$E_C = \frac{n \cdot E_{C,atom} - E_T}{n} \tag{2.2}$$

where $E_{C,atom}$ is the energy of single carbon atom in ground state, E_T is the total energy of each graphyne unit cell, and n is the number of carbon atoms in each unit cell. Density functional theory (DFT) calculations suggest that the cohesive energies of α-, β-, and γ-graphyne are 6.93, 7.01, 7.21 eV atom^{-1}, respectively (Table 2.1). It is seen that the stabilities of α-, β-, γ-graphynes are decreased as the percentages of acetylenic linkages increase. Among them, γ-graphyne that maintains the benzene ring is the most energetically stable structure.

The optimized lattice parameters and bond lengths of α-, β-, and γ-graphyne using different calculation methods are listed in Table 2.1. The shortest bond in α-, β-, and

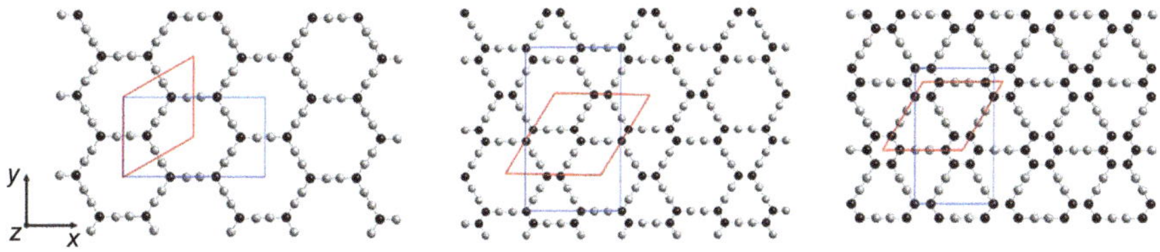

(a) (b) (c)

Figure 2.2 (a) α-, (b) β-, and (c) γ-graphyne structures with the armchair edge along x-axis direction and the zigzag edge along y-axis direction. The sp- and sp^2-carbon atoms are depicted in gray and black colors, respectively. The rhombohedral (red) and rectangular (blue) represent two types of unit cells. Source: Ruiz Puigdollers et al. [26]. © 2016, Elsevier.

Table 2.1 Cohesive energies and geometric parameters. Lattice constants and C−C bond lengths at PBE and PBE-D2 levels.

Method	α-Graphyne	β-Graphyne	γ-Graphyne
Cohesive energy (eV/atom)	6.93	7.01	7.21 (7.95[a])
Lattice constant ($a = b$) (Å)			
PBE	6.966 (7.01[b], 6.97[c], 6.9812[d])	9.480 (9.48[c], 9.5004[d])	6.890 (6.86[a], 6.89[c][f], 6.8826[d], 6.83[g])
PBE D2	6.964 (6.957[e])	9.480	6.888 (6.877[e])
Bond lengths (Å)			
PBE			
sp–sp	1.230 (1.24[b], 1.23[c], 1.2317[d])	1.232 (1.23[c], 1.2343[d])	1.223 (1.223[c], 1.2214[d])
sp–sp^2	1.396 (1.40[b], 1.397[c], 1.3995[d])	1.389 (1.39[c], 1.3922[d])	1.408 (1.408[c], 1.4070[d])
sp^2–sp^2		1.457 (1.46[c], 1.4633[d])	1.426 (1.426[c], 1.4237[d])
PBE-D2			
sp–sp	1.230 (1.229[e])	1.232	1.222 (1.221[e])
sp–sp^2	1.395 (1.394[e])	1.389	1.407 (1.406[e])
sp^2–sp^2		1.456	1.426 (1.422[e])

a) Reference [2] (with LDA).
b) Reference [27].
c) Reference [28].
d) Reference [29].
e) Reference [30] (with Cooper's exchange functional).
f) Reference [31].
g) Reference [32].
Source: Ruiz Puigdollers et al. [26]. © 2016, Elsevier.

γ-graphynes corresponds to the triple bond between sp–sp carbon atoms where the electron density is the most localized. The bond lengths are 1.23 Å for α-, β-graphyne, and 1.22 Å for γ-graphyne, respectively, which are almost the same as the bond length of acetylene in ground state (i.e. 1.21 Å) [33]. The longest bond corresponds to the aromatic bond between sp^2–sp^2 carbon atoms, which are 1.46 and 1.43 Å for β- and γ-graphyne, respectively, similar to the bond length in graphene (i.e. 1.42 Å) [34]. The lengths of single bond between sp–sp^2 carbon atoms are 1.39 and 1.41 Å for β- and γ-graphyne, respectively, which are between the lengths of triple bond and aromatic bond. These results are well consistent with the variation of bond lengths for different types of bonding, such as single, aromatic, or triple bond.

The acetylenic chains in graphynes form abundant pores, which are much larger than the densely packed honeycomb pores in graphene. The large pores in graphynes are critical for the membrane permeability, heterogeneous catalysis, lithium/hydrogen storage, etc. [35]. There are different pore structures in the three graphynes, for example, the hexagonal pores in α-graphyne, the truncated triangular pores in γ-graphyne, and the simultaneously existing hexagonal and truncated triangular pores in β-graphyne. It is estimated that the pore area of α-graphyne is about 42.0 $Å^2$ by regarding C atoms as single points, or 29.7 $Å^2$ by considering the covalent radius of C atoms [36], or 24.9 $Å^2$ by considering the Bader volumes of C atoms [37]. Similarly, the truncated triangular pores area of γ-graphyne is 17.8 $Å^2$ (or 10.0 $Å^2$, or 5.9 $Å^2$) per unit cell. The hexagonal and truncated triangular pores areas of β-graphyne are 41.8 $Å^2$ (or 29.5 $Å^2$, or 23.5 $Å^2$) and 18.0 $Å^2$ (or 10.2 $Å^2$, or 4.5 $Å^2$) per unit cell, respectively. Besides, when regarding C atoms as single points, the corresponding concentrations of hexagonal and truncated triangular pores in graphynes are 2.4×10^{14} and 5.1×10^{14} pores cm^{-2}, respectively.

Different types of pores in graphyne structures lead to different planar packing densities (mg m^{-2}), which is defined as the number of atoms or mass per unit cell. For example, there are eight carbon atoms per unit cell in α-graphyne, corresponding to a planar packing density of 0.379 mg m^{-2}. Similarly, there are 18 and 12 carbon atoms per unit cell in β- and γ-graphyne, giving a packing density of 0.461 and 0.582 mg m^{-2}, respectively. The planar packing density exhibits a reverse tendency with respect to the concentration of acetylenic chains in these graphynes. For example, the planar density in α-graphyne structure is lowest, but the concentration of acetylenic chains is highest. However, the planar density in γ-graphyne structure is highest, while the concentration of acetylenic chains is lowest. As known, graphene consists of only sp^2-hybridization (i.e. 0% of acetylenic chains), but it has a high planar density of 0.756 mg m^{-2}, which is almost double that of α-graphyne containing 100% of acetylenic chains.

2.2 Electronic Structures

2.2.1 Dirac Cones in α-, β-, and 6,6,12-Graphynes

It is generally considered that only graphene with hexagonal symmetry features in the unique Dirac cone structure, thereby inducing the novel electronic properties, such as enormous carrier mobility and charge transport. Gorling and coworkers [5] proved that the characteristics of Dirac cones not only exist in the band structures

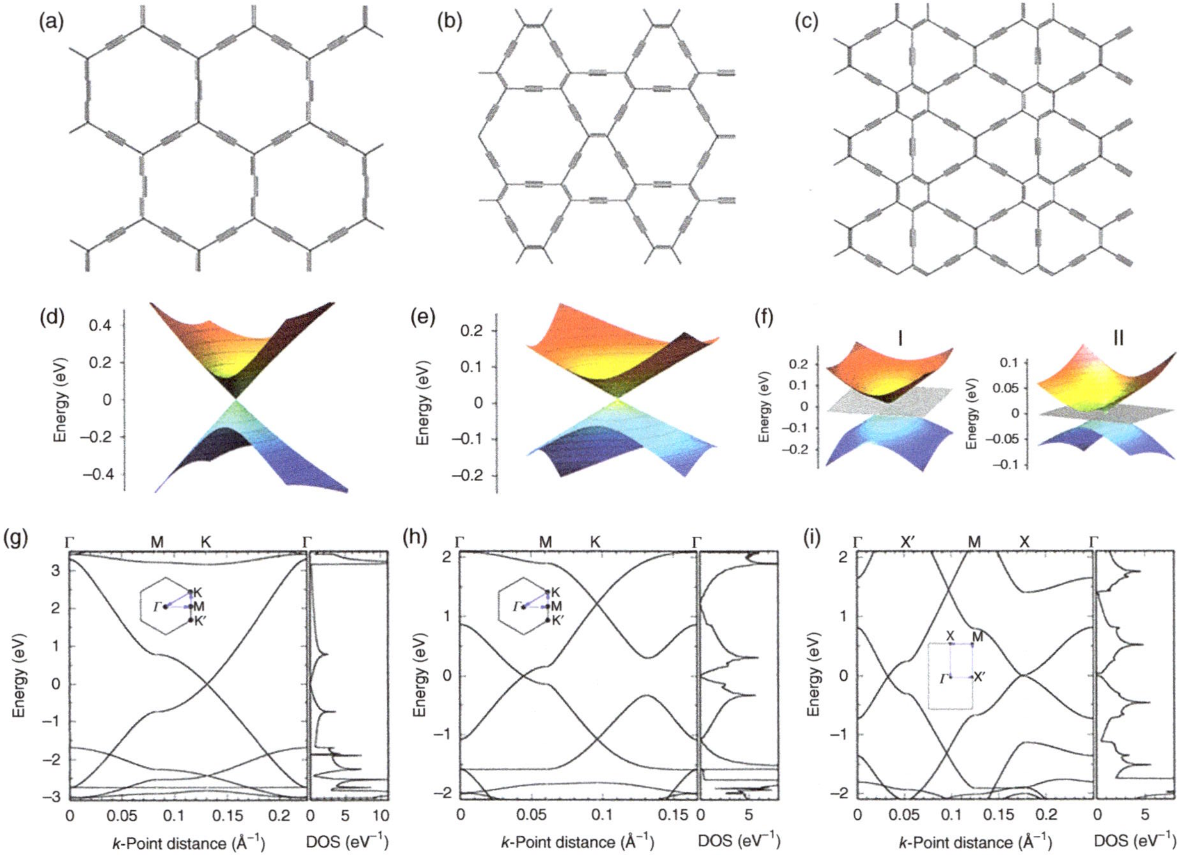

Figure 2.3 Structures of (a) α-, (b) β-, and (c) 6,6,12-graphyne. Dirac cones formed by the valence band and conduction band in the vicinity of Dirac point are displayed in the middle panels for (d) α-, (e) β-, and (f) 6,6,12-graphyne. For 6,6,12-graphyne, two different Dirac cones are named I and II, and the gray planes indicate Fermi level. Electronic structures, including the band structures and density of states (DOS) of (g) α-, (h) β-, and (i) 6,6,12-graphyne. First Brillouin zone with letters designating special points and lines are also displayed. Source: Malko [5]. © 2012, American Physical Society.

of graphynes with hexagonal symmetry, such as α-graphyne (Figure 2.3a) and β-graphyne (Figure 2.3b), but also exist in the band structures of graphynes with rectangular symmetry, such as 6,6,12-graphyne (Figure 2.3c). Therefore, the graphyne materials possess amazing electronic properties similar to those of graphene.

α- and β-Graphyne exhibit very similar Dirac cone structure to that in graphene, as shown in Figure 2.3d,e, respectively. Differently, the Dirac cone is located at the K point of Brillouin region in α-graphyne (Figure 2.3g), while at the line between Γ and M points in β-graphyne (Figure 2.3h). For 6,6,12-graphyne, there are two self-doped nonequivalent distorted Dirac cones (Figure 2.3f) at different points of Brillouin region (Figure 2.3i), indicating the existence of inherent charge carrier. The intrinsic hole and electron mobility in 6,6,12-graphyne are predicted to be 4.29×10^5 and 5.41×10^5 cm^2 V^{-1} s^{-1} at room temperature, respectively, which are larger than those in graphene (approx. 3×10^5 cm^2 V^{-1} s^{-1}) [38]. Besides, the 6,6,12-graphyne shows different electron transport properties, depending on the direction of disturbances such as the applied voltage or strain.

To assess the functionalities and properties of the graphyne materials, it is useful to understand their intrinsic electronic transport properties. The transport properties of the above-mentioned three graphynes were simulated through a combination of DFT with the nonequilibrium Green's function method (NEGF-DFT) [39]. As shown in the left panels of Figure 2.4a–c, the transport setup is constructed using the unit cell defined by the dotted box for α, β and 6,6,12-graphyne. It is seen that there are two distinct transport directions (X_1 and X_2) for these 2D materials. For 6,6,12-graphyne, the X_1 and X_2 directions are not equivalent, namely the transmission function (the right panel of Figure 2.4a) is different, leading to the direction-dependent transport properties of such system. For α- and β-graphynes, only the transports along the X_1 direction with periodicity in X_2 are presented (the right panels of Figure 2.4b,c) owing to the two directions (X_1 and X_2) giving the same results. The electronic transmission functions around Fermi level for all these graphyne structures show similar characteristics to the band structure with Dirac cone, such as for graphene. Since the transmission functions in graphynes are always larger than those in graphene, as indicated in the right panels of Figure 2.4a–c, the electronic current (I_{ds}) flowing on these graphyne systems is always higher than that on graphene (Figure 2.4d). Besides, the current for 6,6,12-graphyne along the X_1 (X_2) direction is almost 3 (4) times larger compared to other structures such as α, β-graphyne, and graphene (Figure 2.4d).

2.2.2 Semiconductor Properties of γ-Graphynes

Interestingly, γ-graphyne is a semiconductor with natural direct band gap. First-principles calculations at the local density approximation (LDA) or generalized gradient approximation (GGA)-Perdew-Burke-Ernzerbof (PBE) levels indicate that the γ-graphyne monolayer gives a direct band gap of 0.46–0.52 eV at the high symmetry point of M in Brillouin zone [2, 27, 30, 40, 41]. However, the band structures calculated at the (Heyd-Scuseria-Ernzerhof) HSE06 level predict a band

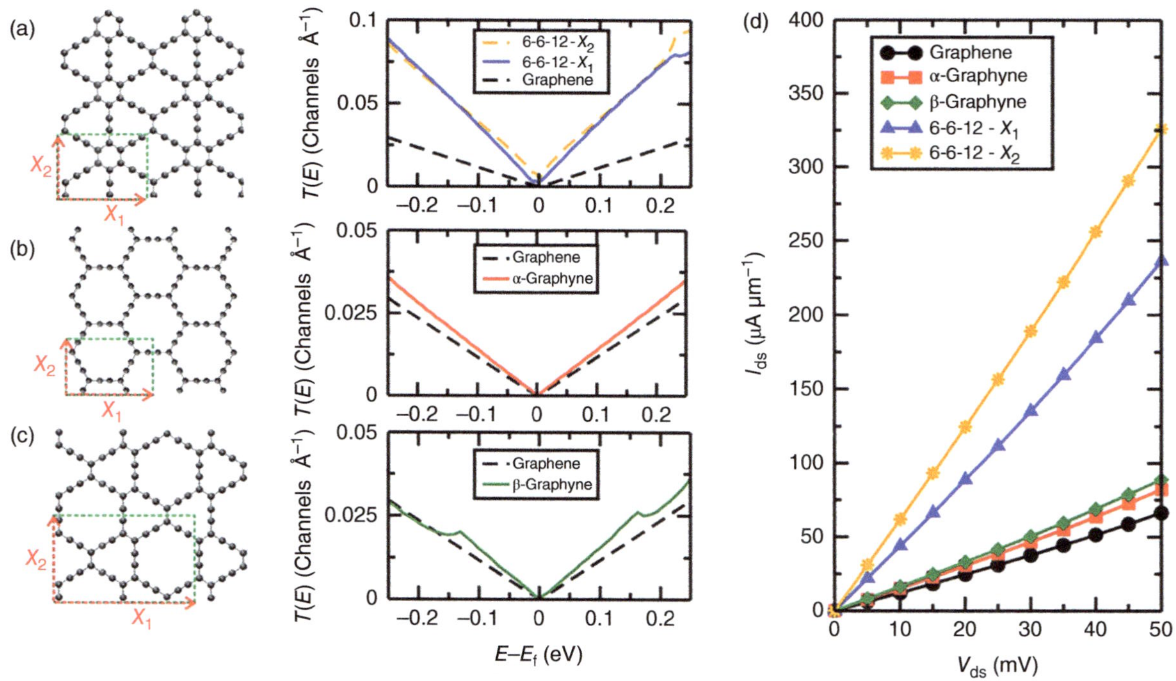

Figure 2.4 Schematic representation of unit cell defined by the vectors X_1 and X_2 (left panel) and the transmission functions (right panels) for (a) 6,6,12-graphyne, (b) α-graphyne, and (c) β-graphyne, (d) The $I_{ds} \times V_{ds}$ for α, β, and 6,6,12-graphyne. For comparison, the transport properties for graphene monolayer are shown. Source: Padilha et al. [39]. © 2014, American Chemical Society.

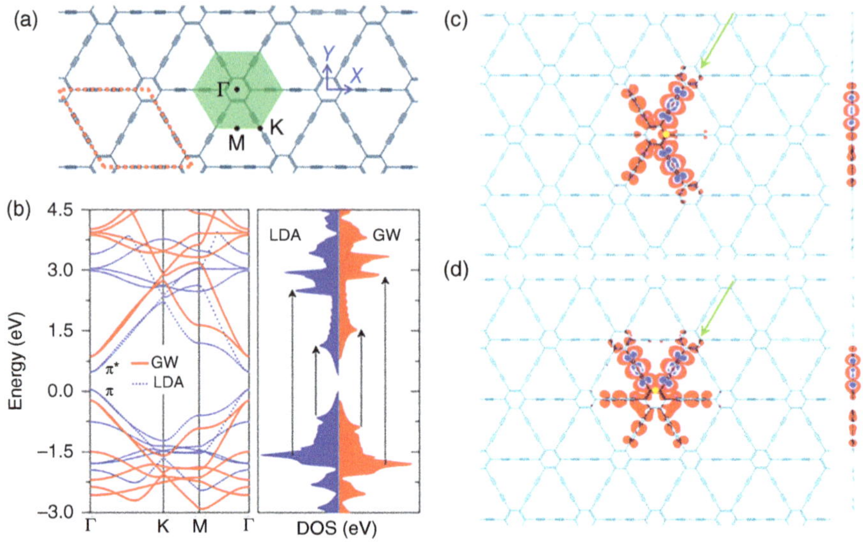

Figure 2.5 (a) Geometrical structure (red diamond dashed) and first Brillouin zone (green hexagon) of GDY unit cell. The structure possesses the $D^1{}_{6h}$ space group with an optimized unit cell length of 9.42 Å. (b) Band structures and density of states (DOS) of GDY calculated at the LDA and GW levels. Optical transitions between the first two Van Hove singularities are indicated. Top and side views of electron densities of the lowest doubly degenerate bright excitons: A_1 (c) and A_2 (d), respectively. The hole is fixed at the yellow spot, and the green arrow indicates the direction of side view. Source: Luo et al. [42].; © 2011, American Physical Society.

gap of 0.96 eV, which is almost double than that of LDA or GGA-PBE [40]. Actually, except that the estimated bandgap values are different, the band structures given by different functional show the same characteristics.

γ-GDY shows similar band structures and density of states (DOSs) to γ-graphyne [42]. The band gap of GDY (Figure 2.5a) is calculated to be 0.44 eV at the LDA level and is noticeably increased to 1.10 eV at the GW level (Figure 2.5b). Such a 1.5-time quasiparticle correction arises from the coulomb interaction in GDY, and the quasiparticle band gap is very close to that of silicon [43]. However, the properties of GDY are probably superior to silicon because it has the direct band gap characteristics, which is quite appropriate for the semiconductor and optoelectronic devices. Figure 2.5c,d show the electron density of the lowest doubly degenerate bright excitons (A_1 and A_2) while fixing the hole (yellow spot) at the density maximum of π_z orbitals. In both cases, the excitons are tightly bound to the GDY plane, which is different from the separation behavior in graphane network [44]. It is deduced that the excitons in GDY have both the Wannier–Mott and Frenkel exciton characteristics due to the significant spatial extent and large binding energy.

2.2.3 Electronic Structures Comparison of GDYs

The electronic structures of α-, β-, and γ-graphynes were compared by calculating the band structures and DOSs [26]. The band structures indicate the existence of

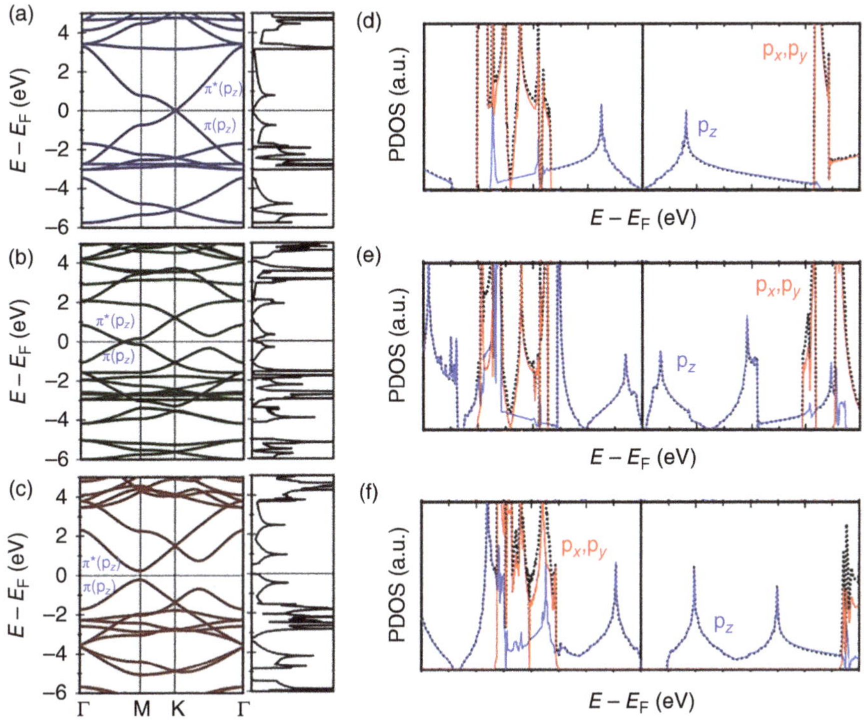

Figure 2.6 Band structures and total density of states (DOS) for (a) α-, (b) β-, and (c) γ-graphyne. Projected density of states (PDOS) for (d) α-, (e) β-, and (f) γ-graphyne. Black dotted lines represent the total DOS, blue lines represent the PDOS on p_z orbitals, and red lines represent the PDOS on p_x and p_y orbitals. Source: Ruiz Puigdollers et al. [26]. © 2016, Elsevier.

Dirac points in α- and β-graphyne, which appear at the high symmetry point of K in Brillouin zone (Figure 2.6a) and at the line between the high symmetry points of Γ and M (Figure 2.6b), respectively. Differently, γ-graphyne has a direct band gap, which is located at the high symmetry point of M (Figure 2.6c). Due to the presence of sp- and sp^2-carbon atoms simultaneously, there are different types of bonds in the graphyne structures. For example, a double bond (i.e. $\sigma + \pi$) formed between two sp^2-carbon atoms; a triple bond (i.e. $\sigma + 2\pi$) formed between two sp-carbon atoms; a single bond (i.e. σ) formed between sp- and sp^2-carbon atoms.

The contributions of different atomic orbitals to the band structures were analyzed. Considering that the graphynes surface lie in the xy plane, s and p_x–p_y orbitals contribute to the σ bond, p_z orbitals contribute to the π bond of both sp^2–sp^2 and sp–sp. The additional π bond of sp–sp bonding is contributed by the in-plane p_x–p_y orbitals. According to the projected DOSs, the band structures of α- (Figure 2.6d), β- (Figure 2.6e), and γ-graphyne (Figure 2.6f) are divided into different regions. The conduction band of α- and β-graphyne meets the valence band at Fermi level, where the DOS is exactly zero. In the region around Fermi level (−1.5–3 eV), only the contributions of p_z orbitals are found for the three graphynes, corresponding to the π and π* states of sp–sp and sp^2–sp^2. The contribution of p_z orbitals can be extended to the lower energies in valence band (less than −4 eV) and the higher energies in conduction band (more than 4 eV). In the region between −3 and −1.5 eV, the contribution of p_x and p_y orbitals becomes dominant. The p_x and p_y orbitals mixed with the p_z orbitals in this narrow region form the additional π bond of sp–sp. Similarly, the bands above 3 eV in conduction band correspond to the additional π* bond of sp–sp.

It is worth noting that the π and π* bands contributed by p_z orbitals cover a wider energy intervals than those contributed by p_x–p_y orbitals, indicating the delocalization of π(p_z) and π*(p_z) states is much higher. Therefore, this type of π bond can appear in both sp^2–sp^2 and sp–sp. In other words, there exists a large p_z–p_z overlap between all adjacent sp- and sp^2-carbon atoms. Contrarily, the π(p_x, p_y) and π*(p_x, p_y) bands are only present between sp–sp carbon atoms, which are separated by sp^2-carbon atoms in the lattice of α-graphyne, by sp^2–sp^2 bonding in the lattice of β-graphyne, and by benzene rings in the lattice of γ-graphyne, respectively.

To compare the conductivity of graphynes with graphene, the Fermi velocities around Dirac points were calculated. Among these graphynes, α-graphyne shows a symmetric Fermi velocity of 6.76×10^5 m s^{-1}, while β-graphyne shows an asymmetric Fermi velocities of 5.07×10^5 m s^{-1} in the Γ → M direction and 3.80×10^5 in the opposite M → Γ direction, respectively. The Fermi velocities in α- and β-graphynes are all lower than those in graphene (i.e. 8.3×10^5 m s^{-1}). This can be expected since the π and π* states in α- and β-graphyne are partly contributed by p_x–p_y orbitals (only localized in sp–sp bonding) due to the existence of sp–sp bonding. Hence, compared to graphene, the π and π* states in graphyne have much higher localization degree since they do not overlap with the orbitals of neighboring sp^2-carbon atoms, which gives rise to much flatter bands.

The effective masses for electrons in valence band (m_v^*) and holes in conduction band (m_c^*) of graphynes were studied. There are $0.21m_0$ and $0.22m_0$ for m_c^* and m_v^* in the M → Γ direction, respectively, and $0.087m_0$ and $0.083m_0$ for m_c^* and m_v^* in the M → K direction, respectively. It is seen that the effective masses for electron and hole are similar, but are different in the two directions. Note that the effective masses of charge carrier are virtual mass, so the closer to zero of effective mass, the higher velocity of charge carrier. For example, the effective masses in graphene, α-, and β-graphyne are strictly zero due to the linear energy dispersion in Dirac points, inducing the amazing conductivity properties of these materials. For γ-graphyne, the effective mass in the M → K direction is approximately the half of that in the perpendicular Γ → K direction, indicating the carrier velocity is higher along the M → K direction. Moreover, the results show that the anisotropy in effective masses govern the electrical properties of γ-graphyne.

2.2.4 Structure and Size-Based Electronic Properties

The absent energy gaps such as for α-, β-graphyne or the small energy gaps such as for γ-graphyne limit their practical applications in electronic devices. It is found that the construction of nanoribbons can effectively open up the band gaps of graphynes. Hence, the investigations on one-dimensional (1D) graphynes nanoribbons will help us to understand the possible structural characteristics and electronic properties of the new carbon allotrope [32]. Various graphyne nanoribbons with armchair or zigzag edges can be obtained by cutting the graphynes sheet in different directions.

First-principles calculations show that all the nanoribbons of graphynes (such as graphyne and GDY) possess the semiconductor properties. Moreover, different edge morphologies and sizes have great influences on the electronic properties of graphynes nanoribbons. For the graphyne nanoribbons, a suitable band gaps of 0.59–1.25 eV (with armchair edge) or 0.75–1.32 eV (with zigzag edge) are obtained. For the GDY nanoribbons, similar band gaps of 0.54–0.97 eV (with armchair edge) or 0.73–1.65 eV (with zigzag edge) are obtained. The band gaps of these nanoribbons are decreased as the widths increase. Particularly, the band gaps of nanoribbons with zigzag edge show a unique "step effect" instead of a smooth gap decrease with the widths increasing. Besides, the band gaps occurr at the Γ point of Brillouin zone for all the GDY nanoribbons, while at the boundary of Brillouin zone for all the graphyne nanoribbons, depending on the numbers of acetylenic linkages (odd or even) between two nearest-neighboring aromatic rings. The large and adjustable band gaps of graphynes nanoribbons make them great application potentials in the electronic devices.

Shuai and coworkers [45] studied the electronic structure and intrinsic carrier mobility of GDY sheet and the corresponding nanoribbons (graphdiyne nanoribbon [GDNRs]) using first-principles DFT coupled with the Boltzmann transport equation with the relaxation time approximation. There are two major ways to cut the GDY sheet into nanoribbons, namely along the direction of nearest-neighbor carbon hexagons (A → B), resulting in an armchair-like edged GDNR (AGDNR, Figure 2.7a-I), or along the direction of next nearest-neighbor carbon hexagons (A → C), resulting in a zigzag-like edged GDNR (ZGDNR, Figure 2.7a-II, III). For ZGDNRs, there are two different configurations: with uniform width (Figure 2.7a-II) or with nonuniform width (Figure 2.7a-III) by cutting at different sites. Figure 2.7b–f illustrates five types of GDY nanoribbons.

The electronic properties of GDNRs with various edge widths are quite different based on the self-consistent field crystal orbital (SCF-CO) investigations [46]. In view of the calculated cohesive energies, all the 1D GDY nanoribbons are more stable than the 2D GDY monolayer. Moreover, the ZGDNRs are more stable than the AGDNRs. This is quite different from the graphene nanoribbons. Similar to the GDY sheet, both the ZGDNRs and AGDNRs are semiconductors with a direct band gap at the Γ point, independent with the edge structures. This is also different from the graphene nanoribbons, for which the semiconducting or metallic properties are dependent on their edge shapes. For example, all the graphene nanoribbons are semiconductors in the armchair direction, while exhibiting the metallic properties in the zigzag direction [47].

Among these GDNRs, the smallest band gap can reach a value of ~0.8 eV (for A_2), larger than that of GDY [45]. This is a useful feature for employing GDNRs as the semiconducting channels in field effect transistors. Huang and coworkers [46] give a relationship between the band gaps (E_g) and widths (W) for GDNRs: $E_g = aW^{-b}$, where index b reflects the sensitivity of band gaps to the nanoribbons widths. It is

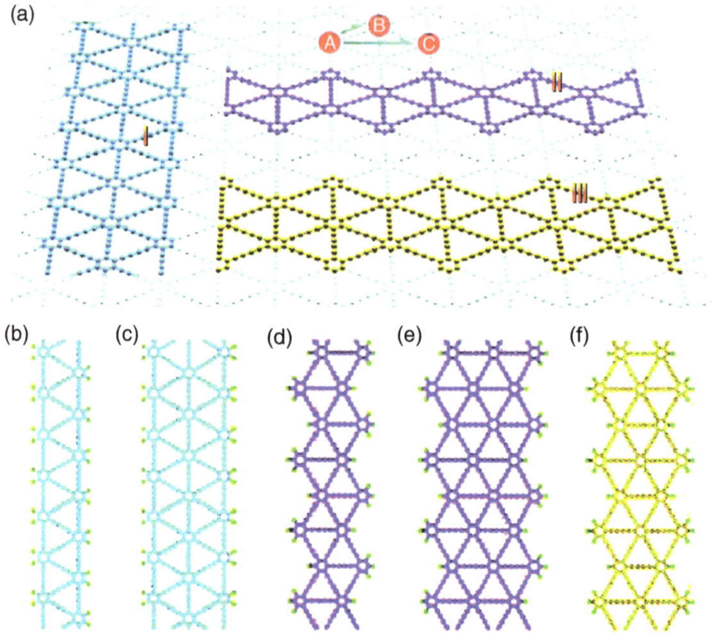

Figure 2.7 (a) Schematic representation of three GDNRs: (I) an armchair-like edged GDNR, (II) a zigzag-like edged GDNR with uniform width, (III) a zigzag-like edged GDNR with alternating width, (b–f) five types of GDNRs structures: (b) and (c) represent the AGDNRs with two (A1) and three (A2) carbon hexagons in width, respectively, (d) and (e) represent the ZGDNRs with two (Z1) and three (Z2) carbon hexagons in width, respectively, (f) represents the ZGDNR with two carbon hexagons at the narrow site and three hexagons at the broad site in width (Z3). Source: Long et al. [45]. © 2011, American Chemical Society.

found that the band gaps of ZGDNRs and AGDNRs are obviously decreased as the widths increase. For example, the band gap of ZGDNRs and AGDNRs with a width of $n = 1$ is 0.971 and 1.538 eV, respectively. As the widths of nanoribbons increase to $n = 12$, the band gaps of AGDNR and ZGDNR are decreased to be about 0.5 eV, close to the band gap of 2D GDY monolayer. The tendency is completely different from the E_g–W relationship for graphene nanoribbons [48–50]. In addition, the b values of ZGDNRs and AGDNRs are 0.302 and 0.593, respectively, which are smaller than the index b of graphene nanoribbon (i.e. 0.872–1.097) [49], indicating that the band gaps of GDY nanoribbons change more smoothly than those of graphene nanoribbons with the widths increasing. These features enable us to precisely modulate the band gaps of GDNRs by adjusting their nanoribbons widths.

The carrier mobility of electron (μ_e) and hole (μ_h) for GDY and GDNRs was calculated using the deformation potential (DP) theory and effective mass approach. The calculated in-plane electron mobility of GDY sheet can reach 2×10^5 cm^2 (V s)$^{-1}$ at room temperature [45], comparable to that of graphene [51]. Whereas the hole mobility is about 2×10^4 cm^2 (V s)$^{-1}$, an order of magnitude lower than the electron mobility. The GDNRs are also potential transport materials with high carrier mobility in a range of 10^2–10^6 cm^2 V^{-1} s^{-1} at room temperature [46], which increase as the

Table 2.2 Calculated band gap, effective mass (m_h* and m_e*), DP constants for valence band (VB) (E_v) and conduction band (CB) (E_c), stretching modulus (C), and carrier mobility (μ) at 300 K for the AGDNRs (A1, A2) and ZGDNRs (Z1, Z2, Z3).

	A_1	A_2	Z_1	Z_2	Z_3
Band gap (eV)	0.954	0.817	1.205	0.895	1.015
m_h^* (m_0)	0.086	0.087	0.216	0.149	0.174
m_e^* (m_0)	0.081	0.086	0.281	0.174	0.207
E_v (eV)	7.406	6.790	4.386	4.786	4.776
E_e (eV)	2.006	1.730	1.972	2.000	2.054
C (10^{10} eV cm^{-1})	1.244	1.864	1.035	1.787	1.420
μ_h (10^3 cm^2 (V s)$^{-1}$)	1.696	2.088	0.755	1.815	1.194
μ_e (10^3 cm^2 (V s)$^{-1}$)	18.590	34.241	2.692	9.127	5.329
μ_h^* (10^3 cm^2 (V s)$^{-1}$)	0.711	1.253	0.426	1.073	0.679
μ_e^* (10^3 cm^2 (V s)$^{-1}$)	10.580	19.731	1.418	5.015	2.829

Source: Long et al. [45]. © 2011, American Chemical Society.

nanoribbon widths increase for both AGDNRs and ZGDNRs (Table 2.2). Moreover, the electron mobility of GDNRs with different widths can reach 10^4 cm^2 (V s)$^{-1}$ at room temperature, which is always significantly larger than the hole mobility (Table 2.2).

The larger mobility of electron than hole for GDY and GDNRs can be explained by the obvious different relationships: $\mu_e = pW^q$ for electrons, while $\mu_h = kW - \mu_0$ for holes [46]. The lowest unoccupied molecular orbital (LUMO) of GDY and GDNRs (both AGDNRs and ZGDNRs) are more extended than the highest occupied molecular orbital (HOMO) (Figure 2.8), which can also qualitatively explain the much larger electron mobility compared to hole mobility. In addition, the carrier mobility of AGDNRs is significantly larger than that of ZGDNRs with the same nanoribbons widths. This is owing to the LUMO for the armchair-edged GDNRs exhibiting much more delocalized characteristics in the axis direction of nanoribbons than those for the zigzag-edged GDNRs (Figure 2.8).

Graphyne nanotubes (GyNTs) were regarded as novel 1D semiconducting carbon allotropes that can be used in nanoelectronics [52–54]. One-dimensional carbon allotropes, including α-, β-, γ-GyNTs and carbon nanotubes (CNTs), are obtained by rolling up the graphene and graphyne sheets into the nanotubes. Figure 2.9 depicts the geometric structures of GyNTs and CNTs with an initial length of approximately 13 nm and a diameter of approximately 15.5 Å [52]. The structural and electrical properties of GyNTs with different diameters and chirality were investigated using the density functional tight binding (DFTB) method [53]. All the GyNTs are fully relaxed along the nanotube axis. The bandgaps of GyNTs with zigzag edge (i.e. ZGyNTs) under negative strain and GyNTs with armchair edge (i.e. AGyNTs) under all strain demonstrate a damped oscillatory behavior, while the ZGyNTs under positive strain show a uniform damped behavior. In addition, the bandgaps of ZGyNTs are

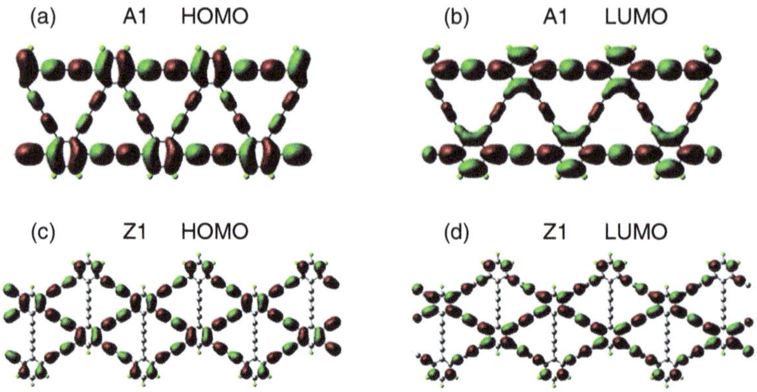

Figure 2.8 Γ-point HOMO (a) and LUMO (b) for the GDNRs with armchair edge (A1). Γ-point HOMO (c) and LUMO (d) for the GDNRs with zigzag edge (Z1). HOMO and LUMO show the antibonding and bonding features between carbon hexagons and diacetylenic linkages, respectively. The coupling strength between hole and acoustic phonon is thereby larger than that between electron and acoustic phonon. Source: Long et al. [45]. © 2011, American Chemical Society.

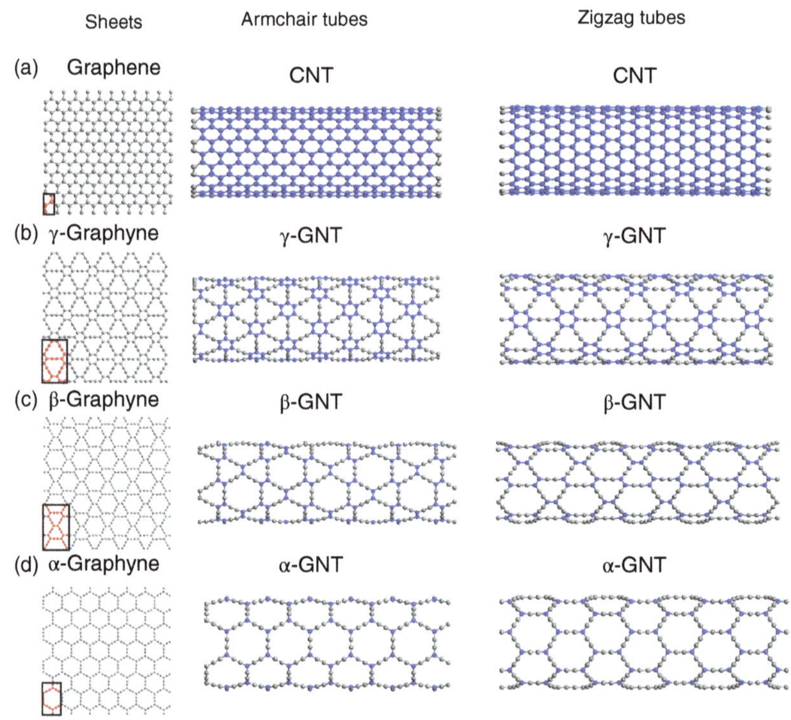

Figure 2.9 The graphene and graphyne nanotubes with zigzag and armchair edge. The geometric structures of different types of nanotubes for: (a) graphene, (b) γ-graphyne, (c) β-graphyne, and (d) α-graphyne. The sp- and sp^2-carbon atoms are depicted in gray and blue colors, respectively. The rectangular (red) represent the unit cells. Source: Azizi et al. [52]. © 2020, Elsevier.

smaller than those of AGyNTs. As the chirality (n) increases, the diameter of nanotubes increases and the in-plain strain on the GyNTs surface decreases, leading to the properties of GyNTs tending toward those of graphyne sheet. The bandgaps of both ZGyNTs and AGyNTs decrease with the increase of diameters. Besides, the variation of bandgaps for ZGyNTs under positive strain is greater than that under negative strain. However, it is contrary for AGyNTs under positive and negative strains. Such large and tunable bandgaps for GyNTs under homogeneous strain make them probably applied for the design of novel optoelectronic and nanoelectronic devices.

2.2.5 Strain-Dependent Electronic Properties

As mentioned above, the surface strain affects the electronic properties of graphynes. Since the 6,6,12-graphyne presents the directional-dependence feature, it is feasible to tune its band structures by applying an anisotropic strain in the system [39]. Figure 2.10a shows the band structures of 6,6,12-graphyne that is stretched (top panel) and compressed (bottom panel) along the X_1 direction. When the 6,6,12-graphyne is stretched along the X_1 direction, the dispersion relations around Dirac cone I decrease. The opposite behavior is observed when the 6,6,12-graphyne is compressed along the X_1 direction. As shown in Figure 2.10c, the Fermi velocity at Dirac cone I decreases from $\sim 5.2 \times 10^5$ to $\sim 2.3 \times 10^5$ m s^{-1} with the strain varying from -6% to 6%. However, the Fermi velocity at Dirac cone II is always increased, independent with the sign of applied strain along the X_1 direction.

When stretching the 6,6,12-graphyne along the X_2 direction (Figure 2.10b), a band gap is opened around the X point of Brillouin zone for Dirac cone II. As shown in Figure 2.10d, the gap opening is continuously tuned (circle blue curve), which can be attributed to the decoupling of acetylenic group, as evidenced by the linear increased distance (i.e. Δ_d) between the acetylenic group and the hexagon (square green curve). In addition, the Dirac cone I remains unaltered and the Fermi velocity is almost constant around 5×10^5 m s^{-1} with this type of strain. When compressing the system along the X_2 direction (Figure 2.10b), the band gap is not opened since there is no decoupling of the acetylenic group. The linear dispersion relations around Dirac cone II increase, and the Fermi velocity increases from 1×10^5 to 4×10^5 m s^{-1}. For the α- and β-graphyne structures under strain, the electronic band structures are different, which remain unchanged even with the applied strain up to 6%.

The change of electronic structures under different strain conditions (varying from -6% to 6% in both X_1 and X_2 directions) will manifestly affect the electronic transport properties of 6,6,12-graphyne. Figure 2.11a depicts the current along the X_1 direction when applying strain on the same direction. It is seen that the current is almost unaltered when the system is stretched, whereas the current is increased by approximately 37% when the system is compressed. However, when the current remains along the X_1 direction but the applied strain is along the X_2 direction, a different behavior is observed (Figure 2.11b). The current is decreased when the system is stretched due to the band gap opening on Dirac cone II and the consequent decrease in the number of quantum channels in this direction. However, the decrease of current is not so expressive; hence, the main contribution arising from

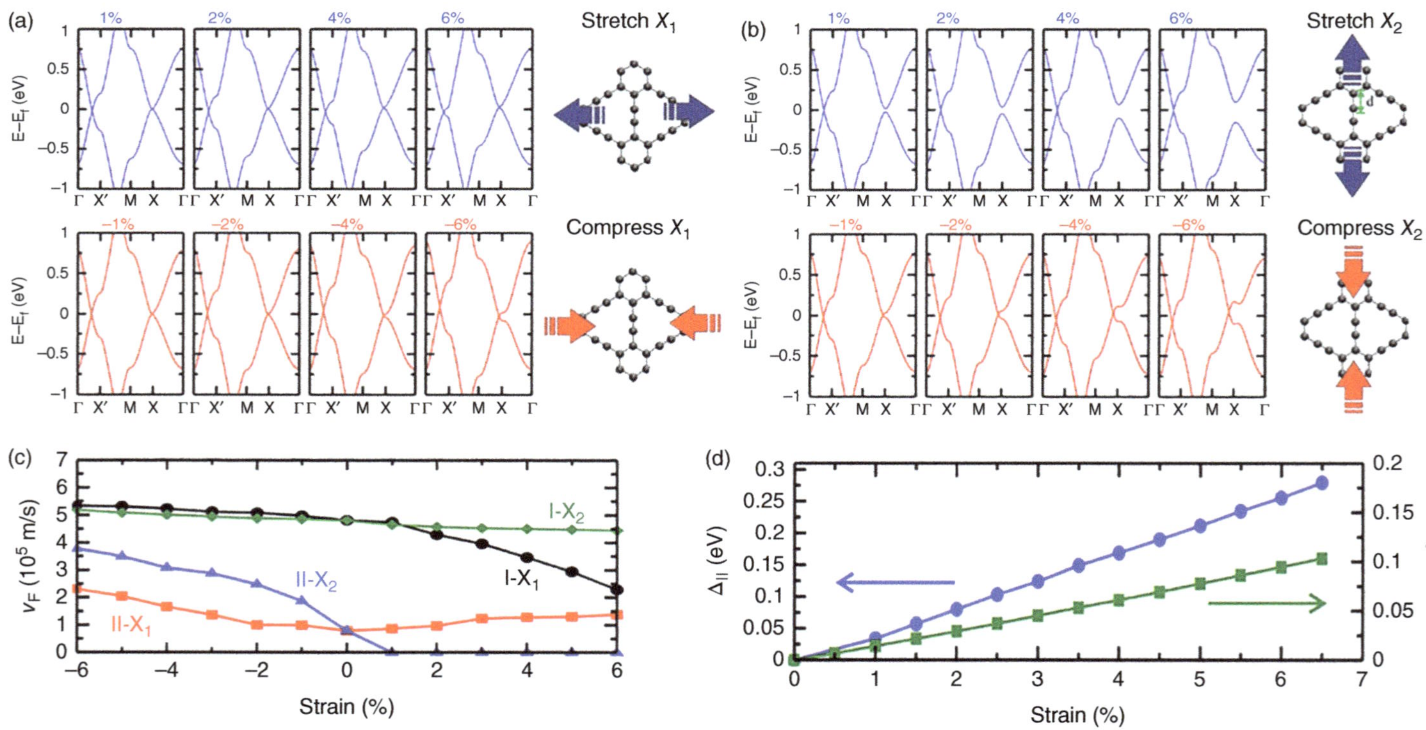

Figure 2.10 Band structures for the stretched (top panel) and compressed (bottom panel) 6,6,12-graphyne along the (a) X_1 and (b) X_2 directions, respectively. (c) Evolution of the Fermi velocity for the Dirac cones I and II as a function of strain. (d) Evolution of the band gap of the second Dirac cone and the distance as a function of strain along the X_2 direction. Source: Padilha et al. [39]. © 2014, American Chemical Society.

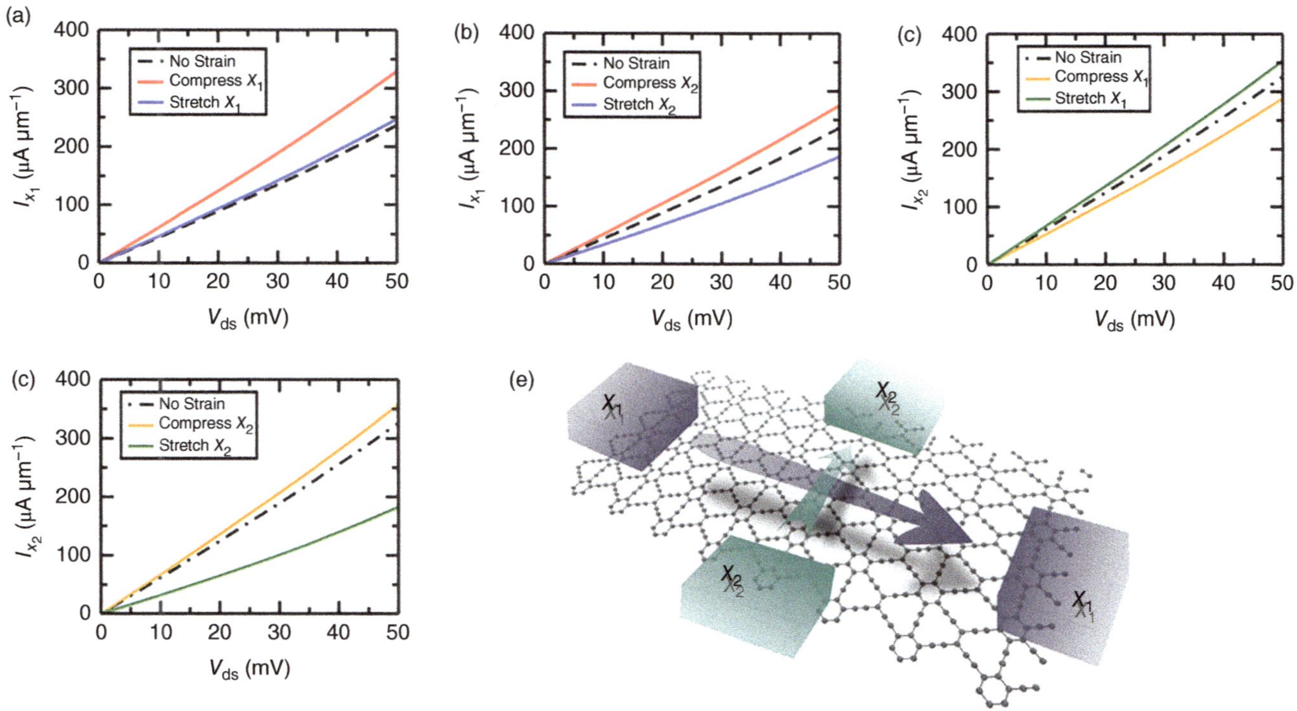

Figure 2.11 Current voltage characteristics for (a) transport on the X_1 direction with strain on X_1, (b) transport on the X_1 direction with strain on X_2, (c) transport on the X_2 direction with strain on X_1, (d) transport on the X_2 direction with strain on X_2, (e) schematic representation of the device used, showing the X_1 and X_2 directions. All the strength of strain are 1%. Source: Padilha et al. [39]. © 2014, American Chemical Society.

Dirac cone I remains unaltered. When compressing the system, the DOS close to Fermi level is increased, resulting in a slight increase of current.

Figure 2.11c,d presents the current along the X_2 direction with the strain applied on X_1 and X_2. The current is increased (or decreased) when the system is stretched (or compressed) along the X_1 direction (Figure 2.11c) due to the increase (or decrease) in the DOSs close to Fermi level, which are the results of changes in the Dirac cone I (II). When compressing along the X_2 direction (Figure 2.11d), there is also a small increase in the current due to the increase in the DOSs. When stretching the system along the X_2 direction, the current is significantly decreased since the main contribution comes from the Dirac cone II when the current is along X_2, and the stretch causes a band gap opening at this point. In this case, the current is dropped by ~45% with a tensile strain of 1%. As discussed above, the current–strain behavior could be very useful in the design of electronic devices that could be controlled by applying strain.

2.3 Mechanical Properties

2.3.1 Mechanical Properties of GDYs

The graphyne family (GDYs) has attracted wide attention and interests of scientists due to its excellent mechanical properties. Zhang et al. [55] studied the mechanical properties of four types of graphynes (i.e. α, β, γ, and 6,6,12-graphynes) with different acetylenic repeats using molecular dynamics (MD) simulations with the adaptive intermolecular reactive empirical bond order (AIREBO) potential. The fracture stresses of graphynes family observe the sequence: γ- > 6,6,12 - > β- > α-graphyne, ranging from 32.48 to 63.17 GPa, which are about 1/3 to 1/2 of those of graphene [56–58]. As shown in Table 2.3, the fracture stress and strain of graphene and graphynes in the zigzag direction are higher than those in the armchair direction. The anisotropic properties can be explained by the difference in the orientation of bonds.

Among the four graphynes, the 6,6,12-graphyne displays the most obvious anisotropy in the fracture stress and Young's modulus. This can be attributed to that all the graphynes and graphene structures having hexagonal symmetry except for 6,6,12-graphyne [5]. Besides, the fracture stresses of the four graphynes structures significantly depend on the percentage of acetylenic repeats, which decrease as the percentage of acetylenic linkages increases (Table 2.3). This is because the increase of acetylenic linkages percentage induces much lower atom densities in these graphynes structures. For example, from γ-graphyne (with 33.33% of acetylenic linkages) to α-graphyne (with 100% of acetylenic linkages), the atom density drops from 29.61 to 18.92 atoms nm^{-2}. Contrarily, the fracture strain of graphynes displays a reverse trend with respect to the percentage of acetylenic linkages; owing to that, the presence of acetylenic linkages in graphynes makes them flexible and thereby enhances the fracture strain. Similar to the fracture stress, Young's modulus of graphynes is also decreased as the percentage of acetylenic linkages increases.

Zhang et al. [55] compared the elastic properties of graphene, α-, β- and γ-graphynes using DFT calculations. It is found that the elastic constants of

Table 2.3 Fracture stresses, strains, and Young's moduli of graphynes and graphene.

Model	Percentage of acetylenic linkage	Atom density (atoms nm^{-2})	Stress (GPa)		Difference in stresses (%)	Strain		Difference in strains (%)	Young's modulus (TPa)	
			x	y		x	y		x	y
α	100	18.92	36.36	32.48	10.69	0.178	0.156	12.37	0.12	0.119
β	66.67	23.13	46.26	38.06	17.72	0.162	0.130	19.54	0.261	0.26
6,6,12	41.67	28.02	61.62	39.06	36.61	0.147	0.116	21.54	0.445	0.35
γ	33.33	29.61	63.17	49.78	21.20	0.148	0.112	24.09	0.505	0.508
Graphene	0	39.95	125.2	103.6	17.27	0.191	0.134	29.93	0.995	0.996

Source: Zhang et al. [55]. © 2012, AIP Publishing.

these graphyne allotropes show more anisotropy compared to those of graphene. The sparse carbon arrangement and the directionality of acetylenic linkages in graphynes lead to the internal intensification of directionality that depends on the applied load, resulting in a nonlinear stress–strain behavior. A.R. Puigdollers et al. [26] investigated the mechanical properties (e.g. uniaxial strains, homogeneous and heterogeneous biaxial strains, in-plane stiffness, Poisson's ratio) of α-, β-, and γ-graphynes based on DFT calculations (Figure 2.12). The three graphynes exhibit isotropic elastic uniaxial strain, α-graphyne being the most easily deformable structure due to the lowest planar packing density (Figure 2.12a). The homogeneous biaxial strain is the most energy-demanding property for the three graphynes materials, being four times higher than the uniaxial strain for α- and γ-graphyne (Figure 2.12b). For the heterogeneous biaxial strain, a softer response compared to the homogeneous biaxial strain is observed, which is mainly due to a compensation that originated from the types of applied strain (Figure 2.12b).

The calculated in-plane stiffness for the three graphynes observes the following order: γ-graphyne (165.51 $N\,m^{-1}$) > β-graphyne (73.07 $N\,m^{-1}$) > α-graphyne (21.98 $N\,m^{-1}$), which is almost half or less than the value of graphene (341.09 $N\,m^{-1}$), indicating these graphyne materials are much softer than graphene. The fact can be attributed to the reduction of average coordination numbers and the decrease of planar packing densities in graphynes compared to graphene [59]. For example, γ-graphyne has the highest in-plane stiffness among the three graphynes, while the average coordination number and planar density value in γ-graphyne are 2.5 and 0.582 $mg\,m^{-2}$, which are still smaller than those in graphene (i.e. 3.0 and 0.756 $mg\,m^{-2}$, respectively). Therefore, it can be concluded that the inclusion of acetylenic linkages into the graphene lattice decreases the bonding number and the planar density, reducing the rigidity of materials and making them much softer. The properties make graphynes potential candidates for a wide variety of applications that need soft materials such as the membrane separations.

Contrarily, Poisson's ratio of graphynes and graphene observes an opposite trend: α-graphyne (0.87) > β-graphyne (0.67) > γ-graphyne (0.42) > graphene (0.18), where

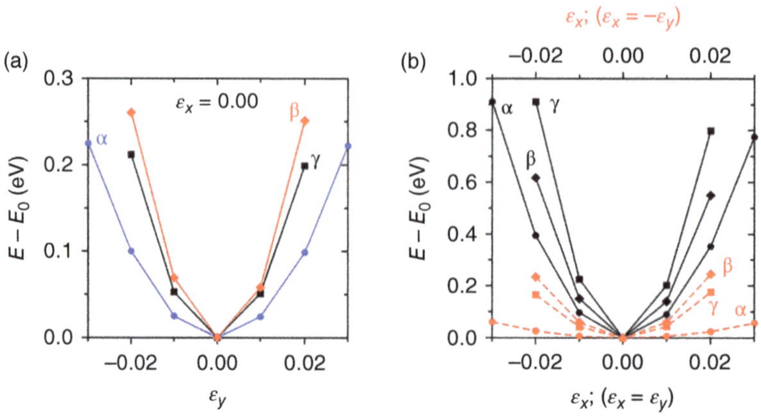

Figure 2.12 (a) Energy curves of uniaxial strains on y-axis maintaining $\varepsilon_x = 0.00$ fixed; the uniaxial strain on x-axis maintaining $\varepsilon_y = 0.00$ fixed is coincident and only the former is shown. (b) Energy curves of symmetric homogeneous (solid lines and bottom x-axis) and heterogeneous biaxial (dashed lines and top x-axis) strains. Circles, rhombus, and squares symbols stand for α-, β-, and γ-graphyne structures, respectively. Source: Ruiz Puigdollers et al. [26]. © 2016, Elsevier.

the value of α-graphyne is almost five times larger than that of graphene. The larger Poisson's ratio of graphynes with respect to graphene can be explained by the available large contractions originating from the sparse in-plane atoms in graphynes. The large Poisson's ratio of graphynes means that it can be highly deformed along the perpendicular direction to the one that performs the strain. Besides, it is worth noting that Poisson's ratio of γ-graphyne (0.42) is the closest to the value of 0.5, corresponding to a perfect incompressible material in all directions. Hereinafter, only the mechanical properties of γ-graphynes are mainly discussed.

2.3.2 Mechanical Properties of γ-Graphyne

The mechanical properties of γ-graphyne monolayer under various strains were studied using DFT based first-principles calculations [60]. Graphyne monolayer can bear large nonlinear elastic deformations up to an ultimate strain of 0.2, then the strain softens until failure. In this process, it is more vulnerable to rupture for the single bond than the triple bond and aromatic bond. The elastic properties of graphyne such as the deformation, the failure behavior, and the ultimate strength are anisotropic with respect to the direction of compressive and tensile strains. According to the orientation of hexagonal crystalline lattice, the edge structures of graphyne are divided into the armchair or zigzag edge. The ultimate strengths of graphyne monolayer under biaxial strain are about 2.81 and 1.81 $N\,m^{-1}$, larger than those under the strain in the armchair and zigzag directions, respectively. The elastic properties of graphyne are described by explicitly determining the elastic constants via fitting the stress–strain curve. The influences of pressure on the elastic constants, in-plane Young's modulus, and Poisson ratio are predicted. It is observed that the variations of local pressure introduced by external stress could be used to modulate the velocity of sound waves in graphyne. Therefore, the graphyne-based materials with controlled local strain can be applied for the next-generation electronic devices such as the nanodevices and the surface acoustic wave sensors.

Buehler and coworker [61] characterized the mechanical properties of γ-graphyne monolayer using first-principles based ReaxFF molecular dynamics simulations. To encompass the deformation of graphyne monolayer until fracture, two uniaxial tensile strains up to 25% are applied. The resulting stress–strain relationships in armchair and zigzag direction are plotted in Figures 2.13a,b, respectively. When applying the uniaxial strain along the armchair direction (x-axis), an ultimate stress of approximately 48.2 GPa and a maximum strain (ε_{ult}) of approximately 8.2% are yielded (Figure 2.13a). It is noted that the ultimate strain does not represent the complete fracture of graphyne sheet, but merely the local rupture and subsequent drop in load capacity. Owing to the acetylenic groups being directly subjected to the applied strain in the armchair direction, both the stress and strain capacity of graphyne are reduced. In contrast, applying the uniaxial strain along the zigzag direction (y-axis) results in a stiffening behavior (Figure 2.13b). The initial stiffness of graphyne is 700.0 GPa (or 224.0 N m^{-1}) by only considering a strain of 5%. Upon applying additional strain, the acetylenic groups undergo alignment to the direction of strain, and the stiffness is increased to 888.4 GPa (284.3 N m^{-1}). With more engaged carbon bonds participating, an ultimate stress of 107.5 GPa and a failure strain of 13.2% are eventually determined, respectively (Figure 2.13b). Different from graphene, the sparser carbon arrangement and the direction-dependent acetylenic groups in γ-graphyne lead to the strong anisotropic and nonlinear stress–strain behavior.

The bending modulus of γ-graphyne is further investigated (Figure 2.13c). To avoid the boundary effects at the fixed edges, only the elastic energy for the interior portion of graphyne sheet is chosen (Figure 2.13d). The effective bending modulus can be regarded as the derived stiffness originated from the graphyne structure that deviates from the initial imposed ideal curvature. Figure 2.13d depicts the temperature-equilibrated and energy-minimized structure. The bending modulus is calculated by fitting the energy-curvature data to the following expression:

$$U_{\text{bend}} = \tfrac{1}{2}\, \mathrm{D}\, \kappa^2 \tag{2.3}$$

where U_{bend} is the system strain energy per unit basal plane area, D is the bending modulus per unit width, and κ is the specified beam curvature. Curvatures in a range of 0.002–0.012 Å^{-1} are imposed on the graphyne system and the minimized energies are plotted versus κ (Figure 2.13e). The calculated bending stiffness of γ-graphyne is ~1.68 eV, which is comparable to that of graphene (i.e. 1.4–1.5 eV) [62, 63].

Young's modulus of corresponding γ-graphyne nanoribbons (GyNRs) with different widths were analyzed using density functional calculations within generalized gradient approximations [64]. The atomic configurations of GyNRs under uniaxial compressive strain and tensile strain are depicted in Figures 2.14a–f, respectively. Within the particular asymmetrical critical compressive and tensile strains, GyNRs will undergo a reversible deformation. According to the energy–displacement relations (Figure 2.14g), the two-dimensional Young's modulus is obtained, which is

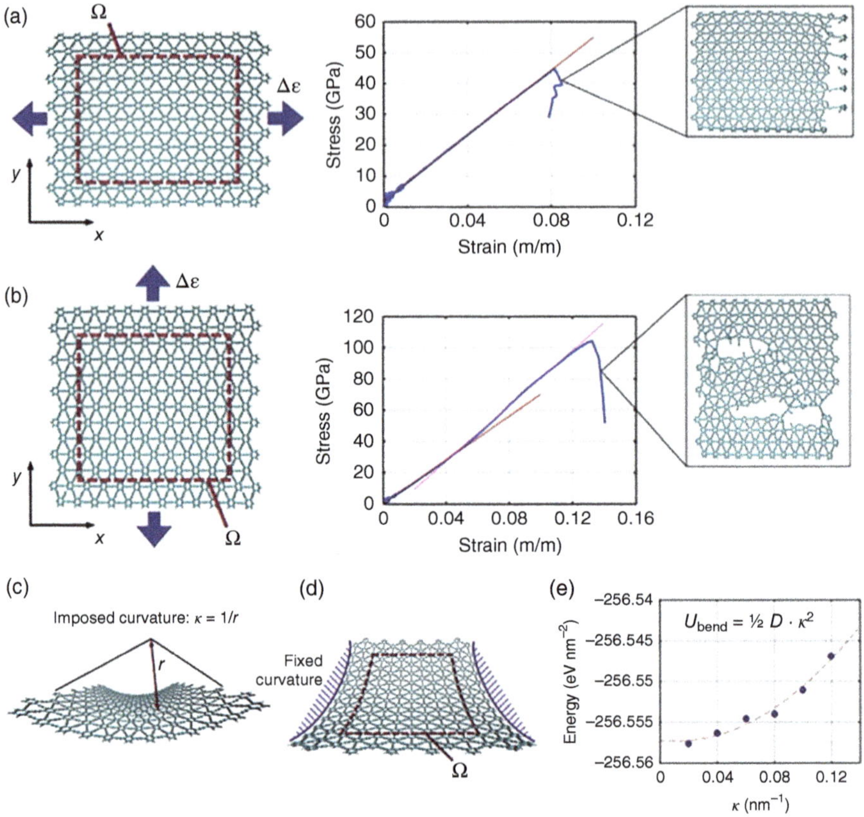

Figure 2.13 Schematic and stress–strain results of uniaxial tension tests in (a) armchair direction (*x*-axis) and (b) zigzag direction (*y*-axis). Determination of out-of-plane bending stiffness: (c) a graphyne sheet bent into a section of cylinder with an imposed curvature throughout the basal plane. (d) Depiction of boundary conditions, where the edge of bent sheet is fixed and the bulk of bent sheet is allowed to relax. (e) The energy-curvature plotted versus κ. Curvatures of 0.002–0.012 $Å^{-1}$ (cylindrical radii of approximately 83–500 Å) are imposed on the graphyne system. Source: Cranford et al. [61]. © 2011, Elsevier.

increased as the widths increase. When the compressive strain is beyond the critical value, the unidirectional corrugations perpendicular to the strain direction are formed in the narrower nanoribbons, while the transverse and longitudinal corrugations are formed in the wider nanoribbons. When the tensile strains exceed the critical value, all the nanoribbons undergo longitudinal corrugations before fracture. The corrugation wavelength does not depend on the applied strain but on the width of nanoribbons. All these nanoribbons are semiconductors with controllable band gaps ranging from 0.139 to 1.216 eV, depending on the width and applied strain. Furthermore, the band gaps of GyNRs are sensitive to the tensile strain, which is decreased steadily as the strain increases. Since the bands near Fermi level are the $2p_z$ states that are mainly composed of π orbitals of benzenes in graphyne sheet, the band gaps can be continuously modulated regardless of the influences of critical strain.

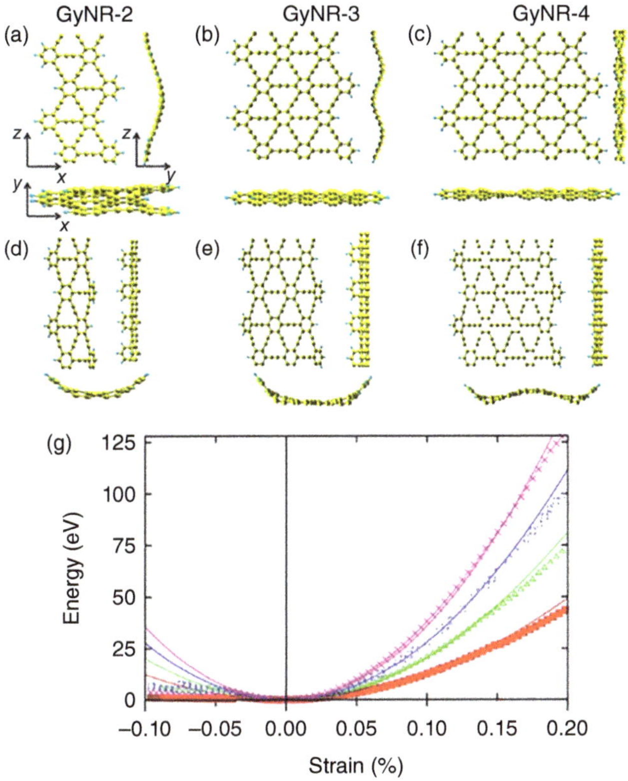

Figure 2.14 Atomic corrugated configurations of GyNRs with a width of $n = 2$ for (a) and (d); $n = 3$ for (b) and (e); $n = 4$ for (c) and (f). Panels (a)–(c) are the configurations under compressive strain, while (d)–(f) are the configurations under tensile strain. The atomic corrugated configurations of 5-GyNR are the same as those of 4-GyNR. (g) Schematic illustration of the strain energy as a function of uniaxial strains for 2-(red, square), 3-(green, triangle), 4-(blue, dot), and 5-GyNRs (pink, star). The data are self-consistently calculated points, and the lines are fitted with the least-squares method. Source: Wang et al. [64]. © 2014, American Chemical Society.

2.3.3 Mechanical Properties of γ-Graphdiyne

The nanoindentation processes of experimental accessible γ-GDY film were demonstrated using MD simulations [65]. The simulated elastic modulus and strength of GDY are about 489.04 and 33.95 GPa, respectively. The GDY film begins to fail when the indentation depth reaches the maximum value of 42.08 Å, corresponding to the critical stress of 40.36 nN. By contrast, the fracture depth and the maximum stress of monolayer graphene film are 55.95 Å and 655.08 nN, respectively, indicating the maximum force that graphene film can bear is much larger than that of GDY film under the same indentation depth. Hence, the strength of GDY is lower than that of graphene.

Figure 2.15 depicts the fracture behavior of GDY film. When the indenter is close to GDY surface (smaller than 3.4 Å), the film is convex due to the attraction interaction

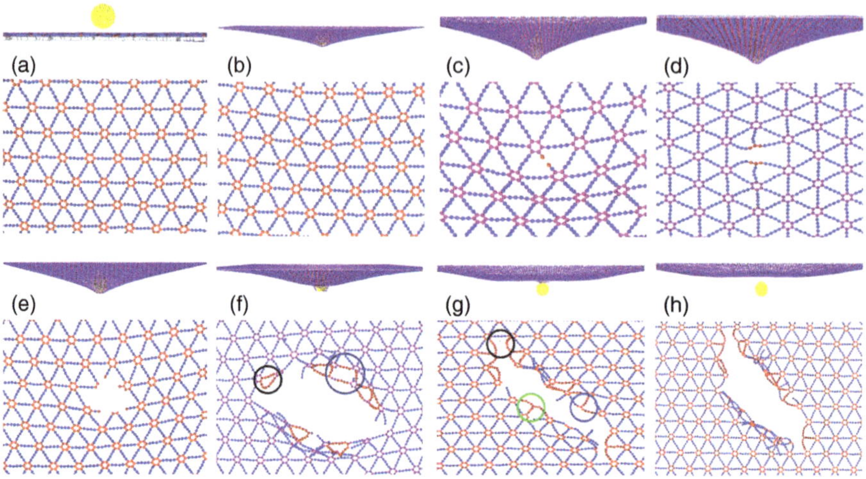

Figure 2.15 Evolution of lattice fracture of graphdiyne film: (a) initial structure, (b) prolonged lattice structure, (c) broken of acetylenic bonds, (d) broken of −C=C− bonds, (e) broken of graphdiyne film, (f) recombined structure of −C−C− bonds and −C≡C− bonds, (g) fracture manners of −C=C− bonds, and (h) final broken behavior. Source: Xiao et al. [65]. © 2019, Elsevier.

(Figure 2.15a). With the increase of indentation depth, the force between indenter and film changes from the attraction to repulsion. Moreover, the carbon bonds in GDY become much longer (Figure 2.15b). As the indentation further increases, the bond lengths of atoms around indenter begin to change. When the indentation depth exceeds the critical value, the π bond in acetylenic bonds (—C≡C—) fractures first. This is because the binding energies of triple bonds are lower than those of single and double bonds. In this process, GDY film is greatly expanded (Figure 2.15c–e). Then, the π bond in some double bonds (—C=C—) breaks. Finally, some single bonds (—C—C—) appear to break owing to the σ bond is much stable than the π bond.

Interestingly, the broken single bonds and acetylenic bonds in GDY can be recombined to sustain further deformation of GDY film (Figure 2.15f), forming the triangular and quadrangular structures, as shown with the highlighted black and blue circles, respectively. The broken double bonds can also be partially bound with the adjacent carbon atoms (Figure 2.15g), forming the unique microstructures and failure patterns, such as the trilateral structure that is highlighted with circles. Once the GDY film is completely destroyed, the force between the indenter and GDY film will suddenly drop to zero, resulting in an irreversible bond fracture. After the indenter completely passes through the GDY film, some broken atoms are separated from the film, forming asymmetrical lathy crack (Figure 2.15h). The behavior is completely different from the symmetrical fracture of graphene [66].

Figure 2.16a gives a representative stress–strain plot for depicting the mechanical properties of GDY, which are subjected to the uniaxial tensile load in both the reclined chair and zigzag directions [3]. It is quantified that there is a modulus of 470–580 GPa and an ultimate strength of 36–46 GPa for GDY, depending on the edge direction (armchair or zigzag). The relationships between the effective moduli (Y)

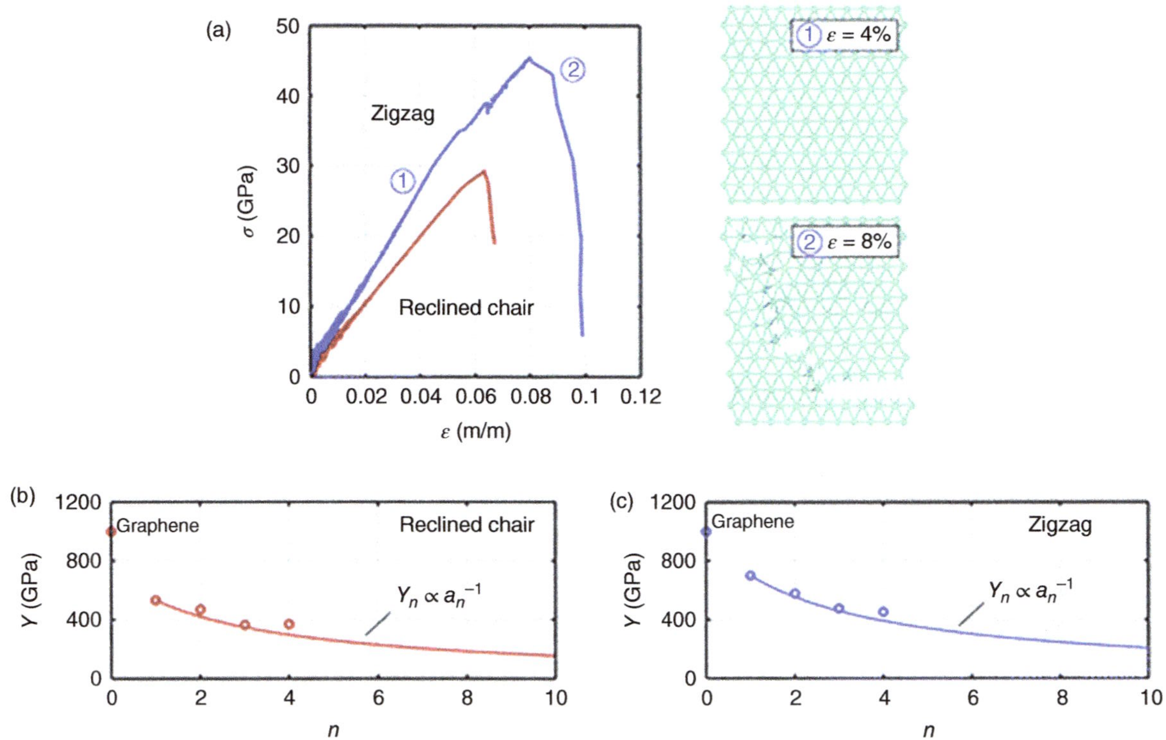

Figure 2.16 (a) Representative stress–strain response for graphdiyne ($n = 2$) along the reclined chair (red) and zigzag (blue) directions, with the snapshots of graphdiyne under deformation at ~4% and ~8% strain (the right insets). Decay of modulus, Y, as a function of acetylenic groups, n, for both (b) reclined chair and (c) zigzag directions. Source: Cranford et al. [3]. © 2012, Royal Society of Chemistry.

and the acetylenic groups (n) for all the extended graphyne models are plotted in Figure 2.16b,c, which are along the reclined chair direction and the zigzag direction, respectively. It is observed that the moduli of these graphynes are continuously degraded with the addition of acetylenic linkages. There is a decrease on the order of 30% in the reclined chair direction and 35% in the zigzag direction from graphyne ($n = 1$) to graphtetrayne ($n = 4$).

2.3.4 Mechanical Properties of γ-Graphynes Family

The mechanical properties (such as the fracture strain and ultimate stress) of γ-graphynes with extended acetylenic repeats are different. Table 2.4 summarizes the mechanical properties of graphene ($n = 0$) and the extended graphynes ($n = 1$–4) models based on the MD simulations [3]. To characterize the mechanical properties of γ-graphynes with extended acetylenic repeats [67], a full atomistic model with 10 nm × 10 nm was constructed (Figure 2.17a). These graphynes allotropes ($n = 1$–5) are named by graphyne, GDY, gaphyne-3, graphyne-4, and graphyne-5, respectively (Figure 2.17b). Based on the orientation of crystalline lattice, the edges of graphynes are designated as armchair or zigzag shape. As depicted in Figure 2.17c, there are three kinds of bonds in these graphynes, namely aromatic bonds (types A and B), single bonds (types C, C′, E, and E′), and triple bonds (types D and D′). Among, types C (C′) and E (E′) represent the single bonds adjacent to aromatic rings and inside the acetylenic chains, respectively.

At the beginning of load test, an obvious structural rupture was observed. This is because the angle between two acetylenic chains (Figure 2.17b) is significantly reduced when the sheet is stretched, which contributes to the release of external stress and the realignment of acetylenic chains to the direction of load. At the latter half of load test, the lengths of carbon bonds become enlarged and the total stress is increased rapidly. Whether upon armchair or zigzag load, the first bond breakage always occurs at the carbon site connecting the aromatic ring with the acetylenic group. The elastic region for all the members of graphynes family extends to ~0.11 in the armchair direction, whereas the ultimate stress averaged on all atoms (σ_c)

Table 2.4 Summary of calculated mechanical properties for extended graphynes.

		Armchair			Zigzag			
Structure	n	Y (GPa)	σ_{ult} (GPa)	ε_{ult} (%)	Y (GPa)	σ_{ult} (GPa)	ε_{ult} (%)	σ_{cr} (GPa)
Graphene	0	1000	130	20	1000	130	20	n/a
Graphyne	1	532.5	48.2	8.2	700.0	107.5	13.2	178.5
Graphdiyne	2	469.5	36.0	6.3	578.6	45.5	8.0	48.1
Graphtriyne	3	365.0	26.8	7.7	476.7	43.7	9.9	19.5
Graphtetrayne	4	370.4	24.8	7.0	453.3	32.5	9.7	9.8

Source: Cranford et al. [3]. © 2012, Royal Society of Chemistry.

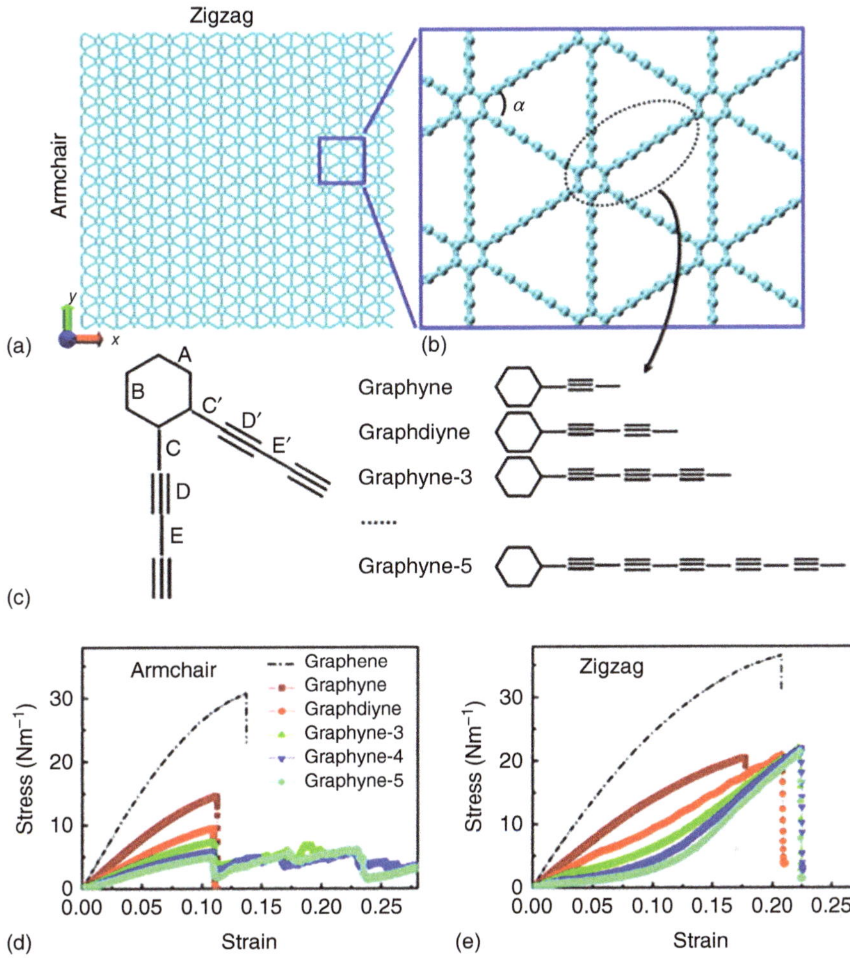

Figure 2.17 (a) A 10 nm × 10 nm γ-graphyne sheet, along with the definition of edges. (b) Schematic illustration of the units of graphynes family with different number of acetylenic repeats between aromatic rings. α corresponds to the included angle of two acetylenic chains. (c) Definition of the types of bonds. The stress–strain relationships for the graphynes family under load along (d) armchair and (e) zigzag direction. The results for graphene with the same size are shown for comparison. Source: Yang et al. [68]. © 2012, Elsevier.

gradually decreases from 14.437 to 4.950 N m^{-1} with the acetylenic groups increasing from one (for graphyne) to five (for graphyne-5). On the contrary, in the zigzag direction, the ultimate strength for the five allotropes nearly remains the same, but the fracture strain is slightly increased by adding more acetylenic groups into the hexagonal grid. These graphynes sheets are found to stretch elastically to 0.177–0.224 and then rupture, with corresponding maximum stress around 21 N m^{-1}.

As discussed above, both the critical strain and strength along the armchair direction are lower than those along the zigzag direction. This can be expected because some bonds are subjected directly to the load during the initial elongation process in the case of armchair deformation, while the external load can be dissipated by the changes of bond length and bond angle in the case of zigzag deformation. In addition, the tensile strengths of graphynes family are less than half of those of graphene sheet, in which the maximum stresses at the fracture point are $30.69\ N\,m^{-1}$ in the armchair and $36.58\ N\,m^{-1}$ in the zigzag load, respectively. Furthermore, it is interesting to note that the graphyne-3, graphyne-4, and graphyne-5 structures exhibit a mechanically strengthening-like behavior in the stress–strain evolution, which is probably derived from the rehybridization of spatial bonds after the first bond breakage, suggesting that more ductile materials can be obtained.

The stress–strain relationships of graphynes family were obtained by applying axially tensile deformation in the armchair and zigzag-oriented monolayer sheets. It is found that these graphynes structures have high strength and stiffness due to the large Young's modulus. In the case of armchair load (Figure 2.17d), the stress–strain curve shows a parabolic increase trend. Moreover, the fracture strain remains almost unchanged, whereas the ultimate stress degrades gradually as the length of acetylenic chains increases (from graphyne to graphyne-5). In the case of zigzag load (Figure 2.17e), the stress increases more slowly in the small strain regime and more rapidly in the large strain regime. Besides, with the increase of chains length, the fracture strain increases slightly, but the ultimate strength remains nearly unchanged. These results show that the fracture strain and ultimate stress for different extended graphynes strongly depend on the loading types, such as along the armchair or the zigzag edge. The unique mechanical properties of graphynes family under armchair and zigzag load can facilitate the design of advanced nanomechanical carbon materials with the direction-dependent properties.

Buehler and coworkers [3] systematically investigated the mechanical properties of extended graphynes using full atomistic first-principles-based ReaxFF molecular dynamics. The increase in the number of acetylenic repeats (n = 0, 1, 2, 3, and 4) has great adverse influences on the elastic modulus (i.e. stiffness), and failure strength of extended graphynes structures. With the accumulation of acetylenic repeats, the ultimate stress is reduced from about 30 GPa for graphyne to less than 10 GPa for graphtriyne/graphtetrayne. The stress–strain response plots show that the stress of graphene and graphynes increases almost linearly with the strain, and then drops sharply during failure (Figure 2.18a). The fracture behavior for graphene is brittle-like, with an ultimate stress of about 60 GPa and a failure strain of about 8% (Figure 2.18a). In contrast, the responses for graphynes are more complex. As visualized in Figure 2.18b, it starts with an initial linear trend of stress versus strain, then followed by a discrete spring network model. Therefore, a unique "two-tier" fracture mode appears due to the mobility of acetylenic groups, resulting in multiple stress peaks. For extended graphynes allotropes with larger acetylenic chain (e.g. $n = 3, 4$), the phenomenon becomes more pronounced.

There is a simple scaling law between the in-plane stiffness (C_n) and the acetylenic linkages number (n) for extended graphynes [31]: $C_n = 166.3/[1 + 0.373\,(n-1)]$. The in-plane stiffness of these graphynes is found to degrade gradually from 166, 123, 102, to $88\ N\,m^{-1}$ with the increase of acetylenic bonds number ($n = 1-4$). Based on the relationships, the stiffness of extended graphyne systems

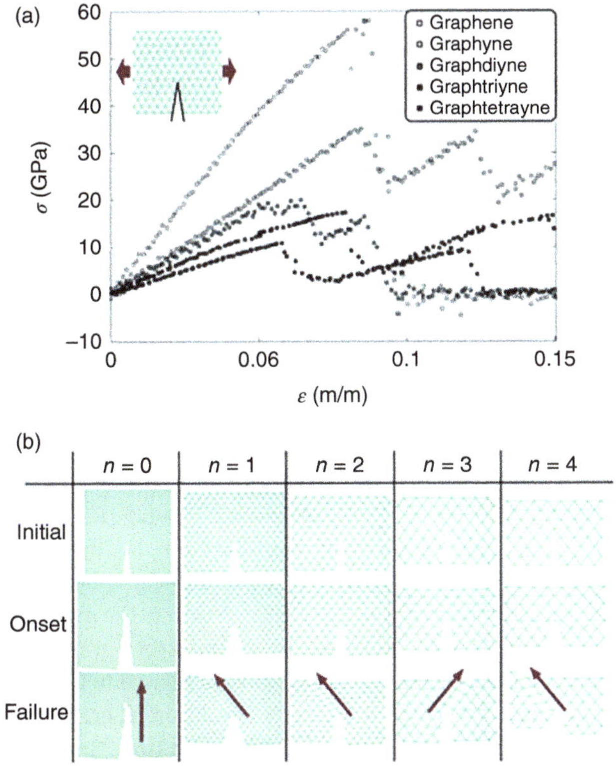

Figure 2.18 Fracture mechanics of graphynes. (a) Stress–strain results for each model. In terms of critical stress (onset of fracture), the observed failure events are consistent with the predictions using the linear spring model. (b) Visualization of failure, which begins at the crack and propagates along the plane defined by the single bonds (C—C) in acetylenic linkages (indicated by arrows). Source: Cranford et al. [3]. © 2012, Royal Society of Chemistry.

with different acetylenic chains lengths can be predicted. The extended graphynes models (such as graphtriyne and graphtetrayne) express significant modulus (>300 GPa) and strength (>20 GPa), which is comparable to the metals and fibers with high performance, indicating the potential applications of graphynes in chemically selective filters, energy storage, or functional coatings/composites scaffolds. Besides, the specific modulus and strength (normalized by areal density) is near-constant, suggesting these extended graphynes can be used in lightweight but structurally robust molecular components.

2.3.5 The Influence Factors on the Mechanical Properties of GDYs

The substitutional doping of nonmetal elements (such as N and B) can regulate the mechanical properties of GDYs (such as α-, β-, γ-graphyne) as revealed by spin-polarized DFT calculations [69]. For example, since the distance of C—N bond is shorter than C—C bond, the presence of N dopant can contract the carbon lattice and increase the in-plane stiffness. However, the length of C—B bond is more than that of C—C bond, thereby the presence of B dopant can expand the lattice and reduce the in-plane stiffness with respect to the pristine graphynes. Hence, the planar packing densities of graphynes increase upon N doping, but decrease upon B doping, resulting in the increase and decrease of corresponding in-plane stiffness, which induces much harder and softer materials, respectively. However, their Poisson's coefficients are almost not changed after doping because they depend on the atomic arrangement rather than the chemical composition in the lattice. Besides, it is found that the doping ratios of N and B impurities in the lattice also affect the mechanical properties of graphynes.

The hydrogen chemisorption can change the mechanical properties of γ-graphyne [70]. Two configurations for the maximum hydrogen adsorption on graphyne surface are considered: the adsorption on the same side or two opposite sides of graphyne sheet. It is indicated that the alternate hydrogen adsorption on two opposite sides of graphyne (H-graphyne) is more stable according to the formation energies. In addition, Poisson's ratio and in-plane stiffness of H-graphyne are 0.23 and 125 $N m^{-1}$, respectively, which are lower than those of pure graphyne. Therefore, the hydrogen chemisorption can significantly weaken the mechanical properties of graphyne.

In fact, the functionalization with hydrogen could greatly deteriorate the mechanical properties of different types of graphynes, such as α-, β-, γ-, and 6,6,12-graphynes [71]. The deterioration degree of mechanical properties is closely related to the percentage of acetylenic bonds in these graphyne structures. For different graphynes allotropes, the higher the percentage of acetylenic bonds in the structures, the larger the reduction in fracture stress (see Table 2.5). For the same graphyne, the reduction in fracture stress depends on the location, distribution, and coverage of hydrogenation. For example, the hydrogenation on the acetylenic chains reduces much greater fracture stress than that on the hexagonal rings. Besides, in the linear hydrogenation, H atoms perpendicular to the tensile direction reduce much larger fracture stress of graphyne than those parallel to the tensile direction. For the random hydrogenation,

Table 2.5 Fracture stresses and strains of graphynes and graphene with and without hydrogenation.

Model	Acetylenic linkage (%)	Density of structure (atoms nm^{-2})	Fracture stress (GPa) H-0%	Fracture stress (GPa) H-100%	Reduction in stresses (%)	Fracture strain H-0%	Fracture strain H-100%	Reduction in strains (%)
α-Graphyne	100	19.12	33.39	10.01	70.00	0.152	0.101	12.37
β-Graphyne	66.67	23.24	38.11	14.77	61.24	0.126	0.088	19.54
6,6,12-Graphyne	41.67	27.81	38.34	16.49	56.99	0.109	0.081	21.54
γ-Graphyne	33.33	28.92	50.25	22.20	55.83	0.111	0.081	24.09
Graphene	0	36.08	104.46	35.44	66.07	0.140	0.095	29.93

Source: Zhang et al. [71]. © 2013, AIP Publishing.

the fracture stress and Young's modulus decrease rapidly at low H-coverage (<10%) and remain stable as the H coverage increases. The reduction in the mechanical properties caused by hydrogenation can be attributed to the formation of weakened out-of-plane C—C bonds at the local hydrogenation region, which leads to the early bonds fracture and the subsequent graphynes fracture.

Besides, the mechanical properties of graphyne and GDY were sensitive to temperature and defects, respectively [72]. For example, the in-plane stiffness and Young's modulus of graphyne decrease as the temperature increases from 300 to 1500 K, while those of GDY decrease as the number of removed carbon atoms

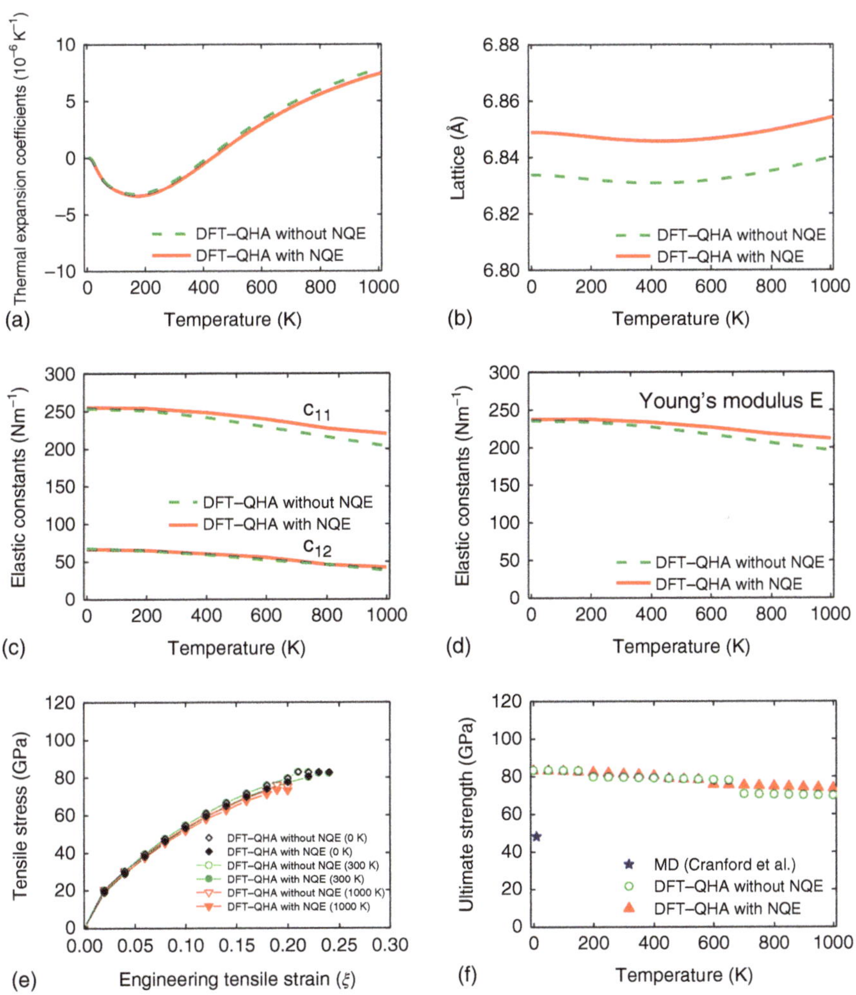

Figure 2.19 (a) In-plane linear thermal expansion coefficient of graphyne, (b) temperature-dependent graphyne lattice, (c) temperature-dependent elastic constants for graphyne, (d) temperature-dependent Young's modulus for graphyne, (e) tensile stress versus strain for graphyne along *x* axis at 0, 300, and 1000 K, (f) temperature-dependent ultimate strength for graphyne along *x* axis. Source: Shao et al. [73]. © 2012, AIP Publishing.

(i.e. defect concentrations) increases. Shao et al. [73] systemically investigated the lattice geometries, thermal expansion coefficients (TEC), elastic constants, and ultimate strength of graphyne based on first-principles calculations. It turns out that these properties are temperature dependent. For the linear TEC, graphyne shows a negative region at low temperature and a positive region at high temperature (Figure 2.19a). As the temperature rose from 0 to 1000 K, the lattice parameters of graphyne decrease first and increase then (Figure 2.19b), while the elastic constants of graphyne decrease as the temperature increase (Figures 2.19c). Besides, graphyne exhibits superior mechanical properties with a Young's modulus of 250.91 $N\,m^{-1}$ and ultimate tensile strength of 81.2 GPa at room temperature, inferior to graphene with a Young's modulus of 350.01 $N\,m^{-1}$ and ultimate tensile strength of 119.2 GPa. As the temperature increases from 0 to 1000 K, Young's modulus of graphyne is decreased by 10.79% (Figures 2.19d), and the ultimate tensile strength is decreased by 12.10% (Figure 2.19f). These indicate that the mechanical properties of graphyne have weaker resistance against temperature. Besides, it is seen that the stress depends on the deformation strain for graphyne at 0, 300, and 1000 K, which increase almost linearly when the strain is less than 0.10 (Figure 2.19e).

2.4 Layers Structure of Bulk GDYs

2.4.1 Stacking Modes for Bilayer α-Graphyne

The electronic properties of multilayer GDY are qualitatively different from those of monolayer GDY. It is shown that the electronic properties of bilayer α-graphyne mainly depend on the stacking modes between two layers [74]. Figure 2.20a–f depicts six nonequivalent stacking modes for bilayer α-graphyne, namely AA, Ab, Aa, AB, ab, and aa. The corresponding band structures of bilayer α-graphyne around *K* point are shown in Figure 2.21a–f. It is seen that the various stacking modes lead to the obviously different electronic band structures. For example, the energy spectrum of AA stacking shows the double Dirac cone structures, one above and the other below Fermi level. However, the energy spectrum of AB stacking (the most favorable stacking mode for bilayer graphene) around *K* point shows the gapless parabolic band structure. These features are similar to the band structures of AA and AB stacked bilayer graphene [75]. Some interesting phenomena appear in the band structures of other stacking modes, such as Ab and ab stacking, where the single Dirac cone is split into two cones. Similar to the monolayer graphene and α-graphyne, both cones for Ab stacking are located at Fermi level, so there are no charge carriers in the neutral state. In other words, the band structure around Fermi level exhibits a linear dispersion.

In addition, the band gap in the Dirac spectrum for bilayer α-graphyne can be opened by an electric field at a rate of 0.3 eÅ. For ab stacking, both Dirac-like electron and hole are located in separate Dirac cones, and the energy displacement between two cones is about 0.1 eV, forming the metallic state. The electronic band structures of Aa and aa stacking modes are similar to those of AA stacking. However, there

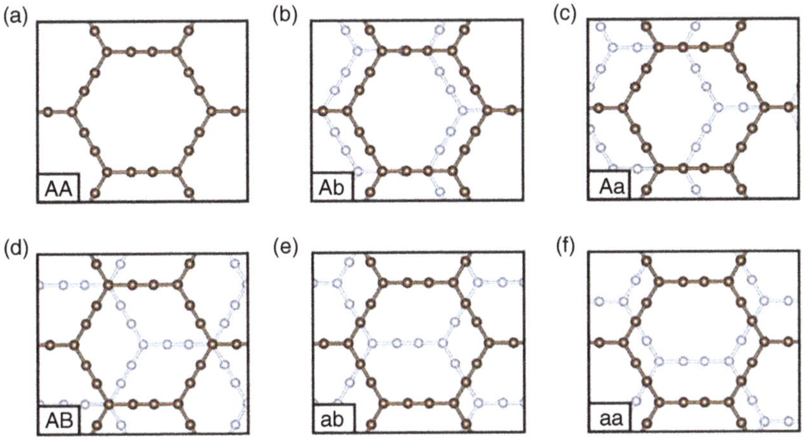

Figure 2.20 (a–f) Six different stacking configurations for bilayer α-graphyne. Two layers are indicated by two colors. Source: Leenaerts et al. [74]. © 2013, AIP Publishing.

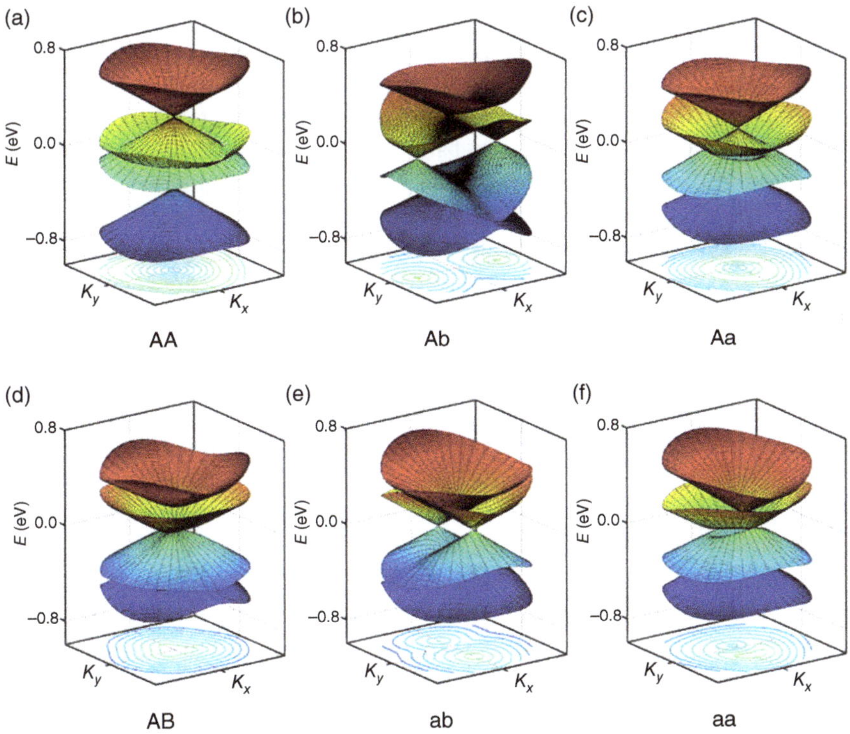

Figure 2.21 (a–f) Band structures near the K point (K_x, K_y) for different stacking styles in Figure 2.20. The Fermi level is taken as the origin of the energy scale and the bands are plotted within a neighborhood of radius $0.09(2\pi/a)$ around the K point. A contour plot of valence band is shown. Source: Leenaerts et al. [74]. © 2013, AIP Publishing.

are still some differences. Compared to AA stacking, the band structures of Aa and aa stacking are opened up wherever two Dirac cones meet. The most stable stacking mode that forms the ground-state configuration of graphyne is related with the employed functionals. For example, the calculated binding energies show that the Ab stacking is the most favorable mode for bilayer α-graphyne at the LDA level, while the AB stacking is the most stable configuration at the optPBE and optB86b levels. Hence, both stacking modes are probably stable.

2.4.2 Stacking Modes for Bilayer γ-Graphyne

Similar to α-graphyne, the bulk γ-graphyne exhibited different electronic structures compared to monolayer γ-graphyne, which was investigated using the first-principles full-potential LCAOs calculations [76]. Narita et al. considered four possible stacking arrangements for bilayers γ-graphyne from the views of symmetry, including two types, such as AA (α) and AB, (β1, β2, β3). The optimized lattice constants and bond lengths are almost the same as those in 2D graphyne monolayer, while the optimized interlayer distance is slightly larger than that in graphite. The binding energy of bulk graphyne is about 90% of that of graphite, which is large enough for the stable synthesis. Besides, the electronic structures of these bilayers γ-graphyne can be classified into two types, namely metallic or semiconducting, depending on the stacking arrangements. For example, β1 and β2 stacked γ-graphynes (the more stable stacking arrangements) show the semiconducting properties with a moderate band gap of 0.19 and 0.50 eV, respectively. However, α and β3 stacked γ-graphynes (the less stable stacking arrangements) show the characteristics of zero band gap semiconductor.

Figure 2.22a shows five different types of stacking modes for γ-graphyne bilayers (i.e. Gyne/Gyne), namely AA, AB1, AB2, AB3, and AB4 [77]. There is no buckling for these Gyne/Gyne structures after full relaxation. In terms of formation energy, the most stable configuration is the AB1-Gyne/Gyne structure with a value of −40.79 meV atom^{-1}. The second stable configuration is the AB4-Gyne/Gyne structure with a value of −39.38 meV atom^{-1}, which is about 1.41 meV atom^{-1} less than that of AB1-Gyne/Gyne. Because the van der Waals (vdW) interaction between two layers is weak, the electronic band structures of Gyne/Gyne along the high symmetry points are very similar to those of Gyne monolayer, and are plotted in Figure 2.22b.

The Gyne monolayer shows a direct band gap of 0.450 eV at the PBE level and 0.875 eV at the HSE06 level. The Gyne/Gyne bilayers have the energy band structures of direct band gap semiconductor, with the conduction band minimum (CBM) and valence band maximum (VBM) located at the M point in Brillouin zone. The band gaps of all the Gyne/Gyne bilayers are smaller than those of Gyne monolayer. Among these stacking structures, the AA-stacked Gyne/Gyne has the smallest band gap of 0.094 eV, in accordance with the previous report [78]. However, a doubly degenerate state emerges in the Gyne/Gyne bilayers due to the vdW force and interlayer interactions. In particular, the obtained band gap values are very

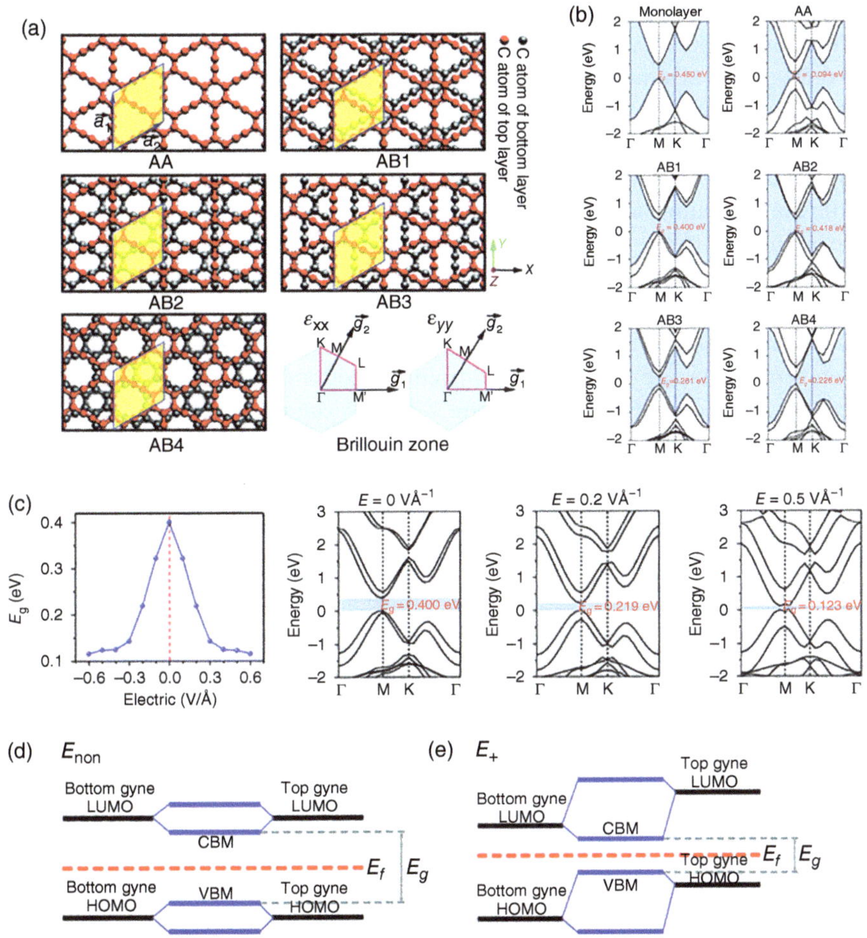

Figure 2.22 (a) Illustration of the Gyne/Gyne bilayers with different stacking modes. The yellow regions indicate the unit cell of bilayer structure. The first Brillouin zone labeling with high symmetry points under strains of ε_{xx} and ε_{yy} is shown, (b) band structures of the Gyne monolayer and Gyne/Gyne bilayers, respectively, (c) the band gap variations for AB1-Gyne/Gyne as a function of electric field and the corresponding band structures under 0, 0.2, 0.5 and 0.6 V Å^{-1}, respectively. Bandgap offset diagrams for Gyne/Gyne under (d) a nonelectric field (left) and (e) an electric field (right) from bottom to top, respectively. Source: Yun et al. [77]. © 2018, Royal Society of Chemistry.

distinct for different stacking modes. This is because the band structures of the Gyne/Gyne bilayers are sensitive to the lattice symmetry, which will be disrupted by different alternate stacking patterns or interlayer interactions [79].

There are great applications potentials for bulk γ-graphyne in the field of nanoelectronic devices and optoelectronic devices. Many reports show that it is an effective method to control the properties of layer structures by applying a gate voltage [80, 81]. The electronic response of the most stable AB1-Gyne/Gyne bilayers was measured under a vertical electric field. Figure 2.22c shows the band gap variations as a function of applied electric field ranging from −0.6 to 0.6 V Å^{-1}. It is found that the band gaps are symmetrically changed under both positive and negative electric fields. When the electric field is smaller than 0.3 V Å^{-1}, the band gaps reduce linearly. However, when the electric field is larger than 0.3 V Å^{-1}, the band gaps change slowly. This is because the position of band gaps is moved from the M point to the Γ–M line in Brillouin zone (Figure 2.22c).

The reduction trend of band gaps under the electric field can be explained according to the molecule orbital model. When there is no external applied electric field, the splits of the LUMO and the HOMO lead to the decrease in the band gap of the Gyne/Gyne bilayers compared to the Gyne monolayer (Figure 2.22d). However, when a vertical electric field is applied from the bottom layer to the top layer, the energy level of the top layer rose (Figure 2.22e). Besides, the splits in the HOMO and LUMO of Gyne/Gyne bilayers are also enlarged. Hence, the energy difference between the CBM of the bottom layer and the VBM of the top layer is further reduced.

2.4.3 Stacking Modes for Bilayer γ-Graphdiyne

Bilayer γ-GDY is a potential candidate for the tunable nanoscale optical devices with high performance [82]. Similarly, different stacking modes obviously affect their electronic properties. Four stacking configurations are considered, namely AA, AB-1, AB-2, and AB-3 (Figure 2.23a). Caused by the relative in-plane transition in these stacked GDY bilayers, the double degeneracy of bands near the Γ point is lifted (Figure 2.23b), leading to the damage of spectra symmetry between X (Figure 2.23c) and Y directions (Figure 2.23d). The spectrum in Z direction is not considered, because the local field effects strongly inhibit the light absorption within the measured energy range. All spectra have three major absorption peaks, which are located below 1 eV, around 1.7 eV, and around 4.02 eV (Figure 2.23c,d), the corresponding electron transitions are indicated in the DOS (Figure 2.23b). It is clearly shown that the adsorption peaks around 4.02 eV are almost the same for all the stacked configurations. However, for the energy below 1 eV, the differences in these spectra are significantly enhanced in the four stacking modes (Figure 2.23c).

For AA stacked configuration, the spectra in the X and Y directions are identical, and the lowest adsorption peak is located at 0.11 eV with the highest intensity, which is thereby easily distinguished from other configurations. The lowest peak of AB-1 stacked configuration is located around 0.15 eV in the X and Y directions. However, the intensity in the Y direction is ~40% greater than that in the X direction. In contrast, the spectrum of AB-2 stacked configuration is very similar to that of monolayer GDY. Finally, the spectrum of AB-3 stacked configuration has two peaks in the X direction, one is located at 0.28 eV and the other is located at 0.49 eV, but there is only one peak located at 0.67 eV in the Y direction. Compared to the spectrum of monolayer GDY, the peaks below 1 eV are generally redshifted in all stacked GDY, which is related to the reduction or disappearance of band gap.

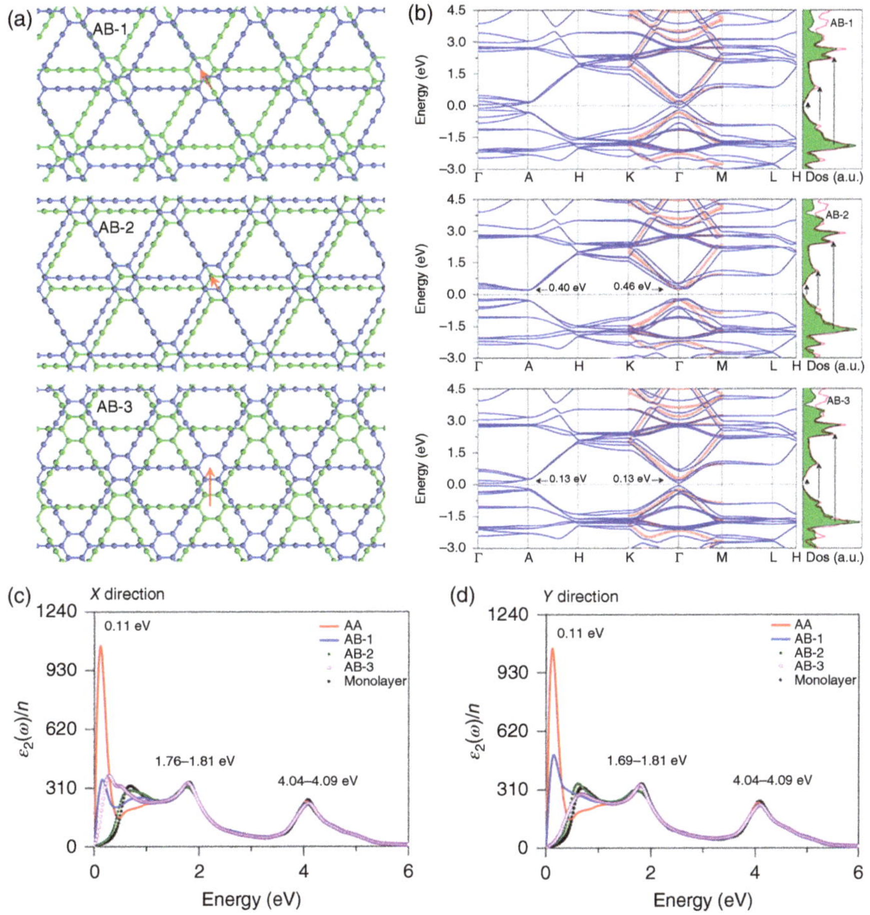

Figure 2.23 (a) Geometrical structures of AB-1, AB-2, and AB-3 stacked graphdiyne bilayers. The red arrows indicate the relative in-plane shifts between two layers. (b) Band structures and DOS of AB-1, AB-2, and AB-3 stacked graphdiyne bilayers, those of graphdiyne monolayer (red lines) are shown for comparison. Fermi level is set to 0 eV. Polarized optical absorption spectra of AA, AB-1, AB-2, AB-3 stacked graphdiyne bilayers and graphdiyne monolayer in the (c) *X* and (d) *Y* directions. Source: Luo et al. [82]. © 2013, American Chemical Society.

2.4.4 Identification on the Stacking Structures of GDY

Numerous studies were performed to identify the most stable stacking structures of bulk γ-GDY [83–85]. The structural and electronic properties of bulk GDY were systematically investigated using DFT calculations with vdW dispersion corrections.

Lu and coworkers [86] reported that both the most stable bilayer and trilayer GDY have their hexagonal carbon rings stacked in a Bernal way, such as the AB- and ABA-style configuration, respectively. The stacked GDY bilayers are either a semiconductor or a metal, depending on their stacking modes. For example, the GDY

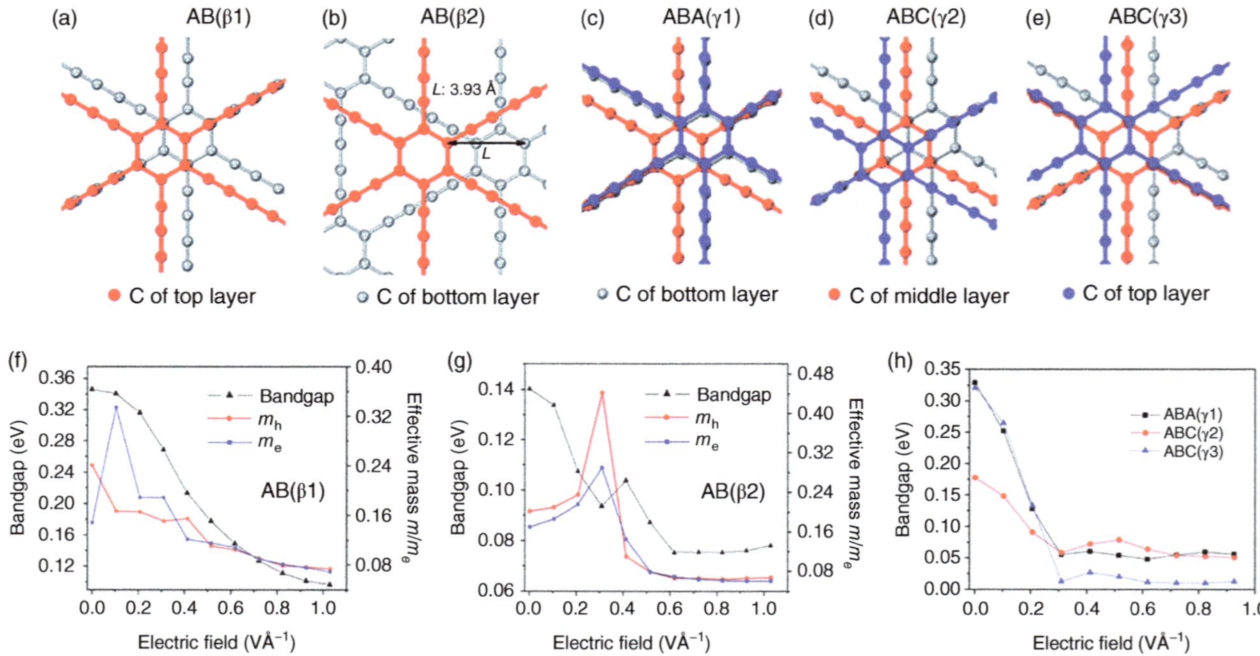

Figure 2.24 Two possible configurations of graphdiyne bilayers from top view: (a) AB (β1) and (b) AB (β2). Three possible configurations of graphdiyne trilayers from top view: (c) ABA (γ1), (d) ABC (γ2), and (e) ABC (γ3). Bandgaps of graphdiyne bilayers of (f) AB (β1), (g) AB (β2), and (h) graphdiyne trilayers as a function of electrical field strength. Source: Zheng et al. [86]. © 2012, Royal Society of Chemistry.

bilayers with the most stable (i.e. β1, Figure 2.24a) and the second most stable (i.e. β2, Figure 2.24b) stacking arrangements are semiconductor with a direct bandgap of 0.35 and 0.14 eV, respectively. By contrast, the AA (α) stacked GDY is the least stable configuration, which is metallic. Since the differences in binding energies and migration barriers between β1 and β2 stacking configurations are very small, both two stacking structures are probably stably existed for GDY bilayers. Therefore, the most stable stacking mode for GDY bilayers is presumably the AB-type [82, 86], either β1 or a mixture of β1 and β2 styles.

For GDY trilayers, the ABA (γ1) configuration (Figure 2.24c) with a bandgap of 0.32 eV is the most stable one among the several stacking modes. The other two GDY trilayers with the stacking configurations of ABC (γ2) (Figure 2.24d) and ABC (γ3) (Figure 2.24e) are less stable, with a bandgap of 0.18 and 0.33 eV, respectively. In addition, the bandgaps of all the semiconducting GDY bilayers and trilayers are generally decreased with the increase of external vertical electric field (Figure 2.24f–h), regardless of the stacking styles. Therefore, it is possible to tune the electronic structures and optical absorption of GDY multilayers by applying an external electric field. It is noted that the CBM and VBM in the AB (β1) configuration are located at the M–Γ line of Brillouin zone. However, in the AB (β2) configuration, the CBM and VBM emerge exactly at the Γ point under field strength of 0.411 V $Å^{-1}$ but move to the M–Γ line of Brillouin zone when the field strength surpasses the value, which accounts for the sudden increase of bandgap in Figure 2.24g.

Despite the great progress made to study the properties of multilayers GDY theoretically, the experimental identification on the stacking style, intrinsic bandgap, and magnetism of bulk GDY are still scarce. Very recent experimental study [87] confirmed that the GDY nanosheet with few layers is exclusively matched with the ABC-type stacking according to the selected-area electron diffraction (SAED) pattern. As shown in Figure 2.25l, the simulated distance (d) of ABC-stacking mode is 0.48 nm [88], much matched with the experimental measured SAED pattern (Figure 2.25f). In comparison, the simulated distance is 0.85 nm for other two AB-type stacking structures, either AA (Figure 2.25j) or AB style (Figure 2.25k). Therefore, the synthesized thinnest graphdiyne nanosheet ever reported in experiments (~2.19 nm thick) is more likely to be the six-layer structure with ABCABC stacking [89], rather than the AA or AB stacking structure. Besides, the optical absorption spectrum measurements show that the bulk GDY has a direct bandgap of c. 0.64 eV [88]. Moreover, it is spin-half paramagnetic at 2 K, which is probably induced by the residual hydroxyl groups on the chains of GDY sheets. The theoretical calculations indicate that the ABC-stacked GDY (Figure 2.25m) is intrinsically nonmagnetic and semiconductor with a direct bandgap of 0.37 eV at the PBE level and 0.73 eV at the HSE06 level (Figure 2.25n). It is reasonable that the hybrid functional can reasonably consider the electron correlation effects, generally yielding much larger bandgap than that calculated by the PBE functional.

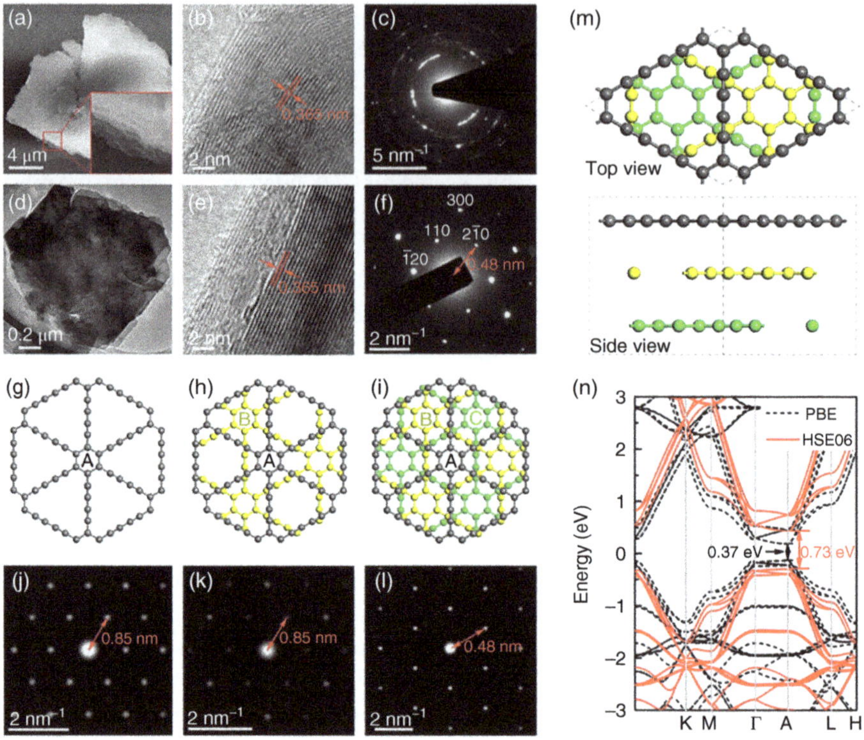

Figure 2.25 (a–c) Scanning electron microscopy (SEM) image, high resolution transmission electron microscopy (HRTEM) image, and experimental SAED pattern of as-GDY; inset in (a) is the image of edge structure. (d–f) Transmission electron microscopy (TEM) image, HRTEM image, and experimental SAED pattern of pristine graphdiyne (pGDY). (g–i) Top view of AA, AB, and ABC stacked GDY multilayers. (j–l) Simulated SAED patterns of AA, AB, and ABC stacked GDY multilayers. (m) Top and side views of the unit cell of ABC stacked GDY multilayers and (n) the corresponding band structures calculated using the PBE functionals (black dotted line) and HSE06 functionals (red solid line), respectively. Source: Reproduced with permission from Lin et al. [88]. © 2019, Elsevier.

2.5 Band Gap Engineering of GDYs

The zero band gap (such as α- and β-graphyne) or the relatively small band gap (such as γ-graphyne) of graphynes family (GDYs) limits the charge transfer from the external regulating systems. To expand the application of GDYs in nanoelectronic devices, it is necessary to further engineer their band gaps.

2.5.1 Influences of Nonmetal Doping

DFT calculations [90] revealed that the nonmetal dopants (e.g. B, N, or O) can effectively modify the electronic properties of α-, β-, and γ-graphynes (i.e. αGy, βGy, γGy). There are two distinct doping sites in graphynes, namely the sp-C site (C1) and sp^2-C site (C2). The dopants with electron less than C atom (such as B atom) will introduce hole into the valence band of graphynes. Therefore, the B-doped graphynes exhibit the p-type semiconductor characteristics because the Fermi level is shifted downwards to the position below the VBM compared to pristine graphynes. In addition, the doping positions (i.e. C1 or C2 site) of B impurity have different influences on the electronic properties of graphynes except for βGy. For example, both B1-αGy and B1-γGy are magnetic, whereas B2-αGy and B2-γGy are nonmagnetic.

On the contrary, the dopants with electron more than C atom (such as N and O atom) introduce additional electron into the conduction band of systems. Therefore, the carrier density is increased. All the N-doped and O-doped graphynes are n-type semiconductor with the conduction band gradually filled up and the Fermi level shifted upwards to the position above the CBM. Differently, the electronic properties of these two kinds of nonmetal doped graphynes are strongly dependent on the doping position of O impurity, but not sensitive to the doping position of N impurity. For example, the O1-αGy is metallic, whereas the O2-αGy is a typical semiconductor; both O1-βGy and O1-γGy are magnetic, whereas O2-βGy and O2-γGy are merely semiconductor and nonmagnetic.

The concentration of dopants (e.g. N and B atom) plays a vital role in engineering the electronic properties of graphynes based on spin-polarized DFT calculations [69]. The impurities with different concentration in the graphynes lattice will lead to the properties transitions from semiconductor to metal, inducing the opening of band gap or the appearance of new Dirac point in the electronic structures, depending on the different parity of electron or hole that is introduced in per unit cell. The band structures of pristine and N-doped graphyne at different concentration show that the π and π* bands are descending as the dopants concentration increase (Figure 2.26a–d). Since the pristine graphyne is a closed-shell system, the doping with an odd number of N atoms per unit cell (i.e. 1 or 3 N) result in a partially filled conduction band structure, inducing the transition from a semiconducting behavior (Figure 2.26a) to a metallic behavior (Figure 2.26b,d). When doping with two N atoms per unit cell (Figure 2.26c), an even number of additional electrons is introduced in the electronic structure and the π* band is completely filled. However, this band indirectly overlaps the next upper band around Γ-point. Such region

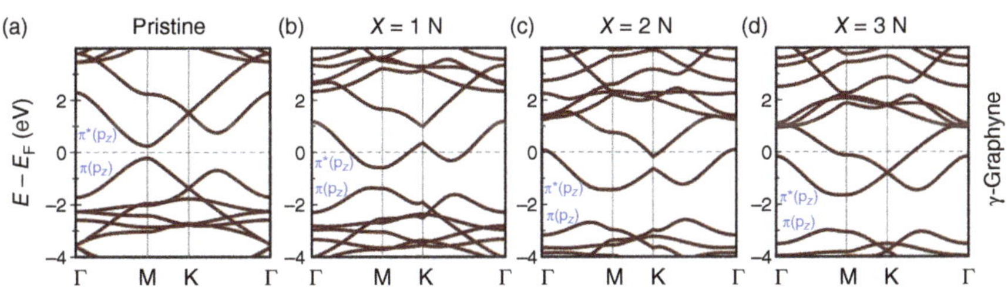

Figure 2.26 Band structures of (a) pristine and N-doped γ-graphyne with (b) 1 atom/cell, (c) 2 atoms/cell, and (d) 3 atoms/cell, in sp^2 positions. The zero of energy corresponds to the Fermi level. Source: Ruiz-Puigdollers et al. [69]. © 2017, Elsevier.

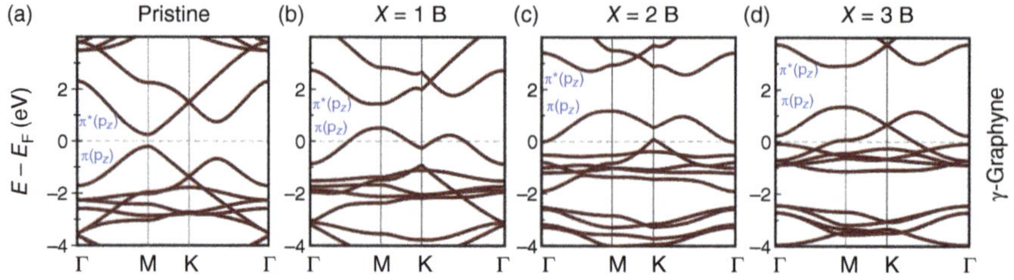

Figure 2.27 Band structures of (a) pristine and B-doped γ-graphyne with (b) 1 atom/cell, (c) 2 atoms/cell, and (d) 3 atoms/cell, in sp^2 positions. The zero of energy corresponds to the Fermi level. Source: Ruiz-Puigdollers et al. [69]. © 2017, Elsevier.

around K-point becomes occupied with part of the electron density from the π* band, causing a partially filled conduction band again.

Similarly, the doping with an odd number of B atoms in per unit cell (i.e. 1 or 3 B atom) removes electron from the valence band, making the highest occupied band half-filled. A transition behavior from semiconductor (Figure 2.27a) to metal (Figure 2.27b,d) is also obtained. When doping with two B atoms in per unit cell (Figure 2.27c), the π band indirectly overlaps with the next lower band around the Γ-point. The π band is not completely empty since a small region around Γ-point remains occupied, resulting in partially filled bands again.

In summary, the introduction of an odd number of electron or hole into per graphyne unit cell (i.e. 1 and 3 heteroatoms/cell) generally leads to the behaviors of spin-polarization and semiconductor-to-metal transition, which are caused by the partially filled bands. When introducing an even number of electrons or holes into per unit cell (i.e. 2 heteroatoms/cell), three different situations in the electronic structures of graphynes appear. For α-graphyne, the doping can separate the HOMO and the LUMO, forming a semiconductor with new band gap. For β-graphyne, a semiconductor with new Dirac cone is formed by connecting HOMO and LUMO through a conical intersection. For γ-graphyne, the HOMO and LUMO overlap, sharing the electron density and resulting in the semiconductor-to-metal transition behavior due to the partial occupation.

The electronic structures of boron-nitride (BN) pair codoped graphyne analog (BN-yne) were investigated using DFT calculations within generalized gradient approximation [42]. The BN-yne consists of hexagonal BN rings connected by carbon chains (Figure 2.28a). The optimized lattice constant of unit cell is 7.00 Å, slightly larger than that of graphyne (i.e. 6.90 Å). It is shown that the charge in BN-yne is mainly accumulated around N atoms and C≡C bonds according to the total electron density diagram (Figure 2.28b), while the C—B and C—N bonds of BN-yne have some ionic characteristics with electron mainly localized around C and N atoms according to the deformation density diagram (Figure 2.28c).

Similar to graphyne, the BN-yne is a direct bandgap semiconductor (Figure 2.28d), with both VBM and CBM located at the M point of reciprocal space according to the band structure and partial density of states (PDOS). For pristine graphyne, the VBM and CBM states are contributed by π and σ* bonding orbitals, respectively. For

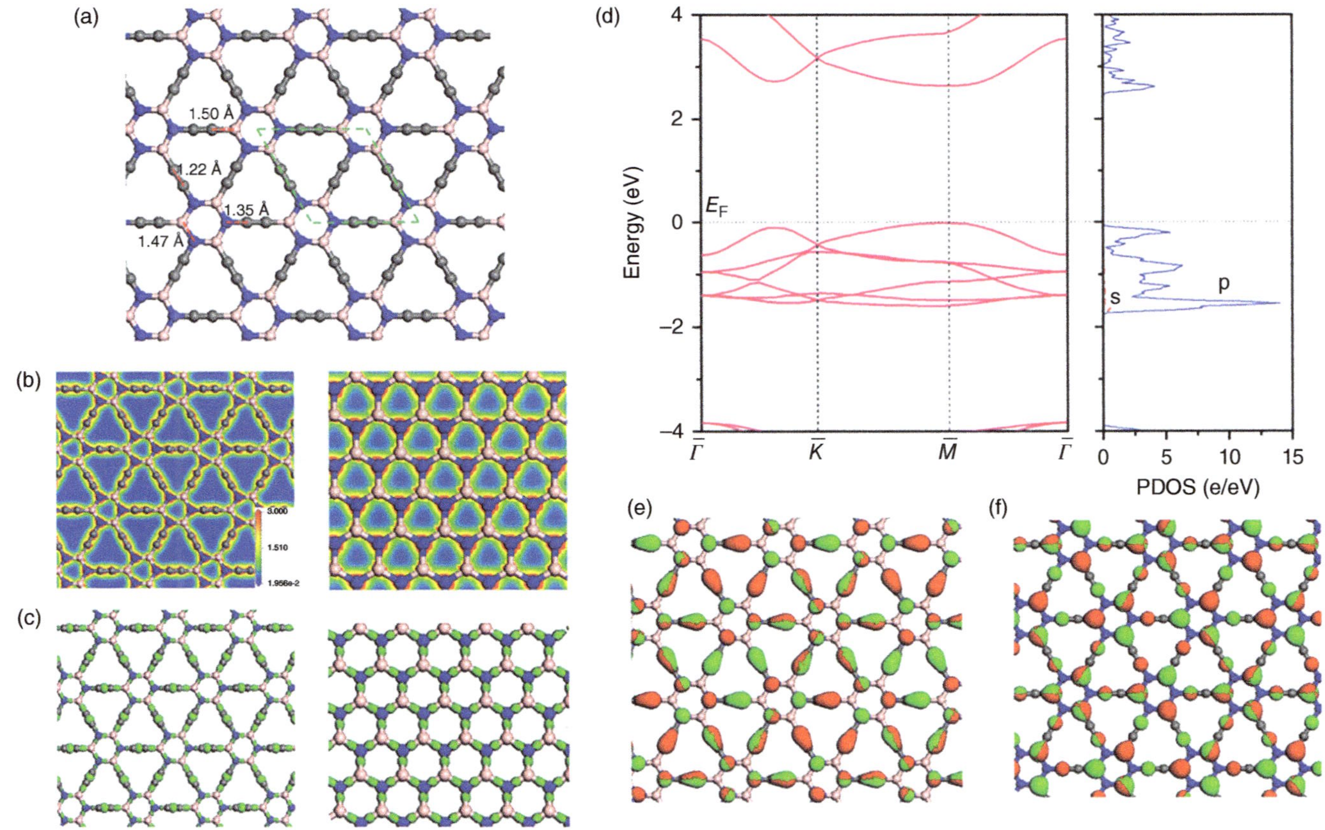

Figure 2.28 (a) Optimized structure of BN pair codoped graphyne (BN-yne). The green dashed lines represent the unit cell. (b) Contour plots of total electron density and (c) deformation electron density of BN-yne (left panel) and graphene (right panel) with an isosurface value of 0.25 electron $Å^{-3}$. (d) Band structure and corresponding partial density of states (PDOS) of BN-yne. (e) Wave function of highest occupied states and (f) lowest unoccupied states in (2 × 2) supercell of BN-yne, with an isosurface value of 0.03 electron $Å^{-3}$. Green and red colors represent positive and negative signs, respectively. Source: Zhou et al. [42]. © 2011, AIP Publishing.

BN-yne, the VBM is localized at the π orbitals of negatively charged N atoms and C≡C bonds (Figure 2.28e), while the CBM is localized at the π orbitals of positively charged B atoms and C atoms of C—N bond (Figure 2.28f). Besides, the bandgap of BN-yne is significantly increased to 2.65 eV compared to that of 0.47 eV for graphyne, providing more flexibility for the energy band engineering of graphyne.

There are two modes for boron–nitrogen (BN) pair doping in GDY [91]. At a low doping ratio ($n \leq 4$), the BN pair preferentially replaces the sp-hybridized carbon atoms in the acetylenic chain, forming a linear BN atomic chain between two aromatic rings. At a high doping ratio ($n \geq 5$), the BN pair first replaces the carbon atoms in the aromatic ring, and then replaces the carbon atoms in the acetylenic chain. The formation energy for the BN doping in GDY is much lower than that in graphene with the same stoichiometry, indicating the doping of BN pair in GDY is much easier. As the ratio of BN pair increases, the band gap first gradually increases, but then abruptly increases up to 4.39 eV (Figure 2.29a–d), corresponding to the transition between the two substitution modes. However, regardless of the doping ratio ($n = 1$–8), the feature of direct gap located at Γ point is preserved in these BN-doped GDY (Figure 2.29a–d).

Unlike graphene, the π-bindings in GDY with small bandgap are not uniform. This is because the π-bindings in the acetylenic chain are different from those in the carbon hexagon. When the BN pairs replace carbon atoms with different hybridized states (sp or sp^2), the heterogeneity of π-bindings will be further increased, and the band gap is inevitably changed accordingly. The spatial distribution of wave functions that correspond to the VBM and CBM also indicates that the bandgap regulation for these BN-doped GDY is derived from the π-bindings between C–$2p_z$ orbitals (Figure 2.29a–d). Besides, the bandgaps of BN-doped GDY can be further modulated by changing the carbon chain length (n), providing more flexibility for the energy band engineering and the application of GDY in nanoelectronic devices.

2.5.2 Influences of Chemical Modification and Functionalization

DFT calculations found that the electronic properties of graphyne (GY) and GDY can be effectively regulated by the chemical modification such as anchoring 3d transition metal (TM) atoms [92, 93]. The electronic structures of TM–GDY/GY are sensitive to the value of on-site Coulomb energy for the 3d orbital of TM, such as V, Cr, Mn, Fe, Co, and Ni [92]. Therefore, it is necessary to adopt DFT + U method, which can accurately account for the electron correlations. The U_{eff} values for all TM–GDY/GY obtained from the linear response method are listed in Table 2.6. As shown, the adsorption of TM atom enables the GDY/GY systems excellent spin-polarized half-semiconductor properties, and the corresponding magnetic moments observe the sequence: Cr > Mn > V > Fe > Co for TM–GDY and Cr > V > Mn > Fe > Co for TM–GY, respectively.

All the spin-polarized energy (ΔE_{spin}) values of GDY/GY are larger than 1.1 eV, indicating the spin-polarized state is stable even at high temperature. It is deduced that the electronic modulation of TM–GDY/GY is mainly derived from the electron transfer between TM atoms and GDY/GY, as well as the electron redistribution

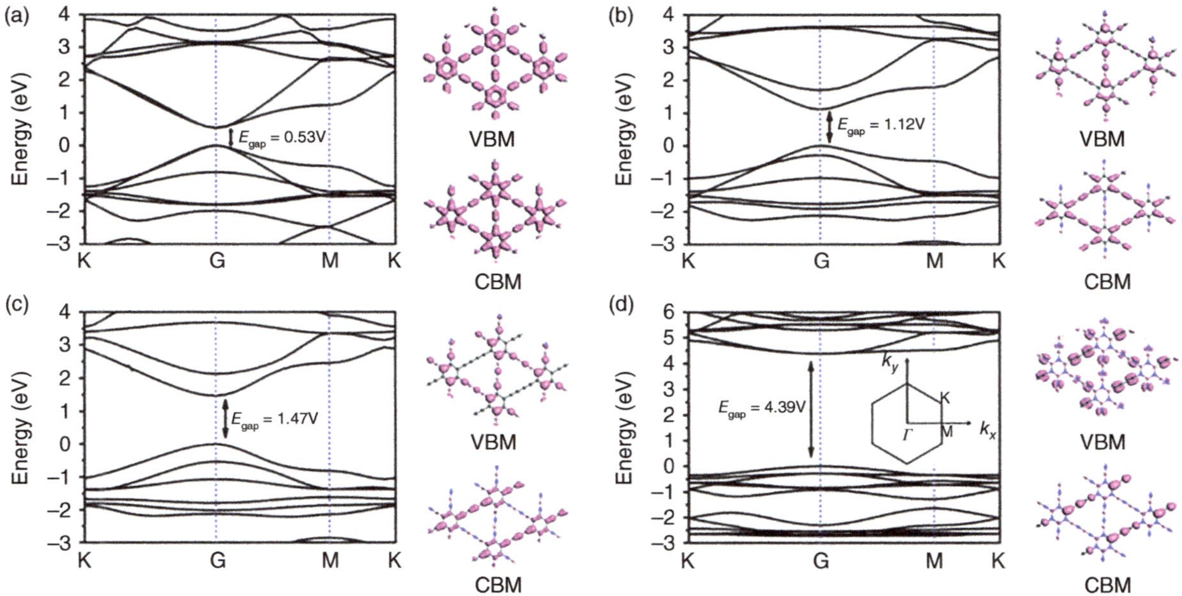

Figure 2.29 The band structures of (a) pristine GDY, (b) 2BN-GDY, (c) 4BN-GDY, and (d) full BN-GDY with the corresponding isosurface of Kohn–Sham state for the VBM and CBM of *n*BN-GDY ($n = 0, 2, 4, 7$) located at Γ point. Source: Bu et al. [91]. © 2012, American Chemical Society.

Table 2.6 Calculated results for TM-GDY/GY.

	U_{eff}	BS	T_e	4s*/3d*/4p*	TM↑/TM↓	μ_T	μ_a	ΔE_{spin}
V-GDY	5.23	M	1.179	0.265/3.164/0.392	3.308/0.513	3.34	2.80	1.119
Cr-GDY	3.36	M	1.023	0.285/4.409/0.284	4.507/0.470	4.85	4.04	3.766
Mn-GDY	4.34	M	1.172	0.236/5.378/0.214	4.957/0.871	3.79	4.09	3.493
Fe-GDY	5.21	M	0.934	0.296/6.474/0.287	5.054/2.003	2.46	3.05	2.788
Co-GDY	6.18	HSM	0.597	0.379/7.582/0.442	4.520/3.883	1.00	0.64	1.397
Ni-GDY	6.63	SM	0.469	0.394/8.493/0.644	4.766/4.766	0	0	0
V-GY	4.84	HSM	1.345	0.175/3.282/0.197	3.493/0.162	3.74	3.33	1.640
Cr-GY	3.55	M	1.017	0.236/4.362/0.385	4.486/0.527	4.85	3.93	3.461
Mn-GY	5.01	HSM	1.250	0.285/5.042/0.423	4.785/0.965	3.00	3.82	2.236
Fe-GY	5.03	M	1.002	0.415/6.035/0.548	4.543/2.455	2.86	2.09	2.701
Co-GY	6.31	HSM	0.593	0.477/7.339/0.591	4.297/3.110	1.00	1.19	1.816
Ni-GY	6.42	SM	0.472	0.145/8.746/0.637	4.764/4.764	0	0	0

T_e is the charge transfer from TM to GDY/GY. "4s*/3d*/4p*" denotes the valence electron numbers according to Bader charge analysis. TM↑/TM↓ is the electron numbers of spin up and spin down in TM atoms. μ_T (μ_B) and μ_a (μ_B) denote the magnetic moment of TM-GDY/GY and TM. ΔE_{spin} is the energy difference between the total energies for nonmagnetic (E_{NM}) and magnetic (E_M) states in the same supercell. "M," "HSM," and "SM" denote metal, half-semiconductor, and semiconductor, respectively.

Source: He et al. [92]. © 2012, American Chemical Society.

between the 4s, 3d, and 4p orbitals of TM atoms caused by the strong coupling. For example, the coupling effects between TM and GDY/GY will produce the TM atoms with less than half-filled d shell (such as V and Cr) in high-spin configurations relative to their freestanding states, and produce the TM atoms with half-filled and more than half-filled d shell (such as Mn, Fe, Co, and Ni) in the low-spin configurations relative to their freestanding state.

The chemical functionalization such as hydrogenation or halogenation can modulate the bandgap of graphyne based on DFT calculations [94]. Hydrogen and halogen atoms prefer to form sp^2 hybridization with sp-bonded carbon atoms, forming an alternate adsorption configuration in two opposite sides of graphyne (Figure 2.30a–c). This is in sharp contrast to the adsorption of these atoms on graphene. For example, fluorine (F) atoms adsorb on graphene, forming sp^3 bonds, while chlorine (Cl), bromine (Br), and iodine (I) atoms do not form any bonds when adsorbing on graphene. The relationships between the band gap of halogenated graphynes and the concentration of halogen atoms were studied. As the concentration of halogen atoms varies, the band gaps of graphyne are broadly increased by ~3 eV (Figure 2.30d), which are comparable to the band gaps of graphene engineered by hydrogenation (~3.4 eV) and fluorination (~2.7 eV), respectively.

The band gap of graphyne can be further regulated by the mixed adsorption of hydrogen and halogen atoms (Figure 2.31a,b), which is more advantageous compared to the separated adsorption of heteroatoms. It is found that hydrogen and

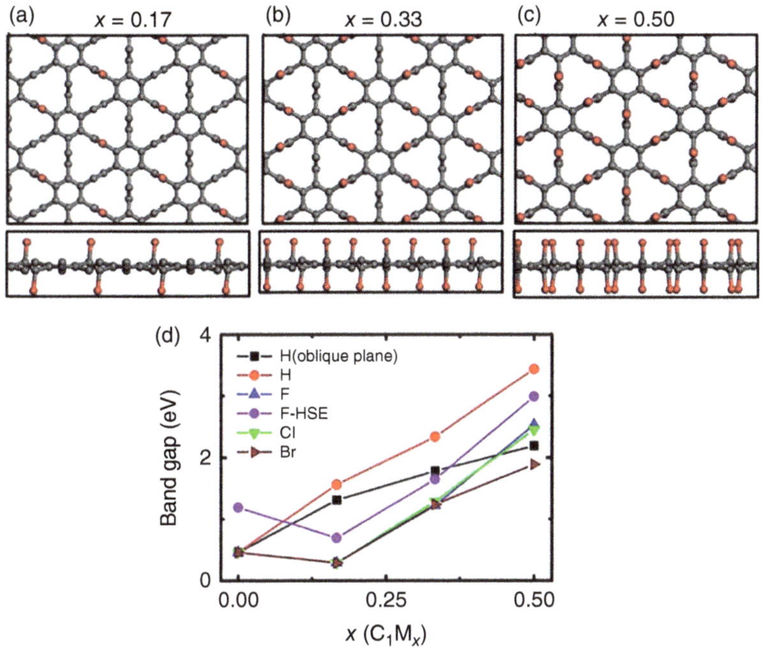

Figure 2.30 Atomic structures of hydrogen or halogen atoms adsorbed graphyne when pairs of heteroatoms are attached to two adjacent sp-carbon atoms at a concentration of (a) $x = 0.17$, (b) $x = 0.33$, and (c) $x = 0.50$ in C_1M_x (M denotes hydrogen or halogen atoms). (d) Calculated bandgaps as a function of the concentration of x. Source: Koo et al. [94]. © 2014, American Chemical Society.

halogen atoms are preferentially attached to both two sides of adjacent sp-bonded carbon atoms alternately (Figure 2.31a,b). At a concentration of $x = 1/3$, the calculated binding energies of adsorbed atoms are 1.88, 0.76, and 0.48 eV atom^{-1} for $C_1H_{1/6}F_{1/3}$, $C_1H_{1/6}Cl_{1/3}$, and $C_1H_{1/6}Br_{1/3}$, respectively (Figure 2.31c), which are slightly larger than or equal to the calculated values for the separated hydrogen or halogen adsorption structures. This can be attributed to the reduced repulsive interaction when the distance between the adsorbed atoms increases. For example, the H–F distance is 3.75 Å in the case of H–F mixed adsorption, while the F–F distance is only 3.55 Å in the case of separated F adsorption. Besides, the band gaps can be tuned by ~1.5 eV as the concentration of hydrogen–halogen mixture varies (Figure 2.31c). The adjustable band gaps may result from the breaking of symmetry due to the mixed adsorption of hydrogen and halogen atoms. It is worth noting that the band gap for $C_1H_{1/3}F_{1/6}$ at the concentration of $x = 1/6$ is abnormal (Figure 2.31d). This is because the adsorbed H atoms with an oblique-plane configuration affect the local geometry of F atoms, which are also changed to the oblique-plane adsorption.

It is interesting noted that the band gap of graphyne is closely related with the sites of functionalization [95]. For example, the band gap of graphyne with fluorination observe the following tendency: graphyne without fluorination

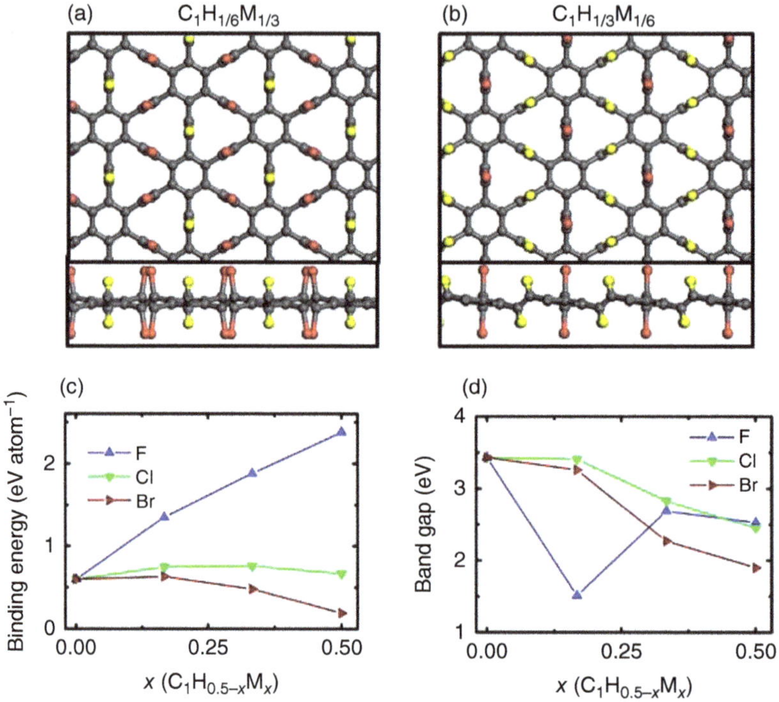

Figure 2.31 Atomic structures of hydrogen and halogen atoms adsorbed graphyne when pairs of heteroatoms are attached to two adjacent sp-carbon atom at a concentration of (a) $x = 1/3$ and (b) $x = 1/6$ in $C_1H_{0.5-x}M_x$, (c) calculated bandgaps as a function of the concentration of x. Source: Koo et al. [94]. © 2014, American Chemical Society.

(0.454 eV) < fluorination at the chain of graphyne (1.647 eV) < fluorination at both chain and ring of graphyne (3.318 eV) < fluorination at the ring of graphyne (3.750 eV). Moreover, the graphynes that are fluorinated at either the chain site or the ring site remain a direct band gap semiconductor, while the graphynes with full fluorination become an indirect semiconductor. It is found that the fluorine atoms mainly contribute to the valence band of graphyne, and the fluorination at different sites activates the s, p_x, and p_y orbitals near Fermi level. Besides, the C–C interaction between neighboring carbon atoms contributes to the bonding state, while the C–F interaction always contributes to the antibonding state near Fermi level.

Similarly, the functionalization such as hydrogenation or halogenation can efficiently regulate the geometry configuration and band gap of GDY [96]. There are two types of adsorption sites on GDY, i.e. sp- and sp^2-carbon sites. Hydrogen or halogen atoms are preferentially adsorbed on the sp-carbon site rather than the sp^2-carbon site, forming sp^2-hybridization when adsorbing one heteroatom or forming sp^3-hybridization when adsorbing two heteroatoms. Due to the difference in the adsorption configurations, the band gap modulation of GDY depends on the types of adsorbed heteroatoms. As shown in Table 2.7, the band gaps of functionalized GDY by adsorbing H, F, Cl, and Br atoms are quite different, which can be

Table 2.7 Calculated bandgap and position of valence band maximum (VBM) and conduction band minimum (CBM) for hydrogenated or halogenated graphdiyne at PBE/HSE levels.

Composites	PBE			HSE		
	VBM	CBM	Band gap (eV)	VBM	CBM	Band gap (eV)
$C_1H_{0.66}$	Γ	Γ	0.66	Γ	Γ	1.10
$C_1H_{1.33}$	Γ	M	3.56	Γ	M	2.63
$C_1H_{1.66}$	Γ	Γ	5.11	Γ	Γ	3.04
$C_1F_{0.66}$	M	M	2.14	K	Γ	2.99
$C_1F_{1.33}$	Γ	Γ	3.10	M	Γ	3.86
$C_1F_{1.66}$	Γ	Γ	4.50	Γ	Γ	5.20
$C_1Cl_{0.66}$	M	M	2.05	K	K–Γ	3.19
$C_1Br_{0.66}$	Γ	M	1.86	K	K–Γ	3.30

Source: Koo et al. [96]. © 2014, Royal Society of Chemistry.

opened to 0.66, 2.14, 2.05, and 1.86 eV at a concentration of $x = 0.66$, respectively. These values are much larger than the band gap of pristine GDY (i.e. 0.49 eV).

The band gap of GDY can be widely increased from ~0.5 to ~5.2 eV by changing the concentration of hydrogen or halogen atoms (Table 2.7). These band gap values are much larger than those of hydrogenated or halogenated graphene [44, 97, 98]. The tunable band gap can be attributed to that the conduction and valence bands of GDY are composed of the vertically arranged orbitals (p_z) of sp- and sp^2-carbon atoms, while the DOSs of p_z orbitals near Fermi level decrease as the concentration of hydrogen or halogen atoms increase. Therefore, the increase in the concentration of heteroatoms leads to the increase in the band gap of GDY.

When forming the sp^3-hybridization structure at a concentration of $x = 1.66$, the band gaps of $C_1H_{1.66}$ and $C_1F_{1.66}$ are 5.11 and 4.50 eV, respectively (Table 2.7). In contrast, sp^3-bonded carbon structures such as diamond have much larger band gap (~5.5 eV). The maximum band gap of GDY can reach up to ~4.6 eV by the hydrogenation and halogenation simultaneously, which is larger than that of graphene (~3.4 eV). Besides, unlike graphane and graphene [44, 97, 98], the aggregation of adsorbed halogen atoms on GDY is energetically unfavorable, suggesting that the hydrogenation or halogenation can be used to modify the electronic properties of GDY in the practical applications, such as nanoelectronics and nanophotonics.

The functionalization with CO or COOH groups could regulate the electronic properties of GY and GDY [99]. The functionalization is realized at the edge of GY/GDY by changing the concentration of functional groups ranging from 5.5% to 33.3%. The adsorption of CO group increases the cohesive energy, while the adsorption of COOH group decreases the stability of systems. The electronic properties, including the range of band gap, the location of Fermi level, and the semiconducting type in the modified GY/GDY, were investigated. For example, the edge functionalization with CO group leads to the lower Fermi level that is toward to the LUMO region. Therefore, the functionalized GY/GDY with CO group becomes electron-rich, forming an N-type semiconductor. The corresponding band gaps are varied in a wide range, such as 0.37–0.72 eV for the modified GY and 0.20–0.67 eV for the modified GDY, depending on the concentration of functional groups. Differently, the band gaps of functionalized GY/GDY with COOH group are not sensitive to the concentration. In addition, the CO functional group is more efficient in tuning the band gap of GY/GDY in a wide range, while the COOH functional group is useful in inducing the semiconducting behavior to achieve the desired band gap, such as 1.33–1.34 eV for GY and 1.40–1.51 eV for GDY.

2.5.3 Tunable Band Gap Under Strain

The band gap (E_g) of graphyne can be continuously modified under uniform strain (ε) [40]. No matter the tensile or compressive strain (Figure 2.32a), the strain energy increases monotonously with the increase of applied strain, indicating that the deformation of graphyne is elastic (Figure 2.32b). Once the stress is removed, it can return to the equilibrium state. The calculated band gaps at the GGA-PBE and HSE06 levels show the same trend (Figure 2.32b). When $-0.10 < \varepsilon < 0.15$, the direct band gap characteristic of γ-graphyne is maintained, and the band gap could be regulated by the strain. Under the tensile strain ($0 < \varepsilon < 0.15$), the band gap increases with the increase of strain. Moreover, under the tensile strain of $0 \sim 0.10$, the band gap almost increases linearly with a slope of 4.15 eV at the GGA-PBE level and 5.13 eV at the HSE06 level, respectively. When ε exceeds 0.10, E_g and ε maintain the linear relationship, but the slope almost doubles, which is 8.10 eV (GGA-PBE) and 10.14 eV (HSE06), respectively. This indicates that the band gap changes more obviously with the increase of tensile strain. Under compression strain ($-0.10 < \varepsilon < 0$), the band gap decreases with the increase of strain, and it is always changed linearly with a slope of 2.97 eV (GGA-PBE) and 3.71 eV (HSE06). In general, when $\varepsilon = 0.15$, the band gap of graphyne is about 1 eV larger than the equilibrium state, while when $\varepsilon = -0.10$, the band gap is about 0.3 eV lower than the equilibrium state.

The band gap variation of γ-graphyne under different strain conditions is related with the width of p_z π and π^* bands, as depicted in Figure 2.32c. When applying the tensile strain, the distance between carbon atoms increases, resulting in the overlap of p_z states between carbon atoms decreases, namely the p_z·π and π^* band get narrow. As a result, the band gap increases with the VBM moving down and the CBM moving up, respectively. When applying the compression strain, the overlap of p_z states increases, namely the p_z·π and π^* band get widened. Therefore, the band gap is decreased with the VBM moving up and the CBM moving down, respectively. To sum up, under the conditions of tensile (compression) strain, the width of p_z band decreases (increases), resulting in the band gap increases (decreases).

Similar to graphyne, GDY is a direct band gap semiconductor without applying strain. It is found that the band gap of GDY increases linearly with the increase of uniform tensile strain using the first-principles calculations combined with the tight binding (TB) approximation [100]. This can be attributed to the fact that the electronic structures around Fermi level are dominantly determined by the

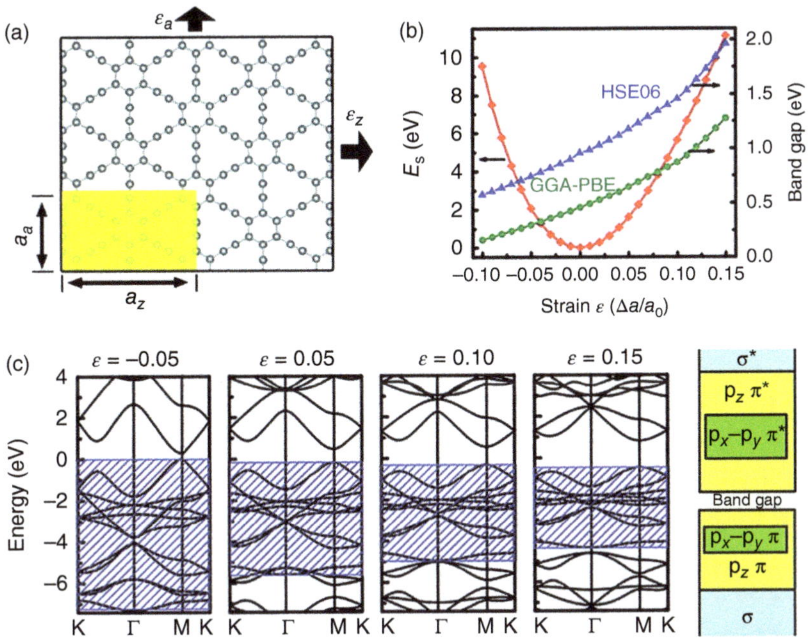

Figure 2.32 (a) Geometry structure of γ-graphyne. a_a and a_z are the lattice constants along the armchair and zigzag directions, respectively. Strain is applied by varying a_z and a_a. (b) Variation of strain energy and band gap of γ-graphyne under uniform strain. The green circles and blue triangles represent the calculated band gap at the GGA-PBE and HSE06 levels, respectively. (c) Band structures of γ-graphyne under different strain. The shaded regions indicate the p_z π bands. The right panel shows the division of energy region according to different bonding types in graphyne. Source: Kang et al. [40]. © 2011, American Chemical Society.

$2p_z$ orbitals of π electrons. Moreover, the orbital overlap between carbon atoms decreases with the strain increasing. Su and coworkers [101] found a similar phenomenon through the first-principles calculations. As the symmetrical biaxial tensile strain increases from 5% to 9–15%, the band gap of GDY is gradually increased to 1.39 eV (Figure 2.33a). Moreover, the bands become much flatter, which means that π electrons become more localized under biaxial tensile strain. However, it will destroy the geometrical symmetry and lift the band degeneracy by applying the asymmetrical uniaxial tensile strain, giving rise to the decrease of band gap. For example, when applying uniaxial tensile strains of 5%, 9%, and 15% along the armchair direction of GDY (Figure 2.33b), the degeneracy of VBM and CBM increases. In other words, the energy level of VBM upshifts and that of CBM downshifts, resulting in the decrease of band gap with the increase of strain. In particular, when the strain reaches up to 15%, the band gap becomes very small (almost zero), forming a Dirac cone-like electronic structure.

However, when applying uniaxial tensile strain along the zigzag direction of GDY (Figure 2.33c), the electronic structures become much complex. As the strain increases from 0% to 9%, the band gap gradually decreases to approximately zero.

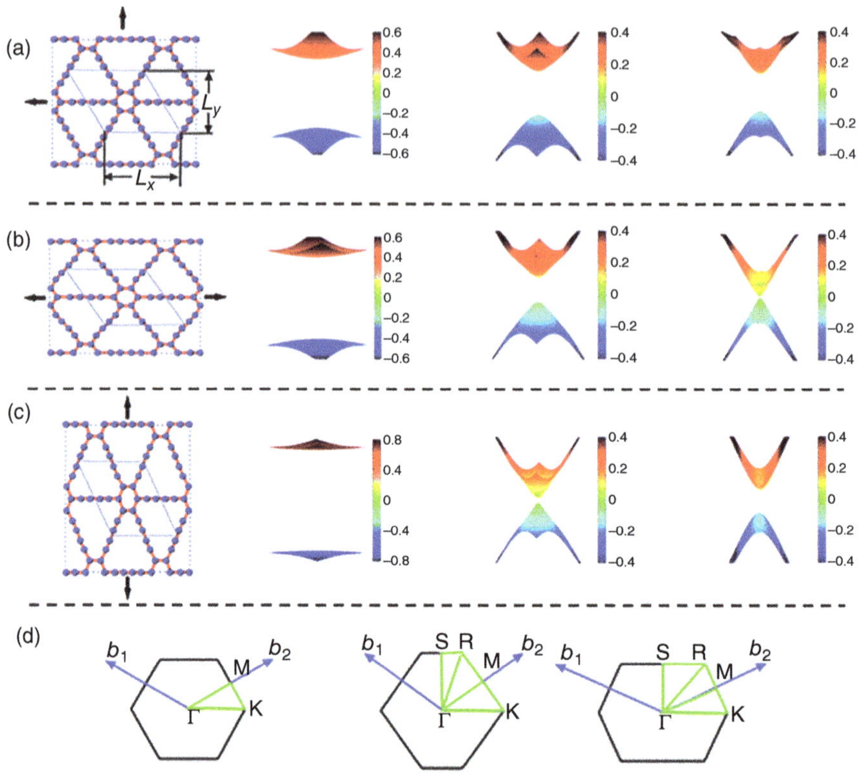

Figure 2.33 Schematic structures of graphdiyne (a) under a symmetrical biaxial tensile strain, and (b) under a uniaxial tensile strain along the armchair direction, and (c) under a uniaxial tensile strain along the zigzag direction. The right panels of (a) and (b) correspond to the three-dimensional energy bands at the Γ point for graphdiyne with a strain of 5%, 9%, and 15%, respectively. The right panel of (c) corresponds to the three-dimensional energy bands for graphdiyne with a strain of 5% and 9% at the Γ point, and 19% at the *K* point, respectively. The parallelogram (dotted line) indicates a unit cell with length of L_x and width of L_y. (d) The corresponding Brillouin zones for each structure. Source: Cui et al. [101]. © 2013, Royal Society of Chemistry.

When the uniaxial strain is between 9% and 17%, the band gap remains close to zero, while the location of band gap is shifted from Γ point to S point. Besides, the Dirac cone around Fermi level becomes much sharper. The results show that the strain brings the properties transition of GDY from semiconductor to semimetal. When the strain is larger than 17%, the band gap that locates at S point is opened again, which is different from the influence of strain in the armchair direction. In this case, the bands around Fermi level form Γ point to S point are obviously uplifted compared with the bands from Γ point to M point. This can also be attributed to the fact that the uniaxial tensile strain destroys the geometric symmetry that supports the band degeneracy, thus changing the electronic structures of GDY. These results show that the electronic properties of two-dimensional GDY can be effectively modulated by applying the external strain.

Table 2.8 Summary of calculated band gaps (in eV) for graphyne sheet and its family under different strain.

Functionals	System	H-strain			A-strain		Z-strain	
		0%	−2%	+10%	−2%	+10%	−2%	+10%
GGA-PBE	Graphyne	0.46	0.40	0.88	0.40	0.17	0.37	0.30
	Graphdiyne	0.48	0.41	0.94	0.39	0.31	0.39	0.21
	Graphyne-3	0.56	0.49	1.07	0.50	0.43	0.47	0.47
	Graphyne-4	0.54	0.46	1.11	0.44	0.45	0.45	0.36
HSE06	Graphyne	0.94	0.87	1.47	0.87	0.56	0.83	0.71
	Graphdiyne	0.89	0.80	1.53	0.78	0.69	0.78	0.56

Source: Yue et al. [31]. © 2013, American Chemical Society.

Based on the first-principles calculations [31], it is found that the band gaps regulation by strain are achievable for the γ-graphyne family with different acetylenic chain lengths (Figure 2.34a). Unlike graphene, the graphyne family are sensitive to almost all types of strain and their band gap can be modified and controlled by applying various strain, as shown in Table 2.8. Since the calculated band gaps at the GGA-PBE and HSE06 levels show almost the same variation trends with the change of strain, only the results at the GGA-PBE level are discussed below.

Under homogeneous biaxial strain (H-strain), the band gaps of graphyne (Figure 2.34c) and GDY (Figure 2.34d) steadily decrease under compressive strain, but increase under tensile strain. The band gaps decrease to 0.40 (0.41) eV at $\varepsilon_H = -2\%$, and increase to 0.88 (0.94) eV at $\varepsilon_H = 10\%$ for graphyne and GDY, respectively. In contrast, under uniaxial strain of A-type, the band gaps decrease monotonically to 0.40 (0.39) eV at $\varepsilon_A = -2\%$ and 0.17 (0.31) eV at $\varepsilon_A = 10\%$ for graphyne and GDY, respectively. Under uniaxial strain of Z-type, the band gaps show a similar variation trend, which decrease to 0.37 (0.39) eV at $\varepsilon_Z = -2\%$ and 0.30 (0.21) eV at $\varepsilon_Z = 10\%$ for graphyne and GDY, respectively. It is seen that the gaps of graphyne and GDY can be altered in a wide range from 0.17 to 0.88 eV and 0.31 to 0.94 eV when applying the tensile A-strain and tensile H-strain, respectively. It is also noted that the tensile A-strain has a stronger ability than the tensile Z-strain in modulating the band gap of graphyne, while the modulating ability is opposite for GDY.

For other γ-graphynes family such as graphyne-3 (Figure 2.34e) and graphyne-4 (Figure 2.34f), the band gaps variation show a similar tendency. The changes in the band structures of graphyne family can be attributed to the shift of energy states near Fermi level when applying the strain. For example, the direct band gaps of graphyne (Figure 2.34c) and graphyne-3 (Figure 2.34e) are located at either M or S point of Brillouin zone, depending on the types of applied tensile strain. However, the direct band gaps of strained GDY (Figure 2.34d) and graphyne-4 (Figure 2.34f) are always located at Γ point, regardless of the types of applied strain. These results suggest that γ-graphyne and its family are promising candidates for fabricating the novel strain-tunable nanoelectronic and optoelectronic devices.

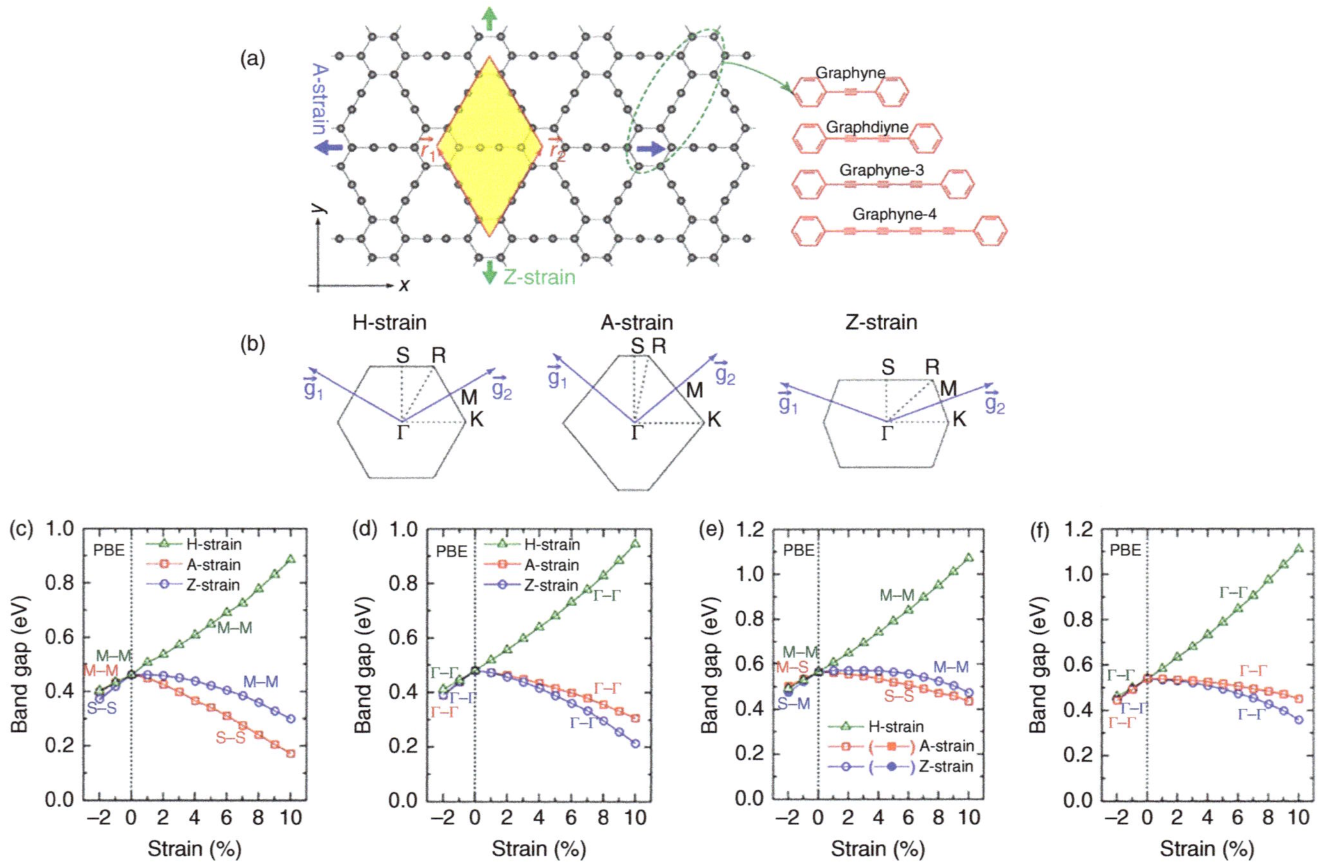

Figure 2.34 (a) Geometric structure of γ-graphyne sheet where the primitive cell is denoted by yellow parallelogram. The green and blue arrows indicate the deformation directions of Z-strain and A-strain. The units of the γ-graphyne family that are constructed by acetylenic linkages between hexagons are schematically illustrated in right panel, (b) Brillouin zone labeled with high-symmetry points under H-strain, A-strain, and Z-strain. Variation of band gaps calculated with GGA-PBE functional versus the strain for (c) graphyne, (d) graphdiyne, (e) graphyne-3, (f) graphyne-4. Open triangle, square, and circle symbols correspond to the direct band gap under H-strain, A-strain, and Z-strain, respectively. Source: Yue et al. [31]. © 2013, American Chemical Society.

2.5.4 Graphyne Nanoribbons Under Strain or Electric Field

There are two types of γ-graphyne nanoribbons by cutting the graphyne sheet along the X direction (a-GNRs) or the Y direction (z-GNRs), respectively, as shown in Figure 2.35a. A series of a-GNRs ($n = 1$–5) and z-GNRs ($n = 1$–4) were investigated using first-principles calculations [102], here n indicates the number of repeated chains of benzene rings. For the strained a-GNR ($n = 1$), the bond lengths of C1–C2 (from 1.37 to 1.45 Å) and C3–C4 (from 1.37 to 1.74 Å) are significantly enlarged with the strain growing (Figure 2.35b), which extremely change the interaction of carbon atoms that connect the stretched bonds, resulting in the obvious change of CBM and VBM states. The band gaps of a-GNRs can be almost linearly modulated in a wide range of 1.36–2.85 eV. On the contrary, for the strained z-GNR with a width of $n = 1$, the band gap is slightly affected even though a wide range of strains (from −16% to +16%) is applied, which remains within a range of 2.68 to 2.92 eV. This is because the strain primarily affects the bond angle (C3–C4–C5) of z-GNR, rather than the bond length (Figure 2.35b). Therefore, the main changes have little influence on the variation of CBM and VBM states.

The band gaps of a-GNRs almost decrease as the width of a-GNRs increases under the same strain (Figure 2.35c). For the wider strained a-GNRs ($n \geq 3$), the band gap generally decreases as the tensile or compressive strain increase. Moreover, the band gap of a-GNRs undergoes the transition from a direct gap to an indirect gap when applying sufficient strain. Differently, for the unstrained z-GNRs, the band gaps show a step-like decrease as the width of z-GNR increases (Figure 2.35c), and the band gap shows a declining tendency under tensile strain, similar to the variation tendency of a-GNRs. For example, under the strain raging from −2% to +16%, the band gaps of z-GNR ($n = 2.5$) decrease monotonically within a range from 0.52 to 0.88 eV. Generally, the band gap of graphyne nanoribbons can be controllably adjusted in a wide range (from 0.05 to 2.92 eV) by changing the applied strain. Therefore, it is an efficient method to engineer the band gap of graphyne by modulating the edge morphology, nanoribbon width, and applied strain.

The transverse electric field can also effectively modulate the band gap of γ-GDY nanoribbons with different edge structures [103], which decrease as the field strength increases. When the electric field is weak, the band gap decreases slowly. However, the band gap shows a linear response for the electric field strength when it is strong. This is because the edge states of valence and conduction bands are strongly localized at the low potential edge and high potential edge of nanoribbons under strong electric field, respectively, showing a giant Stark effect. In recent years, 1D graphdiyne nanowires (GDNWs) with high conductivity and mobility have been successfully synthesized in experiment [67]. The measured conductivity and mobility of GDNWs at room temperature are 1.9×10^3 S m^{-1} and 7.1×10^2 cm^2 V^{-1} s^{-1}, respectively. First-principles calculations show that it is an effective method to open the band gap of 1D GDNWs by applying the electric field [104]. The band gap can be varied in a wide range of 1.76–2.00 eV, and it shows the quantitative relationship with respect to the field: $E_g = 0.024E^2 + 1.760$, indicating that the electronic properties of 1D GDNWs can be precisely regulated

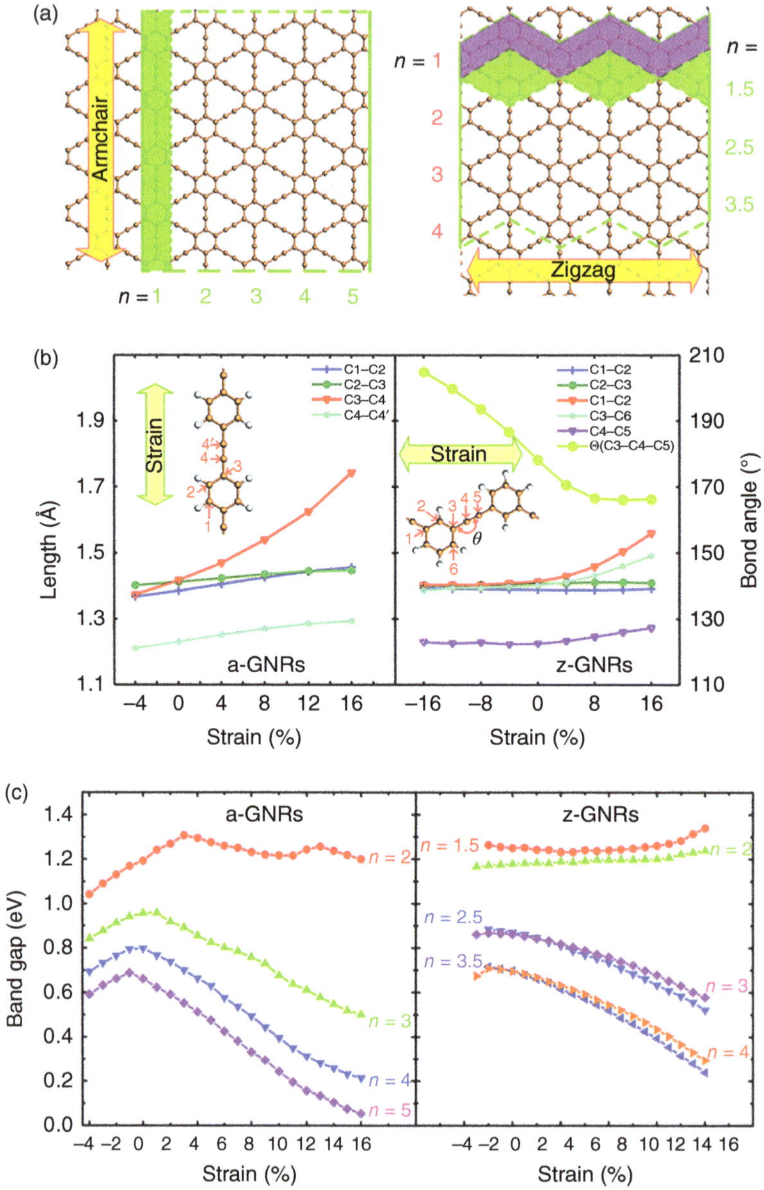

Figure 2.35 (a) Schematic representation of graphyne nanoribbons with armchair edge (a-GNR) along the *Y* direction (left) and zigzag edge (z-GNR) along the *X* direction (right). (b) Structure variations of a-GNR (left) and z-GNR (right) of $n = 1$ under various strains. (c) Band gap variations of a-GNRs (left) and z-GNRs (right) with different widths under strain. Source: Cong et al. [102]. © 2015, Royal Society of Chemistry.

by changing the electric field strength. Actually, the band gaps of nanoribbons of graphynes with different acetylenic chains lengths are all sensitive to the change of electric field. There is an interesting phenomenon: when the acetylenic linkages (—C≡C—) number of graphynes is even, such as GDY and graphyne-4, the band gaps of corresponding graphyne nanoribbons are located at the Γ point of reciprocal space; However, when the acetylenic linkages number is odd, such as graphyne and graphyne-4, the band gaps of corresponding graphyne nanoribbons are located at the boundary of Brillouin zone.

References

1 Baughman, R.H., Eckhardt, H., and Kertesz, M. (1987). Structure-property predictions for new planar forms of carbon - layered phases containing sp^2 and sp atoms. *Journal of Chemical Physics* 87: 6687–6699.

2 Narita, N., Nagai, S., Suzuki, S. et al. (1998). Optimized geometries and electronic structures of graphyne and its family. *Physical Review B* 58: 11009–11014.

3 Cranford, S.W., Brommer, D.B., and Buehler, M.J. (2012). Extended graphynes: simple scaling laws for stiffness, strength and fracture. *Nanoscale* 4: 7797–7809.

4 Enyashin, A.N. and Ivanovskii, A.L. (2011). Graphene allotropes. *Physica Status Solidi B* 248: 1879–1883.

5 Malko, D., Neiss, C., Vines, F. et al. (2012). Competition for graphene: graphynes with direction-dependent Dirac cones. *Physical Review Letters* 108: 086804.

6 Coluci, V.R., Braga, S.F., Legoas, S.B. et al. (2004). New families of carbon nanotubes based on graphyne motifs. *Nanotechnology* 15: S142–S149, Pii s0957-4484 (04)69902-8.

7 Shin, H., Kang, S., Koo, J. et al. (2014). Cohesion energetics of carbon allotropes: quantum Monte Carlo study. *Journal of Chemical Physics* 140: 114702.

8 Li, G.X., Li, Y.L., Liu, H.B. et al. (2010). Architecture of graphdiyne nanoscale films. *Chemical Communications* 46: 3256–3258.

9 Li, J., Gao, X., Liu, B. et al. (2016). Graphdiyne: a metal-free material as hole transfer layer to fabricate quantum dot-sensitized photocathodes for hydrogen production. *Journal of the American Chemical Society* 138: 3954–3957.

10 Gao, X., Li, J., Du, R. et al. (2017). Direct synthesis of graphdiyne nanowalls on arbitrary substrates and its application for photoelectrochemical water splitting cell. *Advanced Materials* 29: 1605308.

11 Xue, Y., Li, Y., Zhang, J. et al. (2018). 2D graphdiyne materials: challenges and opportunities in energy field. *Science China: Chemistry* 61: 765–786.

12 Huang, C., Li, Y., Wang, N. et al. (2018). Progress in research into 2D graphdiyne-based materials. *Chemical Reviews* 118: 7744–7803.

13 Wang, N., He, J., Tu, Z. et al. (2017). Synthesis of chlorine-substituted graphdiyne and applications for lithium-ion storage. *Angewandte Chemie International Edition* 56: 10740–10745.

14 Wang, N., Li, X., Tu, Z. et al. (2018). Synthesis and electronic structure of boron-graphdiyne with an sp-hybridized carbon skeleton and its application in sodium storage. *Angewandte Chemie International Edition* 57: 3968–3973.
15 Wang, F., Zuo, Z., Li, L. et al. (2019). A universal strategy for constructing seamless graphdiyne on metal oxides to stabilize the electrochemical structure and interface. *Advanced Materials* 31: e1806272.
16 Kuang, C.Y., Tang, G., Jiu, T.G. et al. (2015). Highly efficient electron transport obtained by doping PCBM with graphdiyne in planar-heterojunction perovskite solar cells. *Nano Letters* 15: 2756–2762.
17 Xiao, J.Y., Shi, J.J., Liu, H.B. et al. (2015). Efficient $CH_3NH_3PbI_3$ perovskite solar cells based on graphdiyne (GD)-modified P3HT hole-transporting material. *Advanced Energy Materials* 5: 1401943.
18 Jin, Z.W., Yuan, M.J., Li, H. et al. (2016). Graphdiyne: an efficient hole transporter for stable high-performance colloidal quantum dot solar cells. *Advanced Functional Materials* 26: 5284–5289.
19 Li, M., Wang, Z.K., Kang, T. et al. (2018). Graphdiyne-modified cross-linkable fullerene as an efficient electron-transporting layer in organometal halide perovskite solar cells. *Nano Energy* 43: 47–54.
20 Li, S., Chen, Y., Liu, H. et al. (2017). Graphdiyne materials as nanotransducer for in vivo photoacoustic imaging and photothermal therapy of tumor. *Chemistry of Materials* 29: 6087–6094.
21 Li, J., Chen, Y., Gao, J. et al. (2019). Graphdiyne sponge for direct collection of oils from water. *ACS Applied Materials & Interfaces* 11: 2591–2598.
22 Wang, C., Yu, P., Guo, S. et al. (2016). Graphdiyne oxide as a platform for fluorescence sensing. *Chemical Communications* 52: 5629–5632.
23 Parvin, N., Jin, Q., Wei, Y. et al. (2017). Few-layer graphdiyne nanosheets applied for multiplexed real-time dna detection. *Advanced Materials* 29: 1606755.
24 Jin, Z., Zhou, Q., Chen, Y. et al. (2016). Graphdiyne: ZnO nanocomposites for high-performance UV photodetectors. *Advanced Materials* 28: 3697–3702.
25 Lu, C., Yang, Y., Wang, J. et al. (2018). High-performance graphdiyne-based electrochemical actuators. *Nature Communications* 9: 752.
26 Puigdollers, A.R., Alonso, G., and Gamallo, P. (2016). First-principles study of structural, elastic and electronic properties of alpha-, beta- and gamma-graphyne. *Carbon* 96: 879–887.
27 Yue, Q., Chang, S., Kang, J. et al. (2012). Magnetic and electronic properties of alpha-graphyne nanoribbons. *Journal of Chemical Physics* 136: 244702.
28 Deng, X., Si, M., and Dai, J. (2012). Communication: oscillated band gaps of B/N-codoped alpha-graphyne. *Journal of Chemical Physics* 137: 201101.
29 Kim, B.G. and Choi, H.J. (2012). Graphyne: hexagonal network of carbon with versatile Dirac cones. *Physical Review B* 86: 115435.
30 Ducere, J.-M., Lepetit, C., and Chauvin, R. (2013). Carbo-graphite: structural, mechanical, and electronic properties. *Journal of Physical Chemistry C* 117: 21671–21681.

31 Yue, Q., Chang, S., Kang, J. et al. (2013). Mechanical andelectronic properties of graphyne and its family under elastic strain: theoretical predictions. *Journal of Physical Chemistry C* 117: 14804–14811.
32 Pan, L.D., Zhang, L.Z., Song, B.Q. et al. (2011). Graphyne- and graphdiyne-based nanoribbons: density functional theory calculations of electronic structures. *Applied Physics Letters* 98: 173102.
33 Zou, S.L. and Bowman, J.M. (2003). A new ab initio potential energy surface describing acetylene/vinylidene isomerization. *Chemical Physics Letters* 368: 421–424, Pii s0009-2614 (02)01911-5.
34 Ozcelik, V.O. and Ciraci, S. (2013). Size dependence in the stabilities and electronic properties of alpha-graphyne and its boron nitride analogue. *Journal of Physical Chemistry C* 117: 2175–2182.
35 Lu, R., Rao, D., Meng, Z. et al. (2013). Boron-substituted graphyne as a versatile material with high storage capacities of Li and H_2: a multiscale theoretical study. *Physical Chemistry Chemical Physics* 15: 16120–16126.
36 Cordero, B., Gomez, V., Platero-Prats, A.E. et al. (2008). Covalent radii revisited. *Dalton Transactions*: 2832–2838.
37 Tang, W., Sanville, E., and Henkelman, G. (2009). A grid-based Bader analysis algorithm without lattice bias. *Journal of Physics: Condensed Matter* 21: 084204.
38 Chen, J.M., Xi, J.Y., Wang, D. et al. (2013). Carrier mobility in graphyne should be even larger than that in graphene: a theoretical prediction. *Journal of Physical Chemistry Letters* 4: 1443–1448.
39 Padilha, J.E., Fazzio, A., and da Silva, A.J.R. (2014). Directional control of the electronic and transport properties of graphynes. *Journal of Physical Chemistry C* 118: 18793–18798.
40 Kang, J., Li, J., Wu, F. et al. (2011). Elastic, electronic, and optical properties of two-dimensional graphyne sheet. *Journal of Physical Chemistry C* 115: 20466–20470.
41 Zhou, J., Lv, K., Wang, Q. et al. (2011). Electronic structures and bonding of graphyne sheet and its BN analog. *Journal of Chemical Physics* 134: 174701.
42 Luo, G., Qian, X., Liu, H. et al. (2011). Quasiparticle energies and excitonic effects of the two-dimensional carbon allotrope graphdiyne: theory and experiment. *Physical Review B* 84: 075439.
43 Hybertsen, M.S. and Louie, S.G. (1986). Electron correlation in semiconductors and insulators - band-gaps and quasi-particle energies. *Physical Review B* 34: 5390–5413.
44 Cudazzo, P., Attaccalite, C., Tokatly, I.V. et al. (2010). Strong charge-transfer excitonic effects and the bose-einstein exciton condensate in graphane. *Physical Review Letters* 104: 226804.
45 Long, M.Q., Tang, L., Wang, D. et al. (2011). Electronic structure and carrier mobility in graphdiyne sheet and nanoribbons: theoretical predictions. *ACS Nano* 5: 2593–2600.
46 Bai, H.C., Zhu, Y., Qiao, W.y. et al. (2011). Structures, stabilities and electronic properties of graphdiyne nanoribbons. *RSC Advances* 1: 768–775.

47 Nakada, K., Fujita, M., Dresselhaus, G. et al. (1996). Edge state in graphene ribbons: nanometer size effect and edge shape dependence. *Physical Review B* 54: 17954–17961.

48 Wassmann, T., Seitsonen, A.P., Saitta, A.M. et al. (2010). Clar's theory, pi-electron distribution, and geometry of graphene nanoribbons. *Journal of the American Chemical Society* 132: 3440–3451.

49 Barone, V., Hod, O., and Scuseria, G.E. (2006). Electronic structure and stability of semiconducting graphene nanoribbons. *Nano Letters* 6: 2748–2754.

50 Son, Y.-W., Cohen, M.L., and Louie, S.G. (2006). Energy gaps in graphene nanoribbons. *Physical Review Letters* 97: 216803.

51 Long, M.-Q., Tang, L., Wang, D. et al. (2009). Theoretical predictions of size-dependent carrier mobility and polarity in graphene. *Journal of the American Chemical Society* 131: 17728–17729.

52 Azizi, B., Rezaee, S., Hadianfard, M.J. et al. (2020). A comprehensive study on the mechanical properties and failure mechanisms of graphyne nanotubes (GNTs) in different phases. *Computational Materials Science* 182: 109794.

53 Rouzkhash, B., Salehi, A., and Ahmadi, M.T. (2020). Bandgap modulation of low-dimensional gamma-graphyne-1 under uniform strain. *Journal of Computational Electronics* 19: 947–956.

54 De Sousa, J.M., Bizao, R.A., Sousa Filho, V.P. et al. (2019). Elastic properties of graphyne-based nanotubes. *Computational Materials Science* 170: 109153.

55 Zhang, Y.Y., Pei, Q.X., and Wang, C.M. (2012). Mechanical properties of graphynes under tension: a molecular dynamics study. *Applied Physics Letters* 101: 081909.

56 Pei, Q.-X., Zhang, Y.-W., and Shenoy, V.B. (2010). Mechanical properties of methyl functionalized graphene: a molecular dynamics study. *Nanotechnology* 21: 115709.

57 Zhao, H., Min, K., and Aluru, N.R. (2009). Size andchirality dependent elastic properties of graphene nanoribbons under uniaxial tension. *Nano Letters* 9: 3012–3015.

58 Lee, C., Wei, X., Kysar, J.W. et al. (2008). Measurement of the elastic properties and intrinsic strength of monolayer graphene. *Science* 321: 385–388.

59 Zhang, Y.Y., Pei, Q.X., and Wang, C.M. (2012). A molecular dynamics investigation on thermal conductivity of graphynes. *Computational Materials Science* 65: 406–410.

60 Peng, Q., Ji, W., and De, S. (2012). Mechanical properties of graphyne monolayers: a first-principles study. *Physical Chemistry Chemical Physics* 14: 13385–13391.

61 Cranford, S.W. and Buehler, M.J. (2011). Mechanical properties of graphyne. *Carbon* 49: 4111–4121.

62 Kudin, K.N., Scuseria, G.E., and Yakobson, B.I. (2001). C_2F, BN, and C nanoshell elasticity from ab initio computations. *Physical Review B* 64: 235406.

63 Lu, Q., Arroyo, M., and Huang, R. (2009). Elastic bending modulus of monolayer graphene. *Journal of Physics D: Applied Physics* 42: 102002.

64 Wang, R.N., Zheng, X.H., Hao, H. et al. (2014). First-principles analysis of corrugations, elastic constants, and electronic properties in strained graphyne nanoribbons. *Journal of Physical Chemistry C* 118: 23328–23334.
65 Xiao, K., Li, J., Wu, X. et al. (2019). Nanoindentation of thin graphdiyne films: experiments and molecular dynamics simulation. *Carbon* 144: 72–80.
66 Liu, Y., Liu, W.B., Wang, R.G. et al. (2014). Hydrogen storage using Na-decorated graphyne and its boron nitride analog. *International Journal of Hydrogen Energy* 39: 12757–12764.
67 Qian, X., Ning, Z., Li, Y. et al. (2012). Construction of graphdiyne nanowires with high-conductivity and mobility. *Dalton Transactions* 41: 730–733.
68 Yang, Y. and Xu, X. (2012). Mechanical properties of graphyne and its family - a molecular dynamics investigation. *Computational Materials Science* 61: 83–88.
69 Ruiz-Puigdollers, A. and Gamallo, P. (2017). DFT study of the role of N- and B-doping on structural, elastic and electronic properties of alpha-, beta- and gamma-graphyne. *Carbon* 114: 301–310.
70 Mirnezhad, M., Ansari, R., Rouhi, H. et al. (2012). Mechanical properties of two-dimensional graphyne sheet under hydrogen adsorption. *Solid State Communications* 152: 1885–1889.
71 Zhang, Y.Y., Pei, Q.X., Wang, C.M. et al. (2013). A molecular dynamics investigation on mechanical properties of hydrogenated graphynes. *Journal of Applied Physics* 114: 073504.
72 Ahangari, M.G. (2015). Effect of defect and temperature on the mechanical and electronic properties of graphdiyne: a theoretical study. *Physica E: Low-Dimensional Systems and Nanostructures* 66: 140–147.
73 Shao, T., Wen, B., Melnik, R. et al. (2012). Temperature dependent elastic constants and ultimate strength of graphene and graphyne. *Journal of Chemical Physics* 137: 194901.
74 Leenaerts, O., Partoens, B., and Peeters, F.M. (2013). Tunable double Dirac cone spectrum in bilayer alpha-graphyne. *Applied Physics Letters* 103: 013105.
75 Ho, J.H., Lu, C.L., Hwang, C.C. et al. (2006). Coulomb excitations in AA- and AB-stacked bilayer graphites. *Physical Review B* 74: 085406.
76 Narita, N., Nagai, S., Suzuki, S. et al. (2000). Electronic structure of three-dimensional graphyne. *Physical Review B* 62: 11146–11151.
77 Yun, J., Zhang, Y., Ren, Y. et al. (2018). Tunable band gap of graphyne-based homo- and hetero-structures by stacking sequences, strain and electric field. *Physical Chemistry Chemical Physics* 20: 26934–26946.
78 Bhattacharya, B., Sarkar, U., and Seriani, N. (2016). Electronic properties of homo- and heterobilayer graphyne: the idea of a nanocapacitor. *Journal of Physical Chemistry C* 120: 26579–26587.
79 Lu, N., Guo, H., Zhuo, Z. et al. (2017). Twisted MX_2/MoS_2 heterobilayers: effect of van der Waals interaction on the electronic structure. *Nanoscale* 9: 19131–19138.
80 Castro, E.V., Novoselov, K.S., Morozov, S.V. et al. (2007). Biased bilayer graphene: semiconductor with a gap tunable by the electric field effect. *Physical Review Letters* 99: 216802.

81 Padilha, J.E., Fazzio, A., and da Silva, A.J.R. (2015). van der Waals heterostructure of phosphorene and graphene: tuning the schottky barrier and doping by electrostatic gating. *Physical Review Letters* 114: 066803.

82 Luo, G.F., Zheng, Q.Y., Me, W.N. et al. (2013). Structural, electronic, and optical properties of bulk graphdiyne. *Journal of Physical Chemistry C* 117: 13072–13079.

83 Li, Y., Xu, L., Liu, H. et al. (2014). Graphdiyne and graphyne: from theoretical predictions to practical construction. *Chemical Society Reviews* 43: 2572–2586.

84 Jia, Z.Y., Li, Y.J., Zuo, Z.C. et al. (2017). Synthesis and properties of 2D carbon-graphdiyne. *Accounts of Chemical Research* 50: 2470–2478.

85 Ge, C., Chen, J., Tang, S. et al. (2019). Review of the electronic, optical, and magnetic properties of graphdiyne: from theories to experiments. *ACS Applied Materials & Interfaces* 11: 2707–2716.

86 Zheng, Q., Luo, G., Liu, Q. et al. (2012). Structural and electronic properties of bilayer and trilayer graphdiyne. *Nanoscale* 4: 3990–3996.

87 Matsuoka, R., Sakamoto, R., Hoshiko, K. et al. (2017). Crystalline graphdiyne nanosheets produced at a gas/liquid or liquid/liquid interface. *Journal of the American Chemical Society* 139: 3145–3152.

88 Lin, L., Pan, H., Chen, Y. et al. (2019). Identifying the stacking style, intrinsic bandgap and magnetism of pristine graphdyine. *Carbon* 143: 8–13.

89 Li, C., Lu, X., Han, Y. et al. (2018). Direct imaging and determination of the crystal structure of six-layered graphdiyne. *Nano Research* 11: 1714–1721.

90 Kang, B., Shi, H., Wang, F.-F. et al. (2016). Importance of doping site of B, N, and O in tuning electronic structure of graphynes. *Carbon* 105: 156–162.

91 Bu, H., Zhao, M., Zhang, H. et al. (2012). Isoelectronic doping of graphdiyne with boron and nitrogen: stable configurations and band gap modification. *Journal of Physical Chemistry A* 116: 3934–3939.

92 He, J., Ma, S.Y., Zhou, P. et al. (2012). Magnetic properties of single transition-metal atom absorbed graphdiyne and graphyne sheet from DFT+U calculations. *Journal of Physical Chemistry C* 116: 26313–26321.

93 Zhang, Y., Zhu, G., Lu, J. et al. (2015). Graphyne as a promising substrate for high density magnetic storage bits. *RSC Advances* 5: 87841–87846.

94 Koo, J., Huang, B., Lee, H. et al. (2014). Tailoring the electronic band gap of graphyne. *Journal of Physical Chemistry C* 118: 2463–2468.

95 Bhattacharya, B., Singh, N.B., Sarkar, U. et al. (2014). *Tuning of Band Gap Due to Fluorination of Graphyne and Graphdiyne. 5th Young Researcher Meeting.* Bristol: Iop Publishing Ltd.

96 Koo, J., Park, M., Hwang, S. et al. (2014). Widely tunable band gaps of graphdiyne: an ab initio study. *Physical Chemistry Chemical Physics* 16: 8935–8939.

97 Leenaerts, O., Peelaers, H., Hernández-Nieves, A.D. et al. (2010). First-principles investigation of graphene fluoride and graphane. *Physical Review B* 82: 195436.

98 Wei, W. and Jacob, T. (2013). Electronic and optical properties of fluorinated graphene: a many-body perturbation theory study. *Physical Review B* 87: 115431.

99 Mohajeri, A. and Shahsavar, A. (2017). Tailoring the optoelectronic properties of graphyne and graphdiyne: nitrogen/sulfur dual doping versus oxygen containing functional groups. *Journal of Materials Science* 52: 5366–5379.

100 Pei, Y. (2012). Mechanical properties of graphdiyne sheet. *Physica B: Condensed Matter* 407: 4436–4439.

101 Cui, H.J., Sheng, X.L., Yan, Q.B. et al. (2013). Strain-induced Dirac cone-like electronic structures and semiconductor-semimetal transition in graphdiyne. *Physical Chemistry Chemical Physics* 15: 8179–8185.

102 Cong, X., Liao, Y.M., Peng, Q.J. et al. (2015). Contrastive band gap engineering of strained graphyne nanoribbons with armchair and zigzag edges. *RSC Advances* 5: 59344–59348.

103 Kang, J., Wu, F.M., and Li, J.B. (2012). Modulating the bandgaps of graphdiyne nanoribbons by transverse electric fields. *Journal of Physics: Condensed Matter* 24: 165301.

104 Zhu, Y., Bai, H., and Huang, Y. (2016). Electronic property modulation of one-dimensional extended graphdiyne nanowires from a first-principle crystal orbital view. *ChemistryOpen* 5: 78–87.

3

GDY Synthesis and Characterization

Yingjie Zhao, Qingyan Pan, and Hui Liu

College of Polymer Science and Engineering, Qingdao University of Science and Technology, No. 53 Zhengzhou Rd, Qingdao 266042, P. R. China

3.1 Synthesis

As early in 1987, the famous theoretical physicist Baughman predicted that the graphyne (GY) structure could exist stably and was most likely to be chemically synthesized [1]. But, the preparation of this highly ordered two-dimensional (2D) carbon network faces great challenges. It requires smart design of acetylenic monomers and spatial control of the growth from various initiation sites, so that monomers or oligomers can react oriented to obtain crystalline alkyne carbon networks. Therefore, constructing this new carbon allotrope largely depends on precise organic synthesis. During the past decades, scientists conducted a series of early studies on graphyne monomers and oligomers [2–23], which provided a wealth of experimental accumulation for the development of graphyne. Until 2010, graphdiyne (GDY), as a new carbon allotrope, was first prepared on Cu substrate surface through in situ cross-coupling reaction by Li and coworkers [24]. Since then, much of great efforts have been carried out to explore the synthesis of GDYs with different structures and morphologies through various preparation methods. In this section, the synthesis of GDY will be systematically discussed. We will focus on the various synthetic strategies, including Cu surface synthesis, template synthesis, interfacial synthesis, chemical vapor deposition (CVD) method, vapor–liquid–solid (VLS) method, and explosion approach. Different morphologies of the GDY, including films, nanosheets, nanowalls, nanotube arrays, nanowires and nanoribbons, ordered stripe arrays, and 3D frameworks, have been prepared through different synthetic strategies. The contents of this section will be categorized by different synthetic strategies. Basic chemistry for constructing the GDY and the early study for the synthesis of GDY molecular fragments are also mentioned.

3.1.1 Basic Chemistry

Currently, the chemistry involving the synthesis of GDY has focused mainly on the acetylenic coupling reactions. The classical GDY could be synthesized by using the

Graphdiyne: Fundamentals and Applications in Renewable Energy and Electronics,
First Edition. Edited by Yuliang Li.

acetylenic coupling reaction of hexaethynylbenzene (HEB) monomers on specific substrates. To date, two categories of acetylenic coupling reactions have been successfully applied in GDY synthesis.

One type of reaction is the Cu-catalyzed terminal alkyne coupling reaction. This type of reaction includes Glaser–Hay coupling reaction and Eglinton reaction. In 1869, Glaser developed a reaction for the synthesis of symmetric or cyclic diacetylene via the coupling reaction of terminal alkynes in the presence of Cu salts (Scheme 3.1a1). When a mixture of phenylacetylene, CuCl, and ammonium hydroxide in ethanol was exposed in the air, diphenyldiacetylene was smoothly formed [26]. Later, Hay further discovered that other organic bases can be used instead of the ammonium hydroxide. He performed oxidative acetylenic coupling in the presence of a catalytic amount of the bidentate ligand *N*,*N*,*N*′,*N*′-tetramethylethylenediamine (TMEDA) and CuCl as the catalyst (Scheme 3.1a2). The Eglinton reaction was developed based on Glaser coupling reaction, using Cu(II) acetate ($Cu(OAc)_2$) instead of CuCl as the catalyst (Scheme 3.1a3). The Eglinton reaction is also an oxidative coupling reaction of the terminal alkyne, which passes through a free-radical intermediate.

The other type of reaction is the coupling reaction of alkynylsilanes mediated by Cu(I) salt (Scheme 3.1b). The relatively stable terminal alkyne with the trimethylsilyl (TMS) protected group is used as the monomer and CuCl as the catalyst in a polar organic solvent, e.g. *N*,*N*′-dimethylformamide (DMF) and dimethyl sulfoxide (DMSO) [27]. In 2000, Nishihara et al. developed a coupling reaction of the terminal alkyne with a TMS alkyne-protecting group, which was catalyzed by Cu(I) salt and achieved a reaction yield greater than 99% (Scheme 3.1b) [28]. The TMS-protected terminal alkyne can directly yield the alkynyl coupling product without the deprotection step. However, the coupling efficiency is low and the generation of byproducts is unavoidable.

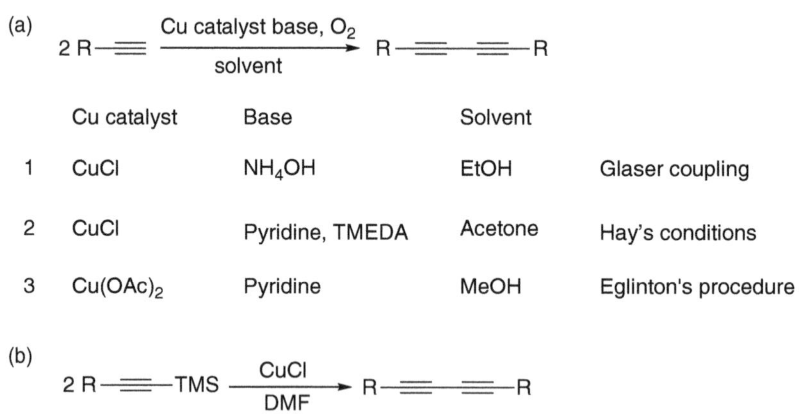

	Cu catalyst	Base	Solvent	
1	CuCl	NH_4OH	EtOH	Glaser coupling
2	CuCl	Pyridine, TMEDA	Acetone	Hay's conditions
3	$Cu(OAc)_2$	Pyridine	MeOH	Eglinton's procedure

Scheme 3.1 Cu-Mediated coupling reaction toward GDY synthesis. (a) Glaser coupling reaction of terminal alkynes, (b) homo-coupling reaction of alkynylsilanes [25]. DMF = *N*,*N*′-dimethylformamide. Source: Modified from Gao et al. [25].

3.1.2 Cu-Surface-Mediated Synthesis

HEB is an ideal precursor for the preparation of GDY by acetylene–acetylene coupling reaction. HEB is extremely unstable and can undergo cross-coupling by itself to form disordered oligomers. Controlling the conditions of the reaction, such as concentration and temperature, can reduce the probability of cross-coupling of HEB to some extent [24]. The catalysts used for acetylene–acetylene coupling reactions are usually Cu salts. For conventional synthesis, these Cu salts are unable to act a flat surface as a "template" for the formation of the film morphology of the GDY during the growth process. The rapid growth process of the GDY can easily bring disorder in structures with many defects. Therefore, the traditional synthetic strategy by using the Cu salts powder as catalysts cannot successfully lead to GDY films. In 2010, the GDY film was first prepared on the Cu foil surface by Li and coworkers (Figure 3.1). The use of Cu foil as a catalyst was a breakthrough for preparing GDY films. The Cu foil is not only an efficient catalyst for a fast alkyne coupling reaction but also affords a flat surface for the growth of the GDY films. The dual roles of catalyst and "planar template" lead to the successful preparation of GDY films. In the presence of pyridine, a small number of Cu ions will be generated on the surface of the Cu foil, and a pyridine–Cu complex formed subsequently, which acts as the catalyst. The growth of the GDY then occurred on the surface of the Cu foil under the catalysis of the pyridine–Cu complex next to the surface. The GDY film grows continuously along the surface, thereby forming a large-area film. The films prepared by this method possess a continuous area of several square millimeters and thickness of hundreds of nanometers.

The Cu-surface-mediated synthesis, as an effective method for preparing GDY films, has also been modified and improved. Zhang et al. developed a synthesis strategy for growing high-quality GDY nanowalls on a Cu surface using a system composed of acetone/pyridine/TMEDA at 50 °C (Figure 3.2a) [29]. The volume ratio of acetone, pyridine, and TMEDA is set to be 100 : 5 : 1. TMEDA acts as a good ligand for Cu. In the presence of pyridine and TMEDA, the Cu can be converted to Cu ions, which functioned as catalytic sites for the synthesis of GDY. First, GDY grew directly at those catalytic reaction sites through the Glaser–Hay coupling reaction at a rapid rate. Then, as the number of dissolved Cu ions increased, new GDY grew along the as-prepared GDY and formed uniform GDY nanowalls on the Cu surface (Figure 3.2b–e). Upon mechanical exfoliation, the thickness of the nanowall can be miniaturized to ~15 nm. Atomic force microscopy (AFM) image of the exfoliated GDY samples indicated the layered structure of the as-prepared nanowalls. The GDY nanowalls have high crystallinity, which was verified by high-resolution transmission electron microscopy (HRTEM) and corresponding selected area electron diffraction (SAED). Notably, the space between the lattice fringes was estimated to be 0.466 nm, which is quite consistent with the theoretical result. Moreover, the curved streaks (space ~0.365 nm) with certain lattice fringe can originate from the layer spacing of GDY.

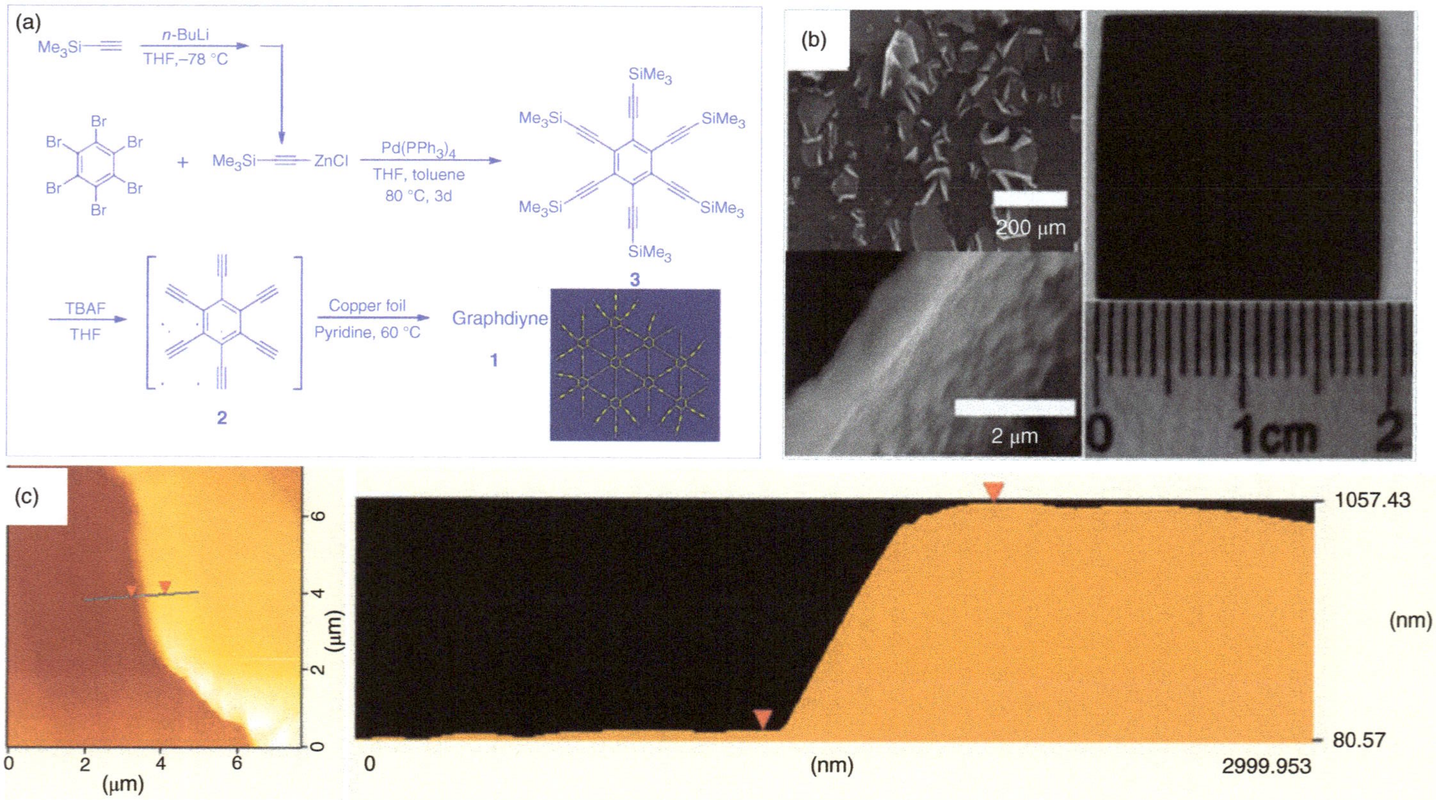

Figure 3.1 (a) Synthesis and structures of precursor HEB and GDY, (b) scanning electron microscopy (SEM) and photographic images, and (c) atomic force microscopy (AFM) images of GDY films [24]. Source: (b–c) Li et al. [24]. © 2010, Royal Society of Chemistry.

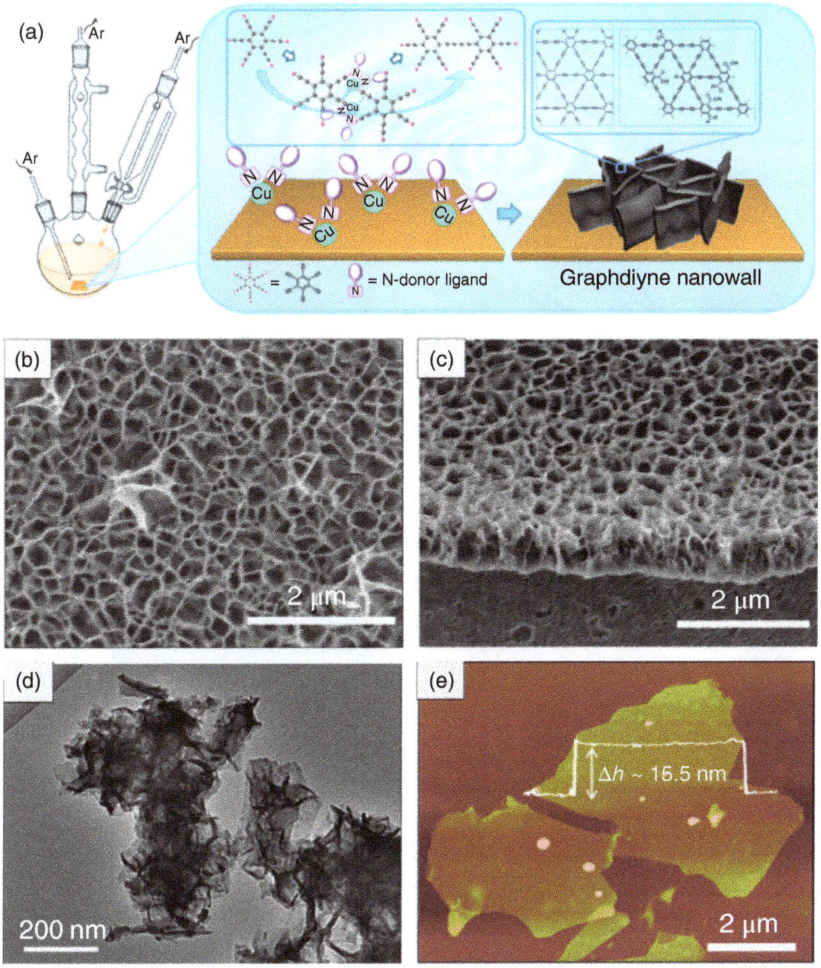

Figure 3.2 (a) Schematic presentation of the synthesis process of GDY nanowalls grown on the surface of Cu foil, (b) top view and (c) cross-sectional view of SEM images for GDY nanowalls, (d) TEM image and (e) AFM image of exfoliated GDY nanowalls [29]. Source: Zhou et al. [29]. © 2015, American Chemical Society.

The Cu-surface-mediated synthesis strategy shows good universality in preparing GDY or its analogs since it was developed by Li in 2010. Subsequently, various surfaces of Cu-containing materials were used as catalysts for the synthesis of GDY. For example, the GDY-based superhydrophobic foam was fabricated by Liu and coworkers (Figure 3.3a) [30]. GDY was directly synthesized on the surface of Cu foam via Glaser–Hay coupling reaction. The Cu foam served as both the robust and porous 3D substrate to support the formation of GDY nanostructures and the catalyst to initiate polymerization. The catalyst distribution and precursor concentration were well conttrolled in the process of the Glaser–Hay coupling reaction to

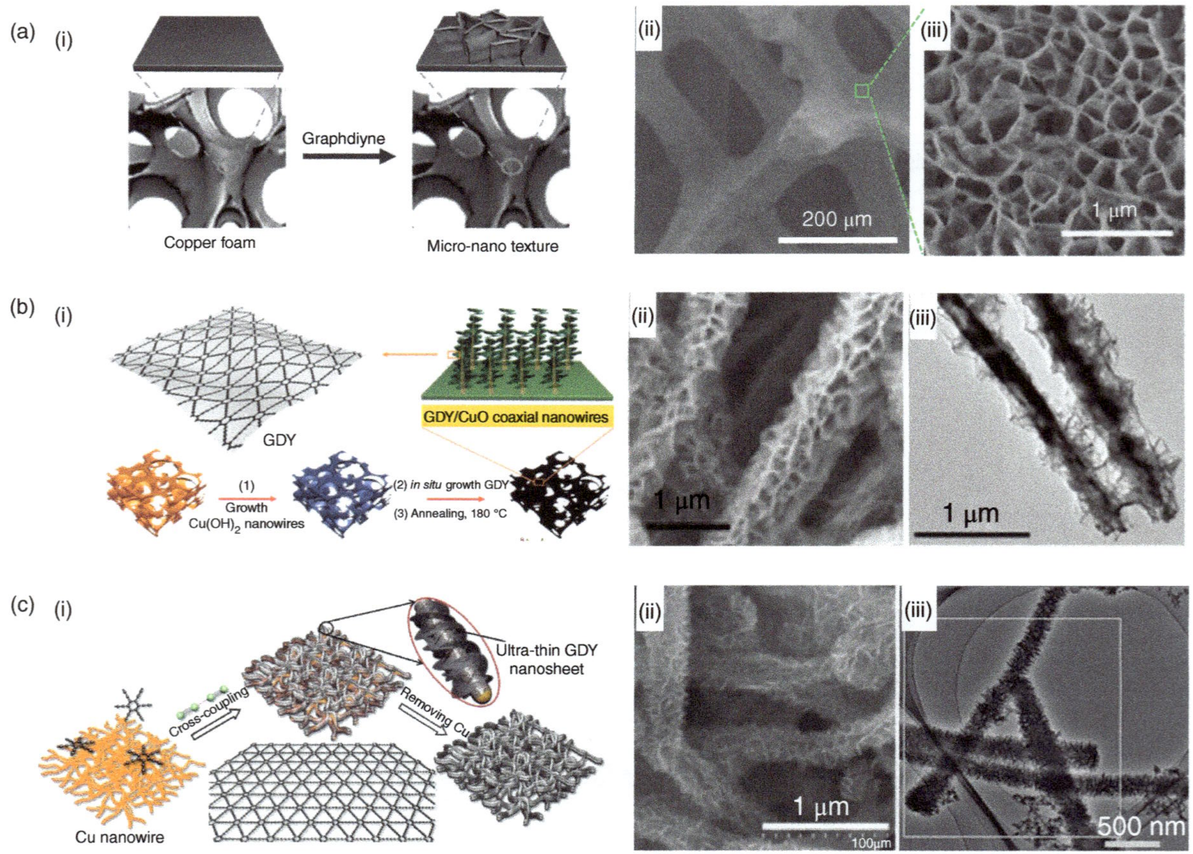

Figure 3.3 (a) Schematic illustration of the GDY-grown Cu foam and its SEM images [30], (b) schematic illustration of GDY/CuO coaxial nanowire and its SEM and TEM images [31], (c) representation of the synthesis of GDY on Cu nanowire and its SEM and TEM images [32]. Source: (a) Gao et al. [30]. © 2015, Wiley-VCH Verlag GmbH & Co. KGaA, (b) Gao et al. [31]. © 2017, American Chemical Society, (c) Shang et al. [32]. © 2017, Wiley-VCH Verlag GmbH & Co. KGaA.

obtain high-quality GDY (Figure 3.3a-i). As a result, honeycomb-like structures, based on GDY, were grown on the Cu foam as shown in the scanning electron microscopy (SEM) images (Figure 3.3a-ii–iii). The nanostructures uniformly possess the entire interconnected 3D scaffolds of Cu, with an average pore size of 150 nm. The cross-section view suggests that the vertical walls of the GDY are about 200 nm high. The Cu foam has a 3D porous structure with a pore size of ≈200 μm and displays approximately the same microlevel structures. Another example of a new structure based on 3D Cu foam, which supported one-dimensional (1D) CuO nanowires anchoring with vertical 2D GDY nanosheets, was developed by the same groups as shown in Figure 3.3b [31]. First, $Cu(OH)_2$ nanowires were synthesized on a Cu foam via the surface oxidation reaction. The hair-like $Cu(OH)_2$ nanowires exhibited diameters in the range of 100–500 nm and lengths in the range of 10–25 μm. Subsequently, vertical GDY nanosheets were in situ synthesized on the surfaces of $Cu(OH)_2$ nanowires via Glaser–Hay coupling reaction. In this way, 1D $Cu(OH)_2$ nanowires were uniformly wrapped by a honeycomb-like structure consisting of vertical 2D GDY nanosheets. Finally, the as-prepared sample was annealed at 180 °C for two hours to complete the dehydration process. Moreover, GDY nanotubes and ultrathin GDY nanosheets, intertwined on free-standing Cu nanowires as the catalyst and substrate, were explored by Li and coworkers (Figure 3.3c) [32]. Different from the previous Cu foil growth, the Cu nanowires could offer more reactive sites for the growth of GDY and improve the quality and size of obtained GDY. From the SEM and transmission electron microscopy (TEM) images (Figure 3.3c-ii–iii), it can be observed that the GDY nanosheets have grown uniformly on the Cu nanowires with a diameter of 200 nm. And after removing the Cu nanowires using a mixed solution of HCl and $FeCl_3$, perfect tubular nanostructures formed with the average thickness of the wrinkles on the nanosheets being about 3.75 nm. The Cu nanowires with defined structure may offer more direct evidence for investigating the growth of GDY.

The characteristic difference of introduced functional groups can be an important positive factor to realize the systematic adjustment of the partial properties of GDY, such as charge density distribution, interlayer distance, and morphology. So the introduction of functional groups into GDY will result in many interesting carbon-based materials, varying the bandgap and the mechanical property, and so on.

One outstanding work is the crystalline multilayer benzene-substituted graphdiyne (Ben-GDY) film with an average thickness of 30 nm developed by Li and coworkers (Figure 3.4) [33]. The π–π/CH–π supramolecular interactions were introduced to control the orientation of the monomers and the growth of the oligomers. The supramolecular interactions could avoid defects and provide high crystallinity in the target networks, which is the key feature of the strategy as illustrated in Figure 3.4a. The vicinity of alkynes was attached with benzene rings to provide the increased stability of the monomer through the bulky protection of alkynes. The π–π/CH–π interactions between the nearby benzene rings in the oligomers can force the alkyne skeleton into a nearly flat surface. The introduction of nearby benzene rings would increase the homo-coupling reaction barrier due

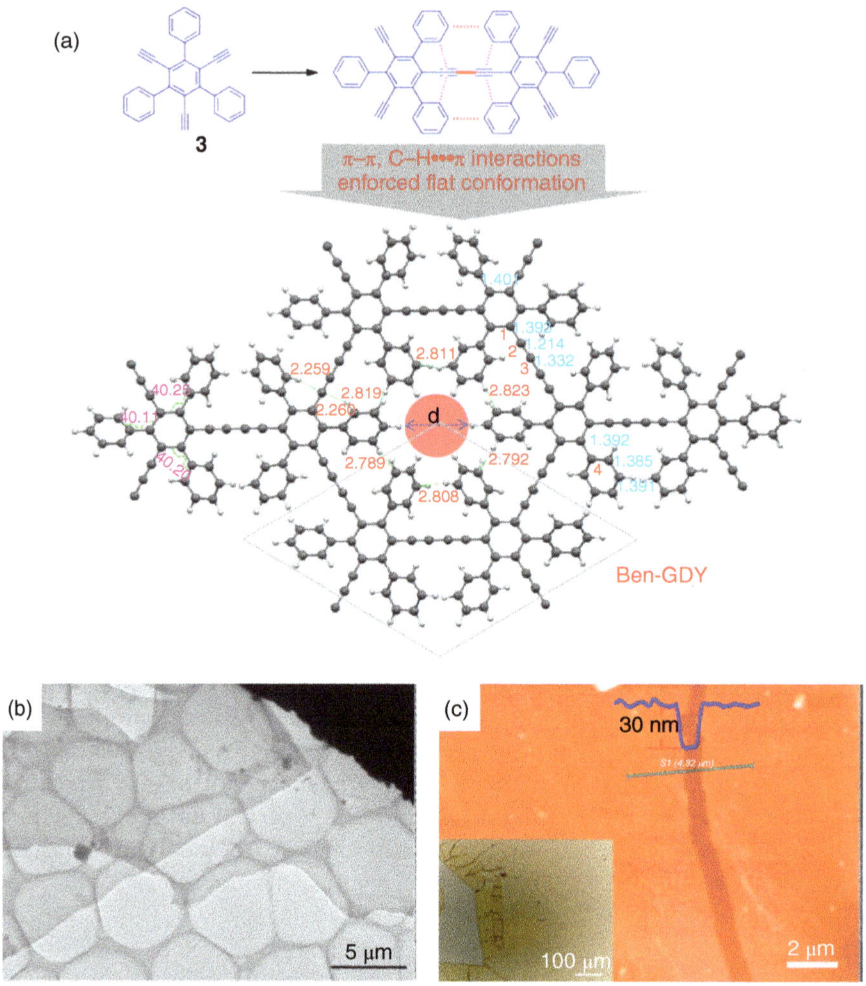

Figure 3.4 (a) The introduction of π–π/CH–π interactions for controlling the flat bonding geometries of Ben-GDY, (b) TEM and (c) AFM image of exfoliated sheets, inset: OM image of BEN-GDY film transferred to a quartz substrate [33]. Source: Zhou et al. [33]. © 2019, American Chemical Society.

to repulsion between benzene rings. Therefore, a higher reaction temperature (110 °C) was adopted for the Glaser–Hay coupling reaction of monomer in pyridine with Cu foil as the substrate and catalyst. A typical low-magnification TEM image, as well as AFM image revealed flat-sheet morphology with a thickness of 30 nm (Figure 3.4b,c). The optical microscopy (OM) of the Ben-GDY film transferred to a quartz substrate also revealed large-scale sheet morphology (inset of Figure 3.4c).

Recently, Huang and coworkers have successfully developed GDYs containing electron-withdrawing/donating groups, such as cyano group (CN), hydrogen-substituted (H), and methyl group (Me) as exhibited in Figure 3.5 [34]. CN-GDY, hydrogen substituted-graphdiyne (H-GDY), and Me-GDY were prepared

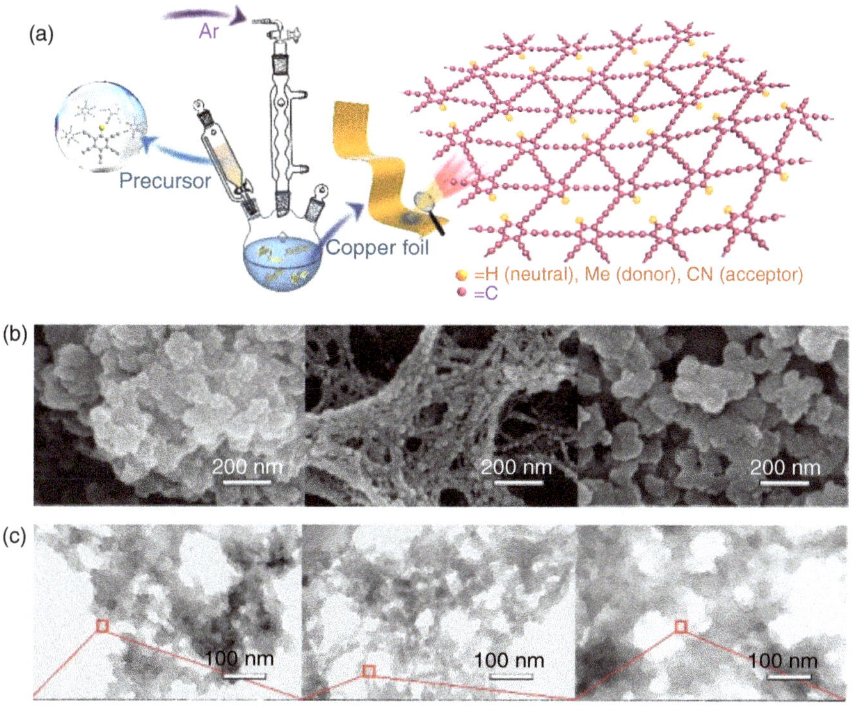

Figure 3.5 (a) Synthetic route of H-GDY, Me-GDY, and CN-GDY, (b) magnified SEM images and (c) TEM images of H-GDY, Me-GDY, and CN-GDY, from left to right, respectively [34]. Source: Xie et al. [34]. © 2020, Wiley-VCH Verlag GmbH & Co. KGaA.

on the surface of Cu foil using the modified precursors. As shown in Figure 3.5b,c, the SEM and TEM images clearly show different morphologies of these three GDYs, all of them are composed of nanoparticles of different sizes (H-GDY, Me-GDY, and CN-GDY, from left to right, respectively).

Except for GDY, many synthetic and theoretical chemists have made great efforts to discover new variants of the GY family from both theoretical predictions and practical experiments in the last few decades. Li and coworkers designed and synthesized a new GY family member-carbon ene-yne (CEY), which is obtained via in situ coupling reaction of tetraethynylethene (TEE) on the Cu surface [35]. As shown in Figure 3.6a, CEY is composed of ene and yne, and possesses black films on the Cu foil. Besides, a highly ordered crystalline graph-tetrayne (GTTY) with precisely controlled and well-defined chemical structure was constructed by He and coworkers. GTTY was obtained via Glaser–Hay coupling reaction of hexa(buta-1,3-diyn-1vl)benzene as precursors on Cu surface under a low temperature of −20 °C (Figure 3.6b–i) [36]. Here, the low-temperature condition not only reduced monomer activity, but also slowed down the growth process. In a diluted monomer solution, the heterogeneous nucleation and crystalline film growth preferred to occur on Cu substrate to form initial nucleation sites to minimize the energy barrier at the interface. Thus, the GTTY formed on the Cu surface. After the

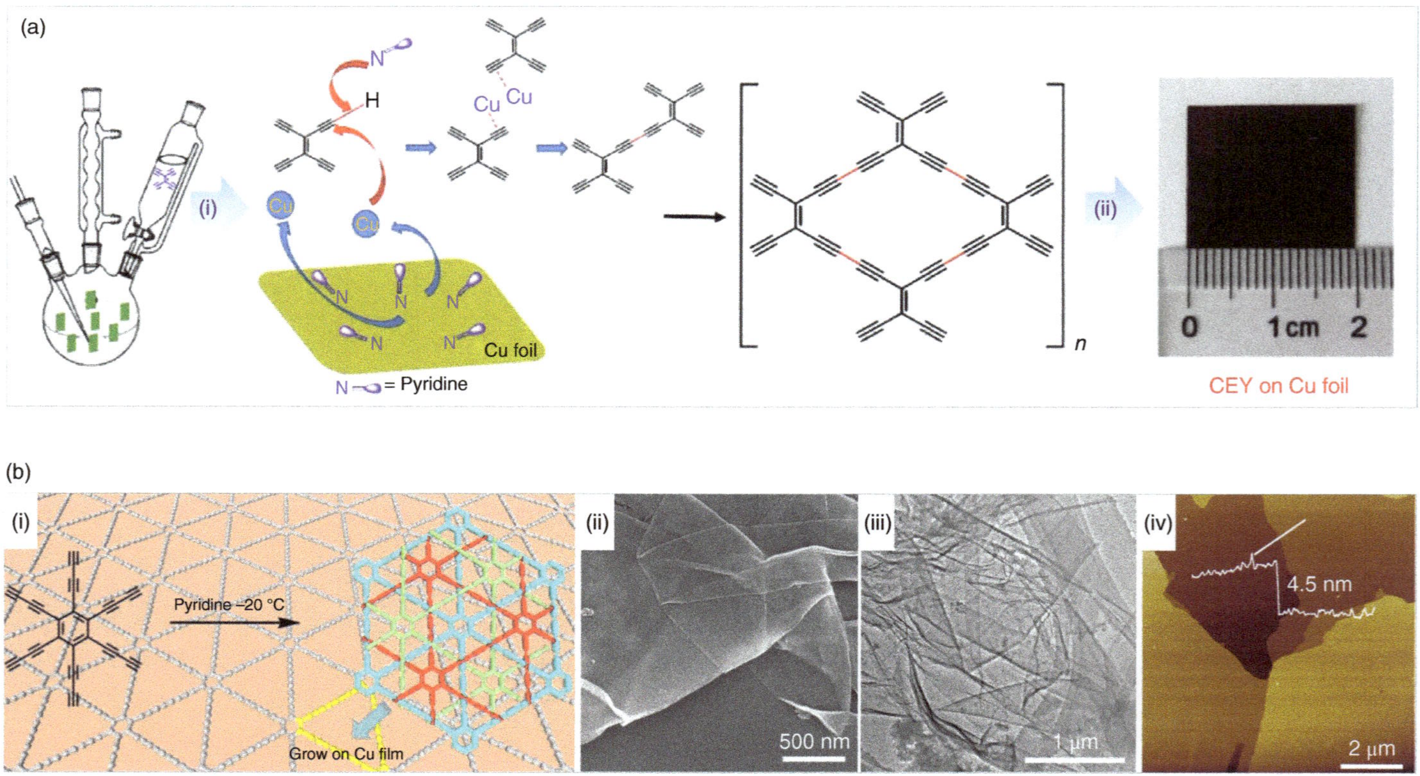

Figure 3.6 (a) Schematic of large-scale CEY synthesis [35]. (i) Synthesis of CEY. (ii) Photograph of large-scale CEY film, (b) schematic of large-scale GTTY synthesis (i) and its morphology images (ii-SEM image; iii-TEM image; iv-AFM image) [36]. Source: (a) Jia et al. [35]. © 2017, Elsevier, (b) Pan et al. [36]. © 2020, Chinese Chemical Society.

reaction, transparent brown GTTY films were observed on the Cu foil surface. The SEM, TEM, and AFM images revealed the ultra-thin film characteristic of GTTY as exhibited in Figure 3.6b-ii–iv.

Many other GDY analogues prepared on Cu surface based on alkyne-yne coupling reaction have also been reported in recent years. A 2D/3D van der Waals (vdW) heterostructure that consists of a crystalline 2D covalent triazine-based framework (TzF) and 3D amorphous component triazine-based graphdiyne (TzG) was prepared on Cu surfaces just like the preparation of GDY film on Cu foil [37]. The TzF/TzG film was grown on a common Cu foil using pyridine as a base and solvent. The authors proposed that the TzF/TzG heterostructure grew through a

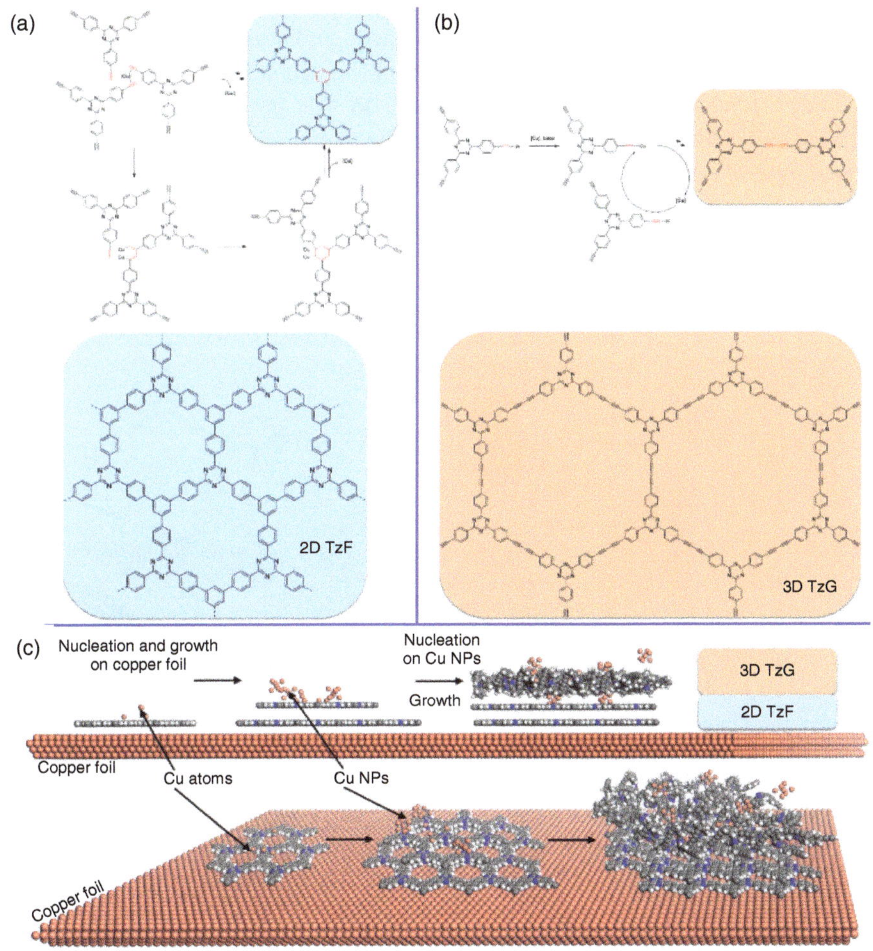

Figure 3.7 (a) Formation of the 2D TzF at the flat Cu interface via Cu-mediated [2+2+2] cyclotrimerization of 2,4,6-tris(4-ethynylphenyl)-1,3,5-triazine, (b) 3D TzG formed by Cu-catalyzed dimerization of the terminal alkynes, (c) reaction scheme for the twinned growth of the TzF/TzG vdW heterostructure on Cu foil surface [37]. Source: Schwarz et al. [37]. © 2017, Wiley-VCH Verlag GmbH & Co. KGaA.

twinned mechanism as shown in Figure 3.7. Initially, Cu catalyzed the formation of the 2D TzF at the flat Cu interface via Cu-mediated [2+2+2] cyclotrimerization (Figure 3.7a). In the later stages of the reaction, Cu-catalyzed dimerization of the terminal alkynes led to 3D TzG (Figure 3.7b). Figure 3.7c exhibits the whole process on the surface of Cu foil. This is different from the preparation GDY on Cu substrate through in situ cross-coupling reaction. TzF/TzG heterostructures can be easily detached from the Cu support for further analysis and processing by a gentle treatment.

Cu-surface-mediated synthesis of poly(1,3,5-triethynylbenzene) (PTEB) nanofibers via Glaser polycondensation of 1,3,5-triethynylbenzene on a variety of conducting (e.g. Cu, graphite, fluorine-doped tin oxide, and titanium) and nonconducting (e.g. Kapton, glass, and silicon dioxide) substrates was explored by Feng and coworkers [38]. In this work, a facile and scalable approach for the synthesis of acetylenic carbon-rich nanofibers via Cu-surface-mediated Glaser polycondensation was developed (Figure 3.8). In this process, both conductive and nonconductive substrates can be uniformly coated with PTEB nanofibers as revealed in SEM and TEM images (Figure 3.8c,d).

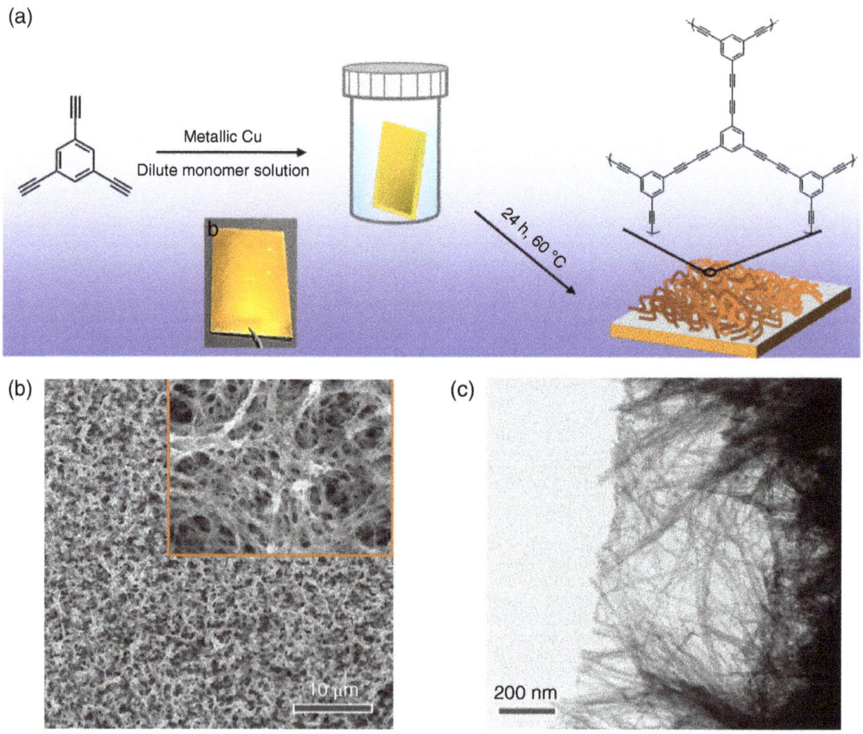

Figure 3.8 (a) Illustration of the synthetic strategy of PTEB nanofibers on the surface of the Cu foil, (b) SEM images of PTEB nanofibers, Inset: magnification image, (c) TEM image of PTEB nanofiber grown on a Cu grid [38]. Source: Zhang et al. [38]. © 2018, Springer Nature.

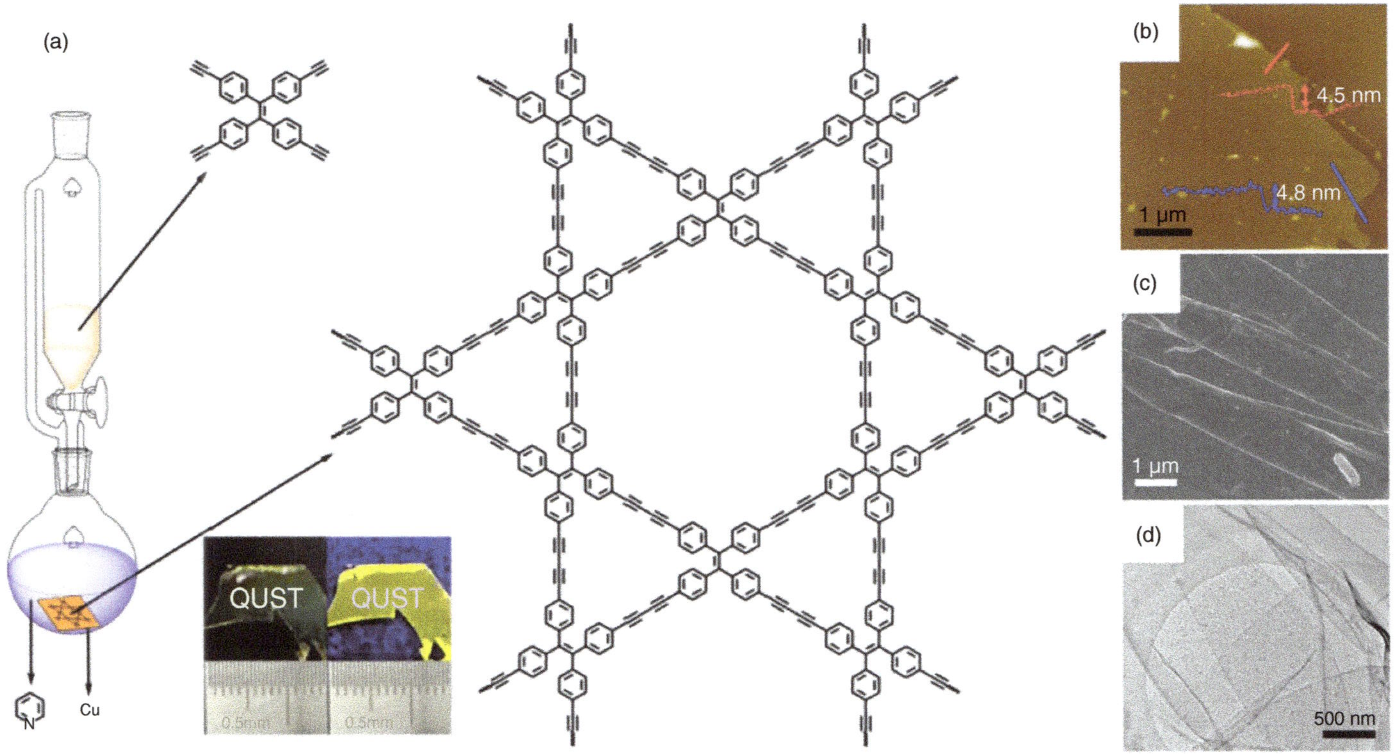

Figure 3.9 (a) Schematic illustration of the synthesis of 2D fluorescent film containing TEPE unit. The inset photo shows centimeter-size film under sunlight and 365-nm UV lamp. Morphology characterizations of the 2D films: (b) AFM image, (c) SEM image, and (d) TEM image [39]. Source: Liu et al. [39]. © 2019, Wiley-VCH Verlag GmbH & Co. KGaA.

Zhao and coworkers developed a fully π-conjugated 2D fluorescent films by Glaser–Hay coupling reaction on the surface of Cu foil (Figure 3.9) [39]. 1,1,2,2-tetrakis(4-ethynylphenyl)ethane (TEPE) was designed as the precursor. Just as the procedure for Cu-surface-mediated synthesis mentioned above, a fluorescent yellow film was formed on the surface of Cu foil (Figure 3.9a). The absolute quantum yield of the fluorescence (Φ_{FL}) of the 2DP-TEPE reaches 30% by using an integrating sphere method. The luminescence was homogeneously distributed across the whole film. The average thickness is around 4.5 nm determined by AFM image (Figure 3.9b). Both SEM and TEM images showed that the obtained 2D fluorescent films exhibited large-sheet morphology (Figure 3.9c,d).

Highly ordered structure of pyrediyne (PDY) nanowall over a Cu surface with much-improved π-conjugation delocalization features and better transport characteristics than that of GDY was synthesized by a modified Glase–Hay coupling reaction using 1,3,6,8-tetraethynylpyrene (TEP) as the precursor in a system composed of acetone/pyridine/TMEDA as presented in Figure 3.10a,b [40]. It has been observed that the polymeric 2D nanowall framework of the PDY can be developed upon Cu plates and be exfoliated to nanosheets. The SEM image in Figure 3.10c shows the layered structure of the transparent nanowalls stacked over one another. The OM image also confirmed the sheet morphology of PDY (Figure 3.10d).

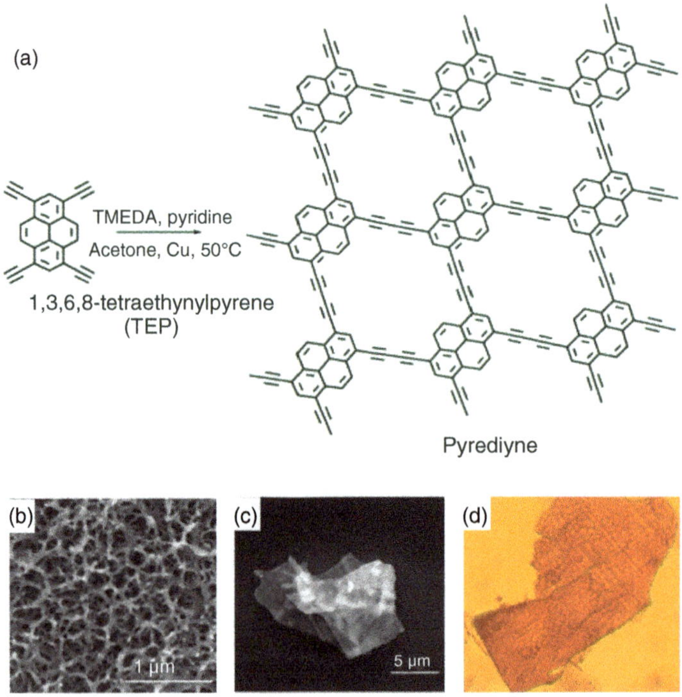

Figure 3.10 (a) Synthetic scheme with the molecular structure of TEP and PDY, (b) SEM image of the 2D PDY nanowalls on the Cu surface, (c) SEM image on a Si(111) wafer, and (d) OM image of exfoliated the transparent 2D nanowalls on a glass substrate [40]. Source: Prabakaran et al. [40]. © 2013, Royal Society of Chemistry.

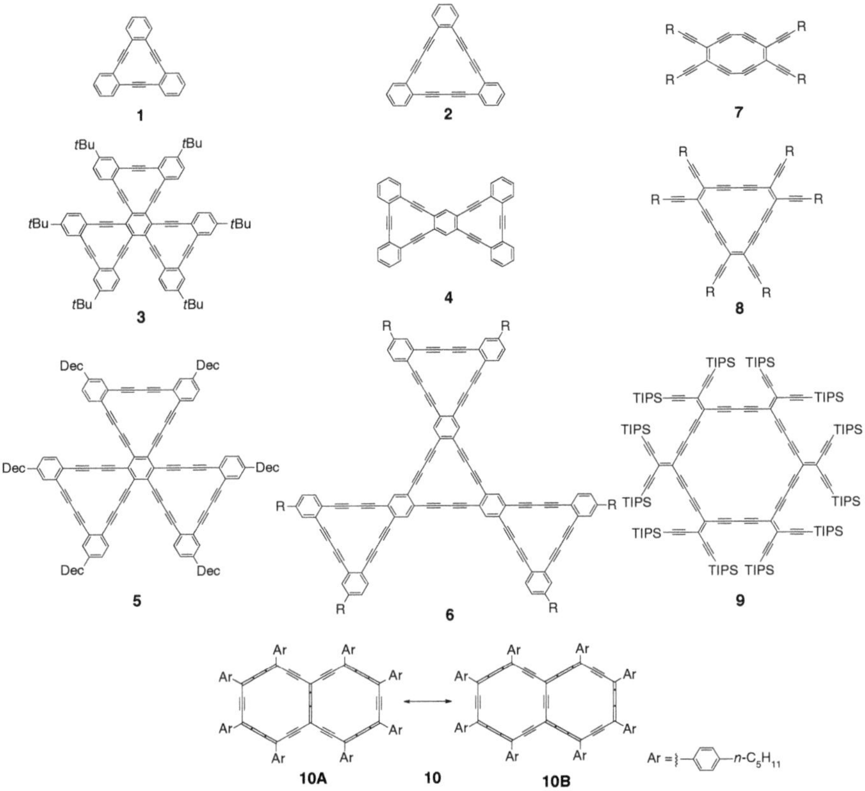

Figure 3.11 Synthesized benzannellated, and perethynylated dehydroannulene-derived substructures and perethynylated expanded radialenes for the construction of GY sheets [5, 41]. Source: Spitler et al. [5]. © 2006, American Chemical Society; Rivera-Fuentes and Diederich. [41]. © 2016, Wiley-VCH Verlag GmbH & Co. KGaA.

Above all, the Cu-surface mediated synthesis is the first successful strategy to prepare GDY. Now, it has become the most widely used strategy for synthesizing GDY, GDY allotropes, heteroatom-substituted GDY, functional groups substituted–GDY, GDY analogues, and so on. The Cu-surface-mediated synthesis method has been proven to be a classical, effective, and universal strategy for the preparation of GDYs films and other 2D materials based on diyne bond formation.

Benefiting from the development of new synthesis methodologies over the past decades, such as metal-catalyzed cross-coupling reactions, alkyne metathesis, and templated synthesis, many monomeric and oligomeric substructures of GY, GDY, and α-GY have been prepared (Figure 3.11). The synthesis details of these GY fragments have already been stated in excellent literatures [5, 41]. These small molecules cannot be polymerized to GY, but the acetylenic coupling is recognized as a versatile ingredient for the rational design of carbon scaffolds and networks.

Learning from the CVD method for fabricating graphene and framework materials, exploring acetylenic coupling reaction on the interfaces is a promising strategy for preparing single-layer GYs. Klappenberger and coworkers reported a GDY analog (only a few ring framework) by sublimating organic monomers onto an Ag(111) surface under the ultra-high vacuum (UHV) system [42]. The monomer of (1,3,5-tris-(4-ethynylphenyl)benzene (Ext-TEB) was initially deposited onto the Ag(111) surface held at 150 K, which results in self-assembled large islands with a long-range periodic structure. Following annealing of the covered surface to 300 K triggered the polymerization. The characteristic dimers were generated, coexisting with some unreacted precursor Ext-TEB species. Further annealing treatment (400 K) realized the formation of six-membered oligomeric cyclic units, which represent the ideal product of a convergent homo-coupling reaction with straight —C≡C—C≡C— connections between the Ext-TEB modules. The scanning tunneling microscopy (STM) images showed a clear hexagonal framework for this GDY analog in Figure 3.12d. Yuan and Xu deposited 1,3,5-tris(bromoethynyl)benzene on Au(111) surface under UHV and observed 2D frameworks by STM as shown in Figure 3.12e [43]. Linear, terminal alkyne-functionalized polyphenylene building blocks were synthesized on a metal surface under UHV, which was named "GDY wire" (Figure 3.12a,c) [19, 43]. Moreover, the carbonitrile (CN)-functionalized terminal alkyne precursors were also used to construct GDY nanowires via surface-assisted reactions under UHV conditions (Figure 3.12b) [44].

More recently, Li and coworkers reported a facile strategy for the controllable synthesis of GDY nanoribbons, which is of great significance for further understanding the intrinsic properties of GDY and the relationship between structure and properties [45]. They designed a two-step strategy to reach the goal of stepwise intermolecular polymerization followed by intramolecular coupling of the ethynyl groups (Figure 3.13). At first, the polymerization of the ethynyl groups in the central part of the monomer guarantees one-dimensional growth. Then, the intramolecular coupling of ethynyl groups on the organized adjacent side chains gave the nanoribbons. The liquid/liquid interfacial assembly and the bulky groups (such as 3,5-di-tert-butylbenzyl group) on the outer edge functioned as a steric, hindering further decrease of the intermolecular coupling. The first chemically synthesized GDY nanoribbons were obtained through this strategy. This nanoribbon was composed of rhomboid benzene units as vertices and butadiyne as sides. A well-defined width of ~4 nm and length of hundreds of nanometers were obtained.

3.1.3 Template Synthesis

A template-assisted approach was explored to control the fabrication of GDY and its analogues on various substrates. On the basis of earlier Cu-surface-mediated synthesis, Li and coworkers prepared GDY nanotube arrays by using anodic aluminum oxide (AAO) as a template catalyzed by Cu foil (Figure 3.14a) [46]. HEB was diluted with pyridine and added slowly to a reactor containing pyridine and AAO template. One side of the AAO was fixed on a Cu foil at 60 °C under a nitrogen atmosphere. The GDY nanotubes were obtained in the AAO template. As revealed by SEM images (Figure 3.14b,c) and TEM image (Figure 3.14d), the as-prepared GDY nanotube shows a smooth surface and the thickness of the wall

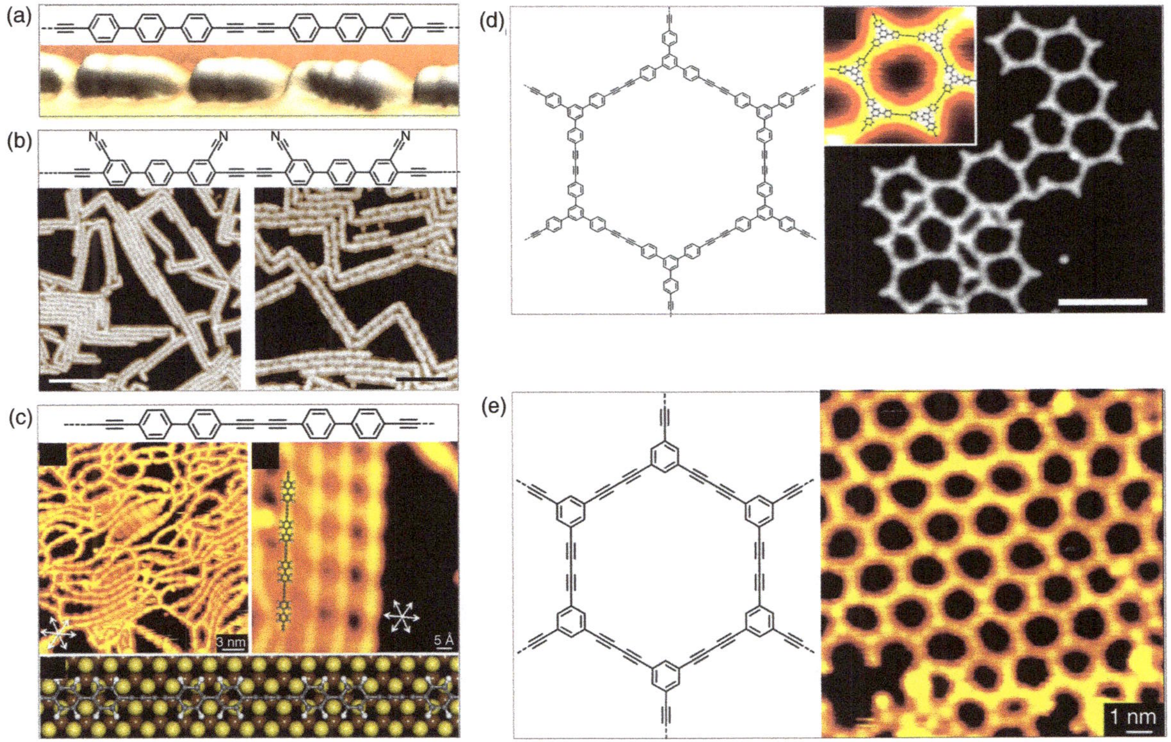

Figure 3.12 (a–c) Large-scale STM images of the formation of molecular chains with acetylenic coupling reaction on a metal surface, (d–e) large-scale STM images of 2D covalent network formed by the homo-coupling reaction of the corresponding monomer on a metal surface [19, 42–44]. Source: Cirera et al. [19]. © 2018, Wiley-VCH Verlag GmbH & Co. KGaA; Zhang et al. [42]. © 2012, Springer Nature; Sun et al. [43]. © 2016, American Chemical Society; Klappenberger et al. [44]. © 2012, Springer Nature.

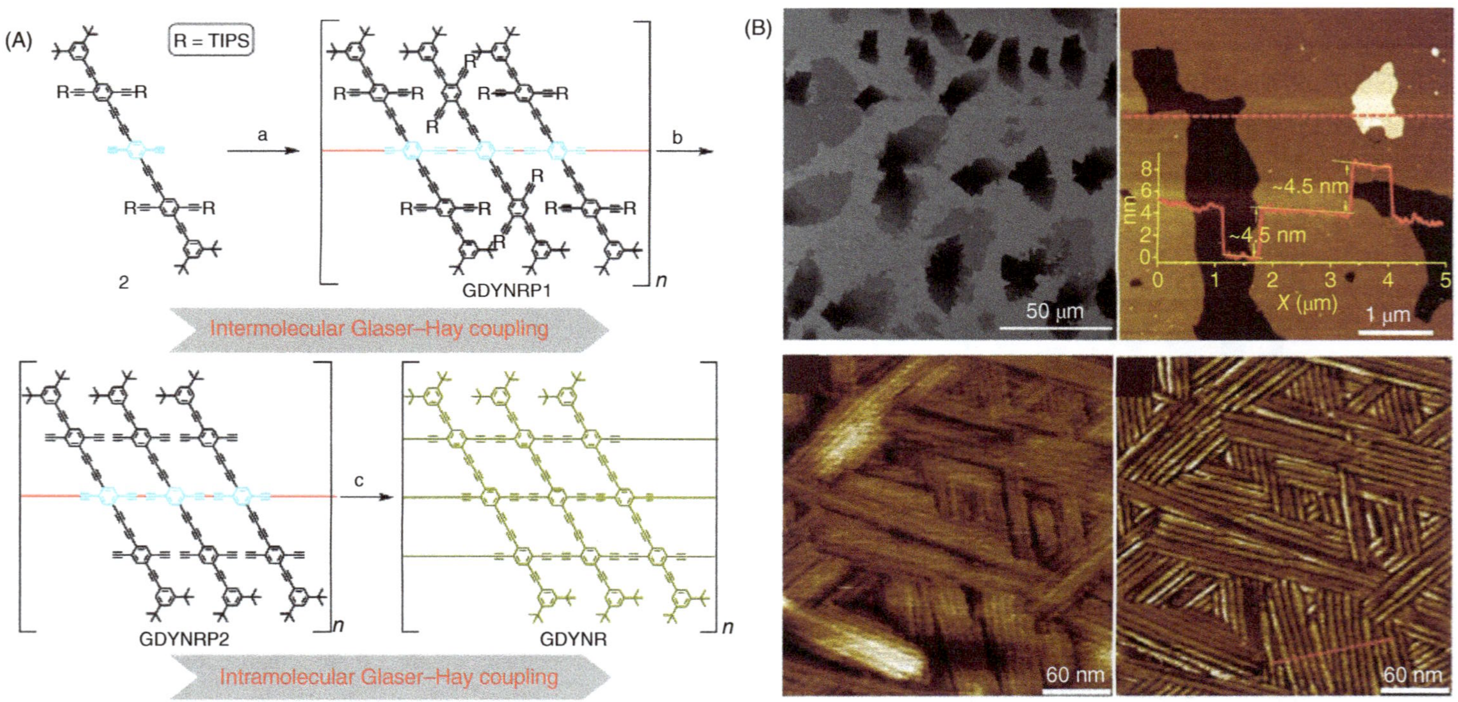

Figure 3.13 (A) The stepwise intermolecular and intramolecular Glaser–Hay coupling reaction for GDY nanoribbons, (B) microscopy characterization of GDY nanoribbons [45]. Source: Zhou et al. [45]. © 2020, Wiley-VCH Verlag GmbH & Co. KGaA.

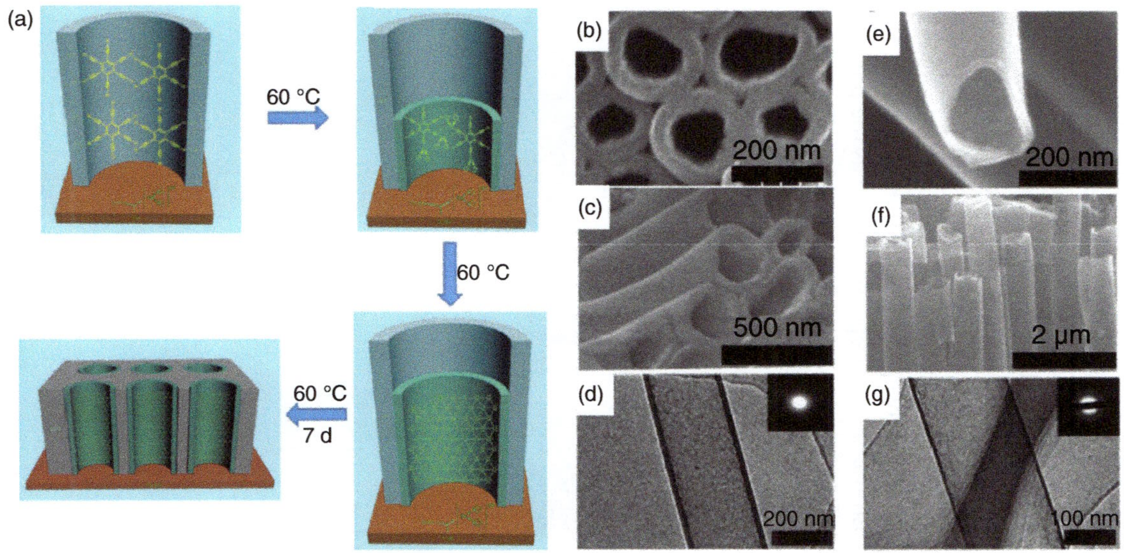

Figure 3.14 (a) The process to fabricate GDY nanotube arrays, (b–d) SEM and TEM images of GDY nanotubes before being annealed: (b) top-view; (c) side-view image; (d) TEM image, the inset is the corresponding SAED patterns, (e–g) SEM and TEM images of GDY nanotubes after being annealed: (e) top-view; (f) side-view image; (g) TEM image, the inset is the corresponding SAED patterns [46]. Source: Li et al. [46]. © 2011, American Chemical Society.

is ~40 nm. After the annealing treatment, the wall thickness decreased to ~15 nm (Figure 3.14e,f).

Wu and coworkers developed a Cu envelope catalysis strategy to synthesize structure-controlled GDY on different substrates. As shown in Figure 3.15, it is a general and simple method for the fabrication of GDY nanowalls on arbitrary substrates [47]. In this method, the target substrates were put into a Cu foil envelope, which ensures that sufficient concentrations of catalyst can diffuse into the interface between the target substrate and solution to induce the in situ growth of GDY (Figure 3.15a). In alkaline solution, Cu tends to transfer to Cu ions in the presence of a catalytic amount of base (pyridine and TMEDA). The coordination of TMEDA to Cu ions formed the Cu-N ligand composites, which act as a "running catalyst" for an acetylenic coupling reaction. The "running catalyst" can diffuse from the Cu envelope to target substrates driven by a concentration gradient. Meanwhile, an acetone solution of HEB was added dropwise, serving as the precursor for the synthesis of GDY. After heating the mixture under an argon atmosphere at 50 °C

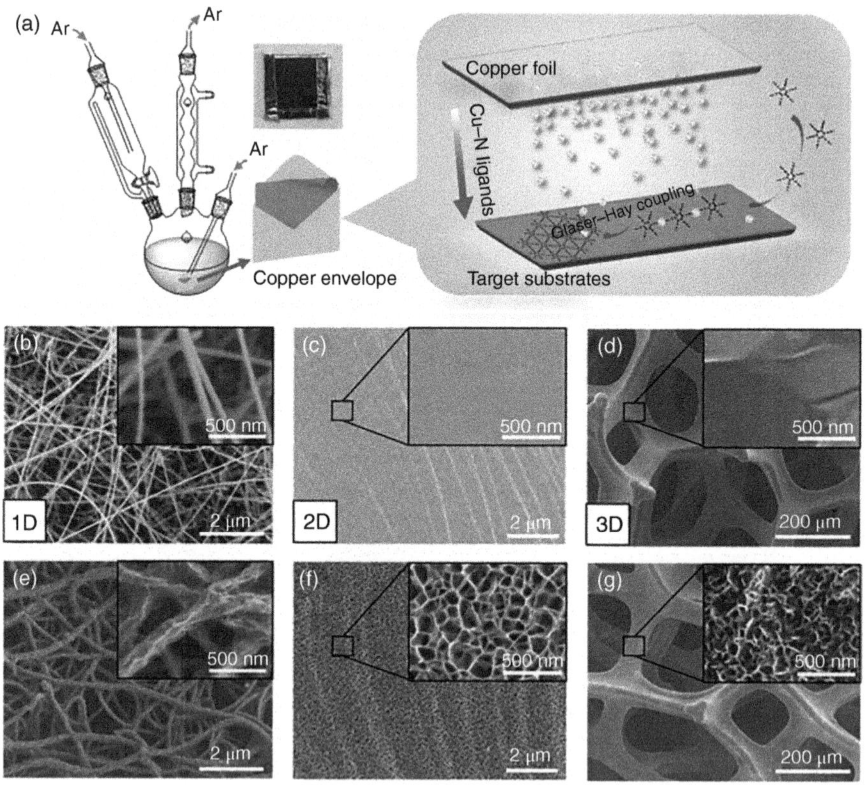

Figure 3.15 Synthesis of GDY nanowalls on arbitrary substrates via Cu envelope catalysis. (a) Schematic illustration of the experimental setup. SEM images of typical substrates before and after growth of GDY nanowalls: (b, e) on 1D silicon nanowires, (c, f) on 2D Au foil, and (d, g) on 3D Ni foam; closeup images are in the corresponding insets [47]. Source: Gao et al. [47]. © 2016, Wiley-VCH Verlag GmbH & Co. KGaA.

for 12 hours, GDY was successfully synthesized on the target substrates where the acetylenic coupling reaction took place at the interface between the target substrate and solution. Through this method, structure-controlled GDY can be grown on 1D Si nanowires (Figure 3.15b,e), 2D Au foils (Figure 3.15c,f), and 3D Ni foam (Figure 3.15d,g).

Later, the growth of GDY under ultra-mild condition was successfully implemented for coating metal oxides (MOs) by Zuo and coworkers [48]. Different dimensional nanostructured MOs on the Cu foils were prepared first by the hydrothermal method. And then the nanostructured MOs loaded on the Cu foils were immersed into a solution containing the precursor of HEB for in situ-weaving GDY nanosheets under room temperature. In this work, the growth of the high-quality GDY nanosheet well conforms to the morphological features of the MOs, and a continuous and seamless conductive network is formed on the interface. As a result, this GDY protection layer significantly enhances the structural and interfacial stabilities and improves the ion and electron migrations over the interface when the MOs are applied as the lithium-ion battery (LIB) anodes.

The synthesized 3D graphene, by employing diatomite as a template, possesses a freestanding texture, high specific area, and electronic conductivity [49]. Inspired by this, it is possible to prepare freestanding 3D GDY by using diatomite as a template. In this method, Cu nanoparticles were primarily absorbed on the surface and in the holes of diatomite through a simple metallic replacement reaction to obtain Cu nanoparticles and diatomite composites (Cu@diatomite). Then, the alkyne coupling reaction occurred on the diatomite surface. The absorbed Cu nanoparticles and diatomite play the roles of catalyst source and substrate, respectively. After the reaction was completed, diatomite was wrapped in GDY flakes (GDY@Cu@diatomite). After carefully removing the residual Cu and diatomite by etching reagents, porous 3D GDY with freestanding structure was obtained [50].

Xue and coworkers developed an in situ growth of GDY and analogues on carbon cloth template (Figure 3.16) [51–55]. To explain it in detail, the closely stacked (the Cu foil on the top and the carbon cloth on the bottom) carbon cloth and Cu foils were put into the reactor containing acetone/pyridine/TMEDA. Cu foils in an alkaline solution can easily release Cu ions into the solution. Subsequently, Cu ligand formed from Cu ions and organic base ligand can easily diffuse onto the template surface and catalyze the coupling of alkynyls. HEB solution was then added to the reaction flask in which carbon cloth was fixed closely to Cu foils. The mixture was kept at a suitable temperature and time under Ar atmosphere. The flexible 3D GDY/carbon cloth with a porous surface was obtained. The 3D GDY/carbon cloth could be further fabricated as atomic catalysts [51, 55] or heterojunction materials [54]. Figure 3.16a shows the synthesis process and morphologies of as-obtained GDY/carbon cloth. Besides, fluorinated-graphdiyne (F-GDY)/carbon cloth was prepared by a similar procedure. Different morphologies on carbon cloth were observed. Figure 3.16b-i–ii shows the uniform film on the carbon fibers [52]. While the nanoporous structure was revealed in Figure 3.16b-iii–iv [53], the F-GDY/carbon cloth as a metal-free catalyst exhibited excellent performance in water splitting and ammonia production.

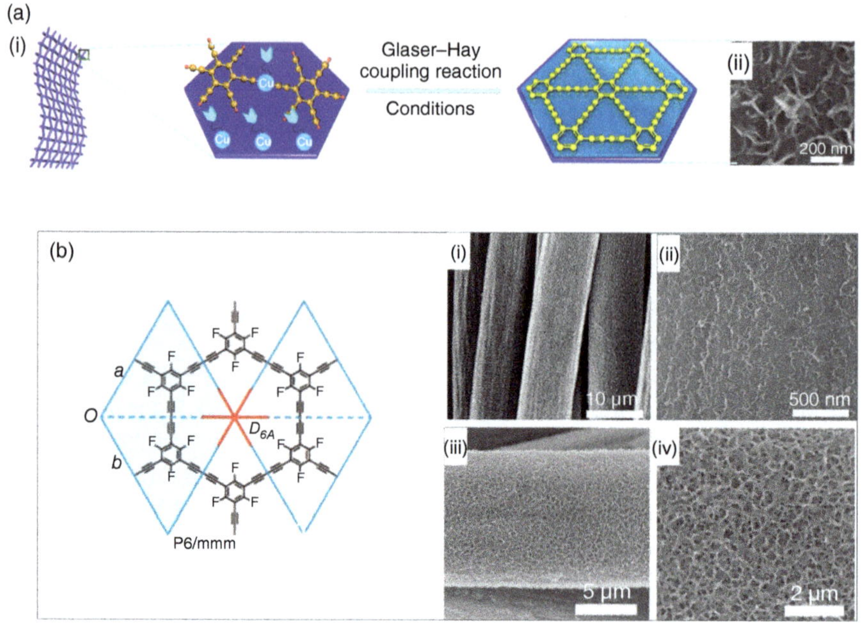

Figure 3.16 Representation of GDY and analogues in situ growth on carbon cloth. (a) GDY in situ growth on carbon cloth and its SEM image [51], (b) F-GDY structure and morphologies of F-GDY/carbon cloth [52, 53]. Source: (a) Xue et al. [51]. © 2018, Springer Nature, (b) Xing et al. [52]. © 2019, Wiley-VCH Verlag GmbH & Co. KGaA; Ref. [53]. © 2020, Royal Society of Chemistry.

Gao and coworkers reported a facile synthesis of an ultrathin single-crystalline GDY film on graphene surface by Eglinton reaction using HEB at a low monomer concentration. As illustrated in Figure 3.17a,b, single-layered graphene on SiO_2/Si acted as the substrate for the synthesis of GDY. The as-prepared graphene substrate was immersed in a CH_2Cl_2 solution of HEB monomers, where the HEB molecules were adsorbed to the graphene surface. Then, a pyridine solution of $Cu(OAc)_2$ was added dropwise, serving as the catalyst of acetylenic coupling reaction. To further control the kinetics of the coupling reaction and avoid the side reactions, the authors adopted a low concentration for both the monomer (0.04 mM) and the catalyst (0.7 mM). After reaction at room temperature for 24 hours, the ultrathin GDY film can be synthesized on the graphene surface. Figure 3.17c,d shows typical OM and SEM images of GDY/graphene films. The GDY film on the graphene surface is continuous with a lateral dimension of 50 mm. AFM analysis (Figure 3.17e) exhibited an ultra-flat surface with a thickness of only 1.74 nm, inclusive of monolayer graphene [56]. Later, Zhang and coworkers extended this template method to hexagonal boron nitride (*h*BN). The same synthesis process was executed with *h*BN as a surface template instead of graphene. This process also worked well on the surface of *h*BN with resembling surface structure, which is a good dielectric layer [57].

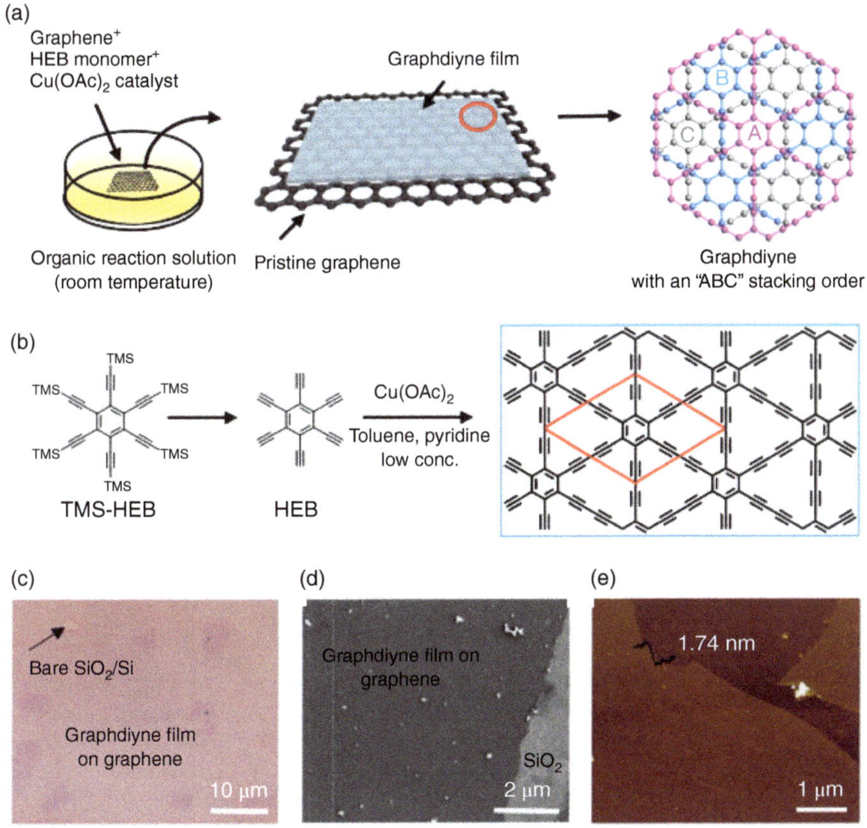

Figure 3.17 Synthetic process of single-crystalline GDY on graphene film. (a) Schematic illustration of the synthetic process, (b) chemical structure of HEB, hexakis[(trimethylsilyl) ethynyl]benzene trimethysily–hexacthynylbenzene (TMS–HEB), and the Eglinton coupling reaction of HEB molecule for the growth of GDY. TMS = $SiMe_3$. Typical OM image (c) and SEM image (d) of GDY film grown on graphene, (e) typical AFM image of GDY/graphene film on SiO_2/Si substrate, showing a thickness of ~1.74 nm (including a single-layer graphene) [56]. Source: Gao et al. [56]. © 2018, AAAS.

Subsequently, by a similar procedure, an ultrathin β-GDY film grew on the graphene surface by the Eglinton coupling reaction using TEE as the precursor. A solution of TEE and CH_2Cl_2 was added dropwise into a two-necked flask containing pyridine, $Cu(OAc)_2$, toluene, and SiO_2 plate supporting single-layer graphene. The reaction mixture was kept at room temperature for 20 hours under argon atmosphere. Then, the β-GDY film grown on the graphene surface was obtained. The as-synthesized β-GDY film presents smooth and continuous morphology and has good crystallinity. The β-GDY film was also directly grown on SiO_2/Si-based *h*BN substrate surface and served as a device with a conductivity of $1.30 \times 10^{-2}\,S\,m^{-1}$ [58].

Precise patterning of GDY could be achieved on the predefined silicon template with different shapes in the microlevel (Figure 3.18) [59]. A similar procedure as

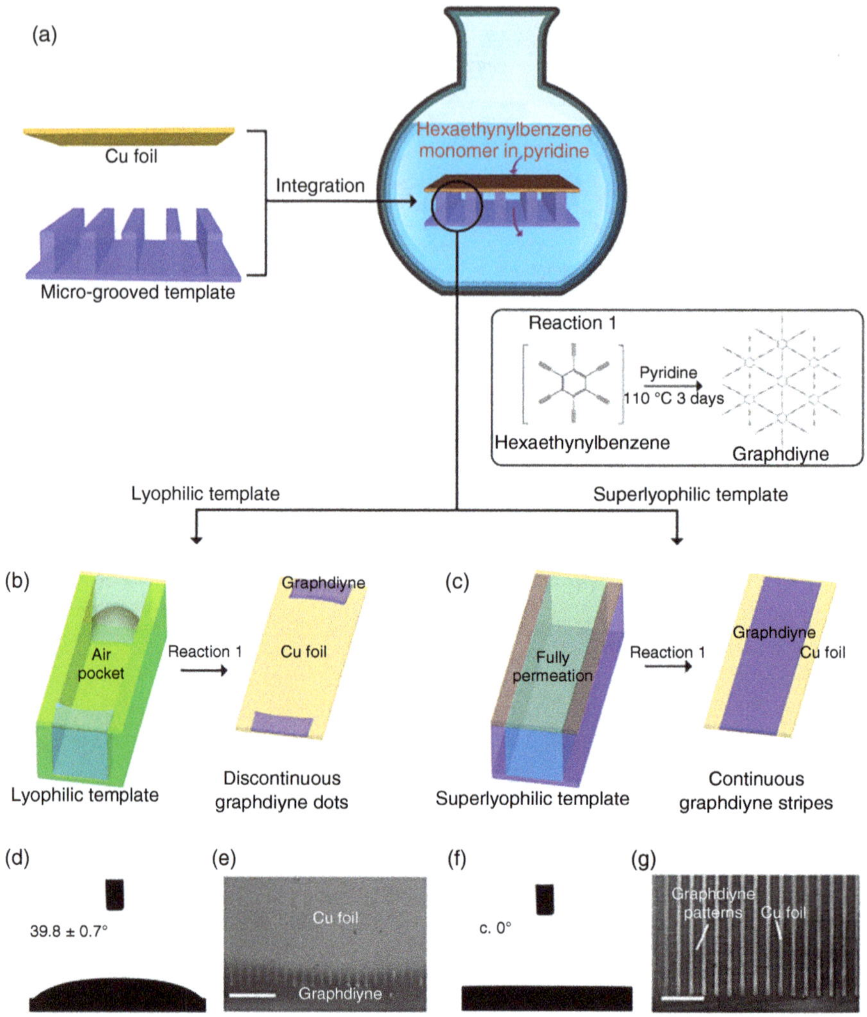

Figure 3.18 (a) Schematic illustration of the in situ synthesis of GDY patterns, (b) the growth of GDY on a lyophilic template. (c) the growth of GDY on a superlyophilic template, (d) the contact angles of pyridine ($\gamma = 39.82\ \text{mN m}^{-1}$) droplet on the lyophilic and (f) superlyophilic grooved templates, (e, g) the corresponding SEM images of GDY growth upon the Cu foil [59]. Source: Wang et al. [59]. © 2016, Wiley-VCH Verlag GmbH & Co. KGaA.

mentioned above is adopted for the in situ growth of GDY on the template. Notably, it is crucial to select a template with superhydrophilicity, so that the reactants could continuously diffuse to the surface of the substrate with microscale spacing. The grooved template provided numerous regularly confined spacing at the microscale for the in situ synthesis of GDY, whereas the wettability of grooved templates played a key role in allowing continuous mass transport of raw reactants. After the completion of the cross-coupling reaction in the microscale spacing, precisely patterned GDY stripes could be generated accordingly. The geometry of the confined spacing, reaction mass, and temperature have been tailored to optimize the final GDY patterning. Following this method, GDY patterns with controllable microshapes were achieved, which can be further used in electronic devices.

3.1.4 Interfacial Synthesis

Bottom-up synthesis of GDY crystalline nanosheets at the interface of liquid/liquid or gas/liquid was proposed by Sakamoto and coworkers [60]. The interfaces are expected to offer a favorable environment for the growth of the GDY, such as preferable orientations for growth, limited access to the reaction space, and controlled reaction rates. The $Cu(OAc)_2$ and pyridine were used to catalyze the coupling reaction and dispersed in the upper aqueous layer. The precursor was dissolved in the lower layer of CH_2Cl_2. Between these phases, a pure water layer was added as a buffer layer to prevent the sudden mixing of the reactants and catalysts. The two aqueous layers gradually merged into one phase, and then GDY nanosheets grew gradually at the interface of those two liquid phases (Figure 3.19a). Inspired by Langmuir–Blodgett technology, they also performed the gas/liquid interfacial synthesis. The gas/liquid interface appears ideal as a reaction space for monomers to form monolayers (Figure 3.19b). The gas/liquid interfacial synthesis for few layers of GDY nanosheets was then investigated. A small amount of HEB in a mixture of CH_2Cl_2 and toluene (1 : 10 v/v) was spread gently onto the surface of an aqueous solution of $Cu(OAc)_2$ and pyridine (under an Ar atmosphere at room temperature). The organic solvent was then evaporated spontaneously, and polymerization occurred at the gas/liquid interface to form GDY nanosheets with single-crystalline characteristics. Besides, the as-prepared GDY nanosheets also have a side size of 1.5 μm and a uniform thickness of ~3.0 nm. In their subsequent work, hexaethynyltriphenylene precursor was also coupled to form free-standing films through a similar procedure [61].

Recently, liquid/liquid and gas/liquid interfacial synthetic strategies were also used to prepare ultrathin nanosheets of nitrogen-substituted graphdiyne (N-GDY), tetrathiafulvalene-graphdiyne (TTF-GDY), and CN-GDY by Zhao and coworkers (Figure 3.20a–c) [62–64, 66]. Various structural precursors were dissolved in the organic phase, the aqueous solution of $Cu(OAc)_2$, and pyridine as catalyst was dropwise added on the upper of the organic phase. Under mild temperature, the GDYs are generated at the interface of the two phases. All these GDYs exhibited thin-film morphology and good crystallinity. Later, a liquid/liquid/gel system as a confined interfacial synthesis strategy was used to prepare high crystalline and ultrathin GDY films by the same group. First, the poly(butyl methacrylate-*co*-lauryl methacrylate) (P(BMA-*co*-LMA)) organogel was prepared, and then immersed in chloroform for at least one day to swell fully before use. Then, the organogel was immersed into the aqueous solution of $Cu(OAc)_2$ and pyridine. The solution of HEB in chloroform was introduced onto the gel surface by using a pipette. After the reaction, the free-standing GDY films were obtained on the gel surface. The GDY exhibited thin-film morphology and good crystallinity [65].

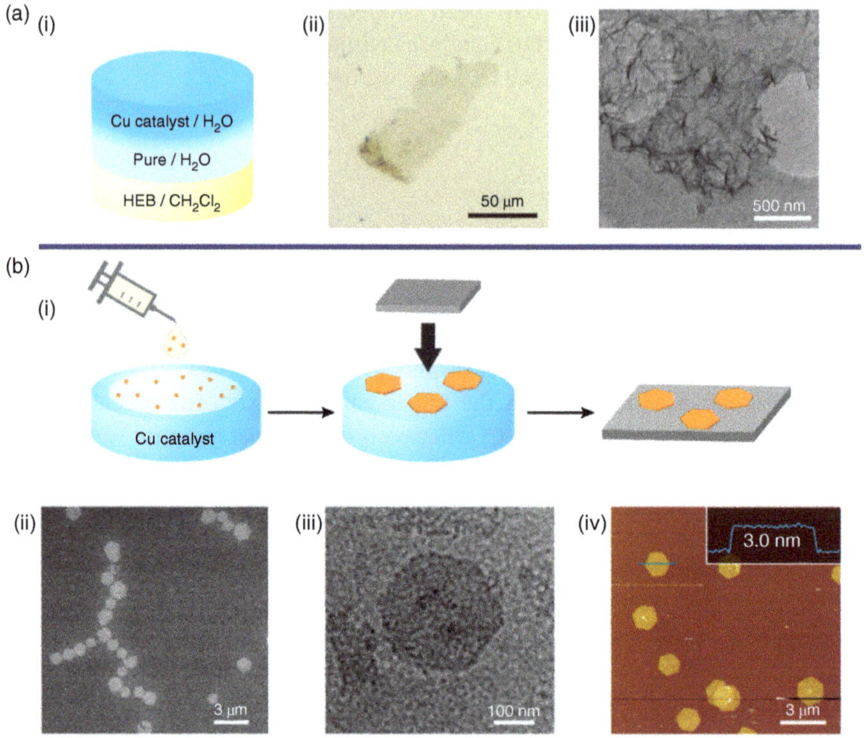

Figure 3.19 Interfacial synthesis of GDY. (a) The liquid/liquid interfacial synthesis and microscopic observations for multilayer GDY (i-liquid/liquid interfacial synthetic procedure; ii-OM image on an hexamethyldisilazane (HMDS)/Si(100) substrate; iii-TEM image of as-prepared GDY), (b) gas/liquid interfacial synthesis and microscopic observations of few-layer GDY (i-schematic illustration of the gas/liquid interfacial; ii-SEM micrograph on HMDS/Si(100); iii-TEM micrograph on an elastic carbon grid; iv-AFM topographic-image on HMDS/Si(100) and its cross-sectional analysis along the blue line) [60]. Source: Matsuoka et al. [60]. © 2017, American Chemical Society.

3.1.5 Vapor–Liquid–Solid (VLS) Growth

GDY films with different numbers of layers can also be prepared through the VLS method [67]. As shown in Figure 3.21, using GDY powder as the vapor source and ZnO nanorod arrays as the substrate, this VLS process was performed by strictly controlling the serving amount of GDY powder and moving the position of the quartz boat accordingly in the heating tube. During the heating process, a spot of ZnO was reduced to Zn droplets, which worked as both catalysts and growth anchor of GDY film. The thickness of the films was approximately 22–540 nm with a smooth surface. This GDY film was continuous over an area greater than 4.8 mm^2. The layer distance of GDY was 0.37 nm, as investigated by HRTEM.

Li and coworkers reported another example of GDY nanowires prepared by the VLS growth process using ZnO nanorod arrays on a silicon slice as a substrate [8].

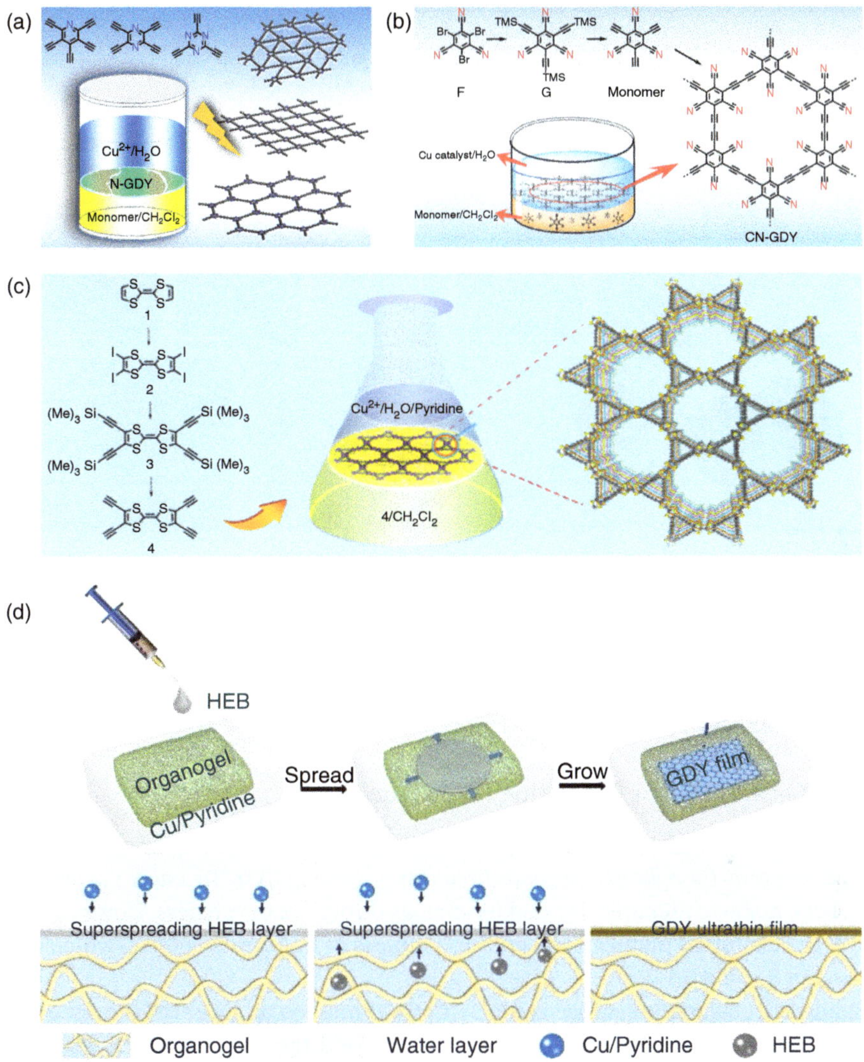

Figure 3.20 Interfacial synthesis of GDY analogues. (a) N-GDYs, (b) CN-GDY, and (c) TTF-GDY [62–64], (d) Schematic process for the confined synthesis of the GDY thin film through a superspreading strategy at the water/oil/organogel interface [65]. Source: (a) Pan et al. [62]. © 2019, American Chemical Society, (b) Pan et al. [63]. © 2019, American Chemical Society, (c) Zhang et al. [64]. © 2020, Royal Society of Chemistry, (d) Kan et al. [65]. © 2020, Wiley-VCH Verlag GmbH & Co. KGaA.

The GDY nanowires are about 0.6–1.8 mm in length and 20–50 nm in diameter. The as-prepared GDY nanowires are excellent semiconductors with a conductivity of $1.9 \times 10^3\ S\ m^{-1}$and mobility of $7.1 \times 10^2\ cm^2\ V^{-1}\ s^{-1}$. This simple, reliable VLS method provides a new approach for preparing high-quality GDY with various nanostructures. It is convenient to fabricate advanced materials with

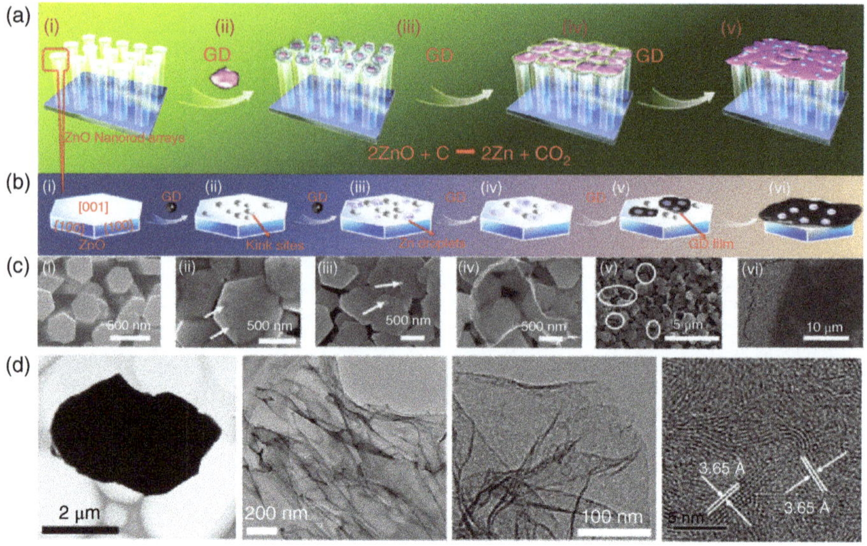

Figure 3.21 (a, b) Illustration of the growth process of GDY film on ZnO nanorods through VLS method, (c) SEM images of the related growth process, (d) TEM and HRTEM images of as-prepared GDY film [67]. Source: Qian et al. [67]. © 2015, Springer Nature.

novel electrical and optoelectronic properties for next-generation electrical and optoelectronic devices.

3.1.6 Chemical Vapor Deposition (CVD) Growth

CVD has achieved great success in preparing high-quality single-layer graphene. It will be naturally thought to be a good choice for preparing GDY. To date, many efforts have been made to prepare GDY and its analogs through CVD process. However, the successful examples mainly focus on the synthesis of fragments of GDY, which can be seen in Section 3.1.2.

Zhang and coworkers synthesized GDY by CVD method on silver foil (Figure 3.22) [68]. The linked carbon monolayer was synthesized by an atmospheric pressure chemical vapor deposition (APCVD) performed in a tubular furnace equipped with a 1-in.-diameter quartz tube. After annealing on the silver foil, samples were transferred onto various substrates by bubbling. The silver-foil enclosure was put in the center of the heating zone and annealed at 890 °C under 300 sccm H_2 and 300 sccm Ar for 30 minutes. After the furnace was cooled to a desired growth temperature (150 °C), 150 sccm Ar was introduced into the quartz tube. Then, a homemade bubbler loaded with the acetone solution of HEB was equipped with the main CVD system. The HEB vapor was introduced into the furnace by bubbling 150 sccm Ar. The typical growth time was two hours. After the growth, the silver enclosure was opened and all the subsequent characterization was carried out on the film grown on the inner surface. AFM and TEM images showed the thickness of film was 0.6 nm. This CVD method provides another simple and possible way of preparing GDY with few layers.

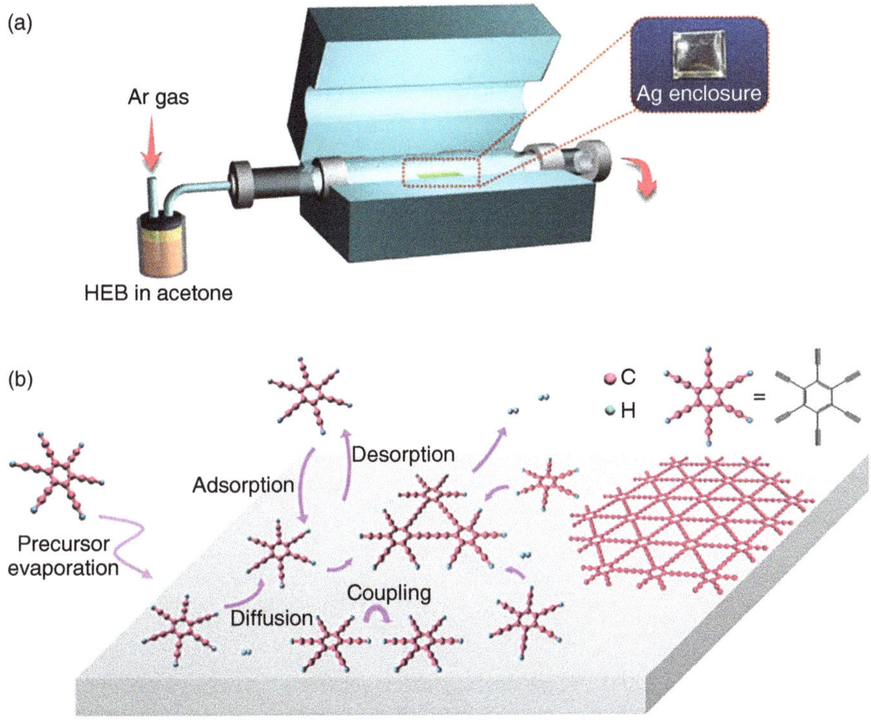

Figure 3.22 (a) Experimental setup of the CVD system for the growth of linked carbon monolayer on a silver surface using HEB as a precursor, (b) schematic view of the surface growth process [68]. Source: Liu et al. [68]. © 2017, Wiley-VCH Verlag GmbH & Co. KGaA.

3.1.7 Explosion Approach

Exploring new ways to prepare a large amount of GDY is an important job since it relates directly to the applications of the GDY. Recently, an explosion method was performed to prepare GDY in a large amount with different nanostructures [69]. As shown in Figure 3.23, three kinds of thermal treatments were carried out to control the cross-coupling reaction of HEB. Following a gradual heating process ($10\,^{\circ}\mathrm{C\,min^{-1}}$) to 120 °C in nitrogen, light yellow HEB becomes dark black without volume variation. When this treatment is applied in air, a popcorn-like explosion phenomenon is triggered once the temperature reaches 90 °C, causing a remarkable sixfold volume increase. This means that oxygen catalyzes the reaction, accelerating dehydrogenation for the coupling reaction. If HEB is directly added to a preheated environment in the air (120 °C), a more violent explosion is immediately initiated, resulting in a 48-fold volume increase. This explosion method was performed without any metal catalyst. GDY nanoribbons, 3D framework, and nanochains can be obtained by controlling the heating process. Such a method could also be utilized to tune the porous structures of GDYs as well as N-configurations of GDYs, exhibiting good controllability [70].

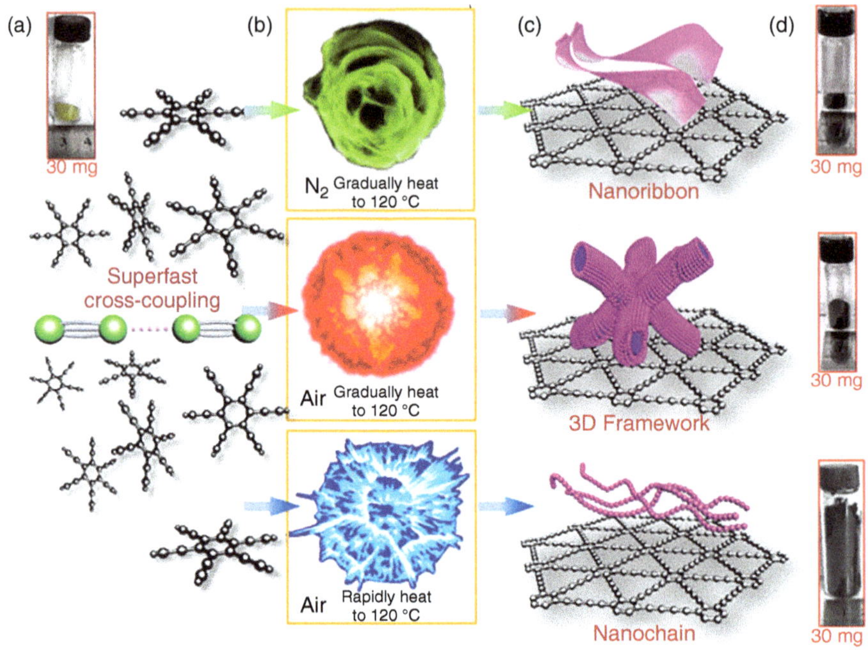

Figure 3.23 Illustrations of the explosion approach preparation processes. (a) A photo of HEB before the reaction, (b) the reaction under three treatments, (c) the GDY morphologies, (d) sample photos showing the volume change after the reaction [69]. Source: Zuo et al. [69]. © 2017, Royal Society of Chemistry.

3.2 Characterization

The characterization of GDY can be roughly divided into two categories. The first one is structure evaluation, which is mainly to characterize the as-prepared GDYs' layers structure and defect to judge the quality of GDY. The other one is performance evaluation, which is mainly to characterize the electronic conductivity, light transmission, mechanical properties, thermal stability, and thermal conductivity of the GDY material to determine its performance. Here we will focus on the structural characterization of GDY. The usual structure characterization methods of GDY mainly include Raman, X-ray photoelectron spectroscopy (XPS), X-ray absorption spectroscopy (XAS), X-ray diffraction (XRD), and microscopy techniques such as SEM, TEM, and AFM.

3.2.1 Raman Spectroscopy

Raman spectroscopy is one of the most common, fast, nondestructive, and high-resolution techniques used to characterize carbon materials. There are six strong peaks of GDY in the Raman spectrum according to the density functional theory (DFT) calculation (Figure 3.24) [30]. The benzene and alkyne-corresponding ring's breathing vibration generated the B peak. The G peak mostly originates

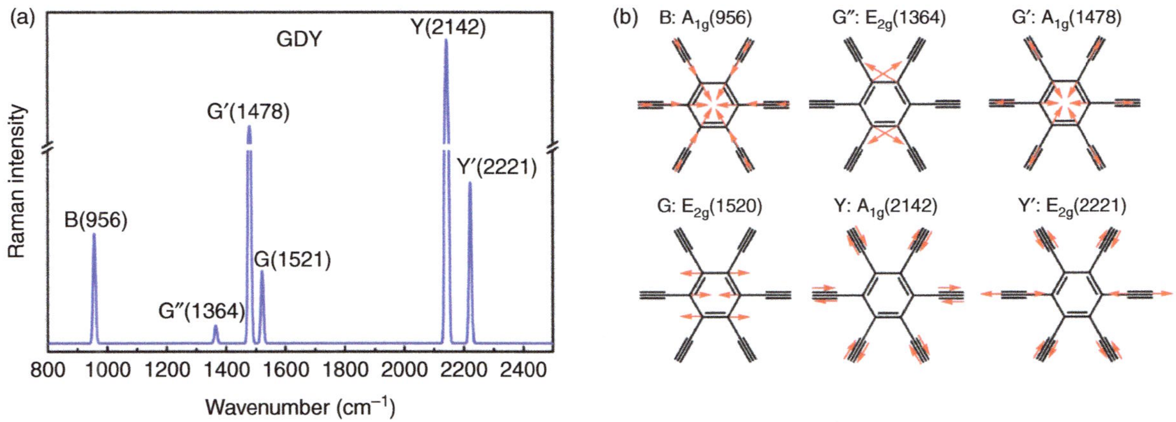

Figure 3.24 Raman spectra and vibrational modes of GDY. (a) Predicted Raman spectrum, whose peaks are with 10 cm^{-1} Gaussian smearing, (b) atomic motions of intense Raman-active modes, in which the red arrows show the motion directions of the main contributors [71]. Source: Ref. [71]. © 2016, American Chemical Society.

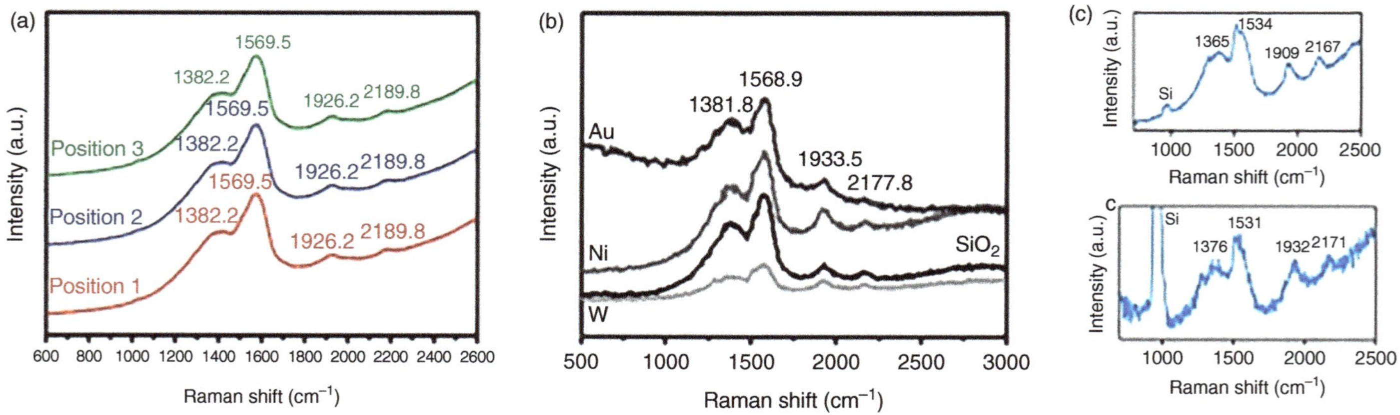

Figure 3.25 Raman spectra of GDY. (a) Raman spectra of GDY films on Cu foil at different positions, (b) Raman spectra of GDY nanowalls on different substrates, (c) Raman spectra of GDY synthesized by the interfacial method [24, 47, 60]. Source: (a) Li et al. [24]. © 2010, Royal Society of Chemistry, (b) Gao et al. [47]. © 2016, Wiley, (c) Matsuoka et al. [60]. © 2017, American Chemical Society.

from stretching of aromatic bonds. The G′ peak comes from the vibrations of C—C bonds between triply coordinated atoms and their doubly coordinated neighbors. The G″ peak is ascribed to the scissoring vibration of atoms in benzene ring. The carbon–carbon triple bonds' synchronous stretching and contracting caused the symmetric mode Y peak. The out-of-phase vibrations of carbon-carbon triple bonds initiated the Y′ peak. On the basis of the theoretical analysis of the above vibrational modes, the G bands were supposed to be useful to detect and characterize the GDY sample, which is due to the fact that it is softened, compared to other sp^2-hybridized carbons such as graphene. The structure variation and strain effect on the Raman shifts of GDY sample were also considered. The inherent relationship between the applied mode of strain and the shift of Raman spectra could also provide useful information for the characterization of GDY-based sample.

In experimental results, the Raman spectra of GDY show four main characteristic peaks at 1382.2, 1569.5, 1926.2, and 2189.8 cm^{-1}, as shown in Figure 3.25a. The strong peak, appearing at 1569.5 cm^{-1} derived from the stretching of aromatic carbon bonds, could correspond to the G band. The peak located around 1382.2 cm^{-1} comes from the carbon atoms vibration in the benzene rings band, which could correspond to the D band. The D band is related to the structural defects of GDY, so the ratio of the D band to the G band can be used to evaluate the quality of the GDY. The smaller ratio of D band to G band indicates fewer defects of as-prepared GDY and higher crystallinity. The peaks of 1926.2 and 2189.8 cm^{-1} can be assigned to the diynyl link (—C≡C—C≡C—) vibrations. Figure 3.25 gives several examples of Raman spectra of the GDY samples synthesized by a template method and interfacial strategy. The 1,3-diyne bonds (—C≡C—C≡C—) vibrations in the Raman spectrum are distinct signs for checking the formation of the diyne bond and have been widely used to evaluate the successful preparation of GDY.

3.2.2 X-ray Photoelectron Spectroscopy (XPS)

XPS can be used to analyze the chemical structure and chemical composition of GDY and its derivatives. The carbon was the main element in GDY, having been verified by the XPS spectra (Figure 3.26). This result was the same as that observed by energy-dispersive spectrometry (EDS). It was possible to divide the fitted C1s peak into four subpeaks, including the sp^2 orbital carbons of benzene rings around 284.5 eV, the sp orbital carbons of C≡C bonds at 285.2 eV, C=O bonds close to 288.5 eV, and C—O bonds near the binding energy of 286.9 eV. Among them, the integral area ratio of C≡C and C=C peaks is 1/2, which is consistent with the chemical structure of GDY. The presence of oxygen was presumed to be the air adsorbed in the GDY sample, as well as some terminal alkyne groups' oxidation.

3.2.3 X-ray Absorption Spectroscopy (XAS)

As an explicit elemental absorption spectroscopic technique, the XAS can be measured by exciting the core electrons to quasi-bound states and is widely applied for determining the electronic structure and local geometry of the complexes. XAS,

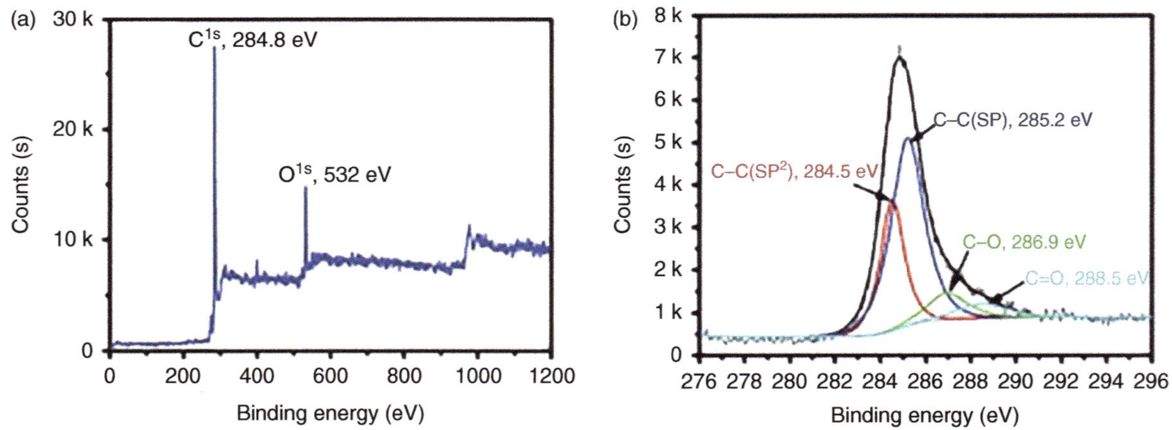

Figure 3.26 XPS spectra of GDY. (a) Elements survey scan, (b) High-resolution core-level XPS spectrum of C1s [24]. Source: Li et al. [24]. © 2010, American Chemical Society.

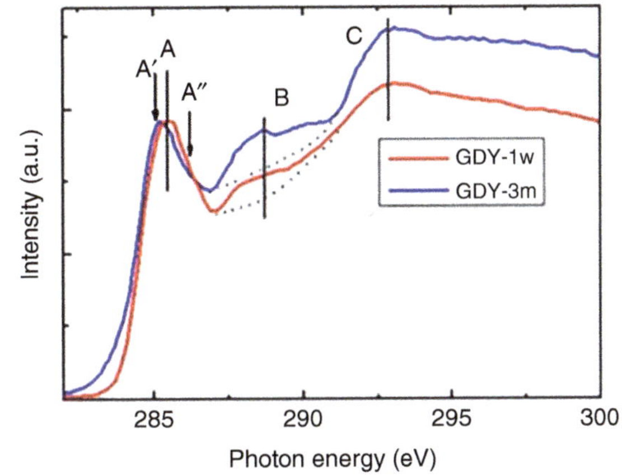

Figure 3.27 Comparison of C K-edge XANES spectra of GDY-1w and GDY-3m normalized to the π* excitation (feature A), GDY exposed to air for one week (GDY-1w) and three months (GDY-3m) [72]. Source: Zhong et al. [72]. © 2013, American Chemical Society.

dealing with the measurement and interpretation of the absorption coefficient of an element in the near-edge region (from just below the edge to ~50 eV above the threshold), is often referred to X-ray absorption near-edge structure (XANES) or near-edge X-ray absorption fine structure (NEXAFS) to be distinguished from extended X-ray absorption fine structure (EXAFS) associated with the modulation of the absorption coefficient in the extended region (50 eV up to as high as 1000 eV above the threshold). XANES probes the unoccupied density of states (DOSs) in a particular chemical environment and is governed by the dipole transition selection rules to a good approximation. Scanning transmission X-ray microscopy (STXM) combines both XANES and microscopy with a spatial resolution of ~30 nm. The combination of XANES with a high spatial resolution down to ~30 nm provides a unique capability to explore electronic and structural information from isolated structures of different compositions in large-area samples. Sham, Sun, and Wang investigated the electronic structure of GDY utilizing XANES and STXM [72]. The K-edge of XANES spectrum of carbon in GDY is mainly derived from two types (Figure 3.27). One is a peak A of around 285.5 eV, which is ascribed to π* excitation of the carbon–carbon bonds in aromatic rings. The other is a peak A″ at 285.8 eV, which can be ascribed to π* excitation that comes from triple bonds of carbon. Thus, they confirm the existence of carbon–carbon triple bonds in GDY. After exposing in the air for three months, feature A shows an obvious lower-energy shift of ~0.3 eV (labeled as A′). Unlike the feature A″, for carbon–carbon triple bonds, the lowering of the energy position suggests the conversion of triple bonds attached to the carbon ring into double bonds. The peak area of feature B assigned to the interlayer states or transitions to sp^3 hybridized states clearly increased significantly due to oxygenated functional groups such as carboxylate.

3.2.4 Microscope Technology

TEM and SEM are usually used to analyze the morphology of the sample, which is combined with X-ray energy spectrometer (EDS) to analyze the composition of the sample. TEM and SEM can be used to check the morphology of the GDY and analyze the chemical element content to facilitate the evaluation of heteroatom doping. A large number of examples of the TEM and SEM of GDY have been reported.

Recently, Lu and coworkers successfully employed low-voltage and low-current-density TEM to directly image an as-synthesized crystalline GDY of ABC stacking with a six-layer thickness [32]. The TEM and HRTEM images of GDY nanosheets are shown in Figure 3.28a,b with its experimental SAED pattern (Figure 3.28c). Three stacking modes, AA, AB, and ABC of GDY, with corresponding simulated SAED patterns, are also presented (Figure 3.28d). The experimental SAED pattern matches well with that of the ABC rather than those of the AA and the AB. Thus, the nanosheet of GDY is confirmed to adopt the ABC stacking mode. To further examine the GDY crystal structure by using Wiener filtering, the noise existing in the HRTEM images was filtered. The HRTEM image by simulation is in accordance with the HRTEM images in the experiment or by the filter. Finally, the GDY nanosheet composed of six layers with a thickness of 2.19 nm, as well as the mode of ABC stacking, has been further confirmed (Figure 3.28d).

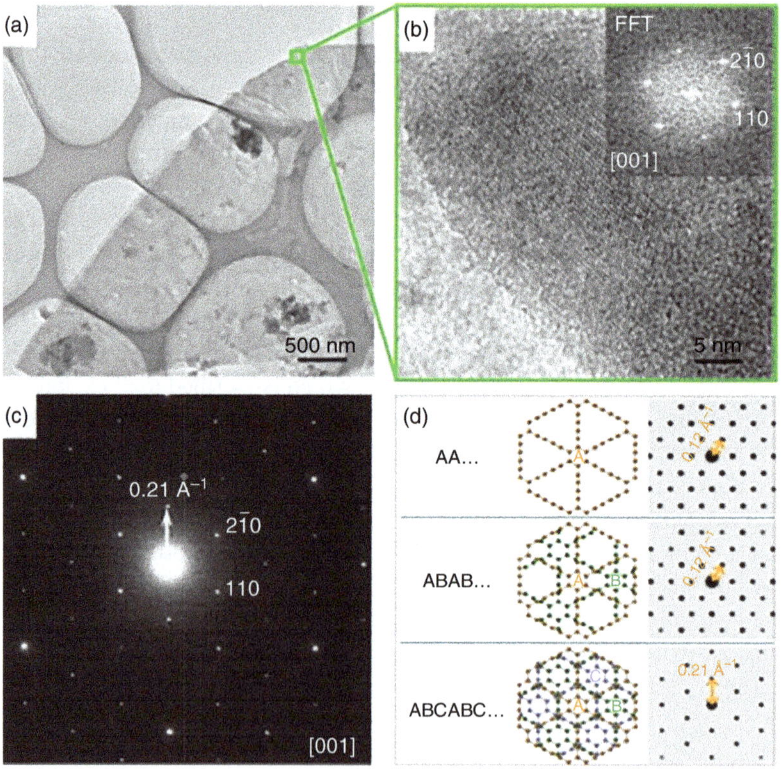

Figure 3.28 Direct imaging of the crystal structure of a GDY nanosheet, (a) TEM and (b) HRTEM, (c) Experimental SAED pattern of the nanosheet, (d) GDY models with AA, AB, and ABC stacking modes and corresponding simulated SAED pattern [32]. Source: Shang et al. [32]. © 2018, Springer Nature.

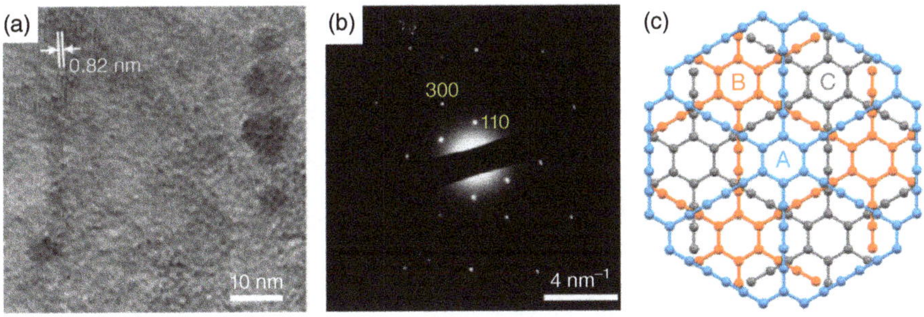

Figure 3.29 (a) HRTEM image of multilayer GDY films, (b) SAED pattern of the nanosheet, (c) graphdiyne lattice of the ABC-stacking configuration [60]. Source: Matsuoka et al. [60].

Sakamoto and coworkers verified the in-plane periodicity of the GDY film obtained at the liquid/liquid interface using TEM and SAED (Figure 3.29), which also exhibited the ABC-stacking configuration [60]. Later, Zhao and coworkers used this method to confirm the stacking structure of N-GDYs, TTF-GDY, F-GDY, CN-GDY, GTTY, Ben-GDY, and so on [36, 53, 62–64].

AFM and scanning force microscope (SFM) are high-resolution characterizations of scanning probe microscope (SPM) with atomic resolution. People can not only check the morphology of GDY, the thickness, and the modulus, but also electronic properties of GDY obtained by various working modes. Some high-resolution atomic structures of the fragments of GDY have been obtained and mentioned in Section 3.1.2.

3.2.5 X-ray Diffraction (XRD) Technique

XRD will be the most effective method for characterizing the structure of GDY. However, the synthesis of large-area long-range ordered GDY is still very challenging. The examples are very rare. As shown in Figure 3.30a, the XRD spectrum of the GDY film has obvious diffraction peaks at 10.59°, 21.72°, 22.21°, and 25.27° [24]. The strong diffraction peak at 10.59 corresponds to the diffraction spacing of 0.42 nm, which is consistent with the lattice fringes as revealed in HRTEM. Besides, a 2D grazing-incidence wide-angle X-ray scattering (2D GIWAXS) assessed the crystallinity and packing structure of few-layer GDY [60]. The 2D GIWAXS pattern (Figure 3.30b) features three diffuse spots along with the horizontal and diagonal directions. This assignment combined with TEM/SAED results to further confirm the crystal structure of GDY.

3.2.6 Others

In addition to these main characterization techniques mentioned above, there are also ultraviolet–visible (UV–vis) absorption spectrum and ultraviolet photoelectron spectroscopy (UPS), which are often used to determine the optical

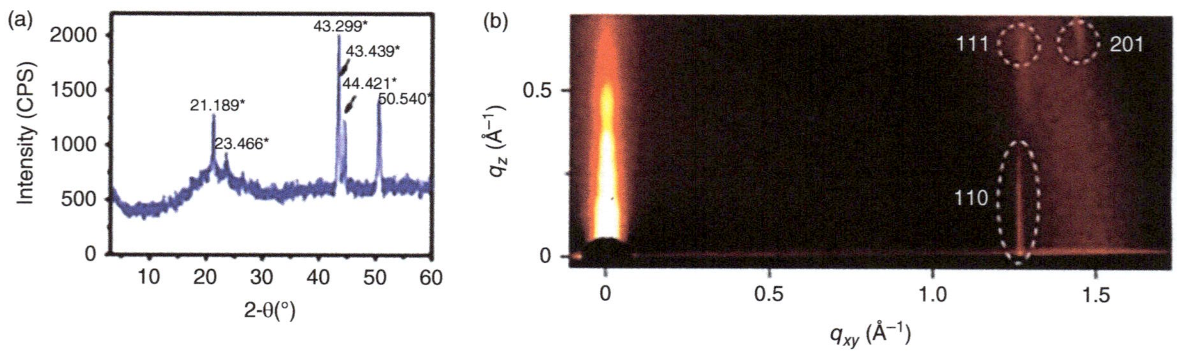

Figure 3.30 (a) XRD pattern of GDY film, (b) 2D GIWAXS pattern of few-layer GDY on Si(100) [24, 60]. Source: Li et al. [24]. © 2010, Royal Society of Chemistry; Matsuoka et al. [60]. © 2017, American Chemical Society.

bandgap and energy level of the valence band of GDYs materials. Fourier transform infrared absorption spectrometer (FT–IR) is mainly used for group structure analysis, qualitative and quantitative analysis of materials, as a complement to Raman spectra. Solid-state ^{13}C NMR (nuclear magnetic resonance) can characterize and analyze the chemical structure and spatial structure of GDY and its derivatives.

The characterization methods can effectively explore the morphology, crystal structure, and bonding mode of GDY. Electron microscopy technology can realize in situ analysis of the morphology and crystallographic properties of GDY. The various energy spectroscopy methods attached to electron microscopy can realize the analysis and characterization of the in situ composition and atomic chemical state of GDY. AFM can be used to study the surface morphology, structure, and thickness. XRD technology can be used to analyze the structure or morphology of the internal atoms or molecules of GDY. Raman, FT–IR, solid-state ^{13}CNMR, XPS, UPS, XAS, and UV–vis can be used to study the composition, chemical structure, and chemical bond states of the elements.

3.3 Summary

Since the first example of GDY was successfully synthesized in 2010 by Li and coworkers, increasing numbers of research groups have been focusing their attention on the preparation of these new promising materials. The requirements of a high content of sp carbon make the synthesis of GDY complicated and challenging. Because of the presence of the triple bond, the traditional high-temperature synthesis strategies are not suitable owing to the unavoidable side reactions. The terminal alkyne coupling reaction of HEB monomers was then considered to be an ideal route. However, controlling the highly ordered growth of the GDY is intractable. In this case, the Cu-surface-mediated synthesis was developed. The dual role of the Cu foil both as substrate and catalyst realizes the successful preparation of GDY film by in situ Glaser coupling on the surface of Cu foil. The reaction conditions have been optimized continually since then. GDYs with higher quality have been prepared. On the basis of this principle, interfacial synthesis has also been developed. The substrate changed from Cu foil to a solution of Cu ion. Various highly crystalline GDYs have been synthesized on the gas/liquid or liquid/liquid interface. Template synthesis is also an important contribution to the preparation of GDY. Arbitrary substrates can be used for the growth of high-quality GDY such as graphene, *h*BN, carbon cloth, metal foam, MOs, diatomite, and so on. Besides these, more synthesis strategies are emerging such as CVD, VLS, and explosion approach. Different morphologies of GDY, including films; nanotube arrays; nanowires and nanoribbons; nanowalls; nanosheets; ordered stripe arrays; and 3D framework, have been successfully prepared. The structures have also been diversified. However, several issues remain. The ultimate goal in GDY synthesis will be the large-scale preparation of large-sized, single-layered GDY single crystal. To touch this grail, new chemistry is definitely needed to be explored especially the dynamic covalent chemistry for constructing high-crystalline GDY. Some ingenious preparation methods assisted by some external factors may help improve the quality of the obtained GDY. The design of new precursors, which could realize the controlled pre-self-assembly process, may lead to the crystalline GDY analogues.

For the characterization, it is particularly important to find a simple and versatile method to quickly judge the quality of GDY. Raman spectroscopy was used to demonstrate the existence of acetylenic linkages in GDY, but this method fails to analyze the percentage of carbon–carbon triple bonds, which is important for judging the structure of graphyne (different kinds of graphynes often have different percentages of acetylenic linkages). In addition, different growth methods often yield samples with different qualities. Therefore, it is necessary to develop an effective way to accurately determine the polymerization degree of GDY samples. Lastly, the accurate structure analysis of GDY is highly needed. The limited crystalline domain size of the obtained GDY so far excluded most of the characterization methods. STM or high resolution atomic force microscope (HR-AFM) techniques may be a possible way to achieve this. Also, it is worth mentioning that SAED technique has shown great potential in characterizing the structure of GDY and analogues since the GDYs have simple layered 2D structures.

References

1 Baughman, R.H., Eckhardt, H., and Kertesz, M. (1987). Structure-property predictions for new planar forms of carbon – layered phases containing sp^2 and sp atoms. *Journal of Chemical Physics* 87: 6687–6699.

2 Diederich, F. and Kivala, M. (2010). All-carbon scaffolds by rational design. *Advanced Materials* 22: 803–812.

3 Cocq, K., Saffon-Merceron, N., Coppel, Y. et al. (2016). Carbo-naphthalene: a polycyclic carbo-benzenoid fragment of alpha-graphyne. *Angewandte Chemie International Edition* 55: 15133–15136.

4 Haley, M.M., Brand, S.C., and Pak, J.J. (1997). Carbon networks based on dehydrobenzoannulenes: synthesis of graphdiyne substructures. *Angewandte Chemie International Edition in English* 36: 836–838.

5 Spitler, E.L., Johnson, C.A. II,, and Haley, M.M. (2006). Renaissance of annulene chemistry. *Chemical Reviews* 106: 5344–5386.

6 Gholami, M. and Tykwinski, R.R. (2006). Oligomeric and polymeric systems with a cross-conjugated π-framework. *Chemical Reviews* 106: 4997–5027.

7 Wiley, J., Youngs, C.A.T., and Bradshaw, J.D. (1999). ortho-arene cyclynes, related heterocyclynes, and their metal chemistry. *Chemical Reviews* 99: 3153–3180.

8 Qian, X., Ning, Z., Li, Y. et al. (2012). Construction of graphdiyne nanowires with high-conductivity and mobility. *Dalton Transactions* 41: 730–733.

9 Baxter, P.N., Al Ouahabi, A., Gisselbrecht, J.P. et al. (2012). Electronic, spectroscopic, and ion-sensing properties of a dehydro[*m*]pyrido[14]- and [15]annulene isomer library. *Journal of Organic Chemistry* 77: 126–142.

10 Desroches, M., Courtemanche, M.A., Rioux, G. et al. (2015). Synthesis and properties of rhomboidal macrocyclic subunits of graphdiyne-like nanoribbons. *Journal of Organic Chemistry* 80: 10634–10642.

11 Ajit Bhaskar, R.G., Haley, M.M., and Goodson, T. III, (2006). Building symmetric two-dimensional two-photon materials. *Journal of the American Chemical Society* 128: 13972–13973.

12 Marsden, J.A., Miller, J.J., Shirtcliff, L.D., and Haley, M.M. (2005). Structure-property relationships of donor/acceptor-functionalized tetrakis(phenylethynyl)benzenes and bis(dehydrobenzoannuleno)benzenes. *Journal of the American Chemical Society* 127: 2464–2476.

13 Tahara, K., Fujita, T., Sonoda, M. et al. (2008). Donors and acceptors based on triangular dehydrobenzo[12]annulenes: formation of a triple-layered rosette structure by a charge-transfer complex. *Journal of the American Chemical Society* 130: 14339–14345.

14 Seo, S.H., Jones, T.V., Seyler, H. et al. (2006). Liquid crystalline order from *ortho*-phenylene ethynylene macrocycles. *Journal of the American Chemical Society* 128: 9264–9265.

15 Tobe, Y., Fujii, T., Matsumoto, H. et al. (2000). [2+2] Cycloreversion of [4.3.2]propella-1,3,11-trienes: an approach to cyclo[*n*]carbons from propellane-annelated dehydro[*n*]annulenes. *Journal of the American Chemical Society* 122: 1762–1775.

16 Wan, W.B. and Haley, M.M. (2001). Carbon networks based on dehydrobenzoannulenes. 4. Synthesis of "star" and "trefoil" graphdiyne substructures via sixfold cross-coupling of hexaiodobenzene. *Journal of Organic Chemistry* 66: 3893–3901.

17 Nishide, H., Takahashi, M., Takashima, J. et al. (1999). Acyclic and cyclic di- and tri(4-oxyphenyl-1, 2-phenyleneethynylene)s: their synthesis and ferromagnetic spin interaction. *Journal of Organic Chemistry* 64: 7375–7380.

18 Belenkov, E.A., Mavrinskii, V.V., Belenkova, T.E. et al. (2015). Structural modifications of graphyne layers consisting of carbon atoms in the sp- and sp^2-hybridized states. *Journal of Experimental and Theoretical Physics* 120: 820–830.

19 Cirera, B., Zhang, Y.Q., Bjork, J. et al. (2014). Synthesis of extended graphdiyne wires by vicinal surface templating. *Nano Letters* 14: 1891–1897.

20 Johnson, C.A. II,, Lu, Y., and Haley, M.M. (2007). Carbon networks based on benzocyclynes. 6. Synthesis of graphyne substructures via directed alkyne metathesis. *Organic Letters* 9: 3725–3728.

21 Kehoe, J.M., Kiley, J.H., English, J.J. et al. (2000). Carbon networks based on dehydrobenzoannulenes. 3. Synthesis of graphyne substructures. *Organic Letters* 2: 969–972.

22 Ferrara, J.D., Tanaka, A.A., Fierro, C. et al. (1989). Synthesis andstructural and theoretical characterization of a nickel(o) complex of tribenzocyclyne (TBC)

and the preparation of a novel organometallic conductor. *Organometallics* 8: 2308–2089.
23 Haley, M.M. (2008). Synthesis and properties of annulenic subunits of graphyne and graphdiyne nanoarchitectures. *Pure and Applied Chemistry* 80: 519–532.
24 Li, G.X., Li, Y.L., Liu, H.B. et al. (2010). Architecture of graphdiyne nanoscale films. *Chemical Communications* 46: 3256–3258.
25 Gao, X., Liu, H., Wang, D. et al. (2019). Graphdiyne: synthesis, properties, and applications. *Chemical Society Reviews* 48: 908–936.
26 Glaser, C. (1869). Beiträge zur kenntniss des acetenylbenzols. *Berichte der Deutschen Chemischen Gesellschaft* 2: 422–424.
27 Ikegashira, K., Nishihara, Y., Hirabayashi, K. et al. (1997). Copper(I) salt promoted homo-coupling reaction of organosilanes. *Chemical Communications*: 1039–1040.
28 Nishihara, Y., Ikegashira, K., Hirabayashi, K. et al. (2000). Coupling reactions of alkynylsilanes mediated by a Cu(I) salt: novel syntheses of conjugate diynes and disubstituted ethynes. *Journal of Organic Chemistry* 65: 1780–1787.
29 Zhou, J., Gao, X., Liu, R. et al. (2015). Synthesis of graphdiyne nanowalls using acetylenic coupling reaction. *Journal of the American Chemical Society* 137: 7596–7599.
30 Gao, X., Zhou, J., Du, R. et al. (2016). Robust superhydrophobic foam: a graphdiyne-based hierarchical architecture for oil/water separation. *Advanced Materials* 28: 168–173.
31 Gao, X., Ren, H., Zhou, J. et al. (2017). Synthesis of hierarchical graphdiyne-based architecture for efficient solar steam generation. *Chemistry of Materials* 29: 5777–5781.
32 Shang, H., Zuo, Z., Li, L. et al. (2018). Ultrathin graphdiyne nanosheets grown in situ on copper nanowires and their performance as lithium-ion battery anodes. *Angewandte Chemie International Edition* 57: 774–778.
33 Zhou, W., Shen, H., Wu, C. et al. (2019). Direct synthesis of crystalline graphdiyne analogue based on supramolecular interactions. *Journal of the American Chemical Society* 141: 48–52.
34 Xie, C., Hu, X., Guan, Z. et al. (2020). Tuning the properties of graphdiyne by introducing electron-withdrawing/donating groups. *Angewandte Chemie International Edition* 59: 13542–13546.
35 Jia, Z., Zuo, Z., Yi, Y. et al. (2017). Low temperature, atmospheric pressure for synthesis of a new carbon ene-yne and application in Li storage. *Nano Energy* 33: 343–349.
36 Pan, Q., Chen, S., Wu, C. et al. (2020). Direct synthesis of crystalline graphtetrayne—a new graphyne allotrope. *CCS Chemistry* 2: 1368–1375.
37 Schwarz, D., Noda, Y., Klouda, J. et al. (2017). Twinned growth of metal-free, triazine-based photocatalyst films as mixed-dimensional (2D/3D) van der waals heterostructures. *Advanced Materials* 29: 1703399.
38 Zhang, T., Hou, Y., Dzhagan, V. et al. (2018). Copper-surface-mediated synthesis of acetylenic carbon-rich nanofibers for active metal-free photocathodes. *Nature Communications* 9: 1140.

39 Liu, H., Zhang, Z., Wu, C. et al. (2019). Interfacial synthesis of conjugated crystalline 2D fluorescent polymer film containing aggregation-induced emission unit. *Small* 15: 1804519.

40 Prabakaran, P., Satapathy, S., Prasad, E. et al. (2018). Architecting pyrediyne nanowalls with improved inter-molecular interactions, electronic features and transport characteristics. *Journal of Materials Chemistry C* 6: 380–387.

41 Rivera-Fuentes, P. and Diederich, F. (2012). Allenes in molecular materials. *Angewandte Chemie International Edition* 51: 2818–2828.

42 Zhang, Y.Q., Kepcija, N., Kleinschrodt, M. et al. (2012). Homo-coupling of terminal alkynes on a noble metal surface. *Nature Communications* 3: 1286.

43 Sun, Q., Cai, L., Ma, H. et al. (2016). Dehalogenative homocoupling of terminal alkynyl bromides on Au(111): incorporation of acetylenic scaffolding into surface nanostructures. *ACS Nano* 10: 7023–7030.

44 Klappenberger, F., Hellwig, R., Du, P. et al. (2018). Functionalized graphdiyne nanowires: on-surface synthesis and assessment of band structure, flexibility. *and information storage potential. Small* 14: e1704321.

45 Zhou, W., Shen, H., Zeng, Y. et al. (2020). Controllable synthesis of graphdiyne nanoribbons. *Angewandte Chemie International Edition* 59: 4908–4913.

46 Li, G., Li, Y., Qian, X. et al. (2011). Construction of tubular molecule aggregations of graphdiyne for highly efficient field emission. *Journal of Physical Chemistry C* 115: 2611–2615.

47 Gao, X., Li, J., Du, R. et al. (2017). Direct synthesis of graphdiyne nanowalls on arbitrary substrates and its application for photoelectrochemical water splitting cell. *Advanced Materials* 29: 1605308.

48 Wang, F., Zuo, Z., Li, L. et al. (2019). A universal strategy for constructing seamless graphdiyne on metal oxides to stabilize the electrochemical structure and interface. *Advanced Materials* 31: e1806272.

49 Chen, K., Li, C., Shi, L. et al. (2016). Growing three-dimensional biomorphic graphene powders using naturally abundant diatomite templates towards high solution processability. *Nature Communications* 7: 13440.

50 Li, J., Xu, J., Xie, Z. et al. (2018). Diatomite-templated synthesis of freestanding 3D graphdiyne for energy storage and catalysis application. *Advanced Materials* 30: e1800548.

51 Xue, Y., Huang, B., Yi, Y. et al. (2018). Anchoring zero valence single atoms of nickel and iron on graphdiyne for hydrogen evolution. *Nature Communications* 9: 1460.

52 Xing, C., Xue, Y., Huang, B. et al. (2019). Fluorographdiyne: a metal-free catalyst for applications in water reduction and oxidation. *Angewandte Chemie International Edition* 58: 13897–13903.

53 Xing, C., Wu, C., Xue, Y. et al. (2020). A highly selective and active metal-free catalyst for ammonia production. *Nanoscale Horiz.* 5: 1274–1278.

54 Yu, H., Xue, Y., Hui, L. et al. (2018). Efficient hydrogen production on a 3D flexible heterojunction material. *Advanced Materials* 30: 1707082.

55 Hui, L., Xue, Y., Yu, H. et al. (2019). Highly efficient and selective generation of ammonia and hydrogen on a graphdiyne-based catalyst. *Journal of the American Chemical Society* 141: 10677–10683.

56 Gao, X., Zhu, Y.H., Yi, D. et al. (2018). Ultrathin graphdiyne film on graphene through solution-phase van der Waals epitaxy. *Science Advances* 4: eaat6378.

57 Zhou, J., Xie, Z., Liu, R. et al. (2019). Synthesis of ultrathin graphdiyne film using a surface template. *ACS Applied Materials & Interfaces* 11: 2632–2637.

58 Li, J., Xiong, Y., Xie, Z. et al. (2019). Template synthesis of an ultrathin beta-graphdiyne-like film using the eglinton coupling reaction. *ACS Applied Materials & Interfaces* 11: 2734–2739.

59 Wang, S.S., Liu, H.B., Kan, X.N. et al. (2017). Superlyophilicity-facilitated synthesis reaction at the microscale: ordered graphdiyne stripe arrays. *Small* 13: 1602265.

60 Matsuoka, R., Sakamoto, R., Hoshiko, K. et al. (2017). Crystalline graphdiyne nanosheets produced at a gas/liquid or liquid/liquid interface. *Journal of the American Chemical Society* 139: 3145–3152.

61 Matsuoka, R., Toyoda, R., Shiotsuki, R. et al. (2019). Expansion of the graphdiyne family: a triphenylene-cored analogue. *ACS Applied Materials & Interfaces* 11: 2730–2733.

62 Pan, Q., Chen, S., Wu, C. et al. (2019). Sulfur-rich graphdiyne-containing electrochemical active tetrathiafulvalene for highly efficient lithium storage application. *ACS Applied Materials & Interfaces* 11: 46070–46076.

63 Pan, Q., Liu, H., Zhao, Y. et al. (2019). Preparation of N-graphdiyne nanosheets at liquid/liquid interface for photocatalytic NADH regeneration. *ACS Applied Materials & Interfaces* 11: 2740–2744.

64 Zhang, Z., Wu, C., Pan, Q. et al. (2020). Interfacial synthesis of crystalline two-dimensional cyano-graphdiyne. *Chemical Communications* 56: 3210–3213.

65 Kan, X., Wang, D., Pan, Q. et al. (2020). Confined interfacial synthesis of highly crystalline and ultrathin graphdiyne films and their applications for N_2 fixation. *Chemistry A European Journal* 26: 7801–7807.

66 Kan, X., Ban, Y., Wu, C. et al. (2018). Interfacial synthesis of conjugated two-dimensional N-graphdiyne. *ACS Applied Materials & Interfaces* 10: 53–58.

67 Qian, X., Liu, H., Huang, C. et al. (2015). Self-catalyzed growth of large-area nanofilms of two-dimensional carbon. *Scientific Reports* 5: 7756.

68 Liu, R., Gao, X., Zhou, J. et al. (2017). Chemical vapor deposition growth of linked carbon monolayers with acetylenic scaffoldings on silver foil. *Advanced Materials* 29: 1604665.

69 Zuo, Z., Shang, H., Chen, Y. et al. (2017). A facile approach for graphdiyne preparation under atmosphere for an advanced battery anode. *Chemical Communications* 53: 8074–8077.

70 Shang, H., Zuo, Z., Zheng, H. et al. (2018). N-doped graphdiyne for high-performance electrochemical electrodes. *Nano Energy* 44: 144–154.

71 Zhang, S., Wang, J., Li, Z. et al. (2016). Raman spectra and corresponding strain effects in graphyne and graphdiyne. *Journal of Physical Chemistry C* 120: 10605–10613.

72 Zhong, J., Wang, J., Zhou, J.-G. et al. (2013). Electronic structure of graphdiyne probed by X-ray absorption spectroscopy and scanning transmission X-ray microscopy. *Journal of Physical Chemistry C* 117: 5931–5936.

4

Functionalization of GDYs

Changshui Huang[1] *and Ning Wang*[2]

[1] *Institute of Chemistry, Chinese Academy of Sciences, Zhongguancun North First Street 2, Beijing 100190, P. R. China*
[2] *Science Center for Material Creation and Energy Conversion, School of Chemistry and Chemical Engineering, Institute of Frontier and Interdisciplinary Science, Shandong University, No. 27 Shanda Nanlu, Jinan 250100, P. R. China*

In recent years, graphdiyne (GDY) has shown great potential in basic science and applied science research in the fields of new energy, catalysis, photoelectric conversion, life science and microelectronics, etc. GDY has a special pore structure and electronic structure, and has important and bright applications potential in the above fields [1–3]. In the past decade, important achievements have been made in both the basic and practical applied research of GDY, making GDY rapidly become a new field in the research of carbon materials [4, 5]. From the view angle of structure, GDY has unique sp and sp^2 co-hybridization, which endows it abundant carbon chemical bonds, natural pore structure, and uneven surface charge distribution. This leads to many outstanding properties and arouses great interest in both scientific and industrial circles. Based on the controlled growth of GDY, aggregation structures with different morphologies can be obtained, such as nanowires, nanotubes, and nanosheets. The unique structure and controllable morphology endow GDY a great potential and wide space to be further modified, so that it can be applied in a wider range of research fields. For instance, the high reaction activity of the acetylene bonding units provides a good platform for its functionalization [6]. In this section, we will introduce the nonmetal heteroatom doping, metal atomic modification, and guest molecular absorption of GDY, and further discuss the related influences on its basic properties and synergistic enhancement of the device performance.

4.1 Heteroatom Doping

Doping heteroatoms on carbon materials is an effective method for the preparation of related carbon-based derivatives [7]. Benefiting from the differences in basic characteristics, such as atom size, electron obit distribution, and electronegativity,

Graphdiyne: Fundamentals and Applications in Renewable Energy and Electronics,
First Edition. Edited by Yuliang Li.

between carbon and other heteroatoms, heteroatom doping can not only improve the porous structure and surface chemical activity of GDY-based materials, but also regulate their electronic structure, conductivity, and supportability [8–10]. As a result, heteroatom doping can greatly broaden the application range of carbon-based materials and open up the application of carbon materials in the applied research fields. GDY owns two-dimensional structures with abundant sp carbon on it, which provides a more suitable environment for the inclusion of heteroatom with different atom sizes and intrinsic characteristics. The existence form of heteroatom in GDY can be multiform, which can be controlled through using different doping methods. Until now, many kinds of nonmetal heteroatoms belonging to different main groups in periodic table have been introduced (Figure 4.1). As a result, a series of doped GDY derivatives, such as nitrogen-doped GDYs (N-GDYs) [11–20], fluorine-doped GDYs (F-GDYs) [21–23], chlorine-doped GDYs (Cl-GDYs) [24, 25], boron-doped GDYs (B-GDYs) [26–30], phosphorus-doped GDYs (P-GDYs) [31], hydrogen doped GDYs (HGDYs) [9, 32–37], and sulfur-doped GDYs (S-GDYs) [38–40], have been successfully prepared. The search results verified that the existence of abundant alkyne bonds in GDY provides more doping sites for various heteroatoms. As a result, some important basic properties of GDY, such as the energy-level band gap and the electron structure, have been systematically adjusted, which opens a new visual angle to improve the performance and widen the application range of GDY in different kinds of practical devices [41, 42].

4.1.1 Nitrogen and Phosphor Doping

N-GDY can be obtained by post-treating GDY under different kinds of nitrogen sources such as ammonia, urea, and melamine. The as-prepared N-GDY exhibits quite a few outstanding properties, for example, three-dimensional porous nanostructures, numerous active sites and diffusion channels and defects, and improved

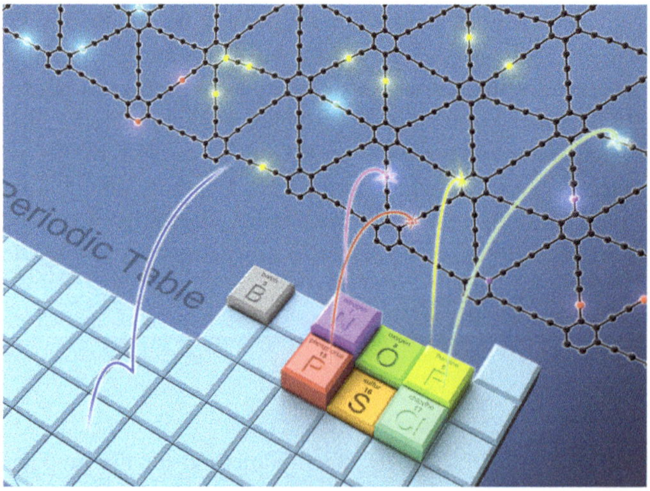

Figure 4.1 Heteroatom doping of GDY.

conductivity [12, 13]. Different from the situation of other carbon-based materials, the nitrogen substitutions of carbon in GDY occur on more possible sites, including the sp hybrid carbon existing in acetylene carbon–carbon triple bond and the sp^2 hybrid benzene ring, which can also be described as pyridine and imidazole nitrogen, respectively (Figure 4.2a). Compared to imidazole nitrogen doping, the substitutions on the ring of benzene induce more defects and larger size pores [13, 17, 45]. As for the experimental conditions, GDY treated under the condition of heating with high-purity ammonia gas in argon can dope N atoms stably into GDY to form N-GDY at the temperature range of 400–600 °C. The optimized post-treating temperature is determined to be 500 °C for the maximum doping amount of 3.70% of heteroatom nitrogen, which can be obtained according to the

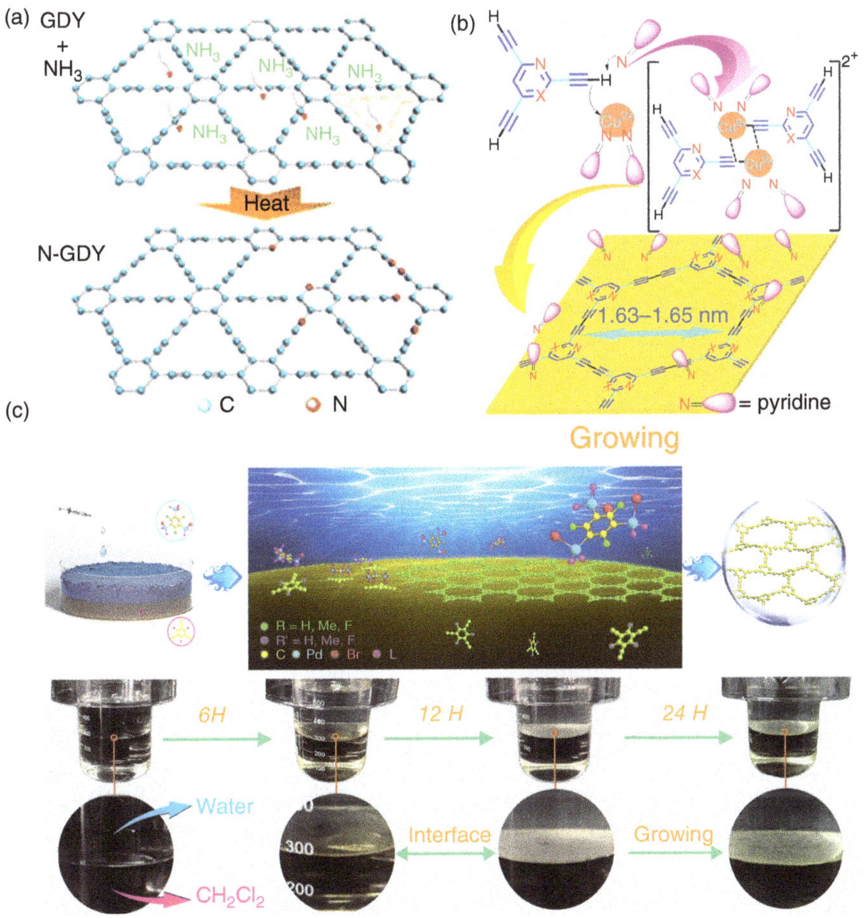

Figure 4.2 Nitrogen-doped GDY prepared through (a) thermal annealing, (b) solvent coupling precursor containing pyrimidine nitrogen, and (c) two-layer interface method. Source: (a) Zhang et al. [12]. © 2016, America Chemical Society. (b) Yang et al. [43]; © 2018, America Chemical Society. (c) Song et al. [44]. © 2019, Royal Society of Chemistry.

experimental results of X-ray photoelectron spectroscopy (XPS) spectrum [11]. Other nitrogen-containing organic compounds such as urea and melamine can also be used as nitrogen source for the nitrogen doping. The above nitrogen-containing compounds can be dissolved in organic solvent. Then, GDY powder can be dispersed in the as-prepared solution. While the solvent is removed under reduced pressure, the organic compound and GDY can be mixed uniformly, which can guarantee the well-proportioned doping of nitrogen in GDY [17, 18, 20].

While nitrogen is induced into the GDY plane, the interplanar distance of N-GDY decreased compared to pristine GDY, for the small atom radius of nitrogen compared to that of carbon, which can be regarded as an efficient way to tune the interlayer distance of GDY powder. Meanwhile, the inclusion of nitrogen also changed the bonding environment of carbon atom adjacent to the heteroatom nitrogen. The electron cloud around the butadiyne linker and benzene in GDY is rearranged, which can not only tune the position of reactive center but also adjust the affinity of GDY to other adsorbate. As a result, the kinetic exciton diffusion and electrochemical process are facilitated due to the electronegativity of nitrogen. Heteroatom active sites and defects are generated in N-GDY, which can lead to enhanced electrochemical properties. Moreover, the resistance between the electrochemical electrode and media was greatly reduced for the improved conductivity of N-GDY. Benefiting from the above specialty, the N-GDY prepared through this post-treating method has been used in various kinds of devices such as energy storage [12, 46], hydrogen purification [47], paramagnetic devices [48, 49], stretchable semiconductors [15], and electrocatalyst devices [12–14]. According to the reports, N-GDY-based devices have shown an improved performance and stability in above kinds of devices. For instance, the negative electrode based on N-GDY for lithium-ion battery (LIB) and sodium-ion battery (SIB) has exhibited excellent electrochemical performance, such as high capacitive, good rate performance, and stable long cycle stability. Furthermore, N-GDY-based lithium ion capacitor (LIC) and sodium ion capacitor (SIC) have also shown good performance. In the case of N-GDY/active carbon-based LIC, at the power density of 112.5 W kg^{-1}, an energy density of 174 Wh kg^{-1} has been obtained. For N-GDY-based SIC, it also exhibited a quite good energy density of 189 Wh kg^{-1}, while the power density is 112.5 W kg^{-1}. Notably, an increase of 58.7% (119 Wh kg^{-1}) in energy density of N-GDY based electrode in LIC has been obtained compared to that of GDY-based electrode [50].

The above-mentioned post-treating method to dope heteroatom has been proved to be an efficient strategy to tune the property and widen the application range of GDY-based materials [13]. Meanwhile, the heteroatom doping method also has a large space to be further updated, for the post-treating method that suffers from the limits of the low doping rate and uncontrollable doping sites, which restricts the unambiguous understanding on the correlation between the structure and properties of doped GDY. Moreover, the application of doped GDY in particular devices could also be restricted, especially while the doping sites and amounts of the heteroatom are needed to be controllable. Therefore, it is of great importance to develop a complementary synthesis method that can precisely control the number and position of heteroatoms on the GDY network. The chemical synthesis strategy of GDY inspires researchers to develop a bottom-up method to fabricate heteroatom doping GDY. The chemical structure of the alkynyl groups containing precursor is easy to be tailored. As a result, the heteroatom could be decorated on the precursors of GDY. Notable, the doping sites and amount of selected heteroatoms are easy to be controlled. Moreover, the doping amount can reach a high percentage, which is difficult to be achieved through the traditional post-treating method. Following this strategy, a series of different kinds of heteroatoms have been doped on GDY such as nitrogen, halogen, hydrogen, and boron. For example, Li and coworkers prepared the N-GDY materials with certain amount of pyridine and imidazole nitrogen heteroatoms. Through the chemical design of GDY precursors, nitrogen heteroatoms were successfully introduced into GDY at 120 °C without solvent and metal catalyst. The as-prepared N-GDY was endowed with high specific surface area and good 3D nanomorphology, which showed high-energy density (8.66 Wh kg^{-1}) and power density (19.3 kW kg^{-1}) in the double-electrode supercapacitor. In addition, this material also shows good performance in catalytic oxygen reduction reaction (ORR) [17]. Notably, the morphology of as-prepared GDY is highly related to the ratio of acetylene group and the N-doping heteroatoms, which strongly influences the reactivity of the selected precursors for the generation of N-GDY. In detail, the reactivity for precursors containing one nitrogen atoms is higher than that containing two or three nitrogen atoms for more acetylenic groups in the former precursor. As a result, the explosion process for the former precursor is more violent than the others, which induces different particle size of generating N-GDY with 3D porous nanostructures. The scanning electron microscopy (SEM) images show that the particle size becomes larger with the increase of the nitrogen heteroatom contents. Based on this report, it can be seen that GDY has outstanding editability in surface area and morphology. The continuous porous nanostructure of N-GDY may offer more ion/electronic migration channels and increased surface area, which can be one of the most important reasons for it to show excellent performance in both LIBs and ORR devices.

In other reports from Huang and coworkers, a series of nitrogen-containing monomers have been designed and prepared, which are used as cross-coupling precursors to fabricate N-GDY through chemical preparing method in the medium of organic solvents such as pyridine and dichloromethane [20, 46, 47, 51, 52]. The optimized doping sites, configuration of selected units in N-GDY, binding energy to the alkali metal atom have been systematically analyzed through both theoretical and experimental methods. For example, the triazine structure that contained GDY, comprising only one type of N atoms with fixed amount, has been synthesized via the bottom-up route. The triethynyltriazine was selected as the starting precursors for the generation of the well-defined 2D carbon network comprising quantitative pyridine-type nitrogen heteroatom and carbon skeleton framework with sp or sp^2 hybridization forms. Moreover, the content of the induced nitrogen is in quite a high determined amount (Figure 4.2b). The atomic ratio of nitrogen to carbon reached 1 : 3. The high content of nitrogen in this kind of N-GDY materials definitely provides more anchoring sites in its hexagonal honeycomb-like pores for the lithium alkaline atoms, which can be the most important factors that make triazine graphdiyne (TA-GDY)-based electrodes exhibit excellent electrochemical performance in LIBs [46]. On the other hand, pyridine nitrogen–containing GDY (PY-GDY) and pyrimidine nitrogen–containing GDY (PM-GDY) were synthesized successfully through modified Glaser–Hay coupling reaction. One of the most important structure features for these two GDY analogues is that composed quantitative pyridine-like N atoms are introduced to GDY framework in a certain amount and fixed doping sites. The 2,4,6-triethynylpyridine and 2,4,6-triethynylpyrimidine were chosen as precursors for the cross-coupling reaction. Interestingly, it can be seen that the amount of heteroatom nitrogen in PM-GDY is exactly twice higher than that in PY-GDY. Impressively, both of these N-GDY can be prepared in the form of transparent films with large size more than 80 cm^2. Moreover, the as-prepared films show good mechanic strength, which can be used as flexible anode materials in LIBs. Combining both the experimental and theoretical results, the excellent performance of this series of N-GDY-based electrodes was mainly ascribed to the high N content, which was endowed with large amount of active sites of the storage of lithium atoms. All the research indicates that the activity of doped GDY-based electrode is closely related to the doping configuration of nitrogen atoms. Notably, pyridyl nitrogen plays a key role for the improvement of both energy storage and electrocatalysis devices' performance, which can be due to its good electronic distribution and affinity [14, 20].

Another efficient method to prepare nitrogen-containing GDY is two-layer interface synthetic method. The interfacial coupling of two-dimensional N-GDY thin films can be performed at either gas–liquid or liquid–liquid interfaces. Following this method, Kan et al. prepared highly ordered N-GDY films with several layer thicknesses [52]. The pyrazine- or triazine-based molecules containing ethynyl group were also selected as precursor for Glaser coupling reactions at interfaces. Compared to the liquid–liquid interface, the gas–liquid interfacial synthesis exhibits great advantages in producing N-GDY film with pyrazine-like structure, for the thickness of as-prepared film is easier to be controlled, and better crystallinity of the GDY film can be obtained. The two-layer interfacial synthetic method provides a simple preparation strategy to control the thickness of N-GDY film. Compared to all-carbon GDY, the hybridization of the N in GDY could induce promising optical and electronic performance in different kinds of practical devices.

In addition to the position and contents of the induced heteroatom nitrogen, the morphology of the as-prepared N-GDY can also be adjusted through optimizing the reaction condition. For example, two-dimensional N-GDY with nanosheet morphology can be synthesized through a two-layer liquid–liquid interface method. As shown in Figure 4.2c, the alkynyl containing small molecule with triazine, pyrazine, and pyridine-based structure is used as monomer [53]. Previous work has proved that N-GDY materials have similar structures with g-C_3N_4. This can be regarded an important factor that endows it impressive semiconductor nature, which makes N-GDY-based practical devices show excellent catalytic, optical, and electronic performance. To further optimize the device performance of N-GDY-based electrode, the morphology, thickness, and interlayer periodicity are considered to be challenging sally ports. The most difficult parts are that all the above three factors should be considered synergetic. To solve this problem, an interfacial preparing strategy has been doped to proceed with the cross-coupling of the as-prepared triazine or pyrazine structure monomers. The reaction can occur at either liquid/liquid or gas/liquid interface. In the case of liquid–liquid interface reaction, $Cu(OAc)_2$ was used as reaction catalyst, which can be dissolved in the aqueous phase. The precursors containing alkynyl groups can be dissolved in dichloromethane phase. During the reaction process, the copper ions, which exist as the catalysis for the cross-coupling reaction, diffuse slowly from aqueous phase to water/CH_2Cl_2 interface. Notably, pyridine is also added to the reaction system to form copper ion complex, which can facilitate the reaction. A brown GDY film can be obtained after removing the aqueous and dichloromethane phase. The experimental results indicate that the band gap of the N-GDY can be tuned through adjusting the reaction conditions such as the reaction temperature and the concentration of the precursors. The N-GDY prepared through this method has good crystallinity, and shows excellent optical and electronic properties. These are crucial factors to improve the performance for the GDY-based nanoelectronics devices. Besides liquid–liquid interface, the gas–liquid interface is also a possible option to prepare N-GDY [52]. One of the most important differences between these two interfacial cross-coupling methods is that the films prepared through gas–liquid method are thinner than those prepared through liquid–liquid method. This can be ascribed that some monomers formed oligomers in dichloromethane solution before they arrive at the interfacial sites for the copper catalysis, and monomers exist in the liquid phase. Following this method, N-GDY films with few layers can be obtained. Notably, it is a big challenge to transfer the GDY film to the substrates for further measurement. Usually, the as-prepared N-GDY films can be transferred through Langmuir–Schäfer procedure. The atomic force microscopy (AFM) images show that the thickness of N-GDY film prepared through gas–liquid interfacial method can be less than 5 nm. This film can show clear selected area electron diffraction (SAED) points, which can provide important structure information for the N-GDY, including the packing mode and interlayer distance.

Compared to the above-mentioned nitrogen atom doping configurations with sp^2-hybridized forms, the doping of sp-hybridized forms of nitrogen could make the adjacent carbon atoms own more positive charge, which is more conducive to the adsorption of oxygen. To realize this prediction, Wang and his coworkers introduced sp-hybridized nitrogen (sp-N) atoms to pericyclically replace the sp carbon atoms in ultrathin GDY. To explain in detail, they firstly obtained the thin layer of GDY from the block GDY products, then mixed it with melamine in a certain proportion, and treated it at high temperature under the inert environment of argon. As shown in Figure 4.3, two possible synthetic routes are represented. The mass spectrum confirmed the existence of the fragment $H_2C{=}N{=}CH_2^+$, which shows a mass-to-charge ratio (m/z) of 42. Finally, sp nitrogen GDY film was obtained through the circumcircle reaction. This work provides a new strategy to

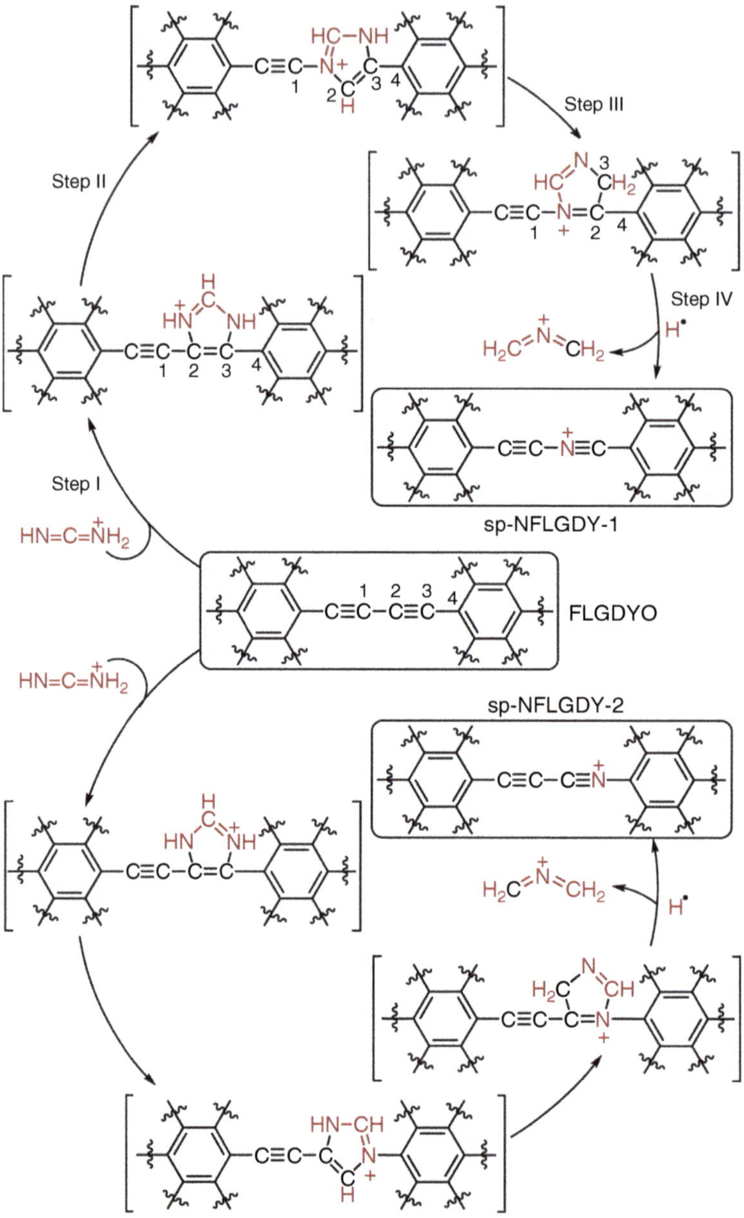

Figure 4.3 Doping of the sp-hybridized nitrogen to GDY. Source: Zhao et al. [18]. © 2018, Springer Nature.

incorporate nitrogen as heteroatom to GDY system. The existence of abundant sp carbon provides sufficient reaction sites for the sp-nitrogen mediate. The inclusion of chemically defined sp-N atoms to GDY significantly enhances its performance in electrochemical catalyst devices. Moreover, the experimental results showed that the catalytic activity increased with the increase of sp-N content. Under alkaline conditions, the ORR activity of the optimized samples was comparable to that of commercial Pt/C catalysts and showed faster reaction kinetics. Under acidic conditions, the catalytic activity of this material is slightly lower than that of Pt/C catalyst, but it is much higher than that of other nonmetallic catalysts [18]. It can be seen that the sp-N-doped GDY catalyst prepared in this way exhibits excellent ORR performance. The density functional theory (DFT) calculations confirm that the high catalytic activity is ascribed to the sp-N dopant, which provide anchoring sites O_2 adsorption. Meanwhile, the electron transfer on the surface also facilitates for the existence of sp-nitrogen. This work indicates that sp-N doped GDY is of great significance for promoting the commercialization of fuel cells.

Based on the above discussion, heteroatom nitrogen can be directly introduced to GDY through post-treatment in high temperature under the atmosphere of nitrogen source. Besides that, nitrogen heteroatom can also be included in the alkynyl-containing precursor, which can be used for the generation of GDY on different substrate. The collected N-GDY products show tunable performance in electrochemical and electronic devices with the changing heteroatom nitrogen contents, doping positions, and existence forms.

In the periodic table of the elements, phosphorus and nitrogen are in the same main group, which indicate that these two elements own the similar electron cloud configuration. Meanwhile, phosphorus also exhibits some different characteristics compared to those of nitrogen, while they are used as heteroatoms to prepare P-GDY materials (Figure 4.4). For example, the existence forms of phosphorus in P-GDY are more complicated. Some phosphorus-containing groups such as P=O, P—O, and P—C bonds could be investigated through Raman spectra, XPS, and Fourier transform infrared (FTIR) spectroscopy. Moreover, sp_3-orbital hybridized configuration of phosphorus-containing groups also exists. As a result, many open edge, distortion, and wrinkled nanostructures are generated. In addition, Brunauer–Emmett–Teller (BET) and Raman spectra analyses indicate that a great number of defects and active sites for alkaline metal have been generated, which induce the existence of micro-mesoporous pores. The existence of the above

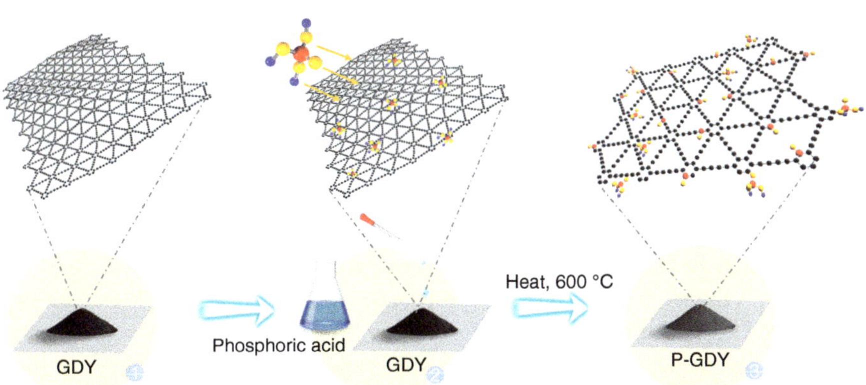

Figure 4.4 Phosphor doping of GDY. Source: Shen et al. [31]. © 2019, Royal Society of Chemistry.

phosphorus functional groups is proved to be good for the thin and dense solid electrolyte interface (SEI) layer formation on the surface of carbon, which can improve electrochemical properties. Besides that, this porous structure of P-GDY is beneficial for the storage of metal Li and diffusion of ions. Especially, the P=O double bonds are proved to be suitable for Li storage, for a stable binding energy is obtained through DFT calculation. As a result, an enhanced electrochemical performance of P-GDY-based electrode is observed compared to that of pristine GDY in LIBs. In the aspect of preparing strategy, a facile thermal treating method was used to prepare P-GDY. The treating temperature is set at 600 °C. The phosphoric acid was selected as phosphorus source, for it can guarantee the sufficient mix of GDY powder and phosphorus source. In this work, a method that combines both the density functional theoretical analysis and experimental results was used to investigate the possible structure and existence forms of phosphorene-containing groups in the doped GDY frameworks. BET investigation of P-GDY reveals that numerous heteroatom defects and hierarchical micromesoporous are generated compared to N-GDY.

4.1.2 Halogen Doping

Halogen element refers to the elements belonging to the seventh main group in the periodic element table. These include iodine (I), bromine (Br), chlorine (Cl), and fluorine (F). The halogen contains seven outermost electrons. The radius of the atom increases from F to I due to the different number of electron layers. Therefore, the nucleus's attraction ability to the outermost electrons decreases, and thus the ability to obtain electrons from the outside atmosphere decreases accordingly. The above characteristic of halogen atoms resulted in different influence on the structure and property while they are induced to GDY framework as heteroatoms.

Until now, a series of heteroatom-doped GDYs have been prepared. Especially, through the method of preparing modified GDY precursor, fluorine and chlorine heteroatoms, which can be only doped on other carbon materials under extreme harsh reaction, have been successfully introduced with controllable contents and doping position [22]. As a result, some specific properties of GDY are significantly improved. For example, while fluorine atoms are introduced to GDY framework, the electrical conductivity of F-GDY-based electrode and its compatibility with electrolyte are greatly improved [23]. The electrode based on F-GDY in lithium ion batteries shows high capacity and stable cycle performance over 9000 cycles at different current densities of 2 A g^{-1}. The excellent device performance of F-GDY-based electrode can be ascribed to the characteristic of fluorine heteroatoms. Besides that, the preparation method has also great influence on the structure and basic properties of F-GDY.

As shown in Figure 4.5a, F-GDY was prepared through bottom-up cross-coupling strategy with 1,3,5-triethynyl-2,4,6-trifluorinebenzene as precursor monomer. The copper foil was selected as the template substrate; meanwhile, the copper ions are also released as reaction catalyst while immersing in pyridine solvent. The fluorine atom is uniformly distributed in the carbon plane. The outstanding feature of F-GDY-based electrode can be ascribed from the good transition reversibility

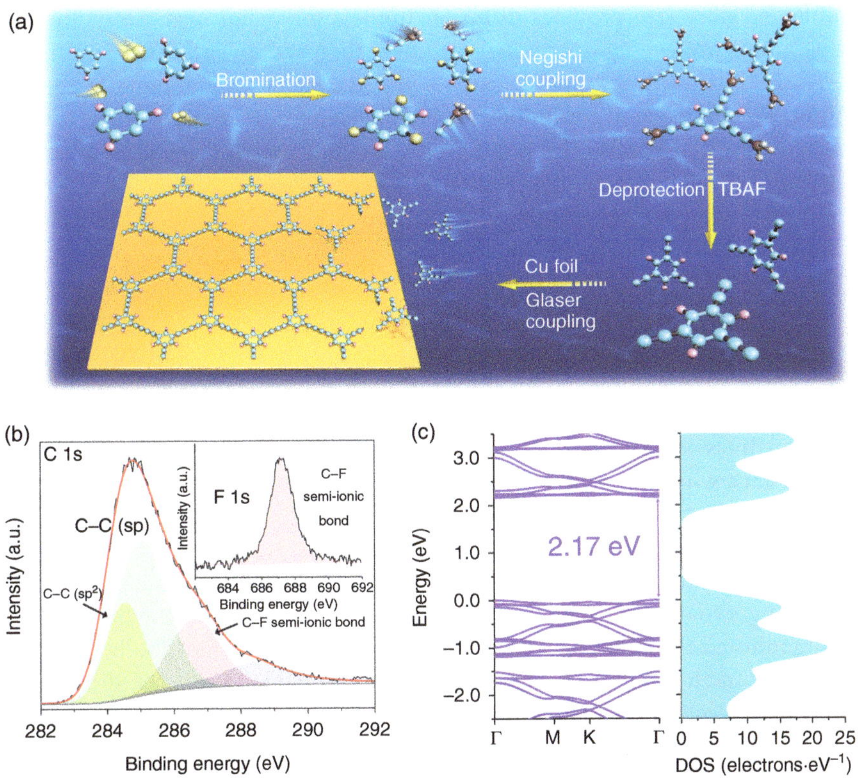

Figure 4.5 (a) Preparation of F-GDY; (b) F1s-XPS spectrum to characterize the C—F semi-ionic bond; and (c) Theoretical investigation for the energy level of F-GDY. Source: He et al. [21]. © 2018, Royal Society of Chemistry.

between the covalent and ionic status of C—F bond, which can be proved through XPS analysis (Figure 4.5b). It can be ascertained that the peak at 687.3 eV in the F 1s-XPS spectrum is assigned to the C—F semi-ionic bond, which stems from the Coulombic attraction between the fluorine and carbon atoms. As shown in Figure 4.5c, a theoretical band gap of 2.17 eV can be estimated through DFT calculation of F-GDY. This value is consistent with the measured optical band gap (2.15 eV). All the structure and energy-level's features lead to a good electrical conductivity of F-GDY ($9.66 \times 10^{-4}\ \mathrm{S m^{-1}}$).

To better understand the function of F in fluorine substituted GDY (FGDY)-based electrodes, we summarized the up-to-date reported experimental and calculation results and discuss the following aspects. From the view of chemical structure, the inclusion of fluorine mainly induces the following changes, which endow the as-prepared F-GDY distinct different basic physical properties. First of all, it can be seen that half the amount of the butadiyne linkers are replaced by F heteroatoms, which obviously enlarge the pore size of the carbon plane. Notably, the conjugation of the whole carbon framework is well maintained, which means the conductivity will not be decreased, for the in-plane transfer route for the carriers is maintained.

The enlarged pore size provides diffusion tunnels for metal ions. Secondly, the fluorine atoms are evenly distributed on the carbon framework. This is difficult to be realized in the case of other carbon materials such as graphene or carbon nanotube, for the reaction conditions for the inclusion of F atoms to these carbon allotropes are quite harsh. The doping position and amount are difficult to be controlled.

From the view of basic properties, the inclusion of F atoms changes the electron distribution of pristine GDY framework. This will generate more active sites for certain electrochemical reaction in F-GDY, which is beneficial for the enhancement of the devices performance with F-GDY as electrodes in both energy storage devices such as LIBs and electrochemical catalysis devices such as ORR process. The calculation results also support the stabilization effect of the F atoms for the Li atoms. As a result, the practical capacity of F-GDY can reach as high as 1700 $mAh\,g^{-1}$, while the current density is set to be 50 $mA\,g^{-1}$. This value is a bit little smaller than the theoretical capacity, which is calculated to be around 1867 $mAh\,g^{-1}$. This can be understood from the fact that the active sites of F-GDY-based electrode are usually not completely occupied by Li atom in the practical devices. Notably, the F-GDY-based electrode shows outstanding stability in LIBs devices. This can be understood from the following two aspects. One is that stable SEI films are formed, which can be in situ monitored through ex XPS spectra. Reversible carbon fluorine bonds with semi-ionic characteristic are observed during the formation process of SEI film on F-GDY based electrodes. The XPS spectra indicate that the potential for the formation of this C—F bond is at about 0.9 V. This value is consistent well with that obtained from the charge–discharge profiles, in which plateaus can be observed. Moreover, while the F-GDY-based electrode is further discharged, this plateau disappeared quickly, which means the electrode is totally covered by surface deposits. The protection effect of these deposits on the F-GDY-based electrode guarantees that the electrochemical process can be performed in atmosphere without the protection of inert gas. This feature greatly increases the practicability of F-GDY-based electrode in soft package or wearable devices. Both the experimental and theoretical analysis results revealed that the semi-ionic bonds play a key role for the ultra-stable long circling charge–discharge process. Besides that, another important feature of F atom is its unique wettability with particular electrolyte. The inclusion of F atoms in GDY will significantly improve the interfacial contact between the F-GDY-based electrodes and electrolyte. The contact resistance is decreased. This can be another important reason for the long stability of F-GDY-based electrode in LIBs.

Besides LIBs, F-GDY-based electrodes also exhibit good performance in the process of HER and OER. Li et al. report a 3D porous F-GDY framework grown on carbon cloth. This metal-free electrode also shows highly stable electrochemical catalysis characteristic. Moreover, the activity of F-GDY-based electrode is also superior to other well-known electrocatalysts in the process of overall water splitting.

In the series reports of halogen atoms–doped GDY from Li and coworkers, 1,3,5-(trimethylsilyl)ethynyl-2,4,6-trichlorobenzene was used as starting monomer for the cross-coupling on the copper foils (Figure 4.6a). Cl-GDY has also shown

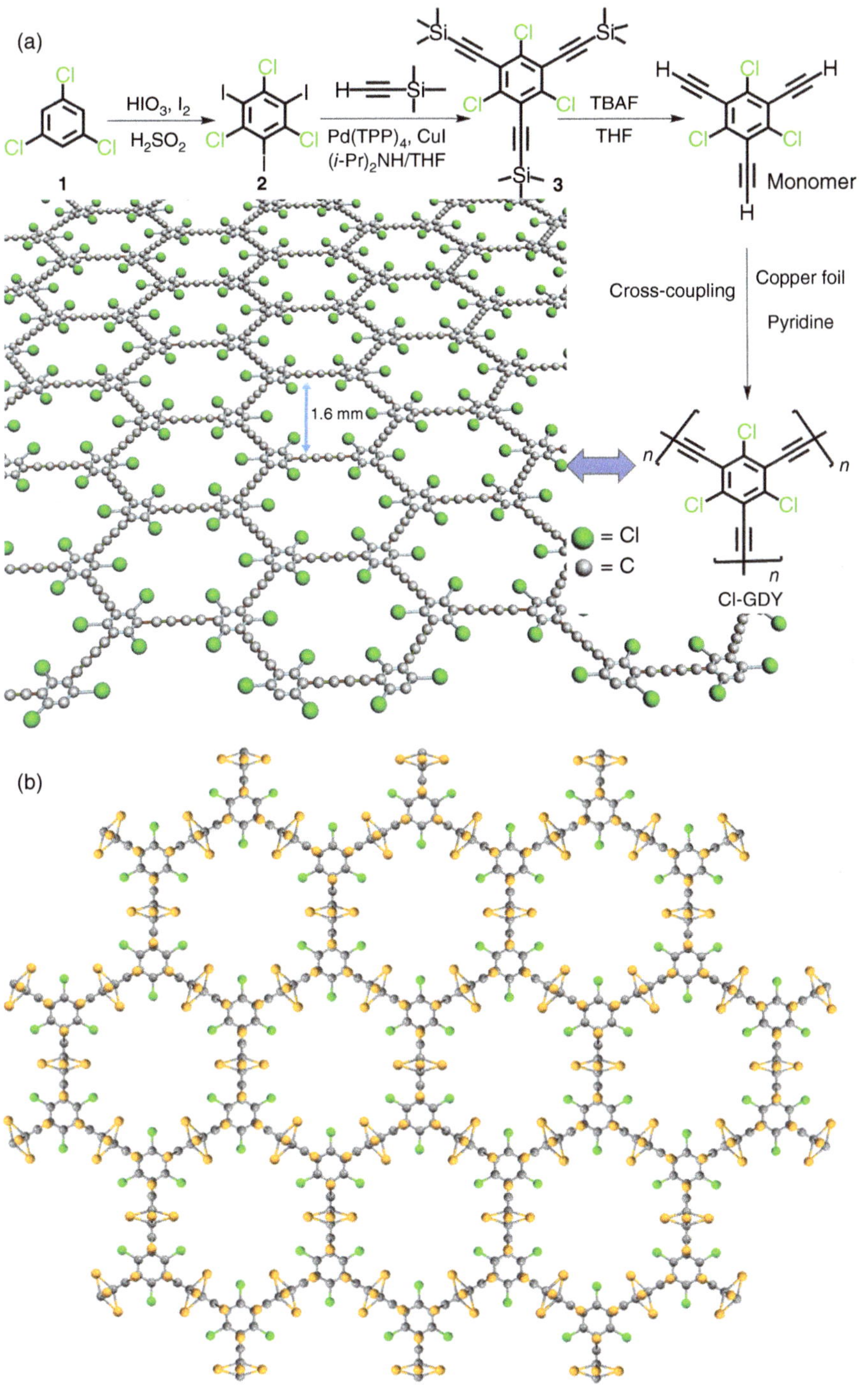

Figure 4.6 (a) Preparation of Cl-GDY. (b) The theoretical analysis of Li storage in Cl-GDY framework. Source: Wang et al. [24]. © 2017, John Wiley & Sons.

some quite different characteristics compared to those of GDY. Obviously, the inclusion of chlorine will definitely change the basic properties of GDY carbon framework. This is quite consistent with the previous reports about the other chlorine-doped carbon materials such as graphene oxide (GO), mesoporous carbon, and reduced graphene, which have also shown quite different properties and device performance compared to corresponding pristine carbon materials. On the other hand, chlorine heteroatom has a milder electronegativity and bigger atom size compared to those of fluorine atoms. These two features make Cl-GDY exhibit quite different properties from the above-mentioned F-GDY. In the aspect of appearance of as-prepared films, the Cl-GDY film exhibits a much deeper color than that of F-GDY. This can be ascribed to the different surface morphology of these two doped GDY films. On the other hand, a synergistically stabilized effect of doped chlorine atoms and adjacent acetylene bonds to anchor the lithium atoms is observed based on the theoretical and experimental results (Figure 4.6b). Profiting from this unique characteristic, Cl-GDY exhibits extraordinary electrochemical properties in lithium ion half-cells [24]. On the other hand, the characteristic affinity of chlorine heteroatoms to the nitrogen also makes Cl-GDY prepared through traditional thermal annealing method be a good candidate material for the toward electrocatalytic nitrogen reduction at ambient conditions [25].

4.1.3 Sulfur, Boron, Hydrogen, and Other Nonmetal Heteroatoms

S-GDY can be prepared through annealing the mixture of GDY and S8 at 600 °C [38]. At this temperature, S8 rings can be decomposed to elementary substance with short chains (S2–S4). During this process, GDY provides excellent adsorbing sites to entrap these polysulfides, which is crucial for the further inclusion of sulfur as heteroatoms to prepare S-GDY. As we mentioned in the previous Chapters 2.1.1, GDY has a porous structure, which is constructed by netted carbon frameworks containing benzene rings and acetylenic linkers. On the in-plane direction, GDY has a two-dimensional carbon plane with uniform pores on it. The size carbon pores are estimated to be 5.42 Å through theoretical calculation. On the out-plane direction, GDY owns a layered structure. The interlayer distance is measured to be 3.65 Å from TEM images. The smaller size of the pores on GDY than that of sulfur chain makes GDY a good container for the decomposed sulfur. Moreover, the strong electrostatic interaction between GDY and sulfur can further result in trapping of polysulfides by chemical adsorption. All the above features guarantee GDY to be a potential conductive carbon material for coating separators that polysulfides can be entrapped by both physical obstruction and chemical adsorption. FTIR spectroscopy spectra of S-GDY confirm the existence of C—S bonds peaked at 671 and 804 cm^{-1}, which cannot be found in the spectra of pristine GDY [54]. The as-prepared S-GDY contains functional sulfur-contained groups, which can be regarded as reactive sites during the electrochemical process. Meanwhile, the introduction of sulfur also improves the electrical conductivities compared to the pristine GDY. The conductivities of GDY and sulfur substituted GDY (SGDY) were measured to be 2.14×10^{-4} and $4.37 \times 10^{-4}\,S\,m^{-1}$, respectively. Besides elemental sulfur, other sulfur-containing organic compounds such as benzyl disulfide (BDS) can also be applied as sulfur source. At high temperature, BDS can generate sulfur radicals during the treating process. The doping process occurs around the acetylenic bonds of GDY, which can be ascribed to the reactivity of sp carbon in GDY (Figure 4.7) [39].

While boron is selected as heteroatom, the sp^2 carbon is easier to be replaced than sp carbon according to the theoretical analysis, for a good stability of the as-prepared B-GDY can be expected for coordination effect [27]. Moreover, B-GDY is supposed to show a good activity for the existence of empty orbitals on boron heteroatoms. As shown in Figure 4.8a, boron heteroatoms uniformly distribute in the alkynyl group's carbon-rich frameworks and are located at the center of three sp-hybridized butadiyne linkers. This doping way is obviously different from the previously reported boron-doped carbon-rich materials and boron carbides. In the experiment, boron substituted GDY (BGDY) composed of sp carbon skeleton and boron atoms was successfully prepared [29]. Notably, the electron-deficient boron centers and the unique bonding environment of the sp-carbon framework generates affinity to metal atoms, which can contribute to extra binding sites. Theoretical analysis confirmed the synergistic stabilization effect of the boron atoms on sodium atoms. Moreover, the diffusion barrier of Na ions along the perpendicular direction of the BGDY molecular plane could be efficiently reduced for the well-distributed pores on the BGDY molecular plane. Compared with the GDY, the excellent performance of BGDY-based electrode could be attributed to the proper size and electronegativity of the molecular pores on BGDY plane. Motivated by this work, theoretical researchers systematically investigate the basic properties of BGDY, including optical, mechanical, and thermal conductivity of single-layer B-GDY. The affinity of BGDY to metal atoms, including Mg, Ca, Li, Na, and K, was also performed. The results indicate that BGDY can exhibit super stretchability. Meanwhile, a band gap of 1.15 eV was predicated through HSE06 method, which is quite consistent with the previous reports (Figure 4.8b,c).

To tune the structure and property of carbon materials, hydrogenation was considered to be an efficient strategy. According to the theoretical analysis, hydrogen is easy to be adopted by GDY for the affinity of sp carbon to hydrogen [36]. Zhao and coworkers have studied the optimized configurations of hydrogenated GDYs [56]. The chemical bonds with lengths of 1.12 Å can be formed between hydrogen atoms and sp carbon. Based on this, it is supposed to break one sp-hybridized carbon atom to change sp-hybridization into sp^2-hybridization. In the experiment, we can realize the hydrogenation of GDY through chemical methods. Most recently, the hydrogen-substituted GDY film is prepared using 1,3,5-triethyltriethynylbenzene as monomers (Figure 4.9a). The conductive matrix with 3D hierarchical porous structure can result in easy transporting and hopping of electrons, which could ensure efficient charge diffusion and collection. The as-designed structure of HGDY therefore provides favorable sites and path to meet the insertion/extraction and diffusion requirements of large-diameter ions. Moreover, the mechanical strength of HGDY is strong, which makes it act as free-standing flexible electrode for both LIBs and SIBs (Figure 4.9b) [57]. The boundary of the HGDY layer is clearly visible in its

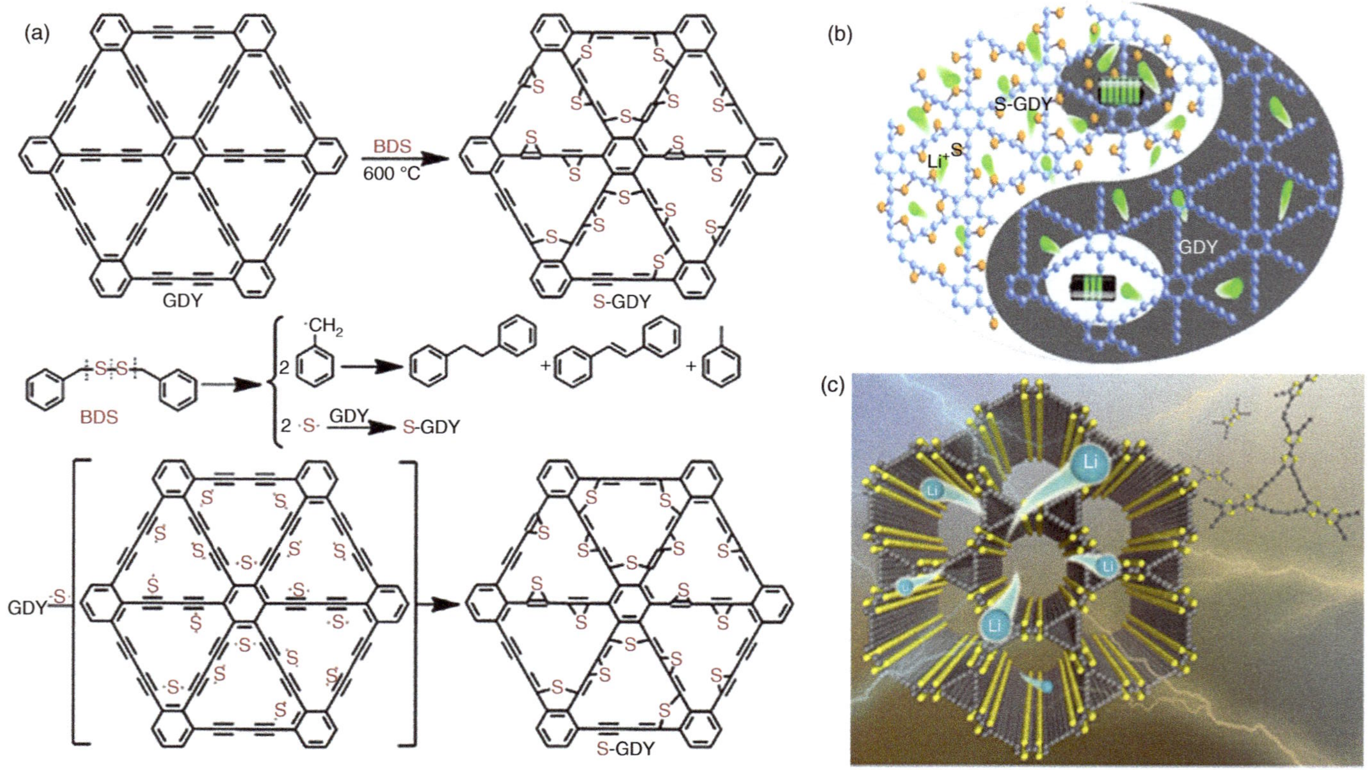

Figure 4.7 (a) Preparation of S-GDY with as BDS as sulfur source. (b) The application of S-GDY in LIBs. (c) Preparation of S-GDY with tetrathiafulvalene-containing compound as precursor. Source: (a, b) Yang et al. [39]. © 2019, John Wiley & Sons, (c) Pan et al. [55]. © 2019, American Chemical Society.

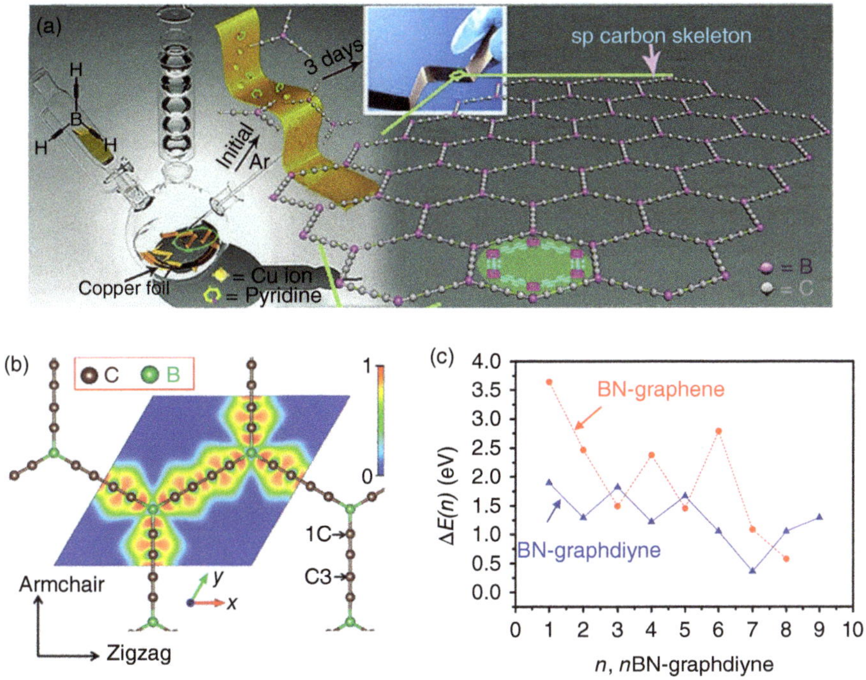

Figure 4.8 (a) Preparation of B-GDY. (b) Atomic structure of C12B2 B-GDY. (c) Isoelectronic doping of GDY with boron and nitrogen. Source: (a) Wang et al. [29]. © 2018, Wiley-VCH Verlag GmbH & Co. KGaA. (b) Mortazavi et al. [28]. © 2018, Royal Society of Chemistry. (c) Bu, H et al. [26]. © 2012, American Chemical Society.

high-resolution TEM images. However, the stacking of HGDY film is difficult to be exfoliated due to the strong interlayer interaction. One interesting phenomenon on HGDY film is that the morphology on the direction perpendicular to the carbon plane is not homogeneous. The morphology of HGDY film close to the copper foil is more compact than that at the side far away from the copper. As we know, in the process of forming HGDY film, the copper foil is not only the provider of the copper ion catalyst for the cross-coupling reaction, but also the substrate supporting the growth of HGDY film. At the beginning of the coupling reaction, the copper ions participate in the reaction immediately after falling off from the copper substrate, which may result in the fast growth of film. As the reaction goes on, the environment for the coupling reaction might be changed gradually and slightly mainly from the following two aspects. (i) The concentration of the copper ions, which can be used as catalyst for the reaction, might be different, for the copper ions need to diffuse through the previous HGDY film, then the HGDY fiber grows on the top of the film. (ii) The subsequent HGDY grows on copper substrate filled into the gap of fibers and/or covered by a layer of previous HGDY film with different thickness, forming thick films. It can be seen that the growth environment for the HGDY film changes gradually as the reaction goes on, which will induce the difference in morphology of the as-growth HGDY film.

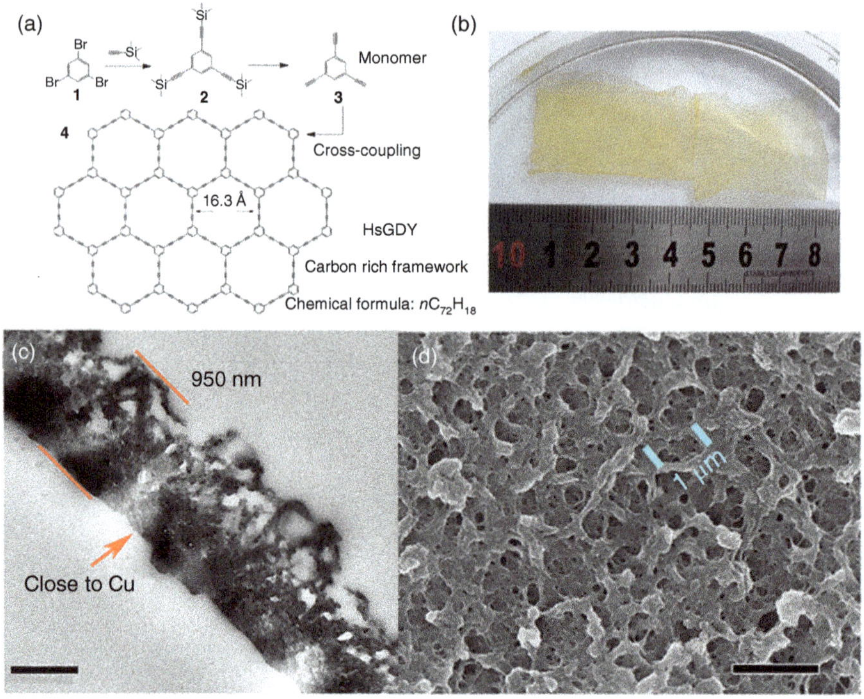

Figure 4.9 (a) Preparation of HGDY through bottom-up method; (b) the images of HGDY film; (c) the cross-section SEM images of HGDY film; and (d) the porous structure of HGDY film. Source: Jia et al. [57]; © 2017, the American Chemical Society.

It is notably displayed in cross-section TEM image of HGDY that the thin film close to the copper side is flat, while that at the far side of the copper is irregular. This could be ascribed to the catalytic activity difference of the copper foil caused by the distance (Figure 4.9c). As it has been reported, the copper foil played the role as both substrate and catalyst supplier during the synthetic process [41, 58]. Therefore, HGDY is grown along the surface of copper foil to form the film. Meanwhile, the subsequent reaction, especially the catalytic process, might be inhibited somehow by the formed HGDY film. As a result, a large amount of micropores exist in the surface of HGDY film. The pore diameter is measured to be in the range from 2 to 15 nm. This length corresponds to about one to eight monomers. One of the reasonable explanations for the existence of these pores is that while cross-coupling occurred on the surface of the HGDY film, some oligomer may form for the lower concentration of copper ions catalyst and change of the reaction templates. Some oligomers are adsorbed on the surface of HGDY film through weak interaction, so that they can be removed while ultrasonically cleaning the HGDY film after reaction. As a result, these microspores retain on the surface of HGDY film. Nitrogen adsorption–desorption studies of this porous HGDY film show that the HGDY film shows a large BET surface area of 667.9 $m^2\ g^{-1}$. The H3-type hysteresis loop indicates the multilayer adsorption occurs in the mesopores and macropores of HDGY film. The surface morphology of HGDY can also be directly observed in SEM images (Figure 4.9d). As shown in Figure 4.5d, the sample also contains large amount of pores, which range from mesoporous to macroporous. This is quite consistent with the above discussion. From this view angle, it can be seen that the pore size of macropores and mesopores can reach from tens of nanometers to even several micrometers, which forms a 3D porous film meso/macropores second-order structure, which is beneficial for fast ionic diffusion. Notably, these pores can also efficiently reduce the length and barrier of lithium diffusion while lithium ions transfer through the basal plane. As a result, the power density of HDGY film–based electrode can be greatly improved. The thickness of HGDY film is around 950 nm, which is in accordance with the observation in TEM image. The thickness of the as-prepared HGDY film is closely related to the loading density of the HGDY-based electrodes. According to the experiment test, the area-loading density of HGDY film with this thickness is about 0.11 mg cm^{-2}. In addition, the thickness of the HGDY films can be tuned via simply changing the amount of precursor. While the film thickness reaches 2.5 μm, the loading density of HGDY-based electrode increases to 0.18–0.20 mg cm^{-2}. Meanwhile, the pore size on the surface becomes larger since less copper ions can reach and induce the cross-coupling reaction.

GDY oxide can usually be obtained by oxidizing GDY using hydrogen peroxide in an acidic environment. Sulfuric acid is usually used as the reaction media [59, 60]. GDY oxide can be regarded as oxygen-doped GDY materials, in which oxygen plays the role of heteroatoms. The XPS spectra confirm that GDY oxide remains in the carbon skeleton of GDY. Benefiting from the existence of hydrophilic oxygen-based functional groups such as ether oxygen bond, hydroxy bond, and carboxy. GDY oxide exhibits a good dispensability in aqueous solution, which makes it possible to be used in the field of humidity response and cytotoxicity detection [60, 61]. In the aspect of humidity response, an ultrafast humidity response of GDY oxide–based devices has been observed. The response speed is measured to be 7 ms. Notably, this value is three times faster than that of GO-based device under the same measurement conditions. Based on the experiment results, sp-hybridized form of carbon existing in GDY oxide plays a key role for the ultrafast humidity response and for the delocalization of electron distribution in GDY oxide frameworks. More active sites, which can bind with water molecule in quite a fast speed, are generated accompanied with the existence of sp carbon. This research indicates the bright application potential of GDY oxide–based humidity monitor devices or sensors. graphdiyne oxide (GDYO) is also a good substrate for small organic molecules. For example, Zhuang reports a hybrid materials system based on Prussian blue (PB) nanoparticles and GDY oxide. The hybrid can be used as a detection electrode toward hydrogen peroxide. The hybrid material was fabricated through anchoring PB nanoparticle on GDY oxide in the presence of $FeCl_3$. Then, $Fe(CN)_6$ was added to the reaction system at room temperature [62]. The PB in nanoparticle form can be anchored and stabilized by the GDYO in neutral or weakly basic solutions. The PB/GDYO hybrid was investigated by TEM, XPS, and X-ray diffraction. The hybrid complex has been measured to show high electrochemical catalytic activity, especially stability for the detection of hydrogen peroxide. Besides that, GDY oxide can also be applied as a good support for

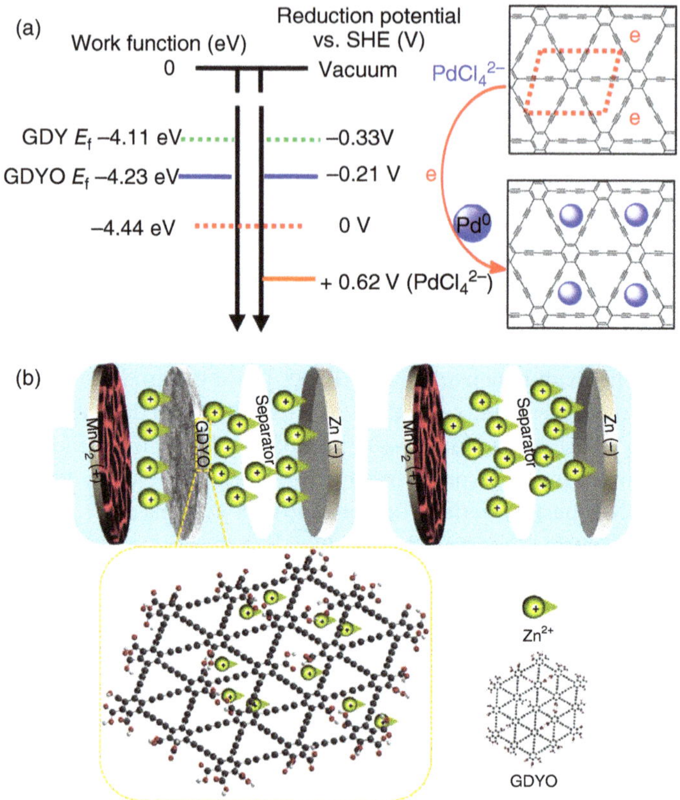

Figure 4.10 GDY oxide used as (a) catalyst supporter for Pd clusters. (b) Aqueous Zn–MnO_2 battery. Source: (a) Qi et al. [63]. © 2015, the American Chemical Society. (b) Modified from Xie et al. [64].

Pd clusters nanocomposite (Figure 4.10a). The as-prepared hybrid catalyst system shows an excellent catalytic property in the reduction reaction of 4-nitrophenol. Based on the experimental result, the high catalytic performance could be ascribed to the synergetic effects that occur at the interfacial phase between Pd cluster and GDY oxide nanocomposite [63]. Besides that, GDY oxide membrane can also be applied in Zn–MnO_2 batteries. A high Coulombic efficiency about 100% and a good stability over 800 hours have been observed for the GDY oxide–based Zn-ion batteries (Figure 4.10b) [65]. Another possible applied field for the GDYO is as chemical or biochemical sensor for the typical substrate [61]. Jia and coworkers also compare the performance of GO and GDY oxide in the application of physicochemistry and mutagenesis. The as-prepared GO and GDY oxide showed similar morphology, oxygen content. However, GDY oxide shows better solubility in selected media. Moreover, GDY oxide shows good separability to cell membrane. In contrast, GO is extended to adhere to cell membrane. Both GDYO and GO exhibit vivo mutagenesis potential, which are bioequivalent at binding to doxorubicin and single-stranded DNA [60].

4.1.4 Dual Heteroatom Doping

As we have introduced in the previous parts, the introduction of heteroatoms, such as nitrogen, phosphorus, halogen, sulfur, boron, and hydrogen, can efficiently tune the structure and properties of GDY-based carbons. In GDY, two kinds of hybridized carbons exist, which make the configuration, doping position, and content of induced heteroatoms versatile. On the other hand, each heteroatom has distinctive characteristic, which can have different influences on the properties of GDY. To tune the properties of GDY more elaborately, two or more kinds of heteroatoms can be doped on GDY in subsequence. As a result, multiple heteroelements doped on GDY usually have more advantages compared to single-element doping [66, 67]. For example, a series of sulfur and sp-hybridized nitrogen atoms dual doped GDY were synthesized by sequential doping method via thermal annealing the sp N-doped GDY and sulfur heteroatom source material at different temperatures in the range of 700–900 °C [68]. The test results indicated the subsequently doped sulfur heteroatoms on sp N-GDY can enhance the current density and improve the catalytic activity and reaction kinetics of oxygen evolution reaction. Notably, the doping position of sp-hybridized N can be verified, the subsequent doping sulfur heteroatom is controllable, which shows the relationship between the stereodistance and synergistic effect of S and N can be investigated distinctly. Besides that, other dual heteroatom doped GDY systems such as B/N [26, 69–72], Fe/N [67], and Co/N [73] have also been prepared and applied in practical devices. The complementary promoting effect stemmed from the individual characteristic of the selected heteroatom partners to make the as-prepared functionalized heteroatom doped GDY materials have a wider application field in various practical devices.

The inclusion of dual heteroatoms to GDY can systematically adjust the structure of GDY, which can obtain the optimized ratio of two heteroatoms. On the other hand, an ideal cathode material for Li-ion batteries needs to have a high capacity of Li storage and a low volume expansion ratio. Therefore, dual heteroatom doping might provide a way to fulfill this requirement. For example, Zhang et al. predicted the Li storage performance of two-dimensional porous GDY as a negative electrode material of Li-ion battery after doping with B and N atoms was studied. The calculation results show that the doping of B in the GDY structure can enhance the adsorption between Li and GDY, and the storage capacity of Li can be increased to 2061 mAh g^{-1}. At the same time, B doping reduces Li's diffusion energy barrier in the direction perpendicular to the GDY plane. On the other hand, N doping reduced the interaction between Li and GDY, but increased the stability site of Li and the storage capacity of Li at 1652 mAh g^{-1}; at the same time, Li's diffusion performance on GDY was greatly improved, and the diffusion barrier in the plane was reduced to 0.37 eV. The charging and discharging properties of N doped GDY have been improved. Therefore, doping B and N can improve the storage capacity of GDY as the cathode material of Li battery. This research can provide a good research idea for developing good Li anode materials and theoretical basis for experimental workers. In the aspect of experiment, a new strategy to introduce two or more heteroatoms to GDY is cross-coupling two different precursors with different kinds of heteroatoms such as hydrogen/nitrogen (Figure 4.11a) and hydrogen/fluorine (Figure 4.11b). The advantage of this method is that contents of two heteroatoms could be simply adjusted through controlling the amount of two precursors in the reaction. For example, in the case of hydrogen/fluorine-doped GDY system, five different ratios of hydrogen and fluorine have been applied in the preparing process. The optimized ratio is determined to be 1 : 1 according to the test results, while the dual doped GDY is used as anode electrode in LIBs.

4.2 Metal Decoration

Recently, the interaction between GDY and metal or metallic compound has drawn great attention of researchers from a broad research field [76, 77]. The theoretical investigation indicates the interaction between the metal atoms and GDY can be regarded as a strong chemical adsorption, which is beneficial from the unique triangular pores structure regularly dispersed on GDY carbon plane [78]. As a result, the doping of metal atoms on GDY can effectively regulate its chemical physical properties such as activity in catalyst [63, 67, 76, 78–83], sensitivity in chembiosensor [84–86], electronic structure [77, 87], and magnetic performance [88–91]. The metal decoration of GDY has been confirmed to be an efficient way to functionalize the GDY, which can provide the basis for its wide application in many kinds of devices [49, 76, 87, 92–96].

4.2.1 Metal Atomic Decoration

GDY shows a full-carbon material with a large π-conjugated system combined by acetylenic bonds and benzene rings [97]. GDY possesses many distinguished properties, including evenly distributed pores, robust chemical stability, and tunable electronic properties. Especially, the porosity of GDY and its active acetylenic bonds endow it an anchor for single-metal atoms [98]. Theoretical calculation has showed that the transition metal elements such as Fe, Co, Ni atoms can be anchored adjacent to the acetylene carbon chain of GDY to form a stable transition metal-GDY heterogeneous system. The doping of transition metals can further improve catalytic performance of GDY [78, 99]. The commonly used transition metals for catalyst are Fe, Co, and Ni. Based on the DFT, the catalytic effect of Fe-, Co-, and Ni-doped GDY on ORR was systematically studied. The results show that the bond energies of the three transition metals are higher than the cohesion energies of the corresponding metals. Moreover, the first-principles calculations have proved that GDY can be used as single-atom catalysts substrate for noble metals. The adsorption energy for Ir, Ru, Pt, Rh, Pd, and Au is estimated to be 5.24, 4.61, 4.31, 3.74, 2.67, and 0.99 eV, respectively [100]. In addition, the adsorption of metal atoms such as Cu, Au, Ni, Fe, and Pt atoms on GDY nanoribbons (GDNRs) also has been proved theoretically. According to the calculation result, except for Au located above the triangle center, the atoms of Cu, Fe, Ni, and Pt are similar to Li in the angle of C triangle [101]. Sun's

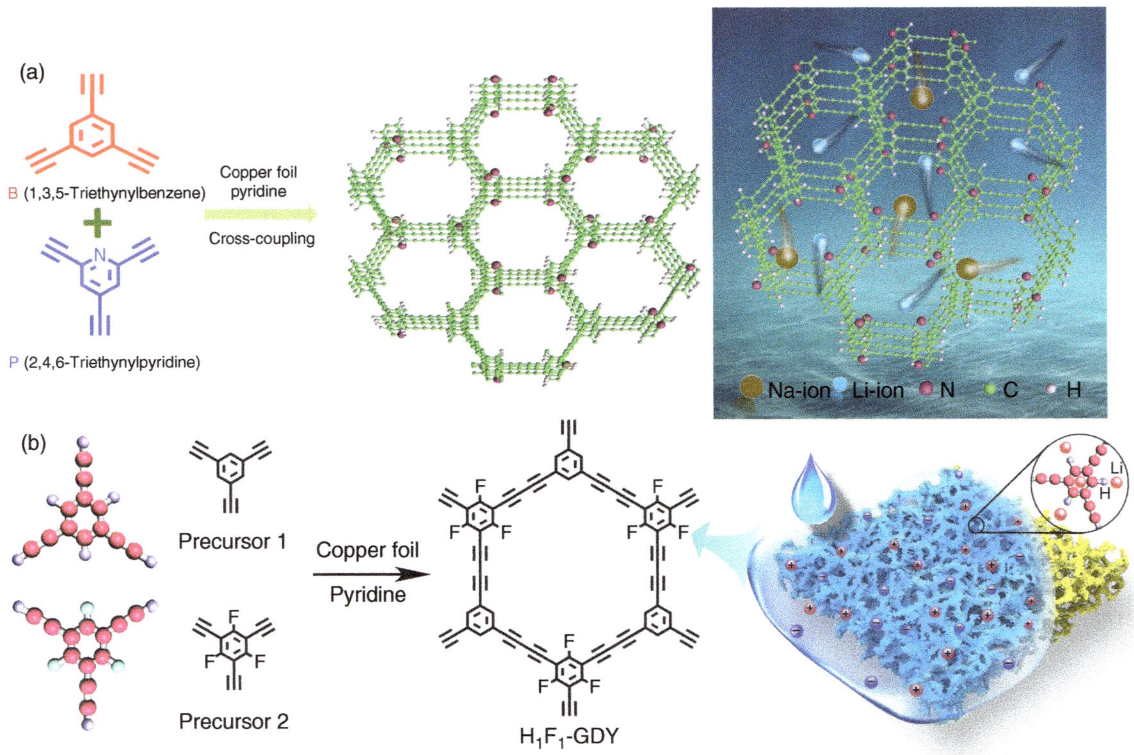

Figure 4.11 Dual heteroatom doping GDY: (a) hydrogen and nitrogen. (b) Hydrogen and fluorine. Source: (a) Gao et al. [74]. © 2020, Royal Society of Chemistry. (b) Lu et al. [75]. © 2020, Elsevier.

group calculated the interaction of transition metal atoms and GDY. This is the most advantageous adsorption site when the transition metal atom is in the alkyne ring of GDY and represents the adsorption energy of transition metal-GDY/graphyne (GY) systems. They also found that the electrons transfer from transition metal to GDY and redistribute between the 4s, 3d, and 4p orbitals of transition metal atoms. By observing the change of band gap, it is found that the structure of the GDY decorated with a transition metal atom has a metallic property in the 18C-hole site [89, 102]. The above results demonstrate that the combination of metal atoms to GDY is pretty stable, which provides a theoretical basis for decorating GDY with metal atoms. Notably, the as-prepared single-atom catalyst can expose the surface area of most metal particles, thus increasing the utilization rate of metal particles. For example, DFT proves that GDY is a good carrier and can support monatomic Fe catalyst [76]. Fe atoms can be firmly embedded in GDY with bond energy of about 4.99 eV and a diffusion barrier of about 1.0 eV. Therefore, Fe atoms can strongly adsorb on GDY to form a monotonically dispersed Fe-alkyne catalyst. It can be seen that the biggest characteristic of GDY is that it anchors the single metal atom through the interaction between the acetylene bond in its own structure and the d-orbital of metal atoms. In the structure of GDY, the existence of sp-hybridized carbon atoms makes the orbitals in the plane rotate arbitrarily in the direction perpendicular to the acetylene bond, so the orbitals of the acetylene bond can always point to the single-metal atom. After the metal atom adsorption on the surface of GDY, metal atom valence electrons not only interact with those in carbon atoms, but also induce hybridization, thus forming a strong combination between acetylene and metal atoms. Moreover, it is found that the single atom of noble metal has the highest adsorption capacity at the center of the single acetylene triangular pore on GDY. Since the noble metal forms a strong covalent bond with the triple bond around the pore after adsorption, it is difficult for metal atoms to migrate, which is conducive to the stable existence of metal atoms.

In the aspect of practicality of experiment, the single-atom catalysts maximize the catalytic efficiency by anchoring the atoms at specific sites on a suitable substrate compared to the corresponding metal clusters. For example, it is well known that Pt single-atom catalysts have higher catalytic efficiency than Pt-based catalysts, which was reduced by utmost utilization of Pt atoms. Lu's group successfully engineered and synthesized two atomically dispersed Pt catalysts anchored on GDY by a wet-chemical route. The existence of abundant isolated Pt atoms on the GDY substrate was confirmed by the HADDF-STEM images. Due to the different coordination environments of Pt atoms, the two single-atom catalysts performance show the differences in catalytic activity on HER, which will be beneficial to better understand the relationship between the catalytic and activity structure in Pt single-atom catalysts (SACs).

The Fe and Ni single-atom catalysts anchored on GDY have been fabricated by Li's group. Under acidic conditions, two GDY-supported atomic catalysts exhibit outstanding high catalytic activity and long-term durability for HER. The onset overpotential of Fe/GDY was measured to be 9 mV, which was close to Pt/C (1.0 mV). To obtain a catalytic current density of 10 mA cm^{-2}, the corresponding overpotential of Fe/GDY and Ni/GDY needed to be 66 and 88 mV, respectively. The

Tafel slopes of two GDY-supported atom catalysts (ACs) were only slightly larger than Pt/C (33.9 mV dec^{-1}). Such excellent performances were attributed to the interactions between the single-atom Ni/Fe and GDY, including strong electronic coupling and chemical interactions. Using first-principle calculations, Du and coworkers investigated the electrocatalytic activity of transition metal atoms loaded on a GDY monolayer. As a result, they found that single atoms are quite stable when located at the corner of the acetylenic ring, which is favorable for single atoms to realize evenly distributed and be well ordered on GDY. Among the researched composites, Ti@GDY and V@GDY are considered the best catalysts for HER, and Pt@GDY and Ni@GDY are considered the candidates of bifunctional catalysts for water splitting. In addition, Cai et al. studied the GDY-supported Fe single-atom catalysts that are utilized for CO reduction. Therefore, the above researches prove that GDY is a candidate support for single-atom metal catalysts.

While alkalis metals such as lithium atoms are introduced to GDY framework, the Li atoms anchored on the GDY framed work exhibit binding energies of −3.86 and −4.47 eV, respectively. These binding energy data can be adjusted through doping heteroatoms on GDY frameworks. For example, while nitrogen heteroatoms are doped on GDY. The binding energy of N-GDY for Li atom is measured to be −4.62 and −4.96 eV, respectively. The shift of the binding energy can be ascribed to the electron-rich characteristic of nitrogen heteroatom. Therefore, pyridine N dopant, which shows lone-pair electrons on nitrogen atoms, is predicted to be highly lithiophilic. These results indicate that the utilization of lithophilic N-GDY can improve the interfacial lithophilic greatly, leading to the optimizations of the Li nucleation.

The experimental results also show that the metal atomic adsorption can not only adjust the electronic structure of GDY, but also introduce GDY excellent magnetic properties, which generate the spin polarization on it. This modulation may be realized by the charge transfer between transition metal atoms and GDY. Another possible explanation for this is the electron redistribution of s, p, and d orbitals between transition metal atoms and sp carbons on GDY. For most adsorbed atomic systems, the magnetic moment is located on metal atoms. But for Cr/GDY, the magnetization of carbon atoms has also been found. The strong bonding between metal atoms and GDY sheets and the charge transfer from the atoms to these sheets lead to a huge shift in the electronic properties of the pristine material. For example, V/GDY, Mn/GDY, Fe/GDY, and Ni/GDY are represented as magnetic semimetals, spin-selective semiconductors, magnetic metals, and nonmagnetic narrow-band gap semiconductors, respectively. Especially, the Co and N dual-doped GDY is prepared using a facile thermal doping method and used as an electrochemical catalyst for water splitting. According to the experiment result, the optimized Co content is measured to be 1%, and the dual doped GDY system shows the most effective catalytic activity in the process of ORR and HER process. Notably, 1% Co-N-GDY shows the recorded best half-wave potential among the reported Co-contained ORR catalyst system. In HER, the 1% Co-N-GDY also exhibits quite a low overpotential, which displays comparable or better performance than many nonprecious metals catalysts as reported [73]. Besides Co-N dual doped system, Fe—N heteroatom dual doped GDY has also been prepared [67]. Similar to above Co—N doped GDY, the

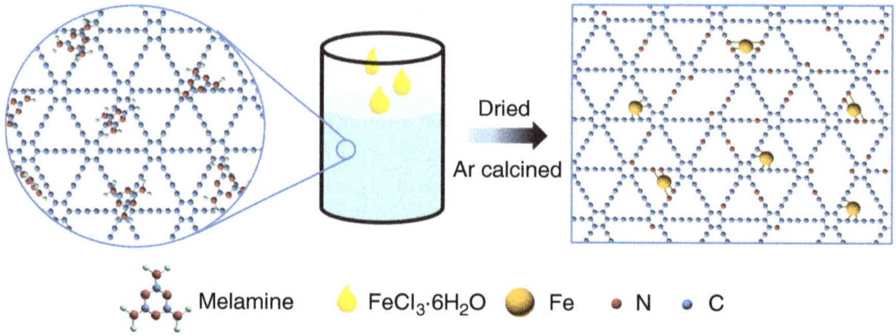

Figure 4.12 Anchoring the single Ni and Fe atom to the GDY. Source: Si, W et al. [67]. © 2019, John Wiley & Sons.

Fe—N co-doped GDY catalyst system also shows the optimized activity while the Fe content is quite low. An onset potential of 0.94 V vs. reversible hydrogen electrode (RHE) electrode has been observed in the ORR process. Meanwhile, the limited current density of 5.4 mA cm^{-2} has been obtained. In addition, the Fe—N co-doped system shows excellent resistance to methanol poisoning. The Fe—N codoped GDY also displays high stability in both alkaline and acidic electrolytes. All those make it as potential application in the field of ORR (Figure 4.12).

4.2.2 Metallic Compounds

Recently, Li and his coworkers grew GDY on the carbon cloth, obtained graphite alkyne foam, and then grew $NiCo_2S_4$ nanowires on the graphite alkyne foam [103]. The catalyst obtained was used as a dual-function catalyst for hydrogen and oxygen evolution, showing excellent catalytic activity and stability. In alkaline electrolyte, while the current density is set to be 10 and 20 mA cm^{-2}, its water electrolysis cell potential was 1.53 and 1.56 V. Its excellent catalytic performance mainly comes from several aspects: (i) the special electronic structure and electronic conductivity of GDY alkyne are conducive to the rapid electron transmission. (ii) $NiCo_2S_4$ nanowires grow directly on graphite alkyne without adding adhesives or conductive additives, which reduce the contact resistance and charge transfer resistance and is conducive to improving the reaction kinetics. (iii) The porous electrode is conducive to exposing more active sites, mass transfer, and bubble removal. (iv) Ni and Co have a strong interaction with the acetylene bond in graphite alkyne, which is conducive to electron transmission and thus improves its catalytic performance.

TiO_2 is a well-known semiconductor material that is often used in different kinds of devices. For example, TiO_2 can be applied to heterogeneous photocatalytic oxygenation of organic compounds. Due to its strong ultraviolet absorption, oxidation performance, nontoxicity, and good photostability, TiO_2 can be considered as a promising photocatalyst and photoelectric catalyst. However, it still has deficiencies such as only absorbing ultraviolet spectrum, and easy combination of photogenerated electrons and holes. In view of the above shortcomings, researchers mainly modified TiO_2 through different methods such as fabricating semiconductor composite, doping, and photosensitizing to optimize its performance. Among them, the composite photocatalyst was constructed by combining carbon-based materials, zinc oxide, iron oxide, and TiO_2, which could effectively inhibit the combination of photogenerated electrons and holes and promote the improvement of photocatalytic performance. Most recently, two kinds of two-dimensional conjugated carbon materials, graphene and GDY, were used to prepare TiO_2–graphene and urea-dispersed TiO_2–graphene and TiO_2–GDY composite photocatalysts by hydrothermal synthesis in situ. The surface morphology and phase structure of the three composite catalysts were characterized, and the photocatalyst properties of the three catalysts were tested. The results showed that the addition of graphene and GDY played a significant role in the dispersion of TiO_2, and the TiO_2 particles were closely bound to graphene and GDY [49]. The addition of urea further reduced the size of TiO_2-GR (graphdiyne ribbon) particles and increased the dispersion degree of TiO_2. Within a certain time under simulated solar radiation, the photocatalyst degradation of organic dye methylene blue and methyl orange light, the results show that the two-dimensional conjugated carbon material graphene and the addition of GDY make composite photocatalyst TiO_2–graphene and photocatalytic properties of TiO_2–GDY for 1.67 times and 1.68 times of pure TiO_2 light catalyst, composite photocatalyst showed better catalytic properties. The outstanding performance of GDY/TiO_2 hybrid system can be ascribed to its unique structures. First of all, it has strong charge separation capability. GDY can capture photogenerated electrons excited by TiO_2, promote charge separation, and inhibit charge recombination. (ii) The existence of impurity levels of GDY/TiO_2 makes it easier for electrons to transfer from valence band to these isolated energy levels (iii) TiO_2 (001)-GDY has higher oxidation capacity. The lower the position of valence band, the stronger the oxidation ability and the higher the photodegradation performance (Figure 4.13).

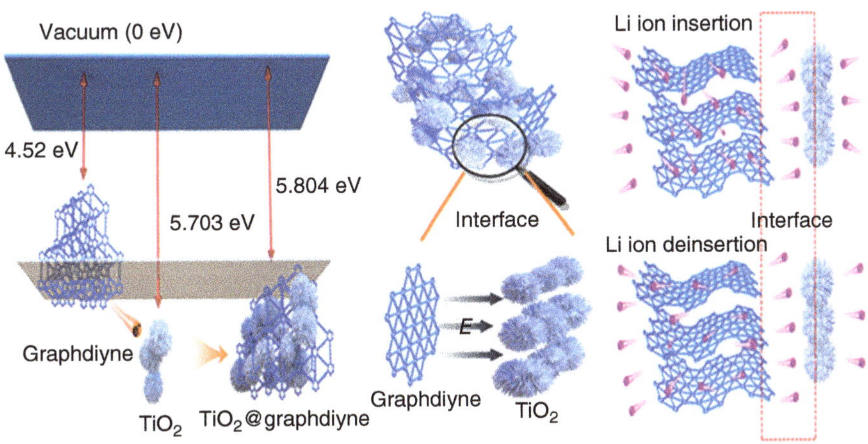

Figure 4.13 The energy-level illustration of TiO_2@GDY composite. Source: Lin et al. [49]. © 2018, Royal Society of Chemistry.

Molybdenum disulfide (MoS_2) is a sandwich structure in which sulfur atoms and molybdenum atoms are covalently bonded, and a single layer of molybdenum atoms is sandwiched between two of sulfur atoms layers to form a special lamella, and the lamella are then weakly connected by van der Waals forces. Based on the different relative positions of sulfur and molybdenum atoms for bonding cooperation, MoS_2 is divided into three crystal systems, named 2H, 3R, and 1T. Due to the weak van der Waals force between layers, MoS_2 has good solid lubricity, showing a low friction coefficient, and is initially used in the field of solid lubrication. With the further study of MoS_2, researchers found that in addition to good solid lubrication, MoS_2 also showed excellent performance in other aspects. For example, in the structure of MoS_2, the Mo—S bond at the edge has a high chemical adsorption capacity and catalytic activity due to insufficient coordination, so it has a wide application potential in ion batteries and catalyst. Recently, GDY-MoS_2 hybrid material has been synthesized and applied in LIBs devices. As shown in Figure 4.14, the flowerlike MoS_2 nanostructure has been prepared via common hydrothermal method. The GDY was in situ grown on the surface of MoS_2 nanoparticles to get the hybrid materials containing GDY and MoS_2. The compact contact through carbon–sulfur bond between MoS_2 can effectively prevent the detachment issue and supply effective transport

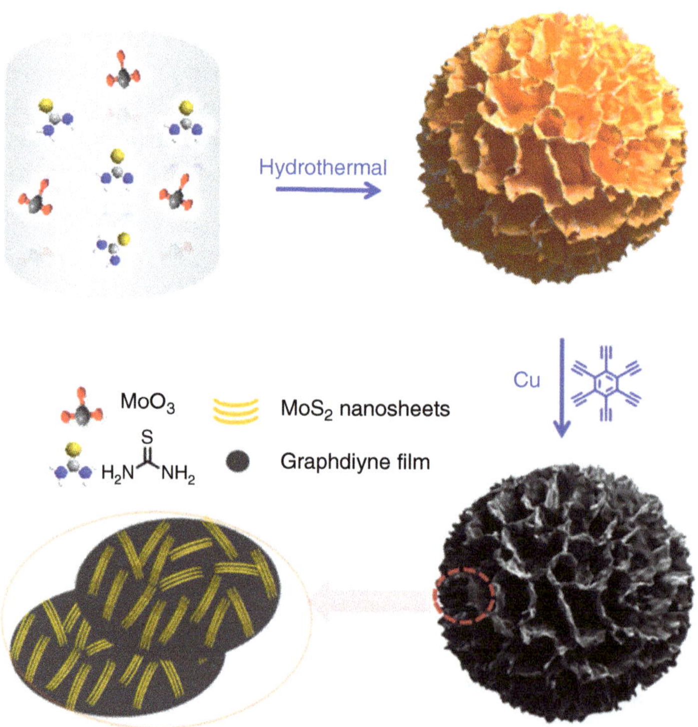

Figure 4.14 Preparing MoS_2@GDY composite. Source: Gao et al. [104]. © 2019, Elsevier.

paths for electron. In this hybrid material system, the important function of GDY layer is to prevent the volume expansion of MoS_2 [104].

4.3 Absorption of Guest Molecules

GDY, a 2D material with a large conjugate structure, due to its excellent electrical conductivity, large specific surface area, and rich unsaturated chemical bonds, has attracted great attention from researchers. The above structure characteristic of GDY makes it a potential host material for several of guest molecules (Figure 4.8). For example, GDY has a good selectivity for hydrogen, which is one of the clean energy sources. As we all know, pore size plays a decisive role in gas molecules selectivity. The side length of the triangle defined by the van der Waals (vdW) surface of the pores on GDY is about 3.8 A [105]. However, the kinetic diameters of H_2, O_2, N_2, CO, and CH_4 are 2.89, 3.46, 3.64, 3.76, and 3.81 A, respectively [106]. Meanwhile, the barriers of a single molecule passing through GDY for H_2, O_2, N_2, CO, and CH_4 are 0.15, 1.09, 1.78, 1.97, 2.63 eV, respectively [107]. Therefore, it is dynamically favorable for H_2 to permeate GDY, which demonstrates that GDY is a good hydrogen separation membrane. The adsorption and diffusion characteristics of GDY monolayer by oxygen and several harmful gases were reported by Lu and coworkers (Figure 4.15). Compared with harmful gases, the overlap of electron densities between O_2 and the pore rim of GDY is small. In a wide temperature range, GDY with well-defined pores exhibits a perfect behavior for oxygen separation from several noxious gases [108].

Compared with graphene, the sp- and sp^2-hybridized GDY carbon networks and their families can provide more active H_2 adsorption space. H_2 adsorption and hydrogenation with acetylene bond to form a new family of carbon allotropes has become a very interesting research direction. Similar to graphene, the main function of GDY sheets is to convert sp^2 and sp carbon atoms to sp^3 or sp^2 carbon atoms by covalently binding to one or more hydrogen atoms. At present, studies on GDY-H system mainly focus on two aspects: the effect of hydrogen adsorption on the regulation. A simplified model of uniformly hydrogenated GDY for the bonding of each sp^2 or sp carbon atom to an H atom has been developed and used to simulate the mechanical properties of the material. It could be found that the

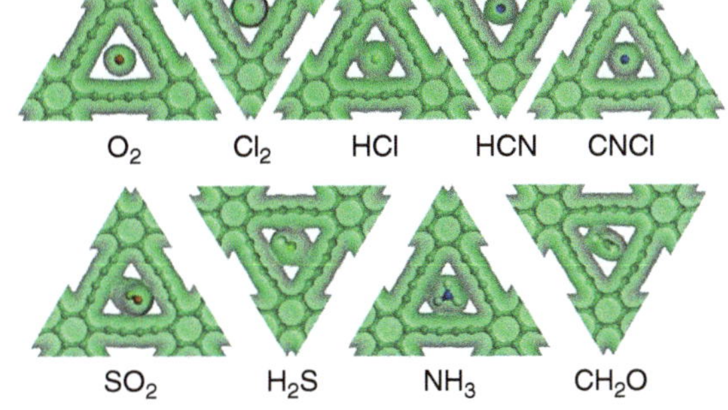

Figure 4.15 Electron densities GDY-adsorbed guest molecules. Source: Meng et al. [108]. © 2016, American Chemical Society.

CH showed a Poisson ratio of 0.23 and an in-plane stiffness of 125 N m^{-1}. So, the in-plane stiffness of the material was lower than that of the original GDY, indicating the negative impact of hydrogen adsorption on the mechanical properties of GDY. According to the coverage (H/C ratio), a more detailed study of GDY showed that an increase in the H/C ratio led to a rapid increase in the band gap. For example, for graphene, if the hydrogen coverage varies from 0 to 1, the band gap varies from 0.45 to 4.43 eV. This provides a new way to regulate the electronic properties of GDY. This also means that the band gap of GDY can be adjusted through the hydrogenation reaction, so as to prepare new two-dimensional hydrocarbon materials.

Using DFT, R. Chandiramouli and coworker researched on the adsorption behavior of NH_3 gas molecules on GDY nanosheets. The adsorption of NH_3 gas results in the increase of the HOMO–LUMO gap in GDY nanosheet. In addition, the physical adsorption of NH_3 gas on GDY nanosheet causes a remarkable change in the conductivity of GDY nanosheet. Therefore, this study demonstrated that GDY nanosheets have important application value in NH_3 gas detection [109]. In addition to adsorbing some gas molecules, GDY also interacts with organic liquid or even solid molecules. The interaction between GDY and amino acids was reported by Zhang's group. They revealed that the amino acid molecules caused a decrease in the conductivity of GDY. This property makes GDY a promising two-dimensional amino acid–sensitive detection material, and may potentially be utilized in biosensors [110]. The adsorption behavior of dimethylamine (DMA) and trimethylamine (TMA) vapor molecules on GDY nanosheet was studied by DFT method. The results showed that the triangular pores of GDY nanosheet were the optimal adsorption sites for DMA and TMA. The electron transfer occurs between the DMA/TMA vapor molecules and GDY nanosheet, which was confirmed by the results of energy band gap and density of states (DOS) spectrum. The variation of energy band gap caused by the adsorption and desorption of DMA/TMA vapor on GDY nanosheet makes GDY nanosheet a potential sensor for detecting DMA/TMA molecules [111]. Few-layer GDY nanosheets possess different affinity to ssDNA and dsDNA and show excellent fluorescence quenching ability, which indicates that GDY NSs could be utilized in biological probes [112].

The adsorption capacity of GDY for small molecules is relatively weak. To further improve the molecular adsorption ability, a method of introducing transition metal atoms (Sc and Ti) into GDY has been developed. Compared to pristine graphene and GDY, Sc- and Ti-decorated GDY exhibited strong adsorption capacity for HCHO molecule. HCHO is considered as an electron acceptor in the molecular doping effect, resulting in a decrease of carrier density in GDY. When Ti-decorated GDY adsorbed HCHO molecules, HCHO takes away almost all the carrier donation of Ti atoms, transforming the system from metal to n-doped semiconductor. Therefore, quantum transport calculations show that the current through the GDY-Ti/HCHO system is significantly lower than that of the GDY-Ti system. The current decrease can be used as a sensitive electrical signal to detect a single HCHO molecule, which suggests Ti-decorated GDY as a promising material for HCHO detection [86].

The mass production of oil has promoted the great progress of the society. Meanwhile, the oil spill also caused serious pollution to the ecological environment.

Accordingly, finding an efficient and simple method of removing organic pollutants is particularly important. Li's group successfully modified GDY to melamine sponge (MS) skeleton by a facile in situ Glaser–Hay coupling. XPS and UV–Vis spectroscopy confirmed that the GDY film was indeed decorated on the MS backbone. Combined with the characteristics of MS and GDY, graphdiyne melamine sponge (GDYMS) exhibits excellent high adsorption capacity (up to 160 times its own weight), unprecedented stability, and outstanding recyclability (up to 100 times). These extraordinary features indicate that GDYMS has a promising application in oil and water separation [23].

In the field of biosensor, the interaction between GDY and organic dye molecules can be used in the field of DNA detection. GDY 2D nanometer sheets (NSs) are a kind of highly efficient fluorescence detection material, which has the ability of adsorption and fluorescence quenching of single-strand DNA labeled with dyes, and can be used as a sensing platform for DNA. DFT simulation calculation shows that graphene diacetylene has a higher ability to adsorb dye molecules than graphene, and GDY NSs have stronger electron capture ability than graphene. GDY NSs are positively electric after adsorption of single-strand DNA, which is more conducive to adsorb negative DNA molecules, so they have a better DNA detection effect. GDY NSs were used to detect DNA in real time. There are two kinds of detection approach. One is fluorescent dye markers of single-stranded DNA first mixed with GDY NSs, GDY NSs and dye molecules form foster resonance energy transfer, dye stimulates the formation of photons absorbed by materials fluorescence quenching, join the target DNA, after the formation of double-stranded DNA weakens the dye-labeled probe and the interaction between GDY NSs, dye-labeled DNA probe from GDY NSs from, fluorescence recovery, complete testing. The target DNA was premixed with the dye-labeled single-stranded DNA and then detected after adding GDY NSs. Taking P1 (H1N1) as an example, after adding GDY NSs, the fluorescence intensity decreased to 2% and the fluorescence was rapidly quenched. After adding target DNA (T1), double-stranded DNA (P1T1) was formed, and fluorescence recovery was achieved.

In summary, significant research progress has been made in the aspects of nonmetal heteroatom doping, metal atomic modification and surface modification, and the photoelectric performance, energy storage and conversion efficiency and catalytic performance of GDY have been improved. In particular, many important results have been achieved in theoretical research, and experimental work has been gradually increased in recent years, but they are still in the preliminary stage of development. Through the design and synthesis of GDY precursor, changing the form of carbon hybridization and the way of connection and the topological structure of the system, synthesizing new two-dimensional carbon GDY and GDY-like alkyne structural materials, fundamentally realizing the regulation of the energy band structure and semiconductor properties of GDY, will be a very important direction in the future. At the same time, the synthesis method of partial graphite acetylene bonds substituted by heteroatoms was developed, which provides a basis for further study on the influence of GDY on its properties through the doping of heteroatoms. It is of great practical significance for the further development of

GDY, a material system with independent intellectual property rights, to carry out extensive experimental research on catalysis and energy application in combination with the results of theoretical research.

References

1 Tang, H., Hessel, C.M., Wang, J. et al. (2014). Two-dimensional carbon leading to new photoconversion processes. *Chemical Society Reviews* 43: 4281–4299.
2 Du, Y., Zhou, W., Gao, J. et al. (2020). Fundament and application of graphdiyne in electrochemical energy. *Accounts of Chemical Research* 53: 459–469.
3 Huang, C., Li, Y., Wang, N. et al. (2018). Progress in research into 2D graphdiyne-based materials. *Chemical Reviews* 118: 7744–7803.
4 Huang, C., Zhao, Y., and Li, Y. (2019). Graphdiyne: the fundamentals and application of an emerging carbon material. *Advanced Materials* 31: 1904885.
5 Wang, N., He, J., Wang, K. et al. (2019). Graphdiyne-based materials: preparation and application for electrochemical energy storage. *Advanced Materials* 31: 1803202.
6 Li, Y. and Li, Y. (2018). Chemical modification and functionalization of graphdiyne. *Acta Physico-Chimica Sinica* 34: 992–1013.
7 Asefa, T. (2016). Metal-free and noble metal-free heteroatom-doped nanostructured carbons as prospective sustainable electrocatalysts. *Accounts of Chemical Research* 49: 1873–1883.
8 Hu, C.G. and Dai, L.M. (2019). Doping of carbon materials for metal-free electrocatalysis. *Advanced Materials* 31: 1804672.
9 Wang, X.Y., Yao, X.L., Narita, A. et al. (2019). Heteroatom-doped nanographenes with structural precision. *Accounts of Chemical Research* 52: 2491–2505.
10 Yu, H., Xue, Y., and Li, Y. (2019). Graphdiyne and its assembly architectures: synthesis, functionalization, and applications. *Advanced Materials* 31: 1803101.
11 Liu, R., Liu, H., Li, Y. et al. (2014). Nitrogen-doped graphdiyne as a metal-free catalyst for high-performance oxygen reduction reactions. *Nanoscale* 6: 11336–11343.
12 Zhang, S., Du, H., He, J. et al. (2016). Nitrogen-doped graphdiyne applied for lithium-ion storage. *ACS Applied Materials & Interfaces* 8: 8467–8473.
13 Lv, Q., Si, W.Y., Yang, Z. et al. (2017). Nitrogen-doped porous graphdiyne: a highly efficient metal-free electrocatalyst for oxygen reduction reaction. *ACS Applied Materials & Interfaces* 9: 29744–29752.
14 Lv, Q., Si, W., He, J. et al. (2018). Selectively nitrogen-doped carbon materials as superior metal-free catalysts for oxygen reduction. *Nature Communications* 9: 3376.
15 Mortazavi, B., Makaremi, M., Shahrokhi, M. et al. (2018). N-graphdiyne two-dimensional nanomaterials: semiconductors with low thermal conductivity and high stretchability. *Carbon* 137: 57–67.

16 Sakamoto, R., Shiotsuki, R., Wada, K. et al. (2018). A pyrazine-incorporated graphdiyne nanofilm as a metal-free electrocatalyst for the hydrogen evolution reaction. *Journal of Materials Chemistry A* 6: 22189–22194.

17 Shang, H., Zuo, Z., Zheng, H. et al. (2018). N-doped graphdiyne for high-performance electrochemical electrodes. *Nano Energy* 44: 144–154.

18 Zhao, Y., Wan, J., Yao, H. et al. (2018). Few-layer graphdiyne doped with sp-hybridized nitrogen atoms at acetylenic sites for oxygen reduction electrocatalysis. *Nature Chemistry* 10: 924–931.

19 Jiang, T., Chen, K., Wang, J. et al. (2019). Nitrogen-doped graphdiyne nanowall stabilized dendrite-free lithium metal anodes. *Journal of Materials Chemistry A* 7: 27535–27546.

20 Yang, Z., Shen, X., Wang, N. et al. (2019). Graphdiyne containing atomically precise N atoms for efficient anchoring of lithium ion. *ACS Applied Materials & Interfaces* 11: 2608–2617.

21 He, J., Wang, N., Yang, Z. et al. (2018). Fluoride graphdiyne as a free-standing electrode displaying ultra-stable and extraordinary high Li storage performance. *Energy & Environmental Science* 11: 2893–2903.

22 Kang, H., Chen, Y., Xu, L. et al. (2019). Top-down strategy synthesis of fluorinated graphdiyne for lithium ion battery. *RSC Advances* 9: 31406–31412.

23 Shen, X., He, J., Wang, K. et al. (2019). Fluorine-enriched graphdiyne as an efficient anode in lithium-ion capacitors. *ChemSusChem* 12: 1342–1348.

24 Wang, N., He, J., Tu, Z. et al. (2017). Synthesis of chlorine-substituted graphdiyne and applications for lithium-ion storage. *Angewandte Chemie International Edition* 56: 10740–10745.

25 Zou, H., Rong, W., Long, B. et al. (2019). Corrosion-induced Cl-doped ultrathin graphdiyne toward electrocatalytic nitrogen reduction at ambient conditions. *ACS Catalysis* 9: 10649–10655.

26 Bu, H., Zhao, M., Zhang, H. et al. (2012). Isoelectronic doping of graphdiyne with boron and nitrogen: stable configurations and band gap modification. *Journal of Physical Chemistry A* 116: 3934–3939.

27 Das, B.K., Sen, D., and Chattopadhyay, K.K. (2016). Implications of boron doping on electrocatalytic activities of graphyne and graphdiyne families: a first principles study. *Physical Chemistry Chemical Physics* 18: 2949–2958.

28 Mortazavi, B., Shahrokhi, M., Zhuang, X. et al. (2018). Boron-graphdiyne: a superstretchable semiconductor with low thermal conductivity and ultrahigh capacity for Li, Na and Ca ion storage dagger. *Journal of Materials Chemistry A* 6: 11022–11036.

29 Wang, N., Li, X., Tu, Z. et al. (2018). Synthesis and electronic structure of boron-graphdiyne with an sp-hybridized carbon skeleton and its application in sodium storage. *Angewandte Chemie International Edition* 57: 3968–3973.

30 Muhammad, I., Wang, S., Liu, J. et al. (2019). Boron-graphdiyne as an anode material for Li, Na, and K ion batteries with high capacities and low diffusion barriers. *Journal of Renewable and Sustainable Energy* 11: 014106.

31 Shen, X., Li, X., Zhao, F. et al. (2019). Preparation and structure study of phosphorus-doped porous graphdiyne and its efficient lithium storage application. *2D Materials* 6: 035020.

32 Psofogiannakis, G.M. and Froudakis, G.E. (2012). Computational prediction of new hydrocarbon materials: the hydrogenated forms of graphdiyne. *Journal of Physical Chemistry C* 116: 19211–19214.

33 Autreto, P.A.S., de Sousa, J.M., and Galvao, D.S. (2014). Site-dependent hydrogenation on graphdiyne. *Carbon* 77: 829–834.

34 He, J., Wang, N., Cui, Z. et al. (2017). Hydrogen substituted graphdiyne as carbon-rich flexible electrode for lithium and sodium ion batteries. *Nature Communications* 8: 1172.

35 Matsuoka, R., Toyoda, R., Shiotsuki, R. et al. (2019). Expansion of the graphdiyne family: a triphenylene-cored analogue. *ACS Applied Materials & Interfaces* 11: 2730–2733.

36 Qiu, H. and Sheng, X. (2018). A first principle study of hydrogenated graphdiyne. *Physics Letters A* 382: 662–666.

37 Zhuo, S., Shi, Y., Liu, L. et al. (2018). Dual-template engineering of triple-layered nanoarray electrode of metal chalcogenides sandwiched with hydrogen-substituted graphdiyne. *Nature Communications* 9: 3132.

38 Du, H., Zhang, Z., He, J. et al. (2017). A delicately designed sulfide graphdiyne compatible cathode for high-performance lithium/magnesium–sulfur batteries. *Small* 13: 1702277.

39 Yang, Z., Cui, W., Wang, K. et al. (2019). Chemical modification of the sp-hybridized carbon atoms of graphdiyne by using organic sulfur. *Chemistry A European Journal* 25: 5643–5647.

40 Zhang, M., Sun, H., Wang, X. et al. (2019). Room-temperature ferromagnetism in sulfur-doped graphdiyne semiconductors. *Journal of Physical Chemistry C* 123: 5010–5016.

41 Jia, Z.Y., Li, Y.J., Zuo, Z.C. et al. (2017). Synthesis and properties of 2D carbon–graphdiyne. *Accounts of Chemical Research* 50: 2470–2478.

42 Xie, C., Wang, N., Li, X. et al. (2020). Research on the preparation of graphdiyne and its derivatives. *Chemistry A European Journal* 26: 569–583.

43 Yang, Z., Shen, X., Wang, N. et al. (2018). Graphdiyne containing atomically precise N atoms for efficient anchoring of lithium ion. *ACS Applied Materials & Interfaces* 11: 2608–2617.

44 Song, Y., Li, X., Yang, Z. et al. (2019). A facile liquid/liquid interface method to synthesize graphyne analogs. *Chemical Communications* 55: 6571–6574.

45 Yang, Z., Liu, R., Wang, N. et al. (2018). Triazine-graphdiyne: a new nitrogen-carbonous material and its application as an advanced rechargeable battery anode. *Carbon* 137: 442–450.

46 Shen, X., Yang, Z., Wang, K. et al. (2018). Nitrogen-doped graphdiyne as high-capacity electrode materials for both lithium-ion and sodium-ion capacitors. *ChemElectroChem* 5: 1435–1443.

47 Jiao, Y., Du, A., Smith, S.C. et al. (2015). H_2 purification by functionalized graphdiyne - role of nitrogen doping. *Journal of Materials Chemistry A* 3: 6767–6771.

48 Liu, R., Zhou, J., Gao, X. et al. (2017). Graphdiyne filter for decontaminating lead-ion-polluted water. *Advanced Electronic Materials* 3: 1700122.

49 Lin, Z., Liu, G., Zheng, Y. et al. (2018). Three-dimensional hierarchical mesoporous flower-like TiO_2@graphdiyne with superior electrochemical performances for lithium-ion batteries. *Journal of Materials Chemistry A* 6: 22655–22661.

50 Shen, X., He, J., Wang, N. et al. (2018). Graphdiyne for electrochemical energy storage devices. *Acta Physico-Chimica Sinica* 34: 1029–1047.

51 Zhang, M., Wang, X., Sun, H. et al. (2017). Enhanced paramagnetism of mesoscopic graphdiyne by doping with nitrogen. *Scientific Reports* 7: 11535.

52 Kan, X., Ban, Y., Wu, C. et al. (2018). Interfacial synthesis of conjugated two-dimensional N-graphdiyne. *ACS Applied Materials & Interfaces* 10: 53–58.

53 Pan, Q., Liu, H., Zhao, Y. et al. (2018). Preparation of N-graphdiyne nanosheets at liquid/liquid interface for photocatalytic NADH regeneration. *ACS Applied Materials & Interfaces* 11: 2740–2744.

54 Kim, H.S., Arthur, T.S., Allred, G.D. et al. (2011). Structure and compatibility of a magnesium electrolyte with a sulphur cathode. *Nature Communications* 2: 1–6.

55 Pan, Q., Chen, S., Wu, C. et al. (2019). Sulfur-rich graphdiyne-containing electrochemical active tetrathiafulvalene for highly efficient lithium storage application. *ACS Applied Materials & Interfaces* 11: 46070–46076.

56 Tan, J., He, X.J., and Zhao, M.W. (2012). First-principles study of hydrogenated graphyne and its family: stable configurations and electronic structures. *Diamond and Related Materials* 29: 42–47.

57 Jia, Z., Zuo, Z., Yi, Y. et al. (2017). Low temperature, atmospheric pressure for synthesis of a new carbon Ene-yne and application in Li storage. *Nano Energy* 33: 343–349.

58 Li, G.X., Li, Y.L., Liu, H.B. et al. (2010). Architecture of graphdiyne nanoscale films. *Chemical Communications* 46: 3256–3258.

59 Wang, C., Yu, P., Guo, S. et al. (2016). Graphdiyne oxide as a platform for fluorescence sensing. *Chemical Communications* 52: 5629–5632.

60 Zheng, T., Gao, Y., Deng, X. et al. (2018). Comparisons between graphene oxide and graphdiyne oxide in physicochemistry biology and cytotoxicity. *ACS Applied Materials & Interfaces* 10: 32946–32954.

61 Yan, H., Guo, S., Wu, F. et al. (2018). Carbon atom hybridization matters: ultrafast humidity response of graphdiyne oxides. *Angewandte Chemie International Edition* 57: 3922–3926.

62 Zhuang, X., Mao, L., and Li, Y. (2017). In situ synthesis of a Prussian blue nanoparticles/graphdiyne oxide nanocomposite with high stability and electrocatalytic activity. *Electrochemistry Communications* 83: 96–101.

63 Qi, H., Yu, P., Wang, Y. et al. (2015). Graphdiyne oxides as excellent substrate for electroless deposition of Pd clusters with high catalytic activity. *Journal of the American Chemical Society* 137: 5260–5263.
64 Xie, C., Hu, X., Guan, Z. et al. (2020). Tuning the properties of graphdiyne by introducing electron-withdrawing/donating groups. *Angewandte Chemie International Edition* 59: 13542–13546.
65 Li, J.F., Chen, Y.H., Guo, J. et al. (2020). Graphdiyne oxide-based high-performance rechargeable aqueous Zn-MnO_2 battery. *Advanced Functional Materials* 30: 2004115.
66 Li, J., Jiu, T., Duan, C. et al. (2018). Improved electron transport in $MAPbI_3$ perovskite solar cells based on dual doping graphdiyne. *Nano Energy* 46: 331–337.
67 Si, W., Yang, Z., Wang, X. et al. (2019). Fe,N-codoped graphdiyne displaying efficient oxygen reduction reaction activity. *ChemSusChem* 12: 173–178.
68 Zhao, Y., Yang, N., Yao, H. et al. (2019). Stereodefined codoping of sp-N and S atoms in few-layer graphdiyne for oxygen evolution reaction. *Journal of the American Chemical Society* 141: 7240–7244.
69 Yang, Z., Zhang, Y., Guo, M. et al. (2019). Adsorption of hydrogen and oxygen on graphdiyne and its BN analog sheets: a density functional theory study. *Computational Materials Science* 160: 197–206.
70 Zhao, J., Chen, Z., and Zhao, J. (2019). Metal-free graphdiyne doped with sp-hybridized boron and nitrogen atoms at acetylenic sites for high-efficiency electroreduction of CO_2 to CH_4 and C_2H_4. *Journal of Materials Chemistry A* 7: 4026–4035.
71 Lin, F., Huang, Y., Shi, C. et al. (2018). Crystal orbital study on one-dimensional beta-graphyne and its BN-substituted derivatives. *Journal of Solid State Chemistry* 265: 402–410.
72 Yun, J., Zhang, Y., Xu, M. et al. (2017). DFT study of the effect of BN pair doping on the electronic and optical properties of graphyne nanosheets. *Journal of Materials Science* 52: 10294–10307.
73 Wang, X., Yang, Z., Si, W. et al. (2019). Cobalt-nitrogen-doped graphdiyne as an efficient bifunctional catalyst for oxygen reduction and hydrogen evolution reactions. *Carbon* 147: 9–18.
74 Gao, J., Wang, N., He, J. et al. (2020). Precise and controllable N/C ratio in graphdiyne for superior Li and Na ions storage capacities. *2D Materials* 7: 025032.
75 Lu, T., He, J., Li, R. et al. (2020). Adjusting the interface structure of graphdiyne by H and F co-doping for enhanced capacity and stability in Li-ion battery. *Energy Storage Materials* 29: 131–139.
76 Xue, Y., Huang, B., Yi, Y. et al. (2018). Anchoring zero valence single atoms of nickel and iron on graphdiyne for hydrogen evolution. *Nature Communications* 9: 1460.
77 Ren, H., Shao, H., Zhang, L. et al. (2015). A new graphdiyne nanosheet/pt nanoparticle-based counter electrode material with enhanced catalytic activity for dye-sensitized solar cells. *Advanced Energy Materials* 5: 1500296.

78 Gao, Y., Cai, Z., Wu, X. et al. (2018). Graphdiyne-supported single-atom-sized fe catalysts for the oxygen reduction reaction: DFT predictions and experimental validations. *ACS Catalysis* 8: 10364–10374.
79 Shen, H., Li, Y., and Shi, Z. (2019). A novel graphdiyne-based catalyst for effective hydrogenation reaction. *ACS Applied Materials & Interfaces* 11: 2563–2570.
80 Li, Y., Liu, Y., Li, Z. et al. (2018). Pd nanoparticles anchored on N-rich graphdiyne surface for enhanced catalysis for alkaline electrolyte oxygen reduction. *International Journal of Electrochemical Science* 13: 12226–12237.
81 Feng, Z., Li, R., Ma, Y. et al. (2019). Molecule-level graphdiyne coordinated transition metals as a new class of bifunctional electrocatalysts for oxygen reduction and oxygen evolution reactions. *Physical Chemistry Chemical Physics* 21: 19651–19659.
82 Shi, G., Yu, C., Fan, Z. et al. (2019). Graphdiyne-supported NiFe layered double hydroxide nanosheets as functional electrocatalysts for oxygen evolution. *ACS Applied Materials & Interfaces* 11: 2662–2669.
83 Xue, Y., Li, J., Xue, Z. et al. (2016). Extraordinarily durable graphdiyne-supported electrocatalyst with high activity for hydrogen production at all values of pH. *ACS Applied Materials & Interfaces* 8: 31083–31091.
84 Liu, J., Shen, X., Baimanov, D. et al. (2019). Immobilized ferrous ion and glucose oxidase on graphdiyne and its application on one-step glucose detection. *ACS Applied Materials & Interfaces* 11: 2647–2654.
85 Dang, Y., Guo, W., Zhao, L. et al. (2017). Porous carbon materials based on graphdiyne basis units by the incorporation of the functional groups and Li atoms for superior CO_2 capture and sequestration. *ACS Applied Materials & Interfaces* 9: 30002–30013.
86 Chen, X., Gao, P., Guo, L. et al. (2017). High-efficient physical adsorption and detection of formaldehyde using Sc- and Ti-decorated graphdiyne. *Physics Letters A* 381: 879–885.
87 Yu, H., Xue, Y., Hui, L. et al. (2018). Controlled growth of MoS_2 nanosheets on 2D N-doped graphdiyne nanolayers for highly associated effects on water reduction. *Advanced Functional Materials* 28: 1707564.
88 Li, J., Xu, J., Xie, Z. et al. (2018). Diatomite-templated synthesis of freestanding 3D graphdiyne for energy storage and catalysis application. *Advanced Materials* 30: e1800548.
89 He, J., Ma, S.Y., Zhou, P. et al. (2012). Magnetic properties of single transition-metal atom absorbed graphdiyne and graphyne sheet from DFT+U calculations. *Journal of Physical Chemistry C* 116: 26313–26321.
90 Chen, X., Gao, P., Guo, L. et al. (2017). Two-dimensional ferromagnetism and spin filtering in Cr and Mn-doped graphdiyne. *Journal of Physics and Chemistry of Solids* 105: 61–65.
91 Zhang, M., Wang, X., Sun, H. et al. (2018). Preparation of room-temperature ferromagnetic semiconductor based on graphdiyne-transition metal hybrid. *2D Materials* 5: 035039.
92 Ku, R., Yu, G., Gao, J. et al. (2020). Embedding tetrahedral 3d transition metal TM_4 clusters into the cavity of two-dimensional graphdiyne to construct highly

efficient and nonprecious electrocatalysts for hydrogen evolution reaction. *Physical Chemistry Chemical Physics* 22: 3254–3263.

93 Yang, N., Liu, Y., Wen, H. et al. (2013). Photocatalytic properties of graphdiyne and graphene modified TiO_2: from theory to experiment. *ACS Nano* 7: 1504–1512.

94 Jin, Z., Zhou, Q., Chen, Y. et al. (2016). Graphdiyne: ZnO nanocomposites for high-performance UV photodetectors. *Advanced Materials* 28: 3697–3702.

95 Xu, J., Li, J., Yang, Q. et al. (2017). In-situ synthesis of MnO_2@graphdiyne oxides nanocomposite with enhanced performance of supercapacitors. *Electrochimica Acta* 251: 672–680.

96 Dong, Y., Zhao, Y., Chen, Y. et al. (2018). Graphdiyne-hybridized N-doped TiO_2 nanosheets for enhanced visible light photocatalytic activity. *Journal of Materials Science* 53: 8921–8932.

97 He, J., Li, X., Lu, T. et al. (2019). Graphdiyne applied for electrochemical energy storage. *Dalton Transactions* 48: 14566–14574.

98 Zuo, Z. and Li, Y. (2019). Emerging electrochemical energy applications of graphdiyne. *Joule* 3: 899–903.

99 Kim, S., Ruiz Puigdollers, A., Gamallo, P. et al. (2017). Functionalization of γ-graphyne by transition metal adatoms. *Carbon* 120: 63–70.

100 Chen, X. and Zhang, S. (2018). Modulation of molecular sensing properties of graphdiyne based on 3d impurities. *Acta Physico-Chimica Sinica* 34: 1061–1073.

101 Nayebi, P. and Zaminpayma, E. (2017). Metal decorated graphdiyne: a first principle study. *Physica B: Condensed Matter* 521: 112–121.

102 He, J.J., Zhou, P., Jiao, N. et al. (2014). Magnetic exchange coupling and anisotropy of 3d transition metal nanowires on graphyne. *Scientific Reports* 4: 4014.

103 Xue, Y., Zuo, Z., Li, Y. et al. (2017). Graphdiyne-supported $NiCo_2S_4$ nanowires: a highly active and stable 3D bifunctional electrode material. *Small* 13: 1700936.

104 Gao, J., He, J., Wang, N. et al. (2019). Robust C—S bond integrated graphdiyne-MoS_2 nanohybrids for enhanced lithium storage capability. *Chemical Engineering Journal* 373: 660–667.

105 Jiao, Y., Du, A., Hankel, M. et al. (2011). Graphdiyne: a versatile nanomaterial for electronics and hydrogen purification. *Chemical Communications* 47: 11843–11845.

106 Pan, Q., Liu, H., Zhao, Y. et al. (2019). Preparation of N-graphdiyne nanosheets at liquid/liquid interface for photocatalytic NADH regeneration. *ACS Applied Materials & Interfaces* 11: 2740–2744.

107 Zhao, W.H., Yuan, L.F., and Yang, J.I. (2012). Graphdiyne as hydrogen purification membrane. *Chinese Journal of Chemical Physics* 25: 434.

108 Meng, Z., Zhang, X., Zhang, Y. et al. (2016). Graphdiyne as a high-efficiency membrane for separating oxygen from harmful gases: a first-principles study. *ACS Applied Materials & Interfaces* 8: 28166–28170.

109 Nagarajan, V. and Chandiramouli, R. (2018). Investigation of NH_3 adsorption behavior on graphdiyne nanosheet and nanotubes: a first-principles study. *Journal of Molecular Liquids* 249: 24–32.

110 Chen, X., Gao, P., Guo, L. et al. (2015). Graphdiyne as a promising material for detecting amino acids. *Scientific Reports* 5: 16720.

111 Nagarajan, V., Srimathi, U., and Chandiramouli, R. (2018). First-principles insights on detection of dimethyl amine and trimethyl amine vapors using graphdiyne nanosheets. *Computational and Theoretical Chemistry* 1123: 119–127.

112 Parvin, N., Jin, Q., Wei, Y. et al. (2017). Few-layer graphdiyne nanosheets applied for multiplexed real-time dna detection. *Advanced Materials* 29: 1606755.

5

Graphdiyne-Based Materials in Catalytic Applications

Yurui Xue[1] and Yuliang Li[2]

[1] *Institute of Frontier and Interdisciplinary Science, School of Chemistry and Chemical Engineering, Science Center for Material Creation and Energy Conversion, Shandong University, 27 Shanda Nanlu, Jinan 250100, P. R. China*
[2] *Institute of Chemistry, Chinese Academy of Sciences, Zhongguancun North First Street 2, Beijing 100190, P. R. China*

The efficient conversion of green energy, such as solar energy, to high-value-added fuels, and chemicals under mild conditions to solve the current global resource shortage and ecological environment deterioration and other problems is still a grand challenge. The catalyst, which determines the catalytic performance and ultimately affects the overall conversion efficiency, plays the key role to achieve high efficiency in the conversion process. Developing new materials with high selectivity and high activity is therefore the key to realize efficient energy conversion and utilization, and is the frontier subject of science.

The successful synthesis of graphdiyne (GDY) has promoted the research on the properties and applications of GDY. The high distribution of the acetylenic bonds in GDY makes the surface charge distribution of GDY extremely uneven, which endows it with more active sites and high intrinsic activity, and thus effectively promotes the catalytic reaction process [1]. The theoretical and experimental results all reveal that GDY has many unique and fascinating properties and applications in various fields, such as catalysis, energy, and electronics. In recent years, a series of original research achievements in GDY-based metal atomic catalysts (ACs), heterojunction catalysts, and metal-free catalysts have been made. Compared with the overall development of metal atomic catalyst, GDY-based ACs are a newly developed research direction, which is still in the preliminary stage. In combination with the previous theoretical research on GDY-based ACs, this chapter will discuss various potential properties and applications of GDY-based ACs in the field of catalysis. It can be predicted that GDY will play an increasingly important role in the field of energy materials.

Graphdiyne: Fundamentals and Applications in Renewable Energy and Electronics,
First Edition. Edited by Yuliang Li.

5.1 Graphdiyne-Based Zero-Valent Metal Atomic Catalysts

Single-metal atomic catalysts, which contain singly dispersed metal atoms and supporting materials, have long been the research frontier in the field of catalysis and energy conversion due to their unique structures and attractive properties, such as maximum atom utilization, specific electronic structure, and high reaction selectivity and activity, and have led to the rapid development of catalytic science [2–7]. During past decades, extensive efforts have been made in the basic and applied research of single-atom catalysts. However, there still remain critical scientific issues and challenges to be addressed in this field, for example, the natural migration and aggregation tendency of metal atoms on traditional supporting materials, the precise structure of the support materials, the valence of the supported metal atoms, the high dispersion of metal atoms, and so on. Atomic catalysts, which feature unique and well-defined chemical and electronic structures, high catalytic activity and selectivity, are of great academic significance for us to clearly understand the anchoring process of single atoms on the support, the interaction of metal atoms with the support, the energy, electron transfer, and conversion behaviors in the system, and the relationship between the catalytic performance and the above scientific issues.

With the purpose of gaining better understanding of the above-mentioned issues, Li and coworkers have successfully anchored transition metal and noble metal atoms on GDY by utilizing the advantages of the electronic and chemical structures of GDY [8–13]. They named the obtained catalysts atomic catalyst (AC). Among the reported single-atom catalysts, zero-valent transition metal ACs have always been a great challenge in the field of catalysis. Scientists have long been waiting for the advent of zero-valent transition metal ACs. But loading zero-valent metal atoms has been an important challenge in the field of catalysis.

5.1.1 Synthetic Strategy for GDY-Based ACs

In 2018, Li and coworkers [8] proposed a novel strategy for rational design and controllable synthesis of zero-valent transition metal atoms by utilizing the rich acetylene bonds in GDY, the interaction between GDY and metal atoms, and combining the spatial scale effect of the porous structure (Figure 5.1). They successfully anchored zero-valent transition metal atoms on GDY (Ni^0/GDY and Fe^0/GDY, with the mass loading of 0.278 and 0.680 wt%, respectively) for the first time across the world, overcoming problems, including the metal atom migration and aggregation and undetermined charge transfer of traditional single-atom catalysts, and realizing the zero-valent metal atom catalysis. The synthetic method has strong universality, expansibility, and convenience in synthesizing atomic catalysts. Yu et al. [10] obtained the first zero-valence noble metal palladium atom catalyst (Pd^0/GDY) based on this method. It is worth mentioning that molybdenum generally exists in the form of compounds with high oxidation state in nature. It is quite difficult to obtain zero-valent molybdenum atom catalysts through traditional method.

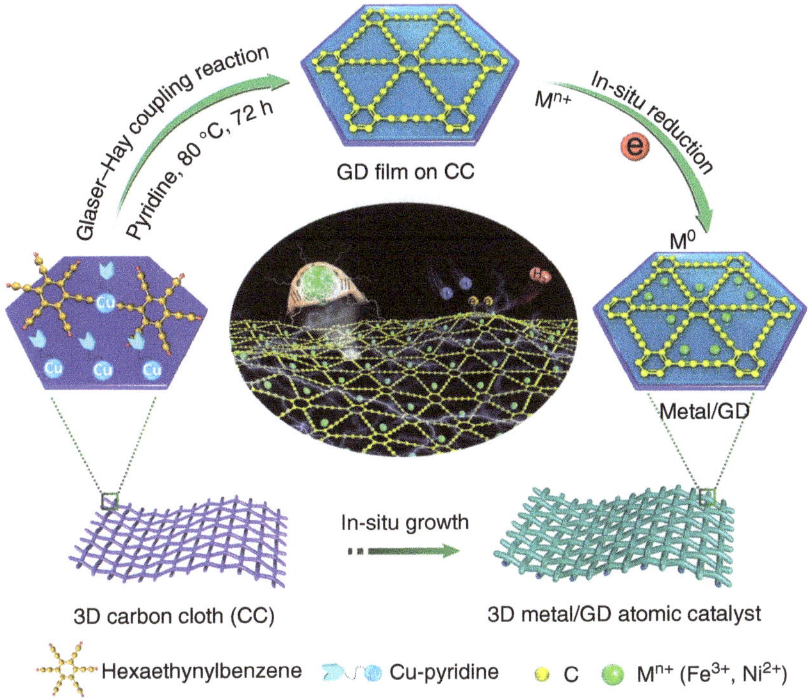

Figure 5.1 Protocols for the synthesis of Ni^0/GDY and Fe^0/GDY. A two-step strategy for anchoring isolated Ni/Fe atoms on GDY, including the in situ growth of GDY layers on 3D carbon cloth (CC) surfaces and the electrochemical reduction of metal ions (Ni^{2+} and Fe^{3+}) into zero-valent metallic species (Ni^0 and Fe^0, respectively). Source: Xue et al. [8]. © 2018, Springer Nature/CC License 4.0.

To this end, Hui et al. reported a facile one-step solvothermal reduction method to reduce high-valence molybdenum atoms to zero valent on GDY, resulting in zero-valent molybdenum ACs (Mo^0/GDY, with a mass loading of up to 7.5 wt%) [9]. More recently, Li and coworkers reported a new metal ion anchoring–electron transfer–self-reduction strategy for anchoring zero-valent palladium atoms on GDY [11]. Up to now, five zero-valent metal atoms (Ni [14], Fe [14], Mo [9], Cu [12], and Pd [10, 11, 15]) have been successfully anchored on GDY, forming zero-valent ACs.

The success of loading zero-valent metal atoms on GDY expands the synthesis methodology of single-atom catalysts and enriches the types of functional catalysts. The emergence of zero-valent catalysts was a real insight into the strategies involved in the development of new single-atom catalysts and opened up a new direction for the development of novel efficient catalysts. GDY–metal atom catalysts have made a series of original breakthroughs in high-efficiency catalysis and energy conversion, and have been first applied in some critical fields related to national economy and people's livelihood, such as splitting water to produce hydrogen and oxygen, and converting nitrogen and water to produce ammonia. GDY has promoted the research on the properties and applications of atomic catalysts.

5.1.2 Adsorption Geometry and Electronic Structures of GDY-Based ACs

There are 18 carbon atoms in a single unit cell of GDY, including six sp^2-hybridized C atoms in the hexagonal ring and twelve sp-hybridized C atoms in the linear alkyne chain. The unique structure of GDY endows it with distinct characteristics, such as natural porous structure and band gap, abundant chemical bonds, uneven charge distribution, and high stability. The abundant sp hybridization of —C≡C— enables the arbitrary angle rotation of π/π^* perpendicular to the axis, making it easier to interact with adjacent metal atoms with larger binding energy than graphene [14], which allows the adsorption of metal atoms more stably to effectively avoid the agglomeration [14]. These are properties that traditional carbon materials, such as graphene, do not have.

The researchers studied the geometry of absorption sites and the hidden information on subtle metal atoms-C bonding, theoretically (Figure 5.2). All possible initial adsorption sites have been considered. A potential energy surface scanning along two possible pathways (routes A and B) was conducted. It was found that the metal atoms are inclined to absorb along route A, and the most favorable adsorption site for metal atoms is A1, as shown in Figures 5.2a,b. The bond length analysis and the presence of strong charge transfer between Ni/Fe and GDY demonstrated their strong chemisorption character (Figure 5.2c). Furthermore, the anchored metal atoms have a significant impact on the electronic structure of GDY, resulting in remarkable charge transfer between GDY and metal atoms, as well as the charge redistribution of the s, p, and d orbitals of metal atoms (Figures 5.2c,d), which will arouse the significant decrease of the free energy and the increase of the number of active sites in the system. These advantages of GDY-based ACs make it have great application prospects in fields of energy conversion and utilization.

5.1.3 Morphology and Valence States of GDY-Based ACs

The high-resolution transmission electron microscopy (TEM), high-angle annular dark-field scanning transmission electron microscopy (HAADF–STEM), X-ray photoelectron spectroscopy (XPS), and X-ray absorption spectroscopy are widely used to determine the local geometric and/or electronic structure of GDY-based ACs. HAADF–STEM images in Figure 5.3 obviously show the isolatedly and uniformly dispersed bright dots, which represent the heavy metal atoms on GDY. This suggested the successful anchoring of metal atoms (Ni, Fe, Pd, and Mo) on GDY. Especially, Figure 5.3o–p gave clear images of the anchoring sites of metal atoms in GDY for the first time, providing evidence for the possible anchoring configurations of the metal atoms on GDY, consistent with the theoretical calculation results.

However, electron microscopy can only characterize a small region of the material. To prove that metal atoms are 100% monatomic dispersed on GDY surface, the researchers further analyzed the local environment surrounding the anchored atoms by the extended X-ray absorption fine structure (EXAFS). As shown in

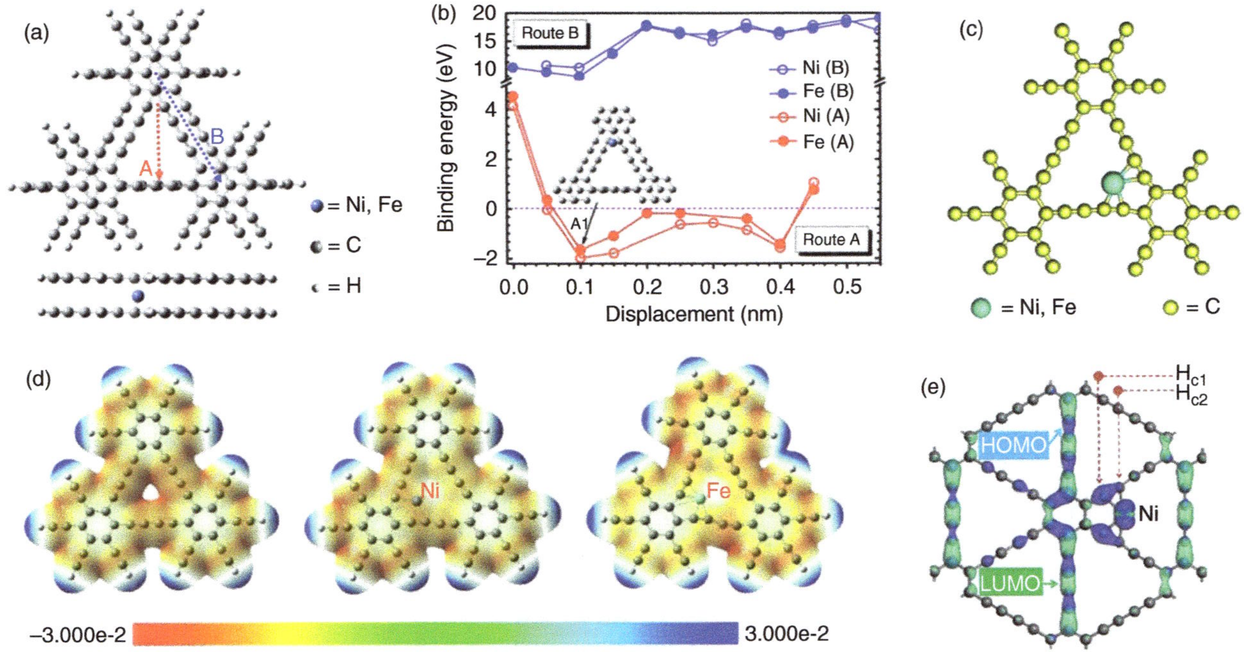

Figure 5.2 (a) Possible adsorption sites for Ni/Fe atoms between GDY layers along routes A and B (upper: top view; lower: side view). (b) Binding energies for Ni/Fe atom adsorption with bilayer GDY as a function of displacement along the A and B directions (a). (c) Adsorption of single-metal atoms on GDY. (d) Electrostatic potential maps. (e) Binding energies, charge transfer amounts, and bond lengths between TM atoms and the nearest carbon atoms. Source: Xue et al. [8]. © 2018, Springer Nature/CC License 4.0.

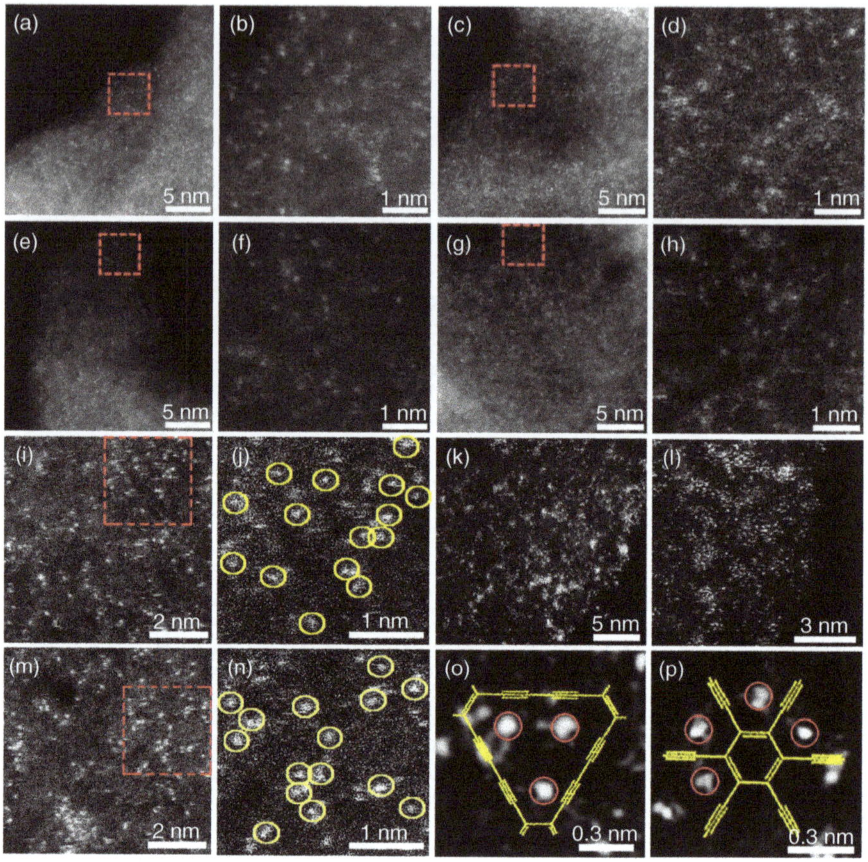

Figure 5.3 HAADF images of (a–d) Ni^0/GDY, (e–h) Fe^0/GDY, (i, j, m, and n) Pd^0/GDY, and (k, l, o, and p) Mo^0/GDY, respectively. The isolated and uniformly dispersed bright dots represent the heavy metal atoms anchored on GDY. Source: (a–h) Xue et al. [8]. © 2018, Springer Nature. Source: (i, j, m, and n) Reproduced with permission from Yu et al. [10]. © 2018, Elsevier. Source: (k, l, o, and p) Hui et al. [9]. © 2019, American Chemical Society.

Figure 5.4, EXAFS spectra of the GDY-based ACs show only the prominent peaks located at smaller distance, which could be assigned to the metal atom–C interactions, and no peak corresponding to metal–metal contributions could be observed. These results further demonstrated the successful anchoring of isolated metal atoms on GDY.

The X-ray absorption near-edge structure (XANES) spectroscopy is capable of detecting the valence changes of the same element with high sensitivity and efficiency and is particularly used as the fingerprint spectroscopy to distinguish different valence states of the same element. Figure 5.4 shows the XANES profiles for Ni^0/GDY, Fe^0/GDY, Pd^0/GDY, and Mo^0/GDY along with corresponding reference samples. Taking Ni^0/GDY as example, pure Ni foil was used as the reference to precisely prove the valence state of metal atoms. For the Ni K-edge spectrum, the

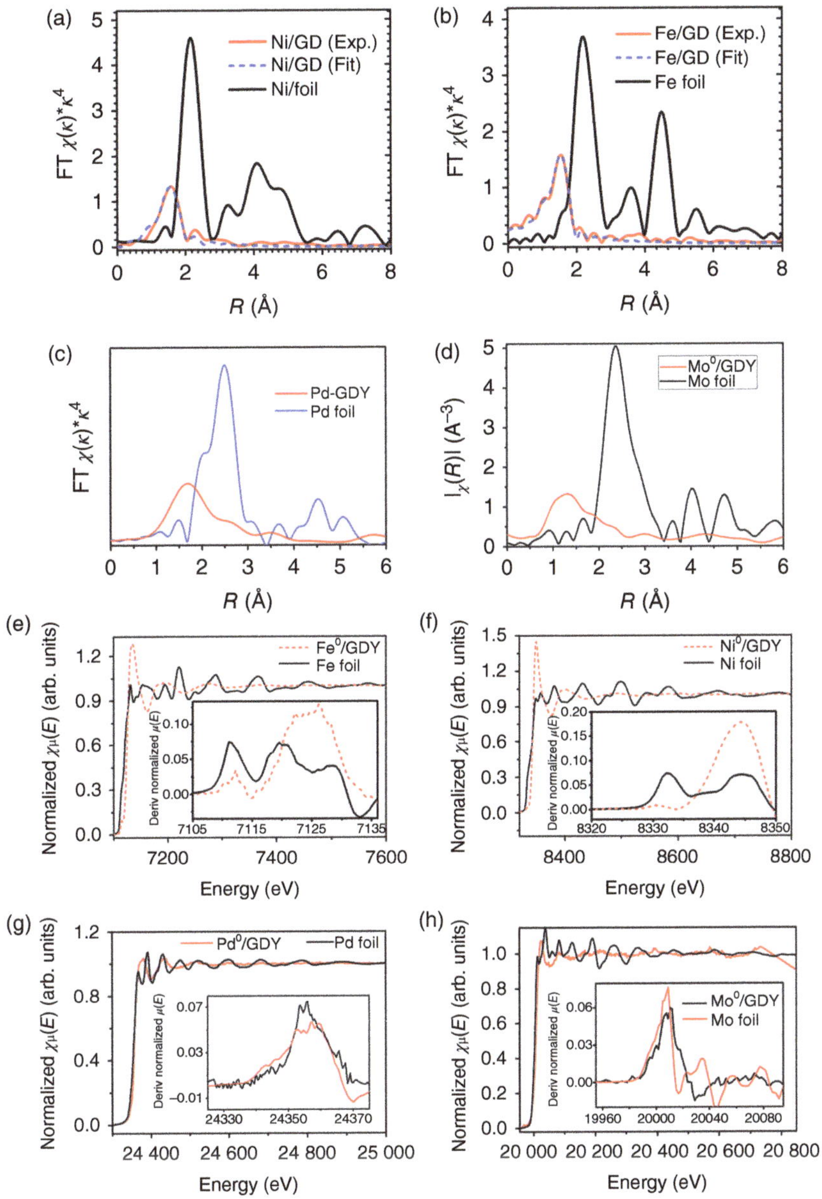

Figure 5.4 Ex situ EXAFS spectra of (a) Ni/GD and Ni foil at the Ni K-edge, (b) Fe/GD and Fe foil at the Fe K-edge, (c) Pd^0/GD and Pd foil at the Pd K-edge. (d) Mo^0/GD and Mo foil at the Mo K-edge. Source: (a, b) Xue et al. [8]. © 2018, Springer Nature/CC License 4.0, (c) Yu et al. [10]. © 2018, Elsevier, (d) Hui et al. [9]. © 2019, American Chemical Society.

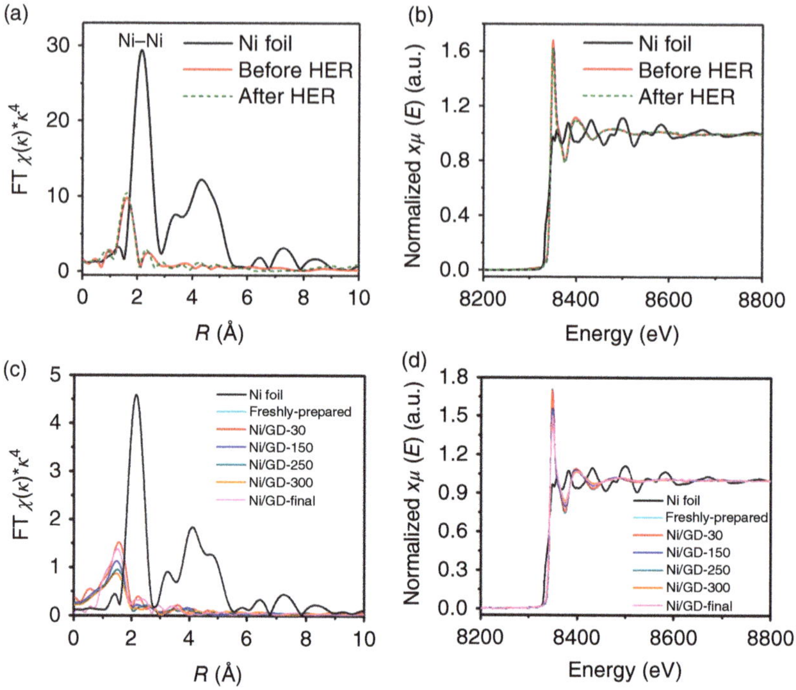

Figure 5.5 (a) Ex situ EXAFS and (b) the normalized XANES spectra at the Ni K edge of Ni/GD obtained before (red line) and after (green dashed line) HER measurements. (c, d) X-ray absorption spectroscopy (XAS) studies of Ni/GD catalysts at different temperatures under 5% H_2/He. (c) Ex situ EXAFS spectra and (d) the normalized XANES spectra at the Ni K edge of Ni/GD at the Ni K-edge obtained at different reduction conditions. Ni foil was measured for comparison. Source: Xue et al. [8]. © 2018, Springer Nature/CC BY 4.0.

absorption position of Ni^0/GDY is located at the same position to Ni foil. Besides, the first-derivative XANES for Ni^0/GDY is similar to that of Ni foil. These results revealed the valence states of Fe atoms in Ni^0/GDY are zero. To gain more insights into the states of metal atoms in the GDY-based atomic catalyst, the in situ XANES and EXAFS measurements were performed on the samples during the catalytic reaction and thermal treatment (5% H_2/He, at different temperatures) processes, respectively. Each temperature step is maintained for 30 minutes. As shown in Figure 5.5, there are no changes in the XANES and EXAFS spectra compared with the freshly synthesized samples, which indicate that the metal atoms anchored on GDY still remain isolated and dispersed, and have zero valence. These results prove the high stability of GDY-based zero-valent metal atomic catalysts.

Simultaneously, theoretical calculations were conducted to further analyze the valence state of metal atoms in Ni^0/GDY (Fe^0/GDY) as well as the interaction between metal atoms and GDY from the perspective of bonding energy. A previously developed method to ab-initially reflect the energy of the targeted on-site orbital especially with electronic occupations under different cases of chemical bonding was used to elucidate the hidden information on subtle metal atoms–C bonding

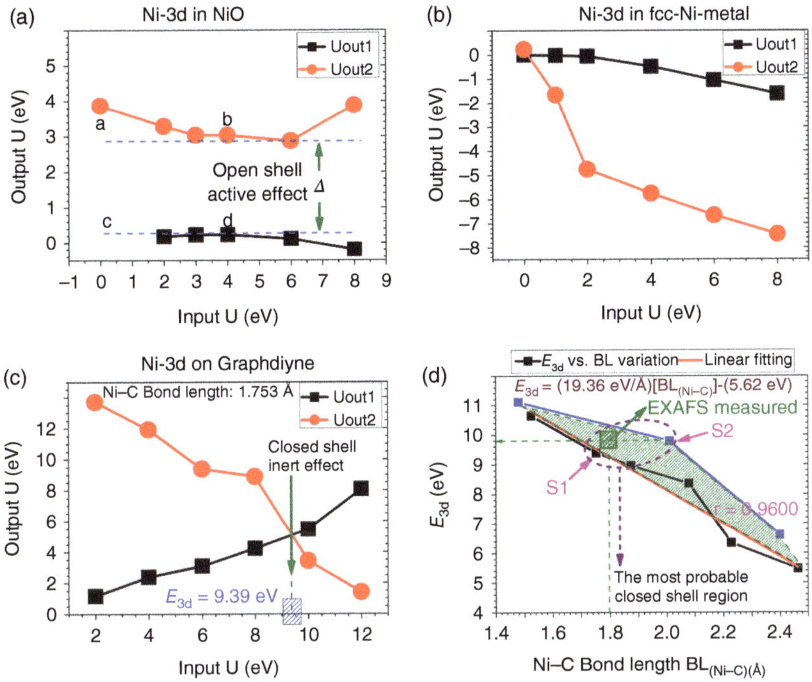

Figure 5.6 (a) The 3d orbital energies for the targeted Ni site in (a) NiO, (b) Ni-on-GD, and (c) Ni-fcc are self-consistently determined via open- and closed-shell charge overlapping. (d) The variation of orbital energy variation with related to the newly formed Ni–C. (The green shaded area denotes the Ni–C interdistance-dependent Ni-closed-shell regions with related to Ni–C interdistance. The purple dashed circle shows the most probable closed-shell orbital regions for Ni to be located on GD system in thermodynamic equilibrium state. The green square tells our experimental EXAFS measured data). Source: Xue et al. [8]. © 2018, Springer Nature/CC BY 4.0.

[16, 17]. Taking the Ni-3d as an example, the open-shell effect (noncrossover) in *cubic*-NiO (Figure 5.6a) and the closed-shell effect (crossover) were found, but with minor 3d–3d overlaps in zero-valent fcc-Ni (Figure 5.6b). As expected, the Ni^0/GDY exhibited an obvious closed-shell effect consistent with that of fcc-Ni (Figure 5.6c), indicating the zero-valent state of the metals in Ni^0/GDY. The larger orbital potential implies stronger orbital overlaps between Ni and neighboring C sites. In addition, the researchers revealed that the theoretically optimized Ni—C bond length (1.753 Å) and Ni-3d orbital energy (9.79 eV) are in agreement with the experimentally measured values (1.8 Å for the Ni—C bond length and 9.89 eV for Ni-3d orbital energy) (Figure 5.6d).

The experimental and theoretical calculation results strongly confirmed that the metal atoms anchored on GDY could stably exist, that is, we have successfully prepared the zero-valence metal atom catalysts. The successful synthesis of the GDY-based zero-valence atom catalyst solved the critical problems that resulted from single-metal atoms supported on traditional materials, such as easy migration and agglomeration and unstable charge transfer. As for GDY-based zero-valence atom catalysts, the anchored zero-valence metal atoms can further activate the carbon atoms adjacent to the anchoring sites, maximizing the electrical conductivity of the system and the number of active sites, as well as its catalytic activity.

Recently, Zhang and coworkers [15] synthesized the GDY-based zero-valent Pd ACs through a wet chemical approach, including the anchoring of single Pd atoms on GDY/G followed by being reduced to zero valence by $NaBH_4$. The XPS spectra of Pd 3d5/2 and 3d3/2 in Pd_1/ GDY/G shift to lower binding energy (335.2 and 340.8 eV) compared to those of Pd1/GDY/G-O (337.4 and 342.7 eV), indicating the formation of zero-valent palladium (Pd^0) complexes.

In addition to the above-mentioned zero-valent atomic catalysts, GDY-based atomic catalysts in which the metal atoms feature high oxidation states have recently been reported, such as Pt-GDY2 [18] Pds-GDY [19], Pd_1/GDY/G [15], and Ru/GDY [13].

5.1.4 Application of GDY-Based ACs

Atomic catalysts (ACs) have long been the research frontier in the field of catalysis because of their unique structures and properties, such as the high atom utilization efficiency, and high reaction selectivity and activity. Compared with the traditional single-atom catalysts, GDY-based metal ACs feature unique and well-defined chemical and electronic structures, highly catalytic activity and selectivity, which is expected to achieve high selectivity, high activity, high stability, and high production yield in catalysis.

5.1.4.1 Applied for Water Splitting

Splitting water into hydrogen (H_2) and oxygen (O_2) provides a feasible and economical approach for energy conversion and storage [20, 21]. The drawbacks of applying these approaches industrially are the sluggish kinetics of the oxygen evolution reaction and hydrogen evolution reaction (OER/HER). However, improvements in the activity, efficiency, and stability of catalysts in the processes still remain intractable challenges. The unique physical and chemical properties of GDY-metal atom catalysts bring new solutions to the problems in the energy conversion process.

Prof. Li and coworkers recently clearly showed the highest occupied molecular orbital (HOMO) and lowest unoccupied molecular orbital (LUMO) charge densities distributions induced by the stabilized metal atoms for GDY-based ACs (for example, Ni^0/GDY and Fe^0/GY [8]). It was found that the fast ($H^+ + e$) charge exchanges between the adsorbed metal atom (Ni^0/Fe^0) and acetylenic carbon is responsible for the electrical conductivity and efficient HER reaction. In addition, the strong interaction between metal atoms and GDY is beneficial to stabilize the metal atoms on GDY. As expected, in acidic conditions (0.5 M H_2SO_4), Fe^0/GDY and Ni^0/GDY exhibited excellent HER activity and stability (Figure 5.7). Fe^0/GDY had the best HER activity with the smallest overpotential of 66 mV at 10 mA cm^{-2}, which is smaller than that of Ni^0/GD (88 mV), pure GDY (578 mV), and other reported ones (Figure 5.7a–c), such as Ni_2P [22], MoS_2 [23], Pt nanowires/single-layered $Ni(OH)_2$ [24], and exfoliated WS_2 [25]. The number of active sites per unit area of

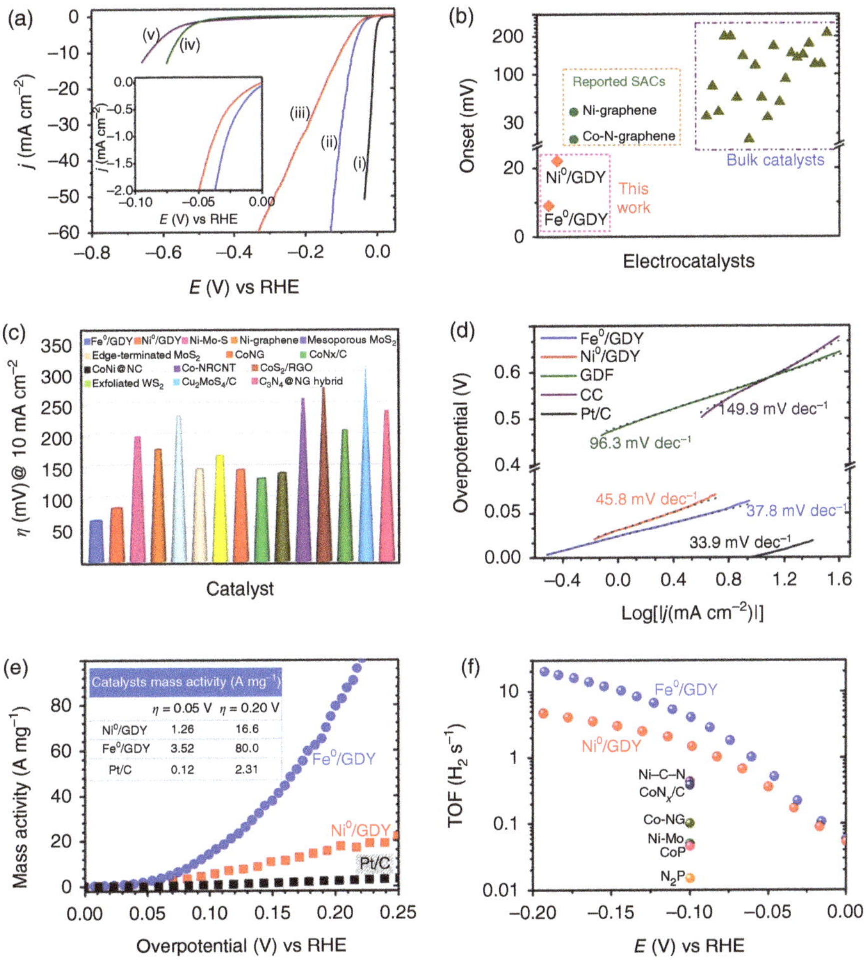

	η = 0.05 V	η = 0.20 V
Ni^0/GDY	1.26	16.6
Fe^0/GDY	3.52	80.0
Pt/C	0.12	2.31

Figure 5.7 HER activities and stabilities of Ni^0/GDY and Fe^0/GDY. (a) Polarization curves of (i) Pt/C, (ii) Fe^0/GDY, (iii) Ni^0/GDY, (iv) GDF, and (v) CC in 0.5 M H_2SO_4 solution (inset: Enlarged view of the linear sweep voltammetry (LSV) curves for Fe/GD and Ni/GD near the onset region). (b) Onset values and (c) overpotentials at 10 mA cm^{-2} of Ni^0/GDY and Fe^0/GDY (red square) along with other nonprecious single-atom HER catalysts (green circle) and several bulk catalysts (olivine triangle). (d) Tafel plots of the presented data in (a). (e) Mass activities of Ni^0/GDY, Fe^0/GDY, and Pt/C (inset: mass activities obtained at overpotentials of 0.05 and 0.20 V). (f) TOF values of Fe^0/GDY (blue dot) and Ni^0/GDY (red dot) together with several state-of-the-art HER electrocatalysts. Source: Xue et al. [8].

Fe^0/GDY (2.56×10^{16} cm^{-2}) and Ni^0/GDY (2.38×10^{16} cm^{-2}) was 17 and 15.8 times greater than that of Pt/C (1.5×10^{15} cm^{-2}). Moreover, the Tafel slopes of Fe^0/GDY and Ni^0/GDY were determined to be 37.8 and 45.8 mV dec^{-1}, respectively, which is close to that of Pt/C (33.9 mV dec^{-1}), suggesting the Volmer–Heyrovsky mechanism was operative during the HER process (Figure 5.7d). The mass activities of Fe^0/GDY and Ni^0/GDY were 34.6 and 7.19 times greater than those of Pt/C, respectively, at overpotential of 0.2 V (Figure 5.7e). Ni^0/GDY and Fe^0/GDY exhibited record-high turnover frequency (TOF) values of 4.15 and 1.59 s^{-1} at 100 mV (Figure 5.7f), respectively, higher than benchmarked electrocatalysts, such as CoP (0.046 s^{-1}) [26] and Ni_2P (0.015 s^{-1}) [22]. These GDY-based ACs exhibited better long-time stability than Pt/C, and Ni and Fe atoms were presented as isolated atoms without any aggregations for Ni/GD and Fe^0/GDY during the long-term cycling operation [8]. Such superior catalytic performance could be ascribed to the synergy effect between the anchored Fe^0/Ni^0 atoms and GDY, which facilitated the efficient charge transfer between the active sites and the carrier, endowing GDY-based ACs with excellent catalytic activity and stability.

Noble metal–based (Pt, Pd, Ru, etc.) materials are still the most efficient HER electrocatalysts so far. Designing and preparing novel noble metal atom catalysts is an ideal approach to maximize the utilization of these noble metal–based electrocatalysts and overcome the limiting factors, including their scarcity and high cost. Lu and coworkers [18] obtained graphdiyne-platinum atom catalyst (Pt-GDY) by reacting GDY with K_2PtCl_4 during which the isolated Pt atoms were anchored on GDY through the coordination interactions between metal atoms and alkynyl C atoms in GDY, forming five-coordinated C_1-Pt-Cl_4 species in Pt-GDY1 and four-coordinated C_2-Pt-Cl_2 species in Pt-GDY2 under different reaction temperatures and times. Their experimental results showed that the coordination environments of Pt atoms in ACs could greatly affect the HER activity. The Pt 5d orbitals of Pt-GDY2 after annealing in Ar atmosphere possessed the highest density of unoccupied states, and these orbitals had significant importance during catalytic reaction process, resulting in a hydrogen adsorption-free energy close to 0 (0.092 eV) on Pt-GDY2. The factors above gave Pt-GDY2 more excellent HER performance in acidic condition, and its mass activity was 26.9 times that of Pt/C. To solve the problem of the high cost of Pt-based catalyst, scientists expect to use cheap metal catalysts to replace Pt. Transition metals that possess higher reserves than Pt have aroused increasing research interests. Li and coworkers [10] experimentally proved that the specific structure and valence of Pd^0/GDY gave it excellent HER performance. It was able to reach a current density of 10 mA cm^{-2} at a very small overpotential (55 mV) with a quite low mass loading (0.2%, only one-hundredth of 20 wt% Pt/C) and exhibited superior mass activity (61.5 A per $mg_{metal}{}^{1}$) and (TOF, 16.7 s^{-1}) than 20 wt% Pt/C. Moreover, Pd^0/GDY showed long-time stability of up to 72 hours. Recently, Yu et al. [13] controllably prepared GDY–ruthenium atom catalyst for the first time. The emergence of a particular charge transfer mechanism between Ru atoms and adjacent C atoms significantly increased the number of active centers of Ru/GDY and its reaction selectivity, activity, and cycle stability in HER and OER in acidic conditions. This was also the first bifunctional metal atomic catalyst, which could perform HER and OER simultaneously, providing a new strategy for the design and synthesis of new class of catalysts.

5.1.4.2 Applied for Ammonia Synthesis at Ambient Conditions

As one of the most fundamental chemical materials in modern industrial and agricultural production, ammonia (NH_3) plays a critical role in the human production and life. However, the process of present industrial ammonia synthesis under harsh conditions with high temperature and pressure (400–600 °C and 20–40 MPa) is energy intensive and will cause serious environmental pollution. The research on the highly efficient synthesis of ammonia, especially at room temperature and atmospheric pressure, has been drawing increasing attention for wide applications of scientists and industries around the world. Developing highly active and selective materials with transformative catalytic performance for ammonia synthesis is still a gigantic challenge.

Li and coworkers [9] reported the first zero-valent atomic catalyst comprising atomically dispersed zero-valent molybdenum atoms on graphdiyne (Mo^0/GDY) with a high mass content of Mo atoms (up to 7.5 wt%) for the synthesis of ammonia at room temperatures and ambient pressures (Figure 5.8a). In the nitrogen atmosphere, Mo site is the optimal N_2 adsorption site and the energetically preferred adsorption configuration is Mo—N≡N. We further interpret the prominent electronic activities given by Mo^0/GDY. From the projected partial density of states (PDOSs) of Mo^0/GDY, the Mo 4d and C–p orbitals dominantly control the electron-transfer activities. The two dominant C-bonding and antibonding orbitals pin the Mo 4d bands in the middle crossover of the E_F. This phenomenon will not only robustly preserve the Mo 4d valence electronic states for the various catalytic steps of the electrocatalytic nitrogen reduction reaction (ECNRR), but also indicate that the Mo site can easily accumulate the active electrons from the C sites for further efficient electron transfer between Mo and N. Additionally, the calculation results show that both C1 and C2 are energetically favored H adsorption sites. The extra chemical adsorption energy indicates that C1 is the optimal site for H adsorption and proton–electron charge exchange (Figure 5.8b). Further examinations of the local structural configurations of the ECNRR process demonstrate that the homopolar N-bond dissociation occurs at the step of [NH_2–NH_3+(H^++e^-)], preventing the overbinding effect of the N intermediate. This result agrees with the trend that the intermediate N—N bonding variation ensures the energetic compensation when the HER is suppressed or H desorption from the local active adsorption sites occurs. The performance of the electrocatalytic nitrogen reduction (ECNRR) of Mo^0/GDY was evaluated in a two-compartment cell separated by a Nafion 211 membrane in different electrolyte solutions (0.1 M Na_2SO_4 and 0.1 M HCl) under ambient conditions (Figure 5.8c). ^{15}N-labeling experiments solidly revealed that the N in produced ammonia comes from the nitrogen gases (Figure 5.8d,e) and no possible byproduct is detected, demonstrating 100% selectivity. Experimental results showed that Mo^0/GDY had higher ECNRR performances in both neutral (0.1 M Na_2SO_4, Figure 5.8f–h) and acidic (0.1 M HCl, Figure 5.8i,j) conditions. For example, in 0.1 M N_2-saturated Na_2SO_4, the highest ammonia yield and Faradaic efficiency of Mo^0/GDY could reach up to 113.4–145.4 $\mu g\,h^{-1}\,mg_{cat}{}^{-1}$ and 15.2–21.0%. There are no obvious variations in the chemical structure and valence of Mo^0/GDY after ECNRR tests, suggesting the robust stability of its structure. According to the experimental and theoretical results, the well-defined chemical/electronic structure and valence states show that the atomically anchored Mo^0 on GDY can create numerous active sites, fast-reversible Mo–C1 charge transfer, and enlarged

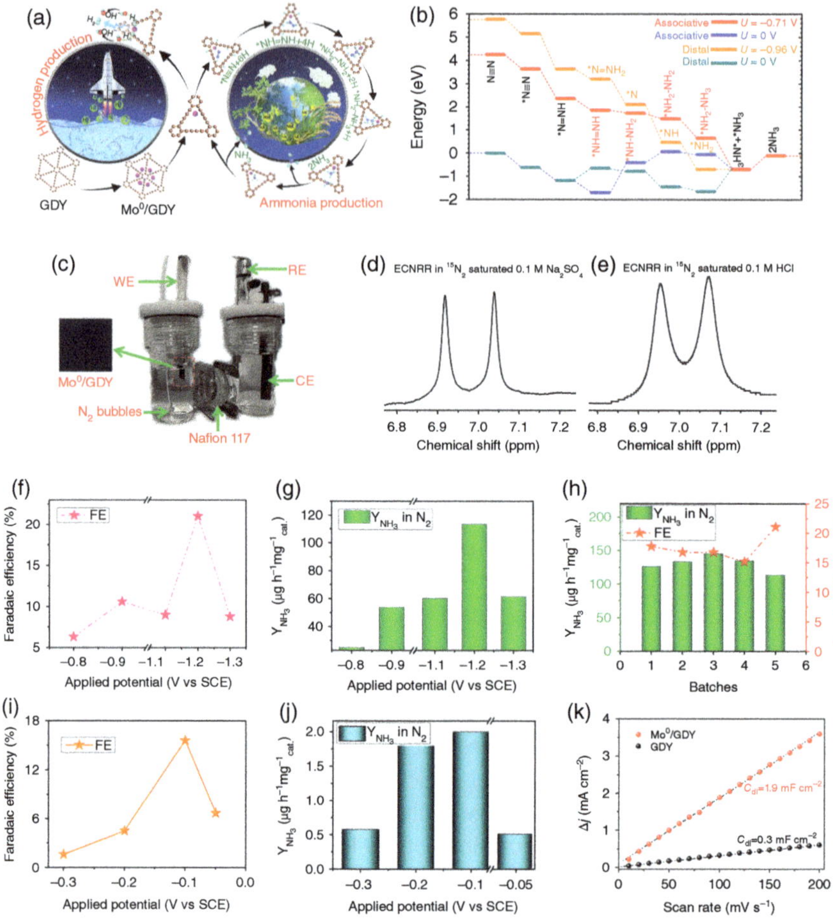

Figure 5.8 (a) Synthesis and structural configuration evolution of catalysis process. (b) ECNRR energetic pathway for Mo^0/GDY. (c) Optical photograph of electrolyte cell for ECNRR. (d, e) Isotope labeling experiments. ^1H NMR (nuclear magnetic resonance spectroscopy) spectra of ammonium obtained after ECNRR in the presence of $^{15}N_2$ saturated (d) 0.1 M Na_2SO_4 and (e) 0.1 M HCl electrolytes. (f) FEs and (g) Y_{NH_3} at different applied potentials in 0.1 M Na_2SO_4. (h) Y_{NH_3} and FEs of NH_3 production of different batches of Mo^0/GDY samples in 0.1 M Na_2SO_4. (i) FEs and (j) Y_{NH_3} at different applied potentials in 0.1 M HCl. (k) C_{dl} measurement of Mo^0/GDY and GDY. Source: Hui et al. [9]. © 2019, American Chemical Society.

electrochemical active surface area ensure such high catalytic performance of Mo^0/GDY.

Exploring new catalysts for ammonia synthesis via nitrogen reduction at ambient pressures and temperatures with ultrahigh ammonia (NH_3) yield and selectivity is still a huge challenge. Recently, by utilizing the unique properties of GDY, Prof. Li proposed a new metal ion anchoring–electron transfer–self-reduction strategy (Figure 5.9a) for anchoring zero-valent palladium atoms [11]. A free-standing

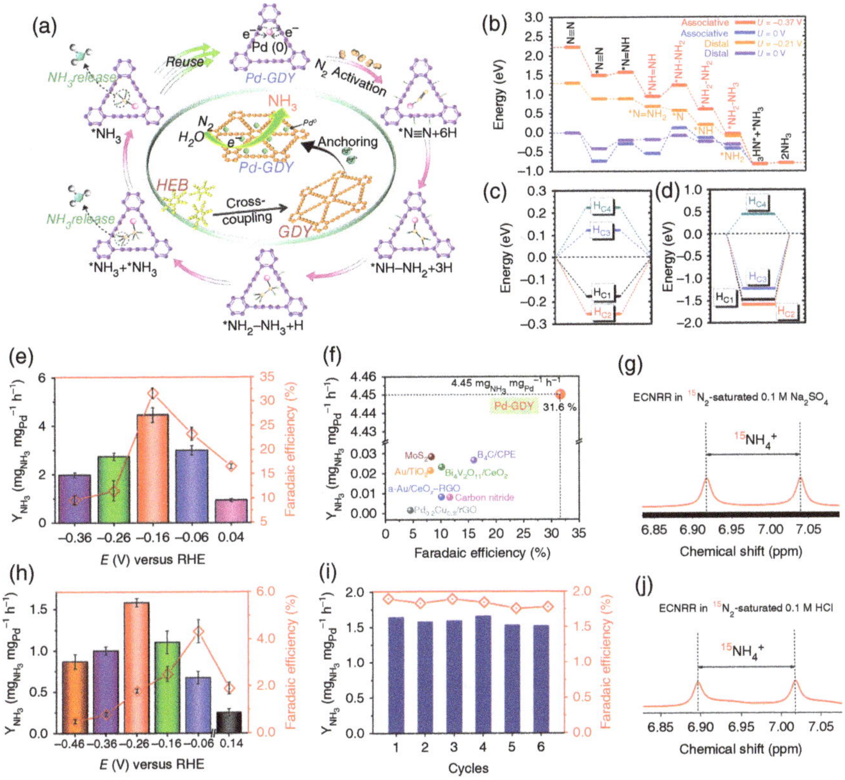

Figure 5.9 (a) Schematic of the synthesis (central green circle) and resuability of the Pd–GDY electrocatalyst for the ammonia production. (b) ECNRR energetic pathway on the GDY–Pd. (c) Formation energies of H-adsorption on C-sites of GDY–Pd. (d) H-chemisorption energies on C-sites of GDY–Pd. (e) Y_{NH_3} and FEs at applied potentials in 0.1 M Na_2SO_4. Error bars represent calculated standard deviation from independent experiments (at least three times). (f) Comparison of the ECNRR performance of Pd-GDY with others. Error bars represent calculated standard deviation from independent experiments (at least three times). (g) ^1H-^{15}N NMR spectra of 0.1 M Na_2SO_4 after ECNRR under $^{15}N_2$ with Pd-GDY as catalyst. (h) Y_{NH_3} and FEs at applied potentials in 0.1 M HCl. (i) Stability test of Pd-GDY at −0.26 V vs. RHE under ambient conditions. (j) ^1H-^{15}N NMR spectra of 0.1 M HCl after ECNRR under $^{15}N_2$ with Pd-GDY as catalyst. Source: Yu et al. [11]. © 2020, Oxford University Press/CC License 4.0.

3D zero-valent atomic catalyst electrode was fabricated and showed high performance in electrocatalysis conversion of nitrogen to ammonia reaction at ambient conditions. This catalyst shows unique advantages such as determined electronic and chemical structure, determined active sites, clear reaction mechanism and process, and excellent properties and performance. Density functional theory (DFT) calculation reveals that strong orbital interaction between Pd and neighboring C sites induces a close shell correlation effect that leads to an equivalent point for repulsion and electronegative activity (ENA) energetic competition. With pinning of 2p-band from (C1, C2) site, the ENA has been more promoted by the negative correlation to overcome the orbital repulsion. This leads GDY–Pd to preserve

Pd^0 state and possesses an even higher ENA than the fcc-Pd^0, which dominates the strong electronegative reduction character for nitrogen reduction reaction (NRR). The downshifted s-band from the electronic structure arises from the elimination of nonbonding lone-pair N_2-2s orbitals through N-hydrogenation, which electronically suppresses the HER. Experimentally, in neutral conditions, the Pd-GDY catalyst has the highest ammonia yield of $4.45 \pm 0.30\ mg_{NH_3}\ mg_{Pd}{}^{-1}\ h^{-1}$, which is orders of magnitude higher than the reported materials, such as Au/TiO_2 ($21.4\ \mu g_{NH_3}\ mg_{cat.}{}^{-1}\ h^{-1}$) [27], $Bi_4V_2O_{11}/CeO_2$ ($23.21\ \mu g_{NH_3}\ mg_{cat.}{}^{-1}\ h^{-1}$) [28], Ru NC ($3.6\ mg_{NH_3}\ mg_{cat.}{}^{-1}\ h^{-1}$) [29], Pd/C ($4.9\ \mu g_{NH_3}\ mg_{cat.}{}^{-1}\ h^{-1}$) [30], and $Pd_{0.2}Cu_{0.8}$/rGO ($2.8\ \mu g_{NH_3}\ mg_{cat.}{}^{-1}\ h^{-1}$) [31], and even compared favorably to those worked at higher temperatures/pressures; in acidic conditions, its catalytic activity can also reach up to $1.58 \pm 0.05\ mg_{NH_3}\ mg_{Pd}{}^{-1}\ h^{-1}$. The ^{15}N isotope-labeling experiments confirmed that the NH_3 was formed from the reduction of N_2, revealing that the Pd-GDY is highly selective and active toward ammonia synthesis. Moreover, both the ammonia production rate and Faradaic efficiency of the catalyst can be maintained for several cycles without decay, confirming its robust stability.

5.1.4.3 Applied for Oxygen Reduction Reaction

The oxygen reduction reaction (ORR), a fundamental step for energy conversion in fuel cells and metal–air batteries, requires favorable catalysts to obtain fast reaction kinetics for practical applications. Single-atom catalysts show excellent superiority for maximizing the used efficiency of metal atoms. However, the major challenge in the synthetic process is finding an appropriate and efficient substrate to anchor the well-defined and highly dispersed single atoms. As described in Section 5.1, GDY is a good substrate for anchoring single atoms to facilitate the mass-transport properties of ORR-relevant species and promote the electron-transfer kinetics of ORR. Recently, Wu and coworkers [32] successfully prepared a GDY-based Fe atom catalyst (Fe–GDY with a mass loading of 0.63%) through in situ reduction of Fe^{3+} on the surface of GDY by utilizing $NaBH_4$. The experimental results showed that Fe–GDY could promote the $4e^-$ pathway of ORR while limiting the $2e^-$ pathway, which was quite consistent with theoretical predictions. In basic condition (0.1 M KOH), the onset potential ($U_{onset} = 0.21\ V$), half-wave potential ($U_{1/2} = 0.10\ V$), kinetic current density ($i_k = 6.70\ mA\ cm^{-2}$ under 0.1 V), and rate constant ($k = 1.47 \times 10^{-2}\ cm\ s^{-1}$) of Fe–GDY are close to commercial Pt/C. Simultaneously, the half-wave potential and rate constant of Fe–GDY after 5000-cycle accelerated stability test (ADT) hardly exhibited variations in comparison with the initial values. Accordingly, for the commercial Pt/C, these parameters showed obvious attenuation, fully demonstrating the excellent long-term stability of Fe–GDY. This study reveals the unique advantages of GDY-based metal atom catalysts in the rational design and preparation of novel highly active ORR catalysts.

5.1.4.4 Applied for Organic Reactions

Inspired by the excellent catalytic performance of GDY-based zero-valent metal atom catalyst, Liu and coworkers [15] prepared the atomic Pd catalysts through anchoring Pd atoms on graphdiyne/graphene heterostructure (GDY/G) by a vdW

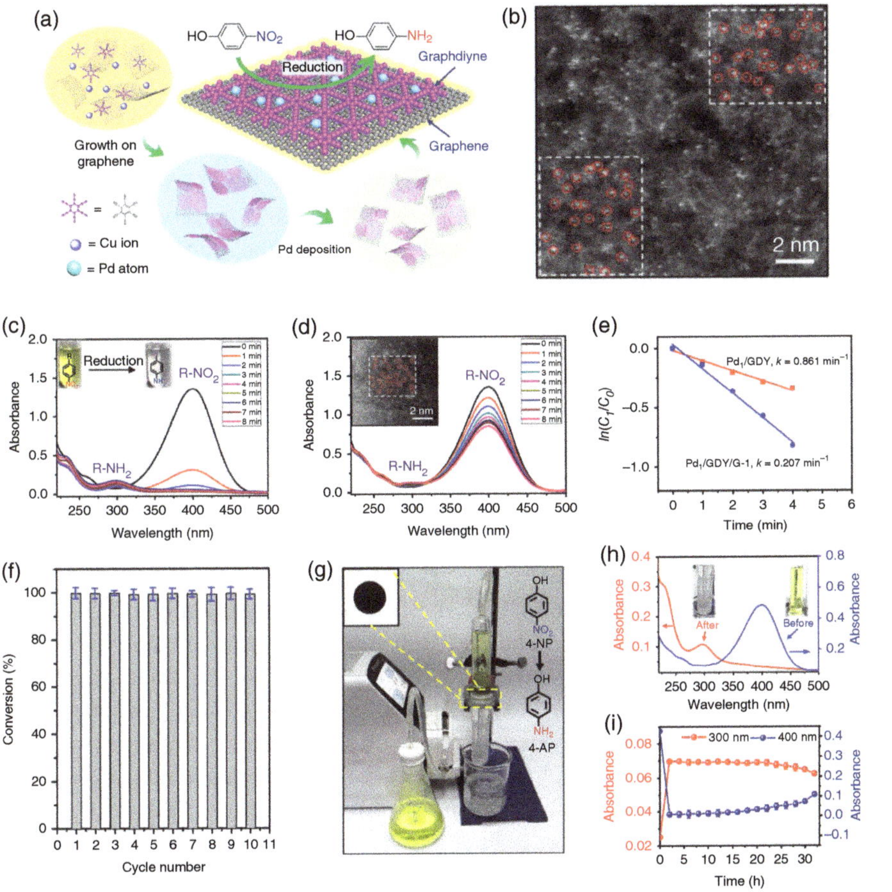

Figure 5.10 (a) Schematic illustration of the synthesis of GDY/G heterostructure through solution-based van der Waals epitaxy, preparation of Pd1/GDY/G, and catalytic process of 4-nitrophenol reduction. (b) Atomic-resolution HAADF–STEM image of Pd_1/GDY/G–O. Time-dependent UV–vis absorption spectra recorded during the 4-NP reduction catalyzed by (c) Pd1/GDY/G and (d) Pd1/GDY. Corresponding atomic-resolution HAADF–STEM image of Pd1/GDY is shown in the inset. (e) Plots of ln (C_t/C_0) as a function of the reaction time for the reduction of 4-NP catalyzed by Pd1/GDY/G-1 and Pd1/GDY. (f) Catalytic performance of Pd1/GDY/G after ten catalytic cycles. (g) Images of the setup of continuous flow to evaluate the catalyst performance in the 4-NP reaction. (h) UV–vis absorption spectra recorded before (blue curve) and after (red curve) reduction in the continuous mode. (i) The absorbances of the lower solution at 400 and 300 nm after reduction as a function of the reaction time. Source: Li et al. [15]. © 2019, Wiley-VCH Verlag GmbH & Co. KGaA.

epitaxy method in the solution phase (Figure 5.10a). The Pd loading is 0.855 wt%. In the presence of $NaBH_4$, Pd_1/GDY/G could achieve highly active and selective catalysis of the reaction from 4-nitrophenol (4-NP) to 4-aminophenol (4-AP), with the rate constant of 0.953 min^{-1}, which was 44 times than that of Pd/C. The TOF was 1762.17 min^{-1}. It was found that the conversion rate of 4-NP still had an over 99% retention after 10-time repeated cycles, suggesting the excellent long-time stability of Pd_1/GDY/G. In addition, DFT calculations showed that graphene in the GDY/G heterostructure played a key role in improving the catalytic efficiency of the electron transfer process. More recently, Lu and coworkers [19] prepared the Pd^s–GDY catalysts by simple wet-chemistry routes. It was observed that the Pd–GDY catalysts showed obvious size-dependent performances on the hydrogenation of phenylacetylene (Figure 5.11a). The Pd^s–GDY exhibited the highest catalytic activity, with TOF up to 6290 h^{-1} at 100% conversion with 99.3% selectivity in hydrogenation of phenylacetylene to styrene, compared to that of GDY-supported Pd nanoparticle catalysts, namely Pd^{NP1}–GDY (with 2-nm Pd NPs) and Pd^{NP2}–GDY (with 12-nm Pd NPs). Their results revealed that Pd^s–GDY had the weakest adsorption to styrene, which is responsible for its high performance. These reports demonstrated the application potential of GDY-based metal atomic catalysts for organic reactions with desirable activity and selectivity.

5.2 GDY-Based Heterojunction Catalysts

A homogeneous catalyst, which has defined active sites, generally exhibits high activity and selectivity in the catalysis. However, homogenous catalysts always contain noble metal and are difficult to be recycled, limiting their large-scale application. It is therefore of great significance for the design and development of new catalytic materials to effectively overcome these problems in homogeneous catalysts. Heterogeneous catalysts have been widely applied in practical uses because of their easy fabrication, high stability, and richness in different catalytically active sites. But their reaction selectivity and activity are relatively low. Exploring the catalysts that feature the advantages of high selectivity, high activity, and long-term stability is thus of great demand.

Carbon materials have emerged as the promising candidates due to their tunable molecular structures, abundance, and strong tolerance to acid/alkaline environments [33–35]. Among all reported carbon materials, GDY features many unique chemical and electronic structures and fascinating physicochemical properties, for example, sp and sp^2-hybridized two-dimensional network structure, richness in carbon chemical bonds, natural hole structure and band gap, and highly intrinsic activity induced by uneven distributed surface charge. Also, GDY can be controlled grown on arbitrary substrate surface at low temperatures and ambient pressures. Such superiorities of GDY have changed the monohybrids, and the harsh synthetic conditions (e.g. high temperature and high pressures) of traditional carbon materials and show huge advantages and advancements in many aspects, including the growth, assembly, and performance modulation; forming new materials with definite chemical and electronic structures; and initiating new phenomenon and new knowledge for materials research. Beneficial from the unique properties, GDY is expected to realize the combination of the advantages of homogeneous and heterogeneous catalysts, achieving new catalysts with high selectivity, activity, long-term stability, and production yields.

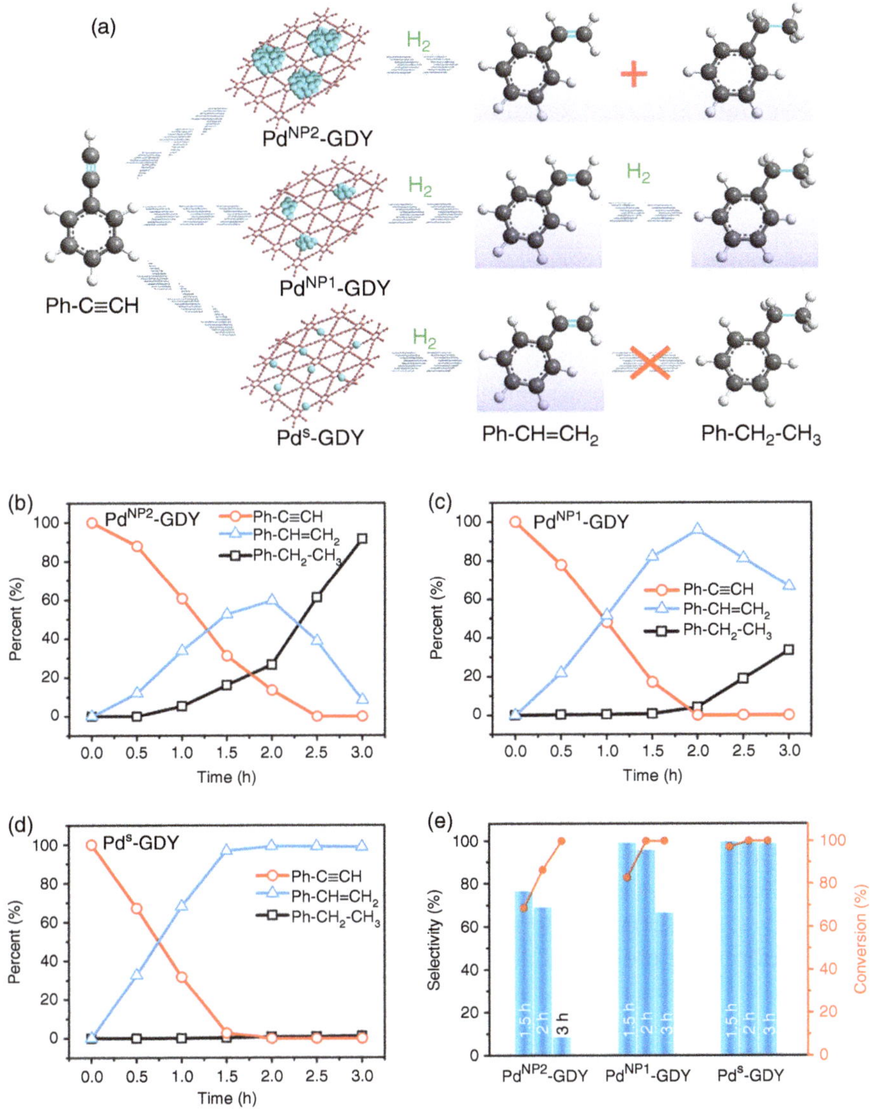

Figure 5.11 (a) Illustration showing different hydrogenation processes of phenylacetylene over different Pd–GDY catalysts. (b–d) The conversion of phenylacetylene and the selectivity toward styrene and phenylethane as a function of reaction time over Pd^{NP2}–GDY (b), Pd^{NP1}–GDY (c), and Pd^{s}–GDY (d). (e) The comparison of catalytic performances for Pd^{s}–GDY, Pd^{NP1}–GDY, and Pd^{NP2}–GDY catalysts. Reaction conditions: 2 mg catalyst, 1.0 mmol phenylacetylene, 2 ml ethanol, 25 °C, 0.2 MPa H_2. Source: Yin et al. [19]. © 2020, Royal Society of Chemistry.

Recent research results have challenged drawbacks of conventional catalysts. GDY has played an important role in the field of catalysis and contributed new ideas to solve the problems in the energy conversion process, and provided a fresh solution for the construction of catalytic system with excellent performances.

5.2.1 Hydrogen Evolution Reaction on GDY-Based Heteros

Li and coworkers for the first time reported a novel protocol for the rational design and controllable preparation of self-supported core–shell nanowires array electrode consisting of GDY as the shell and Cu as the core on Cu foams (Cu@GD NA/CF) through a simple chemical synthesis strategy (Figure 5.12a), including the first growth Cu nanowires on CF, followed by the self-catalyzed growth of GDY films outside the surface of Cu nanowires via Glaser coupling reaction using hexaethynylbenzene (HEB) as the precursor (Figure 5.12b) [36]. In 0.5 M H_2SO_4, Cu@GD NA/CF shows a small-onset overpotential (η) of 52 mV and a Tafel slope of 69 mV dec^{-1}, and requires only 79 mV to reach the current density (j) of 10 mA cm^{-2}. Besides, it has an exchange j of 0.81 mA cm^{-2}, further demonstrating its high intrinsic catalytic activity (Figure 5.12b). This is the first GDY-based electrocatalyst for the highly efficient HER process with excellent catalytic performance and durability. Experimental results showed that GDY could interact with zero-valent copper atoms (Cu^0) to form catalytic active centers, resulting in unique electronic structure and conductivity, which could significantly enhance the charge transfer ability. DFT calculations showed that electrons transferred from Cu enrich the electron density on the carbon atoms near the Cu atoms, resulting in a more favorable proton affinity of the GDY shell. Benefiting from the efficient electron transfer from Cu to carbon atoms in GDY, the catalysts would rapidly promote the adsorbed H* species to final molecular hydrogen. These findings clearly showed that the synergistic effect between metal atoms and GDY could efficiently enhance catalytic activity for HER.

Water splitting includes two half-reactions of HER and OER. At present, OER catalysts usually show excellent performance in alkaline environments. To enhance the integral performance of the overall water-splitting reaction, one of the challenges in this field is to develop an efficient HER catalyst that can be active and stable over a wide pH range. For this, Prof. Li's group has developed a universal method for in situ assembly and growth of heterojunctions on the surface of GDY, and also a series of GDY-based heterojunctions with excellent catalytic performances. In the research work of Yu et al. [37], a film of ultrathin molybdenum sulfide (MoS_2, MDS) nanosheets was in situ grown on the surface of the GDY layer forming a MoS_2–GDY heterojunction electrode (denoted as eGDY/MDS, Figure 5.13a–e). This is the first effort to integrate transitional metal molybdenum disulfide (TMDS) with eGDY to form a highly active interface with definite chemical and electronic structures. Experimental (Figure 5.13f–h) and DFT (Figure 5.13i) results all clearly revealed the electronic interactions occurring at the interfaces between eGDY and MDS, resulting in the facilitated electron transfer kinetics that could lead to an optimal ΔG_H close to thermoneutral, which finally caused the efficient dissociation of the

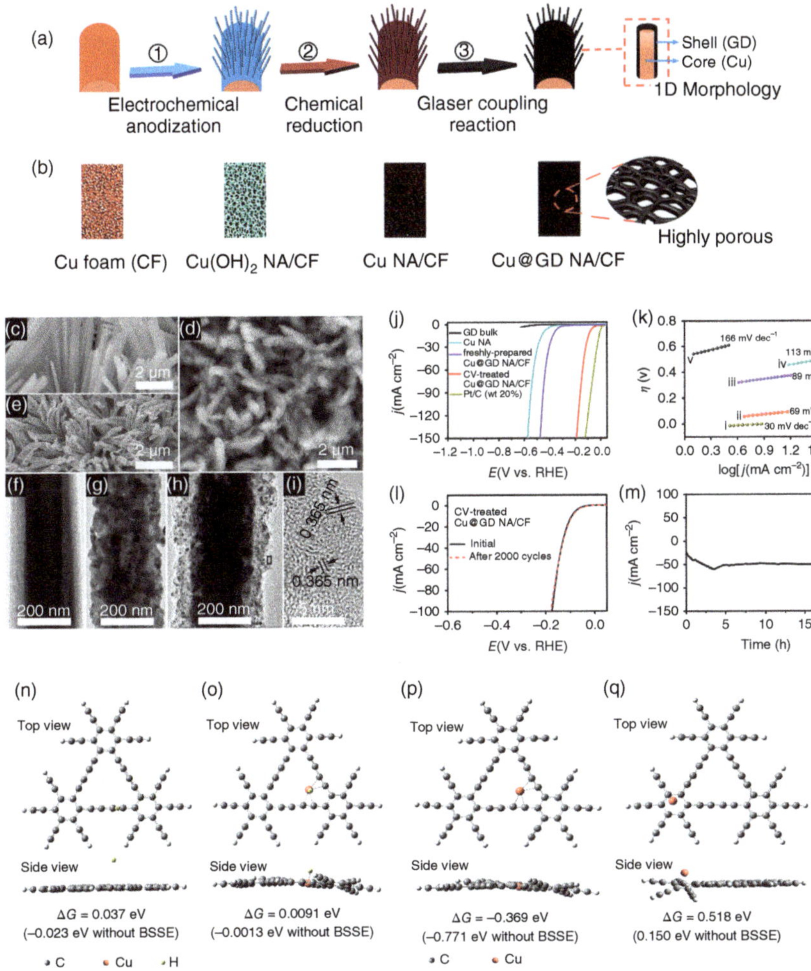

Figure 5.12 (a) Fabrication process of the Cu@GD NA/CF, including enlarged model of a single Cu@GD nanowire. (b) Optical photograph of Cu foam (CF), $Cu(OH)_2$ NA/CF, Cu NA/CF, and Cu@GD NA/CF (from left to right). SEM images of (c) $Cu(OH)_2$ NA/CF, (d) Cu NA/CF, and (e) Cu@GD NA/CF. TEM images of single (f) $Cu(OH)_2$ (g) Cu, and (h) Cu@GD nanowires. (i) high resolution transmission electron microscopy (HRTEM) image taken from the area marked with a black rectangle in (h). (j) The HER polarization curves and (k) corresponding Tafel plots of as-prepared catalysts at a sweep rate of $5\,mV\,s^{-1}$. (l) the HER polarization curves of CV-treated Cu@GD NA/CF before and after 2000 CV scanning between 0 and 0.7 V. (m) time-dependent current density curve obtained at an overpotential of 239 mV for almost 20 hours in 0.5 M H_2SO_4 solution. Sample labels are: i, 20 wt% Pt/C; ii, CV-treated Cu@GD NA/CF; iii, freshly prepared Cu@GD NA/CF; iv, CuNA; and v, GD bulk. Optimized adsorption structures along with the binding energies of H on the (n) pristine and (o) Cu-doped graphdiynes. (p) The Cu atom interacts with the acetylenic bonds (configuration A). (q) The Cu atom is located above the phenyl ring (configuration B). Source: Xue et al. [36]. © 2016, Elsevier.

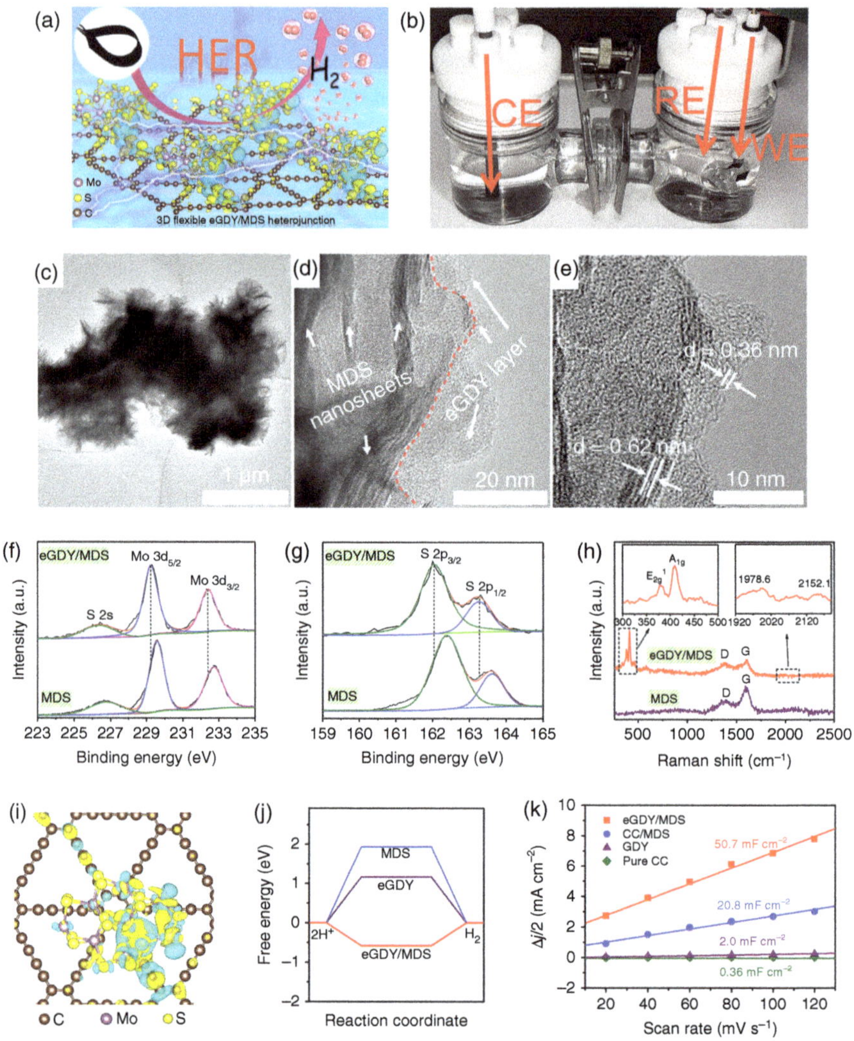

Figure 5.13 (a) Schematic illustration of the hydrogen production on eGDY/MDS. (b) The electrolysis cell (two chambers) using eGDY/MDS as cathode under working conditions. (c) Low- and (d) high-magnified TEM images of eGDY/MDS. (e) HRTEM images of eGDY/MDS. Polarization curves (a) and corresponding Tafel slopes (b) of samples obtained in 1.0 M KOH. (f) Mo 3d and (g) S 2p of eGDY/MDS and MDS. (h) Raman spectra of eGDY/MDS (red line) and GDY (violet line) (insets: enlargements of the selected regions). (i) Charge density difference maps for eGDY/MDS. (j) Hydrogen adsorption-free energy (ΔGH) on eGDY/MDS (red line), eGDY (violet line), and MDS (blue line). (k) The capacitive currents at 0.70 V vs. RHE plotted against the scan rates (20, 40, 60, 80, 100, and 120 mV s^{-1}) for eGDY/MDS, CC/MDS, GDY, and CC, respectively. Source: Yu et al. [37]. © 2018, Wiley-VCH Verlag GmbH & Co. KGaA.

O—H bonds of the H_2O molecule. Interestingly, the eGDY/MDS displayed metallic conductive property, in sharp contrast to the semiconducting nature of MDS and eGDY. In addition, the incorporation of GDY and MoS_2 can effectively increase the electrochemical active surface area (Figure 5.13l). These unique advantages of eGDY/MDS heterojunction make it a highly active and stable HER catalyst from acidic conditions to alkaline conditions, showing small overpotential values (128 mV in 0.5 M H_2SO_4 and 99 mV in 1.0 M KOH) and Tafel slopes (46 mV dec^{-1} in 0.5 M H_2SO_4 and 89 mV dec^{-1} in 1.0 M KOH) at 10 mA cm^{-2}. Especially, the catalytic activity of eGDY/MDS in the alkaline media was even superior to that of Pt/C.

The above-mentioned controllable synthesis method of GDY-supported heterojunction has high universality and has been successfully extended to the preparation of other GDY-supported composite materials, such as MoS_2/nitrogen-doped graphdiyne (NGDY) [38], GDY-MoS_2 NS/CF [39], and WS_2/GDY [40].

However, although above-mentioned GDY-supported HER electrocatalysts exhibit high catalytic activity over a wide pH range from acidic to alkaline, such supported catalysts would be corroded gradually during the catalysis, leading to the changes in structure, morphology, and the reduction of service life. As one of the unique properties for GDY, it is the only one carbon material that can grow on any substrate surface at a low temperature and in a controlled manner. Therefore, it can be easily and effectively grown on the surface of the catalysts, protecting the catalysts from corrosion and enhancing the long-term stability. Moreover, GDY has high intrinsic activity, abundant number of active sites, and large surface area, which are all beneficial to maximizing the catalytic activity of the catalyst.

Based on these considerations, Li and coworkers [41], for the first time, showed the in situ exfoliation and modification of bulk iron–cobalt layered double hydroxides (LDH) nanosheets through a GDY-induced intercalation/exfoliation/decoration strategy (e-ICLDH@GDY/NF) as illustrated in Figure 5.14a. Typically, the self-supported iron–cobalt LDH (ICLDH) nanosheet array was first grown on nickel foam (NF) through the hydrothermal treatment, followed by a cross-coupling reaction. During this process, the HER molecules could enter the LDH interlayer channel and exchange with the anions in the interlayer of ICLDH. Hydrogen bond could then be formed between HEB and LDH structure, and the highly ordered self-assembly occurred on the surface of LDH layer. Afterward, the GDY layer is formed through coupling reaction in the confined state. Owing to the flexibility of both GDY and LDH nanosheets, the stress/strain caused by the close contact of GDY and LDH further expanded the layer spacing, achieving the complete exfoliation of the bulk LDH (~40–80 nm in thickness, Figure 5.14b–e) to obtain a sandwich structure catalyst with a thickness of around 7.0 nm (Figure 5.14f–i) and more wrinkled than the pristine ICLDH nanosheets. Experimental results revealed that the formation of a heterojunction structure between GDY–LDH can significantly reduce the resistance of solution and charge transfer, improve the charge transfer ability, and increase the electrochemical active area, leading to high catalytic activity. As a result, the e-ICLDH@GDY/NF structure displays greatly enhanced electrocatalytic activities and stabilities for the HER in 1.0 M KOH, requiring only

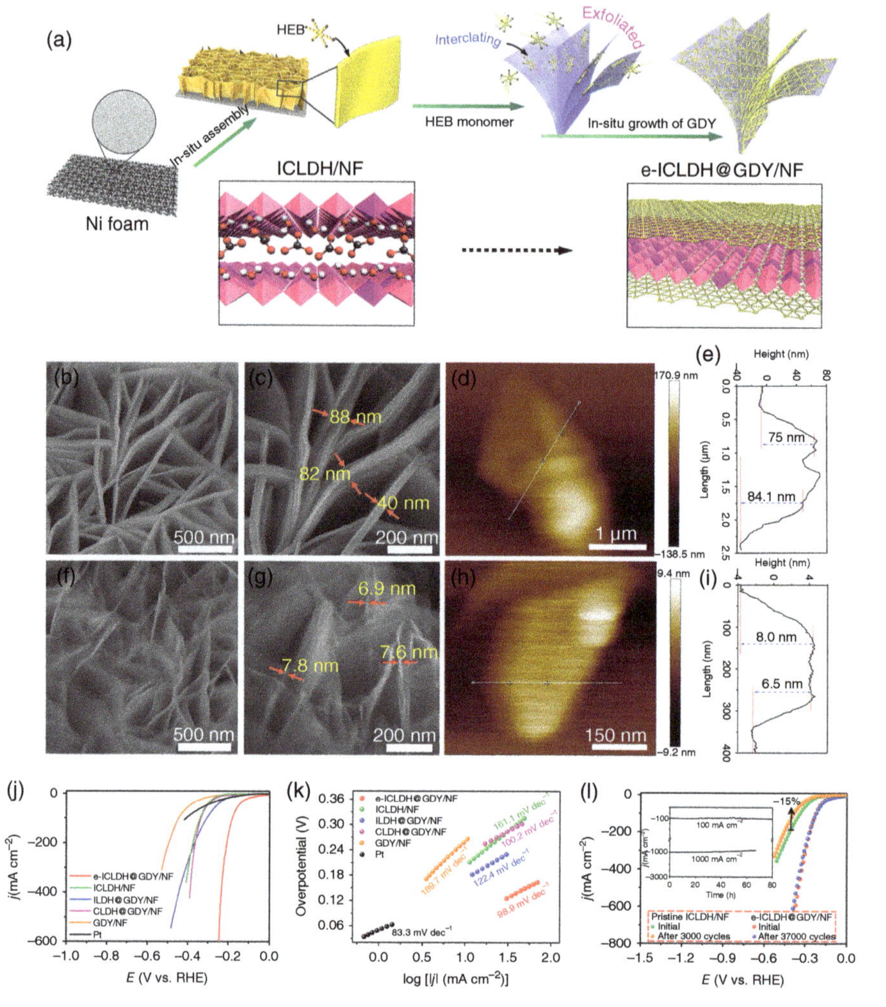

Figure 5.14 (a) Schematic representation of the synthetic strategy for the preparation of e-ICLDH@GDY/NF structures. SEM images of (b, c) pristine ICLDH/NF and (d, e) e-LDH@GDY/NF (recorded at different areas). Atomic force microscope (AFM) images and height profiles of (f, g) ICLDH nanosheets and (h, i) e-LDH@GDY nanosheet, respectively. (j) HER polarization curves and (k) corresponding Tafel plots for the as-synthesized samples. (l) HER polarization curves of e-ICLDH@GDY/NF and pristine ICLDH/NF, recorded before and after 37 000 and 3000 cycles, respectively, of the HER. Inset of (l): Chronopotentiometric curves of e-ICLDH@GDY/NF recorded at 100 and 1000 mA cm^{-2} for both the OER and HER. Source: Hui et al. [41]. © 2018, Springer Nature.

43, 174, 215, and 256 mV at 10, 50, 100, and 1000 mA cm^{-2}, respectively, which are better than those of the pristine ICLDH/NF (202 mV at 10 mA cm^{-2}) and Pt/C. At the overpotential of 200 mV, the TOF value is as high as 8.44 s^{-1}. Moreover, the HER activity of e-ICLDH@GDY/NF exhibits no decay after 37 000 continuous cyclic voltammetry (CV) cycles, far exceeding pure FeCoLDH (35% attenuation after 3000 CV cycles). The research results confirm the GDY can not only greatly enhance the catalytic activity of the catalysts, but also effectively improve the stability of the catalyst. This offers new insights for designing and synthesizing novel non-noble metal-based electrocatalysts with high catalytic activity and high stability. This method has a strong universality. During subsequent exploration and expansion, a series of GDY-coated electrocatalysts with superior catalytic properties were reported, such as GDY-encapsulated metal oxide (NiO-GDY NC [42]), nitrides, (CoN_x@GDY [43]), carbonate hydroxide (FeCH@GDY [44]), and sulfides (GDY–MoS_2 NS/CF [39] and GDY/CuS [45]).

In addition to the electrocatalytic hydrogen production, water splitting by sunlight to produce hydrogen is another promising solution to address the problem of environmental issues and energy shortages [46, 47]. GDY, a novel large π-conjugated carbon material, for the first time, was introduced as the hole transfer layer into a photoelectrochemical (PEC) water-splitting cell. Zhang and coworkers [48] prepared a film of the CdSe QDs (quantum dot)/GDY nanosheets through the π–π stacking between mercaptopyridine and GDY (Figure 5.15a,b). The GDY acted as hole transfer layer for hydrogen production in neutral water. Their experimental results showed that the strong π–π interactions between GDY and mercaptopyridine surface-functionalized CdSe QDs facilitated the hole transportation and photocurrent enhancement, which could promote the photogenerated electron-charge separation efficiency and improve the photocatalytic hydrogen evolution performance. As a result, when being exposed under the illumination of Xe lamp, the CdSe QDs/GDY photocathodes exhibited an enhanced photoactivity and nearly $90 \pm 5\%$ faradaic efficiency for hydrogen production in a 12-hour test. Our results are attributed to the higher hole mobility and stability of GDY. A new GDY-based metal-free 2D/2D heterojunction was synthesized by incorporating GDY with g-C_3N_4 (g-C_3N_4/GDY) by Lu and coworkers [49] and used as a photoelectrocatalyst for water splitting (Figure 5.15c–e). They showed that g-C_3N_4/GDY photocathode exhibited enhanced photocarriers separation because of the excellent hole transfer nature of GDY and the structure of g-C_3N_4/GDY. Compared with g-C_3N_4 (88 μs and $-32\,\mu A\,cm^{-2}$), the electron lifetime and photocurrent density for g-C_3N_4/GDY were significantly increased to 610 μs and $-98\,\mu A\,cm^{-2}$, respectively, at 0 V (vs. normal hydrogen electrode [NHE]) in neutral aqueous solution. The 2D/2D heterojunction of g-C_3N_4/GDY was further applied for photocatalytic H_2 production by Si et al. [50]. It was found that only 1% GDY could realize a high H_2 evolution rate of $454.28\,\mu mol\,h^{-1}$. Recently, Lu and coworkers further [51] combined GDY and CdS nanoparticles to obtain a CdS/GDY heterojunction for photocatalytic hydrogen production (Figure 5.16a). Their results show that GDY can not only stabilize CdS, but also effectively transfer photogenerated holes and effectively prevent photogenerated electron–hole recombination. When the GDY content was 2.5 wt%, the catalytic performance of CdS/GDY heterojunction was 2.6 times higher than that of pristine CdS and the catalyst also showed high recyclability (Figure 5.16b,c). By utilizing the advantage that GDY can be grown on any substrate surface, more recently, Li et al. [53] reported a one-step method to construct GDY-coated CuI composite materials (GDY–CuI) by using cuprous iodide as the catalyst carrier to grow a GDY. The obtained catalysts could be directly applied to photocatalytic hydrogen

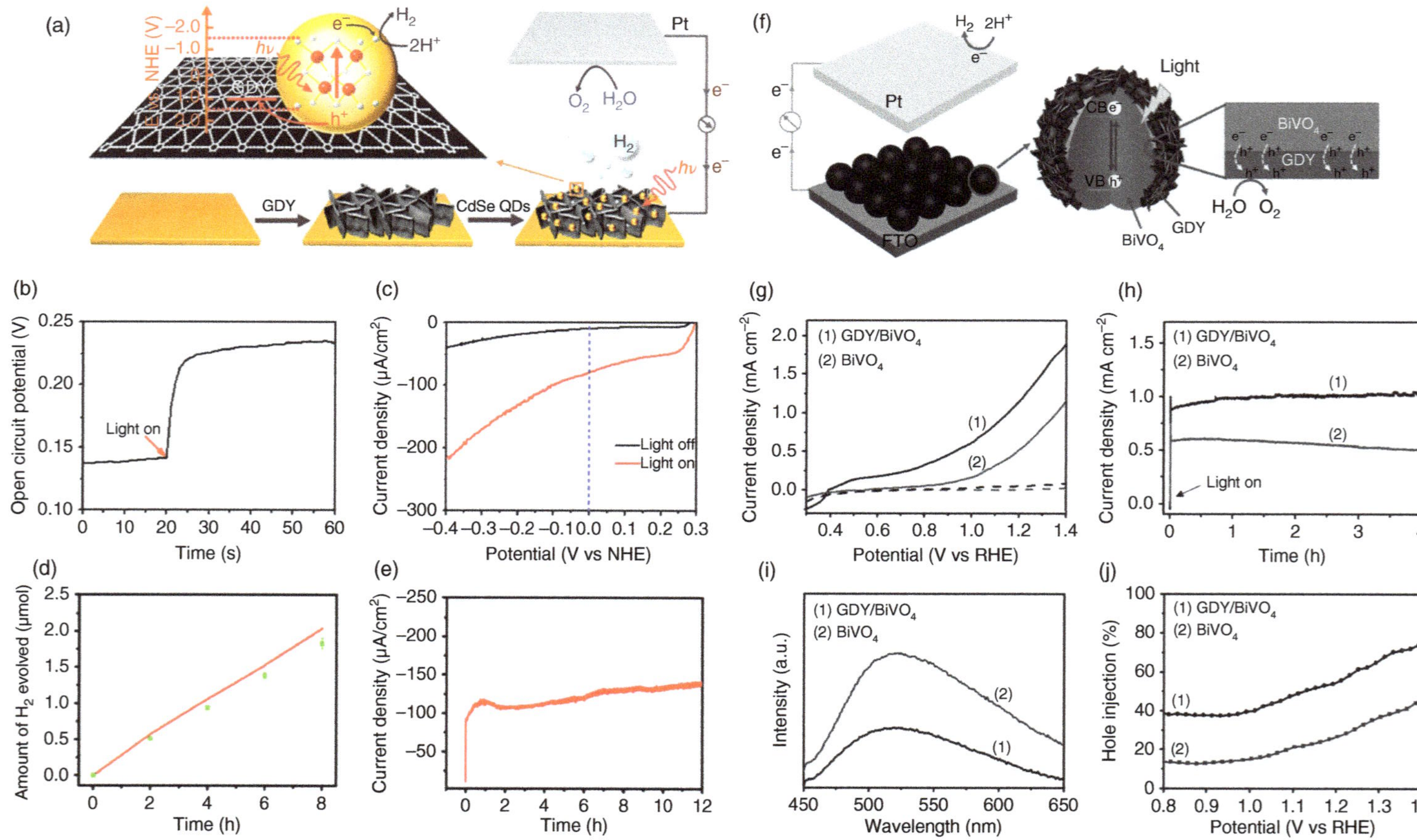

Figure 5.15 (a) Schematic diagram of the PEC cell, consisting of the assembled CdSe QDs/GDY photocathode, Pt wire as counter electrode, and corresponding interfacial migration process of the photogenerated excitons. (b) Open circuit potential response of the CdSe QDs/GDY photocathode under dark and illuminated conditions (300 W Xe lamp). (c) LSV scanning from 0.3 to −0.4 V at 2 mV s^{-1} with light off (black trace) and on (red trace) for CdSe QDs/GDY photocathode. (d) Amount of evolved hydrogen and recorded charge carrier during photoelectrolysis. (e) Controlled potential electrolysis of the CdSe QDs/GDY photocathode during 12-hour test. (f) The schematic illustration for the construction of 2D/2D graphitic carbon nitride/graphdiyne heterojunction on 3D GDY nanosheet array. (g) The open circuit potential (OCP) response of g-C_3N_4/GDY photocathode under dark and irradiated conditions. (h) The electrochemical impedance spectra of g-C_3N_4 and g-C_3N_4/GDY photocathodes measured under light. (i) The Bode phase plots of g-C_3N_4 and g-C_3N_4/GDY photocathodes. (j) Linear sweep voltammetry scanning for different photocathodes measured under dark and light. Source: (a–e) Li et al. [48]. © 2016, American Chemical Society, (f–j) Modified from Han et al. [49].

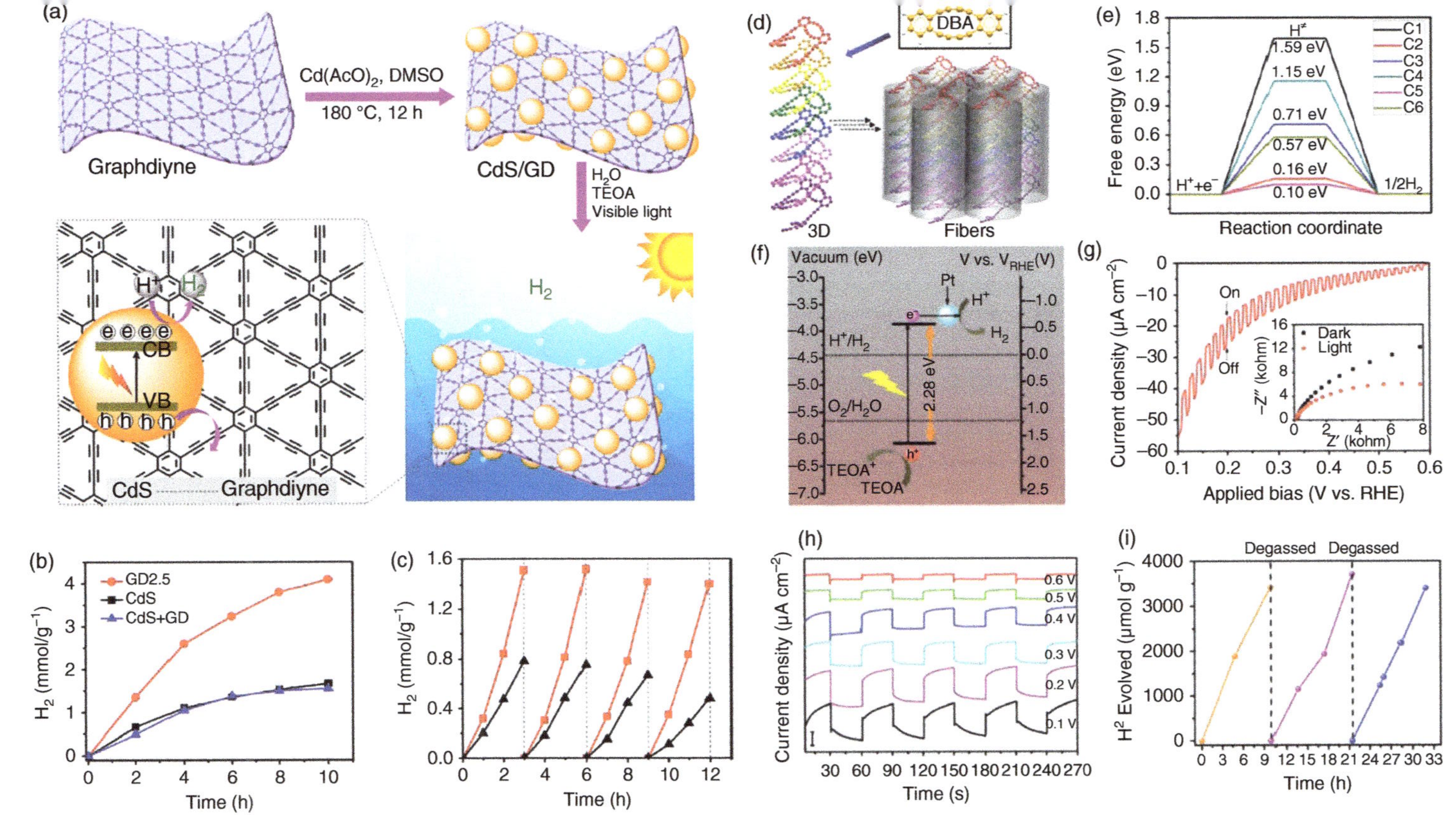

Figure 5.16 (a) Preparation of CdS/GD composite and its photocatalytic process. (b) Photocatalytic evolution of H_2 catalyzed by GDY2.5 (red), CdS (black), and the mixture of CdS and GDY (blue) (97.5/2.5, w/w), respectively. (c) Recycling experiments for H_2 production of GDY2.5 (red) and CdS (black). Light source: 450 nm LED, 200 mW cm^{-2}, illumination area, 0.8 cm^2 in the presence of 2 mg of catalyst and triethanolamine (TEOA) (0.3 M) in 5 ml of aqueous solution at 25 °C. (d) Designed 3D alkyne material based on DBA and its possible aggregate structure; (e) Hydrogen adsorption-free-energy (DGH) at various sites: C1, C2, C3, C4, C5, and C6 denote the different active sites. (f) Band structure diagram of PDBA and the potentials for water splitting, the energy values (in vacuum level) were converted to electrochemical potentials according to a standard reference electrode. (g) Current density–potential curves vs. the bias of PDBA under intermittent illumination. The inset shows electrochemical impedance spectroscopy (EIS) (recorded at 0.50 V vs. RHE) of PDBA under light or in the dark. (h) Transient photocurrents under visible light illumination (l > 420 nm) of PDBA nanofiber films obtained at various bias voltages (vs. RHE) in a 0.2 M Na_2SO_4 aqueous solution, scale bar: 5 mA cm^{-2}. (i) Cycle runs for the photocatalytic H_2 production over PDBA nanofiber suspension under visible light irradiation (l > 420 nm). Source: (a–c) Lv et al. [51]. © 2019, American Chemical Society. (d–i) Shen et al. [52]. © 2020, Royal Society of Chemistry.

production, showing a greatly enhanced photocatalytic hydrogen production activity (465.95 mol/5 hours), which is 15.8 times and 3.0 times that of pure GDY (29.42 mol/5 hours) and CuI (156.49 mol/5 hours), respectively.

Tailoring of the alkyne triple bonds promises effective tuning of the electronic properties, which provides a very wide space for chemical derivation of GDY. These derivations have been expected to promote the development of new kinds of carbon materials with a clear structure, high activity, and stability. Li and coworkers [52] recently reported their efforts on the synthesis and self-assembly of a crystalline dehydrobenzoannulene-based three-dimensional graphdiyne (dehydrobenzoannulene-based three dimensional graphdiyne [PDBA], Figure 5.16d) and its application in photocatalytic hydrogen production. DFT calculations revealed that the C5 site (0.10 eV, carbon atom on the twisted butadiyne of the dehydrobenzoannulene [DBA] unit) and C2 site (0.16 eV, carbon atom on the connecting butadiyne) possessed the higher activity and were favorable for H_2 evolution (Figure 5.16e), confirming the role of alkynes in photocatalysis. As shown in Figure 5.16f, the energy level for the reduction of H_2O to H_2 is below the E_{cb} of PDBA and that for the oxidation of H_2O to O_2 is above the E_{vb} of PDBA. In a three-electrode setup, the PDBA photocathode gave a prompt and reproducible photoresponse under on/off cycles of visible light excitation, and exhibited high ability for H_2 production with a rate of 340 mmol h^{-1} g^{-1} and an apparent quantum efficiency of 4.68% at 420 nm (Figure 5.16g–i).

Experimental and theoretical results reveal that the high catalytic ability can be ascribed to the formation of GDY/semiconductor heterojunction, in which GDY can not only stabilize semiconductor catalysts but also can act as a hole transfer material for efficient electron–hole pairs separation, therefore effectively preventing the recombination of photogenerated electrons and holes.

5.2.2 Oxygen Evolution Reaction on GDY-Based Heterojunction

OER is an important half-reaction of water splitting. While its slow kinetics severely restricts the overall performance and efficiency of the water-splitting process, which brings about a higher overpotential to achieve the desired reaction rate. How to design and achieve the controllable synthesis of high-efficiency and stable OER catalysts with new structure, understand, and interpret the intrinsic catalytic behavior of the material, and finally achieve breakthrough enhancement in electrocatalytic efficiency has always been a crucial issue in this field.

As a pioneering work, Li and coworkers [54] prepared a three-dimensional GDY network as the substrate for the controlled growth of $NiCo_2S_4$ nanowire arrays (Figure 5.17a–i), and obtained the first GDY-based OER catalyst – $NiCo_2S_4$ NW/graphdiyne foam (GDF). Benefiting from the synergistic interaction between $NiCo_2S_4$ and GDY, the $NiCo_2S_4$ NW/GDF showed better OER catalytic performances with much smaller overpotentials at the same current density and Tafel slope than that of pure $NiCo_2S_4$ NW (Figure 5.17j,k). These results confirmed the inconceivable potential of GDY as the key material for high-performance catalysts for OER application. Inspired by this, a number of new GDY-based

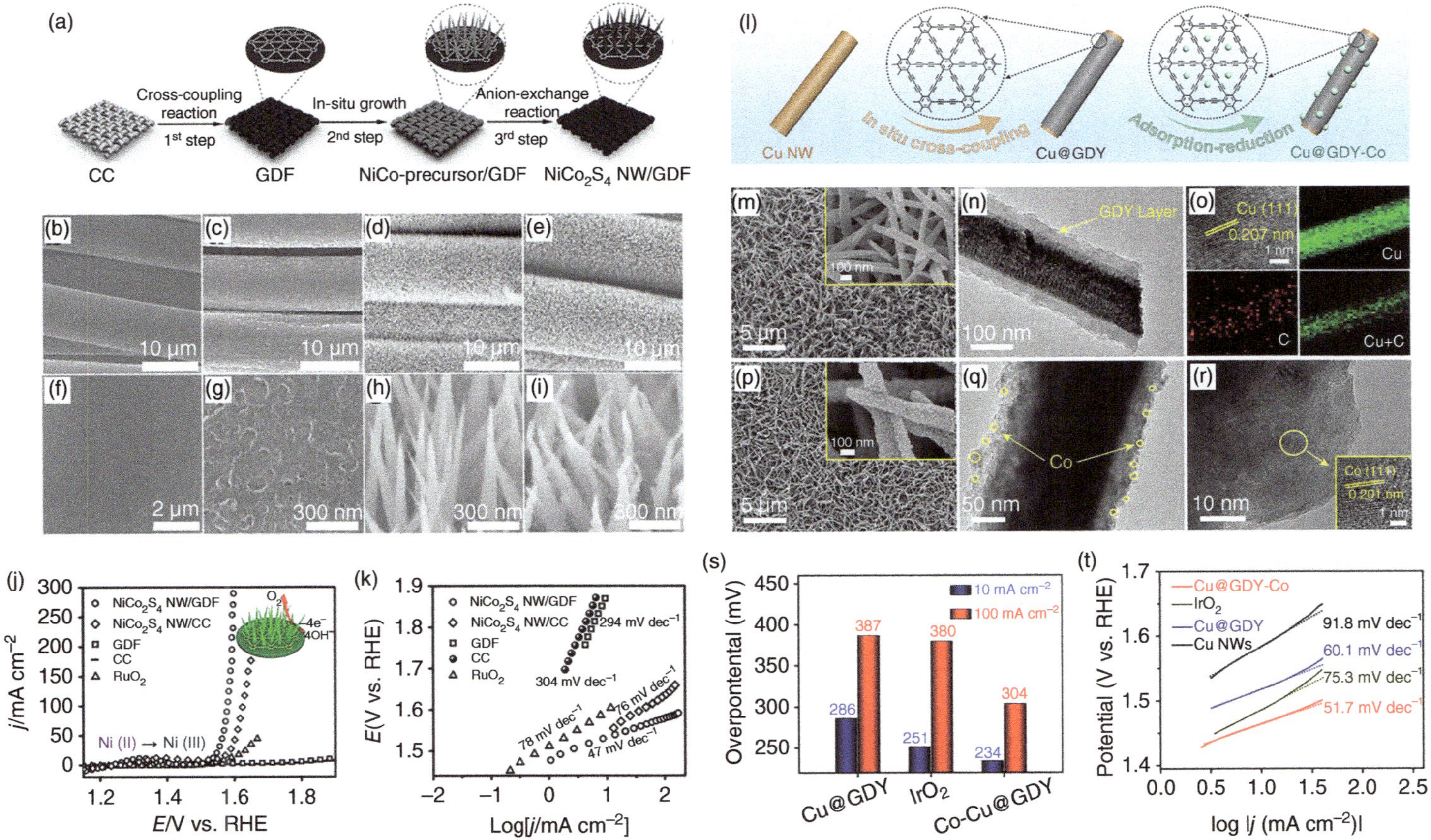

Figure 5.17 (a) Schematic illustration of the fabrication process of the $NiCo_2S_4$ NW/GDF and scanning electron microscope (SEM) images of samples. Low- and high-magnification SEM images of (b and c) CC, (d and e) GDF, (f and g) NiCo-precursor NW/GDF and (h and i) $NiCo_2S_4$ NW/GDF. (j) Polarization curves of for $NiCo_2S_4$ NW/GDF, $NiCo_2S_4$ NW/CC, GDF, CC, and RuO_2 for OER with a scan rate of 5 mV s^{-1}. (k) Corresponding OER Tafel slopes. (l) The schematic illustration for the synthesis strategy of the Cu@GDY-Co. (m) SEM image of Cu@GDY, inset high-resolution image. (n) Low- and (o) high-resolution TEM and corresponding mapping images of Cu@GDY. (p) SEM image of Cu@GDY-Co, inset high-resolution image. (q) Low- and (r) high-resolution TEM image of Cu@G (s) LSV curves, and (t) corresponding Tafel slopes of the samples for OER. Source: (a–k, Xue et al. [54]. © 2017, Wiley-VCH Verlag GmbH & Co. KGaA, (l–t) Shi et al. [55]. © 2020, Elsevier.

electrocatalysts were fabricated for efficient OER processes. Lu and coworkers [56] fabricated Ni–Co–P nanosheets on GDY nanowalls (Ni–Co–P/GDY) for efficient OER process. Moreover, GDY-supported NiFe-layered double hydroxide heterostructures obtained by different synthetic methods, including hydrothermal reaction [57] and electrochemical deposition [58], were reported to have improved OER catalytic activity than pure NiFe-layered double hydroxide. In addition to the above-mentioned GDY-supported metal compounds (e.g. sulfides, nitrides, and hydroxides), GDY-supported metal clusters have also been reported to have high OER performances. Shi et al. [55] anchored Co clusters on electron-rich GDY nanoarrays forming a multidimensional Cu@GDY-Co nanoarray electrocatalyst (Figure 5.17l–r). Both the theoretical and experimental results demonstrated the strong electronic perturbation (p–d charge transfer) that originated from GDY, which played the key role for stabilizing the Co cluster and endowed the catalysts with high electrocatalytic activity requiring overpotentials of 234 and 304 mV to reach the current density of 10 and 100 mA cm^{-2}, respectively. A metalloporphyrin-based GDY analogue (Co-PDY) with a unique p-conjugated structure was innovatively designed and prepared [59], and exhibited excellent OER catalytic performance, including a small overpotential of 270 mV at 10 mA cm^{-2}, a low Tafel slope of 99 mV dec^{-1}, and promising long-term stability and durability.

During the water-splitting process, the reactions take place only at the interface between the catalysts, electrolytes, and gaseous molecules; engineering the interface microenvironment of the catalysts would be beneficial to the identification of the main factors influencing the reaction activity, and facilitating the mass diffusion and electron transfer to promote the catalytic performance. To this end, Zhang and coworkers [60] introduced oxygen-containing groups (–O–, –OH, and –COOH, etc.) to the surface of GDY via air plasma, followed by in situ assembly of the ultrathin CoAl (CO_3^{2-}) LDHs nanosheets on hydrophilic GDY for OER (Figure 5.18). Experiments together with DFT calculations showed that the charge transfer behavior, catalytic activity, and long-time stability were all obviously improved. The impressive OER performance of superhydrophilic CoAl-LDH/GDY could be ascribed to the more negative charge density of hydrophilic GDY surface, the stronger interaction between superhydrophilic GDY and CoAl-LDH, and more sufficient electrolyte–catalysts contact, which are beneficial for improving the loading mass of the catalysts, and the rapid interfacial mass and electron transport [60].

According to the above experimental results, GDY can effectively enhance the catalytic activity of conventional catalysts. However, the materials supported on the surface of GDY would be entirely exposed to electrolytes (strong acid or alkaline solutions) for a long time, which are inclined to be corroded, leading to the destruction of the structure, the reduction of the number of active sites, and therefore the degradation of the catalytic performance. Coating the surface of the catalysts by GDY could be an effective way to enhance the catalytic efficiency and stability. Taking advantage of the characteristic that GDY can grow on arbitrary substrates in a controllable manner, Li and coworkers designed and synthesized GDY-exfoliated and sandwiched iron/cobalt-layered double-hydroxide nanosheet arrays grown on NF (Figure 5.19) [41]. The GDY-exfoliated and -sandwiched LDH

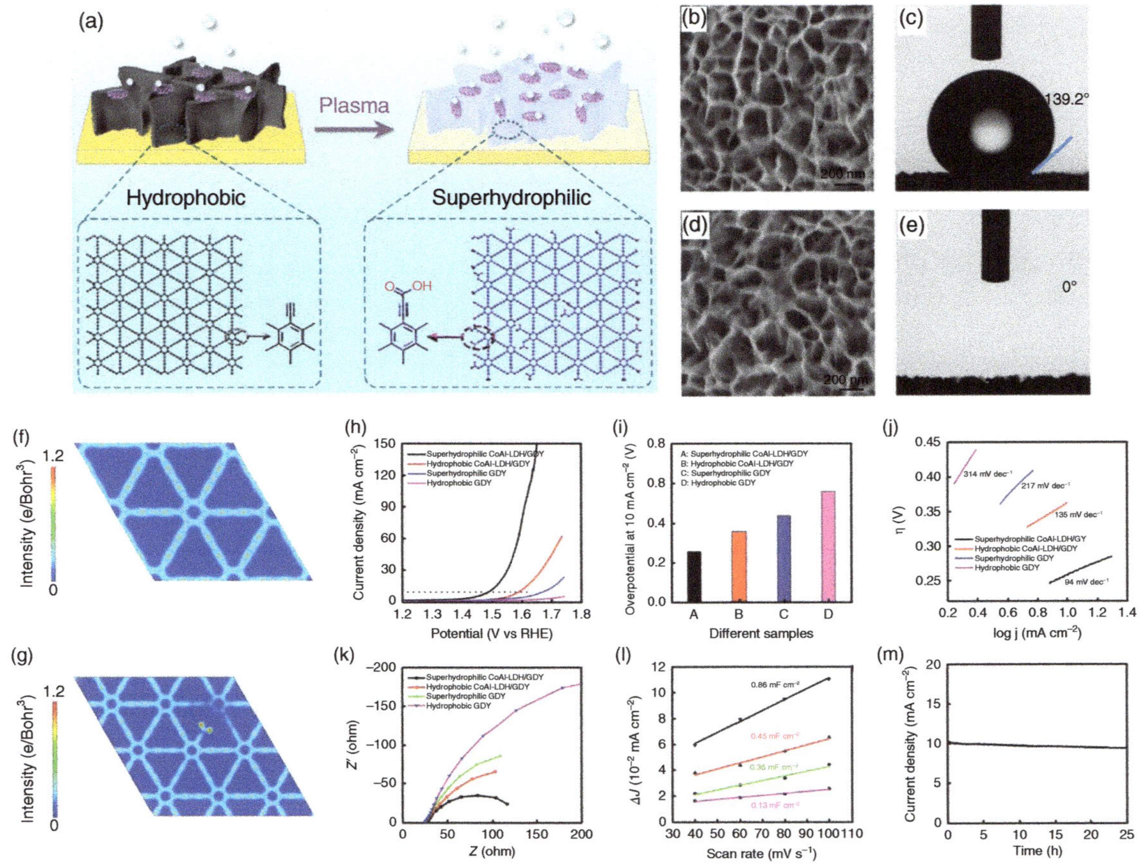

Figure 5.18 (a) Comparison of CoAl-LDH (CO_3^{2-}) assembled hydrophobic and superhydrophilic GDY electrodes. (b,d) SEM images of GDY before and after plasma treatment. (c,e) Static water contact angles of GDY before and after air-plasma treatment. Electron density mapping of (f) hydrophobic GDY and (g) superhydrophilic GDY. (h) LSV curves of hydrophobic GDY, superhydrophilic GDY, hydrophobic CoAl-LDH/GDY, and superhydrophilic CoAl-LDH/GDY electrodes in 0.1 m KOH; (i) required overpotential to reach 10 mA cm^{-2} of the different samples; (j) Tafel plots of the different samples; (k) Nyqyist plots of the corresponding electrodes; (l) double-layer capacitance per geometric; (m) long-time stability test for superhydrophilic CoAl-LDH/GDY electrode. Source: Li et al. [60]. ©2019, John Wiley and Sons.

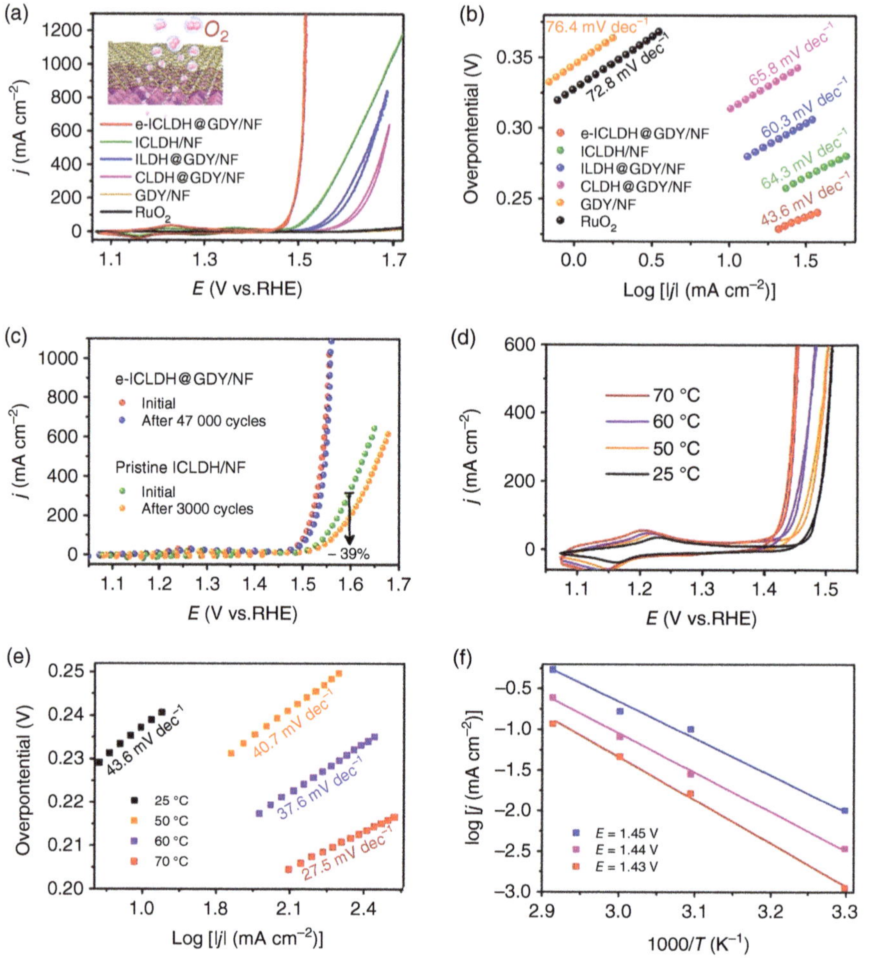

Figure 5.19 Electrocatalytic performance of e-ICLDH@GDY/NF. (a) OER CV curves and (b) corresponding Tafel plots of as-synthesized samples. (c) Polarization curves of e-ICLDH@GDY/NF and pristine ICLDH/NF, recorded before and after performing 47 000 and 3000 cycles, respectively, of the OER. (d) OER CV curves and (e) corresponding Tafel plots for e-ICLDH@GDY/NF, recorded at various temperatures. (f) Arrhenius plots for the OER performed over e-ICLDH@GDY/NF at various potentials. Source: Hui et al. [41]. ©2018, Springer Nature.

exhibited superior OER performance. Under alkaline conditions, the overpotential of eICLDH@GDY/NF at a current density of 10, 50, 100, 500, and 1000 mA cm^{-2} was only 216, 238, 249, 275 and 278 mV, respectively (Figure 5.19a). Impressively, the catalytic activity had almost no degradation after continuous 47 000-cycle CV tests (Figure 5.19c). The relationship between temperature and electrocatalytic activity was examined to determine the kinetic barriers for the OER (Figure 5.19d–f). Arrhenius plots were constructed based on the current densities recorded at different potentials. As the potentials increased from 1.43 to 1.44 to 1.45 V (vs. reversible hydrogen electrode, RHE), the activation energies decreased from 100.9 to 91.52 to 82.33 kJ mol^{-1}, respectively, indicating the potential-determining step of the electrocatalyst. Based on such in situ GDY growth strategy, various GDY encapsulated catalysts were fabricated and applied to water splitting. The GDY-wrapped iron carbonate hydroxide nanosheets array on nickel foam (FeCH@GDY/NF) was first synthesized by Hui et al. [44], and showed excellent OER activity giving an overpotential of 260 mV to drive 10 mA cm^{-2}, outperforming most of the recently reported OER electrocatalysts. The first GDY-based metal nitride OER catalyst was synthesized by Fang et al. [43] by in situ growing GDY layers on the surface of cobalt nitride nanosheets (CoN_x@GDY NS/NF, Figure 5.20a–d). Benefiting from the unique chemical and electronic structures and associated interactions between GDY and CoN_x, the electrocatalyst exhibited greatly enhanced catalytic activity and stability in OER process (Figure 5.20d–f). Besides, Yu et al. [42] fabricated heterostructures of GDY-wrapped NiO nanocube (NiO–GDY NC). The NiO–GDY NC achieved a low overpotential of 278 mV in 1.0 M KOH, at a current density of 10 mA cm^{-2}. The OER activity could be maintained even after 8000-cycle CV tests. Cui et al. [61] fabricated a novel catalyst comprising the 1D/2D nanostructural nanohybrid of CoCu-based zeolitic imidazolate framework nanosheets and GDY nanowires applied in electrochemical oxygen production. The OER activities of above catalysts were all superior to commercial RuO_2.

5.2.3 Photo-/Photoelectrocatalytic Oxygen Evolution Reaction

Apart from applications in electrocatalytic OER, GDY can also be employed as a hole transport layer to construct excellent photo/photoelectrocatalysts for oxygen evolution. For instance, the CoAl–LDH/GDY/$BiVO_4$ composite reported by Zhang and coworkers [60] (Figure 5.18) could be directly used as a photoanode for PEC oxygen evolution and showed enhanced photocurrent density of about 3.15 mA cm^{-2} at the potential of 1.23 V vs. RHE in 0.1 M Na_2SO_4 solution with a three-electrode system under illumination of 100 mW cm^{-2} Xe lamp. Zhang et al. [62] successfully introduced GDY nanowalls to the Si-based PEC system for significantly enhanced water-splitting performance. A SiHJ/GDY/NiO_x heterojunction material was constructed by in situ growth of GDY on the surface of Si substrate followed by controllably plating NiO_x film with a certain thickness on GDY, as confirmed by the scanning electron microscope (SEM) (Figure 5.21a–c), TEM (Figure 5.21d), and Raman (Figure 5.21e) results. As shown in Figure 5.21e, both sides of the n-type single crystalline (280 μm in thickness) were coated by an ultrathin (7 nm in thickness) intrinsic a-Si : H layer, then by an 8-nm-thick n-type a-Si : H layer and 15-nm-thick p-type a-Si : H coating, respectively, on each side. At last, conducive indium tin oxide (ITO) layers were sputtered on both sides, and a silver electrode was also added on the backside. Owing to the superiorities of GDY and NiO_x, the optimized SiHJ/GDY/NiO_x-10 nm exhibited a saturated photocurrent density reaching up to 39.1 mA cm^{-2} (Figure 5.21i–k).

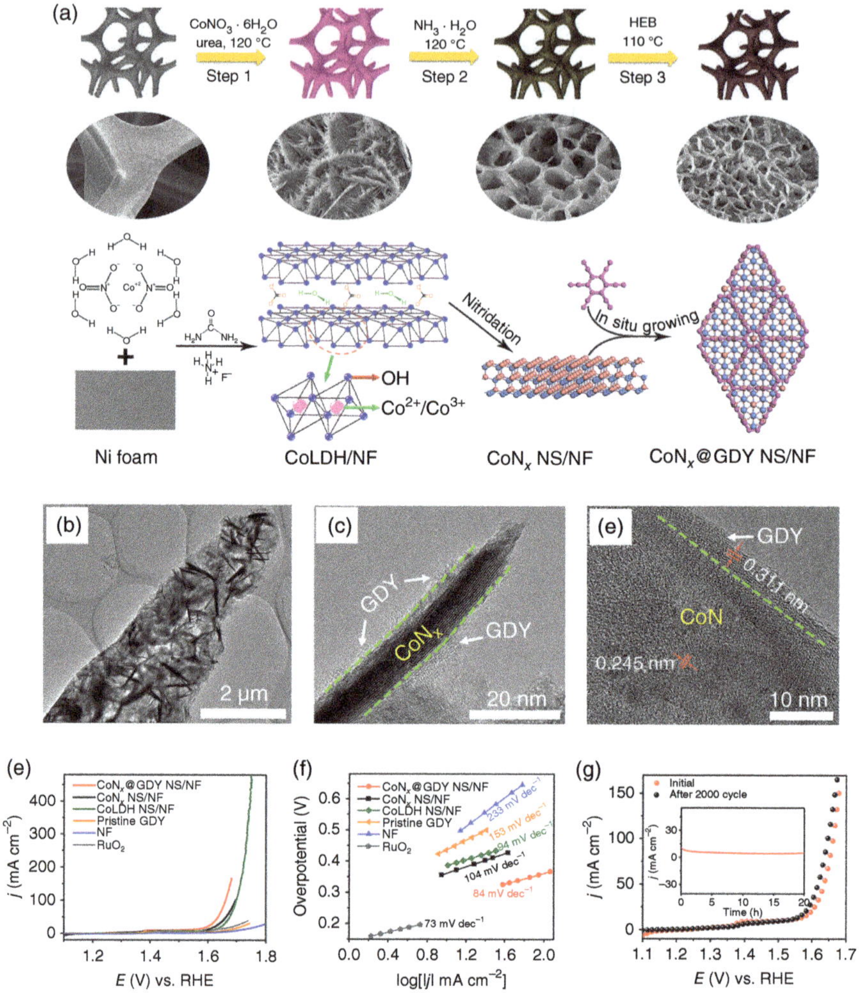

Figure 5.20 (a) Schematic of the synthesis of CoN$_x$@GDY NS/NF via an in-site growth strategy. (b) Low- and (c,d) High-resolution TEM images of CoN$_x$@GDY NS. (e) Polarization curves and (f) Tafel plots of as-prepared catalysts for OER in 1.0 M KOH. (g) Polarization curves of CoN$_x$@GDY NS/NF before and after 2000 cycles (inset: time-dependent current density curve of CoN$_x$@GDY NS/NF at an overpotential of 280 mV vs. RHE for 20 hours. Source: Fang et al. [43]. © 2019, Elsevier.

Ag_3PO_4 (APO) has shown high quantum efficiency and excellent water-splitting activity under visible light illumination [63]. Motivated by the intriguing properties of GDY and Ag_3PO_4, Mao and coworkers [64] reported that hydrophobic GDY was able to form a picking emulsion with hydrophilic Ag_3PO_4 nanoparticles, which can promote the photocatalytic water-splitting process (Figure 5.22A,B). Due to the fascinating properties, such as reduction ability, high stability, hole transfer mediator with its low work function, highly conjugated structure, and high electron mobility of GDY, Ag_3PO_4/GDY-based emulsion possessed a higher apparent rate constant of

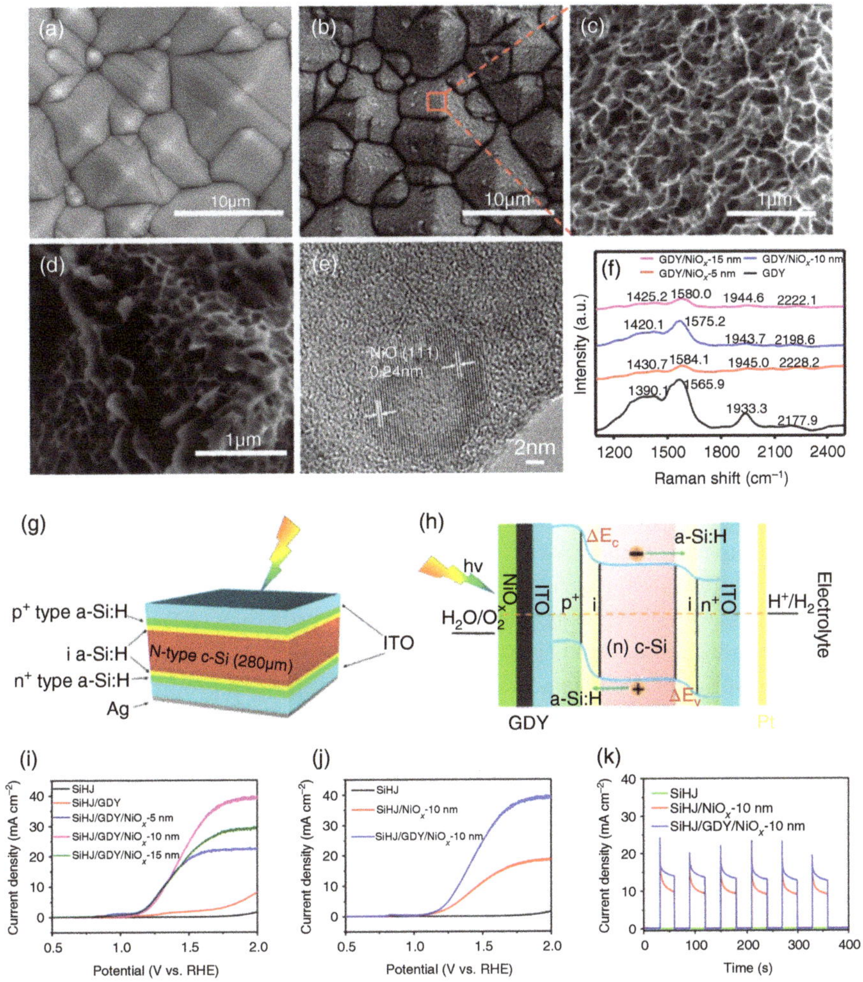

Figure 5.21 SEM images of (a) micropyramidal structure of silicon, (b) silicon substrate loaded with GDY at low magnification, (c) porous structure of the GDY at higher magnification, and (d) GDY loaded with NiO_x catalyst. (e) HRTEM image of NiO_x catalysts on the GDY. (f) Raman spectra of GDY, GDY/NiO_x-5 nm, GDY/NiO_x-10 nm, and GDY/NiO_x-15 nm. (g) Schematic of the SiHJ and (h) energy band diagrams of the SiHJ photoanode. (i) J–V characteristics of GDY modified SiHJ photoanodes with varying thickness of NiO_x as OER catalysts; (j) J–V characteristics of the SiHJ photoanodes loaded with 10-nm-thick NiO_x catalysts with/without GDY modifying. (k) Chronopotentiometry curves of the SiHJ photoanodes loaded with 10-nm-thick NiO_x catalysts with/without GDY modifying at potential of 1.4 V. All the experiments were conducted in 1 M KOH under the illumination of 100 mW cm^{-2} regularly. Source: Zhang et al. [62]. © 2019, American Chemical Society.

about 0.477 min^{-1} for the degradation reaction of methylene blue, and a higher photocatalytic water oxidation activity, which was 1.89 times and 1.75 times higher than that of Ag_3PO_4/CNT and Ag_3PO_4/graphene, respectively (Figure 5.22C,D). Si et al. [65] prepared a novel Z-type Ag_3PO_4/GDY/g-C_3N_4 nanocomposite via electrostatic and π–π stacking interaction for photocatalytic OER activity with a high O_2 yield of 753.1 μmol g^{-1} h^{-1} (Figure 5.22E,F), which was 12.2 times than that of pure Ag_3PO_4 nanoparticles.

5.2.4 Applied for Overall Water Splitting

Among the drawbacks of applying electrochemical or PEC splitting of water into hydrogen and oxygen industrially, there still remains a huge challenge to exploit non-noble metal-based catalysts with excellent activity and stability simultaneously for both HER and OER at the same pH conditions. To address this challenge, Li and coworkers [54] utilized the unique advantages of GDY to obtain the first GDY-based electrocatalyst, $NiCo_2S_4$ NW/GDF (Figure 5.17a), for overall water splitting (OWS). When used as both anode and cathode, this electrocatalyst delivered the 10 mA cm^{-2} at a much smaller cell voltage of 1.53 V (vs. RHE) in 1.0-M KOH than RuO_2||Pt/C couple (1.63 V@10 mA cm^{-2}), and exhibited excellent stability with negligible decrease in current density over 140 hours galvanostatic electrolysis at about 20 mA cm^{-2}. Afterwards, a series of efficient electrocatalysts with higher catalytic performances toward OWS were fabricated by in situ growth of GDY on the surface of hydrotalcite-like compounds [41, 44, 57], metal oxides [42], metal nitrides [43], phosphides [56], metal clusters [55], etc. For example, Li's group fabricated an alkaline electrolyzer using e-ICLDH@GDY/NF as both the anode and cathode for OWS. Their results showed that the assembled electrolyzer had outstanding performances in achieving the current densities of 10, 100, and 1000 mA cm^{-2} at small cell voltages of 1.43, 1.46, and 1.49 V, respectively [41], outperforming the performance of the reported RuO_2||Pt system and most of the reported excellent bifunctional electrocatalysts (Figure 5.23a–c). As shown in Figure 5.23d,e, upon increasing the applied voltages, the current densities increased from approximately 50 mA cm^{-2} to approximately 1000 mA cm^{-2} and stabilized rapidly. The current densities decreased from 1000 to 50 mA cm^{-2} upon decreasing the applied voltage, revealing the system's excellent stability. Moreover, the e-ICLDH@GDY/NF electrolyzer can continuously work for more than 60 hours at 100 mA cm^{-2} (Figure 5.23f), with the current density remaining constant and the Faradaic efficiency of O_2 evolution of 97.40 ± 1.30%. Li and coworkers prepared NiO–GDY NC that was used as both the anode and cathode for OWS (Figure 5.23g) [42]. The assembled electrolyzer exhibited excellent OWS performance having a small cell voltage of 1.52 V at 10 mA cm^{-2} (Figure 5.23h) and could remain almost constant current density over 60 hours (Figure 5.23i). Fang et al. reported the controlled synthesis of GDY-wrapped cobalt nitride nanosheets (CoN_x@GDY NS/NF) and their direct application as a bifunctional electrocatalyst toward electrical splitting of water for the first time [43]. A current density of 10 mA cm^{-2} could be obtained at an ultralow cell voltage of 1.48 V, much smaller than that of Pt/C||RuO_2 couple, with its long-time stability

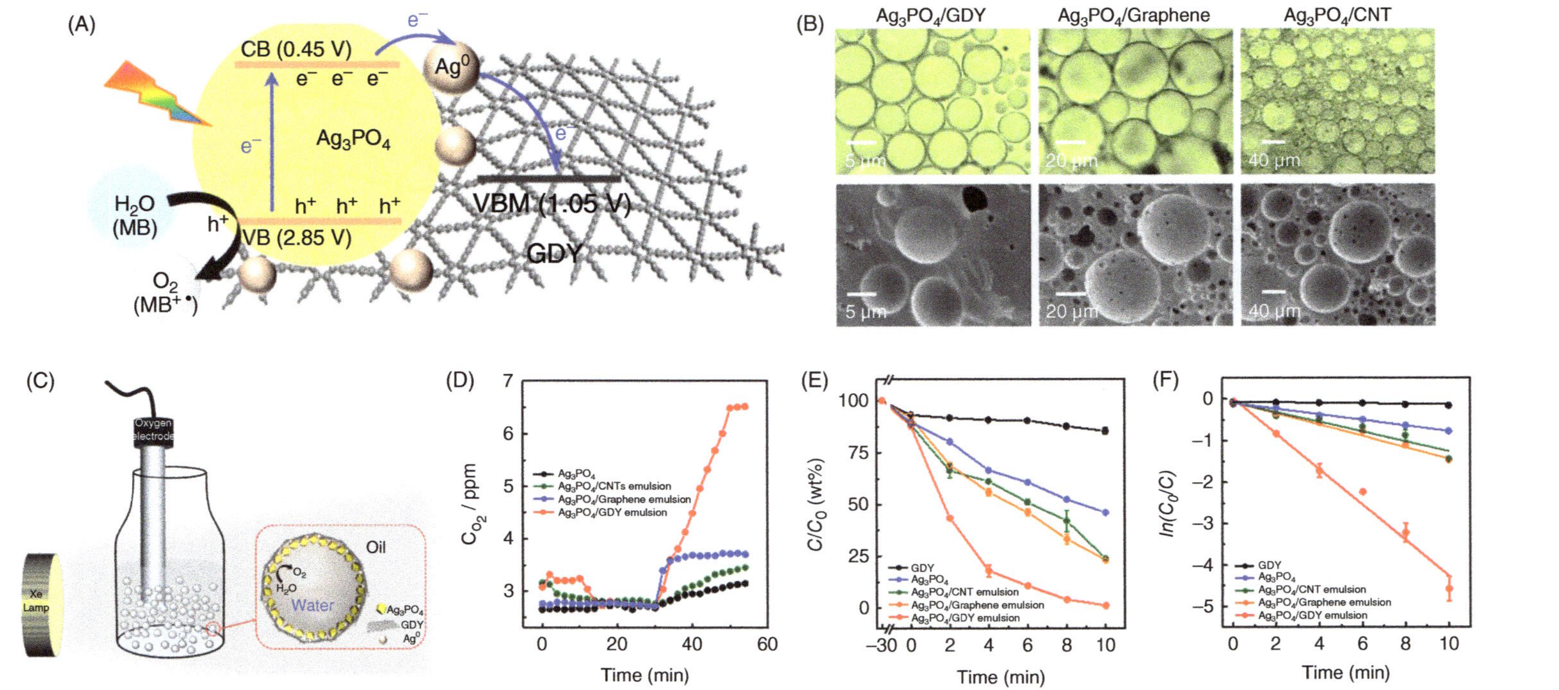

Figure 5.22 (A) Schematic representation of the Ag_3PO_4/GDY-based emulsion for photocatalysis. (B) Optical microscopy images (a–c) and Cryo-SEM images (d–f) of the Ag_3PO_4/GDY-, Ag_3PO_4/graphene-, and Ag_3PO_4/CNT-based emulsions. (C) Schematic illustration of O_2 evolution under visible-light irradiation. (D) O_2 concentration measured with dissolved oxygen meter in Pickering emulsion-based systems (water volume: 15 ml) and Ag_3PO_4 solution-dispersed system (15 ml). (E) MB decomposition and (F) first-order linear transforms of MB decomposition under visible-light irradiation as a function of time in various systems. Source: Guo et al. [64]. © 2019, American Chemical Society.

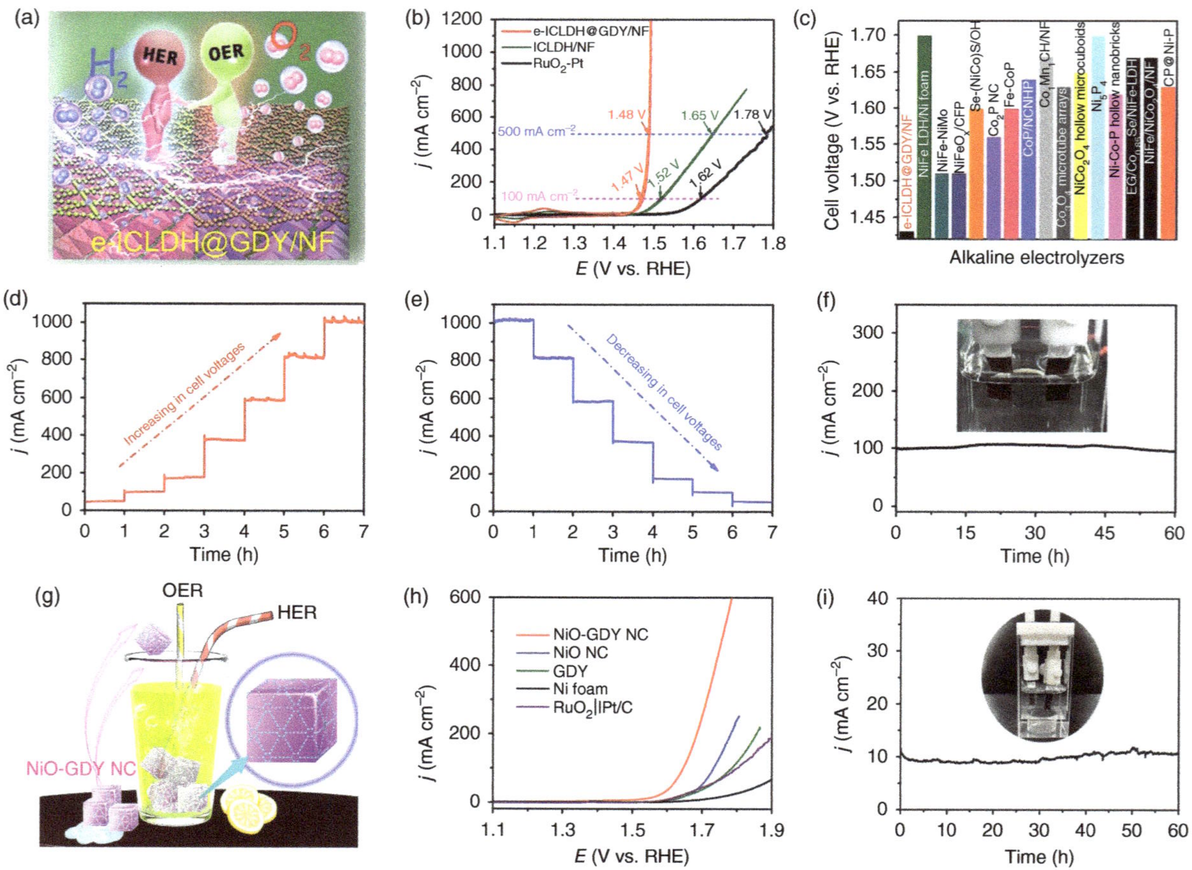

Figure 5.23 (a) Schematic representation of the OWS process of e-ICLDH@GDY/NF. (b) OWS activities of e-ICLDH@GDY, a precious RuO_2-Pt couple, and the pristine ICLDH/NF. (c) Comparison of the OWS activity of e-ICLDH@GDY with previously reported electrocatalysts. Current density–time curves measured with (d) increasing (red line) and (e) decreasing (blue line) cell voltages. (f) Current density of the e-ICLDH@GDY/NF electrode in an alkaline electrolyzer under a constant cell voltage of 1.56 V, measured over 60 hours (inset: the photograph for the two-electrode system). (g) Schematic representation of the OWS process of NiO-GDY NC. (h) CV curves of NiO-GDY NC, NiO, GDY, and RuO_2||Pt/C toward overall splitting in 1 M KOH. (i) Chronopotentiometry curve of NiO-GDY NC at 10 mA cm^{-2} toward overall splitting in the same alkaline condition. Source: (a–f) Hui et al. [41]. ©2018, Springer Nature, (g–i) Yu et al. [42]. ©2019, Elsevier.

and electrochemical activity surface area (ECSA) being improved (Figure 5.8j–l). Moreover, Cu@GDY-Co electrode prepared by Shi et al. exhibited impressive water-splitting activity and stability [55]. The performances of GDY-based electrocatalysts mentioned above were all comparable to the state-of-the-art non-noble metal catalysts for water electrolysis.

Liu and coworkers [66] grew GDY on the surface of $BiVO_4$ to synthesize GDY–$BiVO_4$ heterojunction materials (Figure 5.24a–c) for photoelectrocatalytic water splitting. The energy-wasting and undesirable charge recombination process was significantly suppressed due to the efficient extraction of photogenerated carriers from $BiVO_4$ by the GDY layer. A significant increase in the PEC performance of $BiVO_4$ was achieved benefiting from the unique and fascinating properties of GDY. The photocurrent density of $1.32\,mA\,cm^{-2}$ could be obtained at 1.23 V (vs. RHE), which was nearly twice that of pure $BiVO_4$ (Figure 5.24d,e).

5.2.5 Applied for Other Catalysis

Unique carbon hybridization (sp^2 and sp) endows GDY with an uneven distributed surface charge as compared to carbon nanotubes (CNTs) and graphene [8, 9, 67–69]. This has promoted the advancement of GDY into a highly active electrode material for various applications, including (photo)-electrochemical water splitting, nitrogen reduction reaction, ORR, and batteries. Recently, some explorations have been made for the application of GDY/GDY-based materials for other catalytic reactions. For example, Mao and coworkers [70] showed that the oxygen-containing groups on graphdiyne oxide (GDYO) surface play an important role in controlling the formation of Pd clusters through increasing the anchoring ability of Pd nuclei and avoiding Ostwald ripening following nuclei, and exhibited high reaction activity for 4-NP reduction giving the rate constant about 40-, 11-, and 5-fold higher than that of Pd/multiwalled carbon nanotube (MWNT), Pd/GO, and commercial Pd/C, respectively. Lu and coworkers [71] reported a catalyst comprising Pd subnanometric catalysts (SNCs, an average particle size of 0.83 nm) stabilized on pyrenyl graphdiyne (Pyr-GDY) ultrafine nanofibers (3–10 nm) for the reduction of nitroarenes to arylamines and Suzuki coupling reactions (Figure 5.25a,b). For instance, the catalytic activities for reduction of 4-NP are 300 and 25 times higher than those of commercial Pd/C and Pd/GO, respectively. Tan et al. [74] reported a 2D heterogeneous hybrid nanomaterial (P5A-Au-GD) based on GDY and pillar[5]arene (P5A)-reduced Au nanoparticles (P5AAu) toward efficient reduction of 4-NP and methylene blue with higher catalytic performance than commercial Pd/C catalyst. Li and coworkers [72] utilized the strong interactions between GDY and Pt nanoparticles that could prevent the thermal migration of Pt nanoparticles on the GDY surface to fabricate an ultrastable Pt-GDY catalysts (Figure 5.25c). Their results showed that Pt NPs with size of 2–3 nm exhibited high performance on hydrogenation of aldehydes and ketones to the corresponding alcohols compared with commercial Pt–C (Figure 5.25d).

GDY was also applied for photocatalyzed organic reactions. Wang et al. [73] successfully incorporated titania nanoparticles (P25) with GDY nanosheets

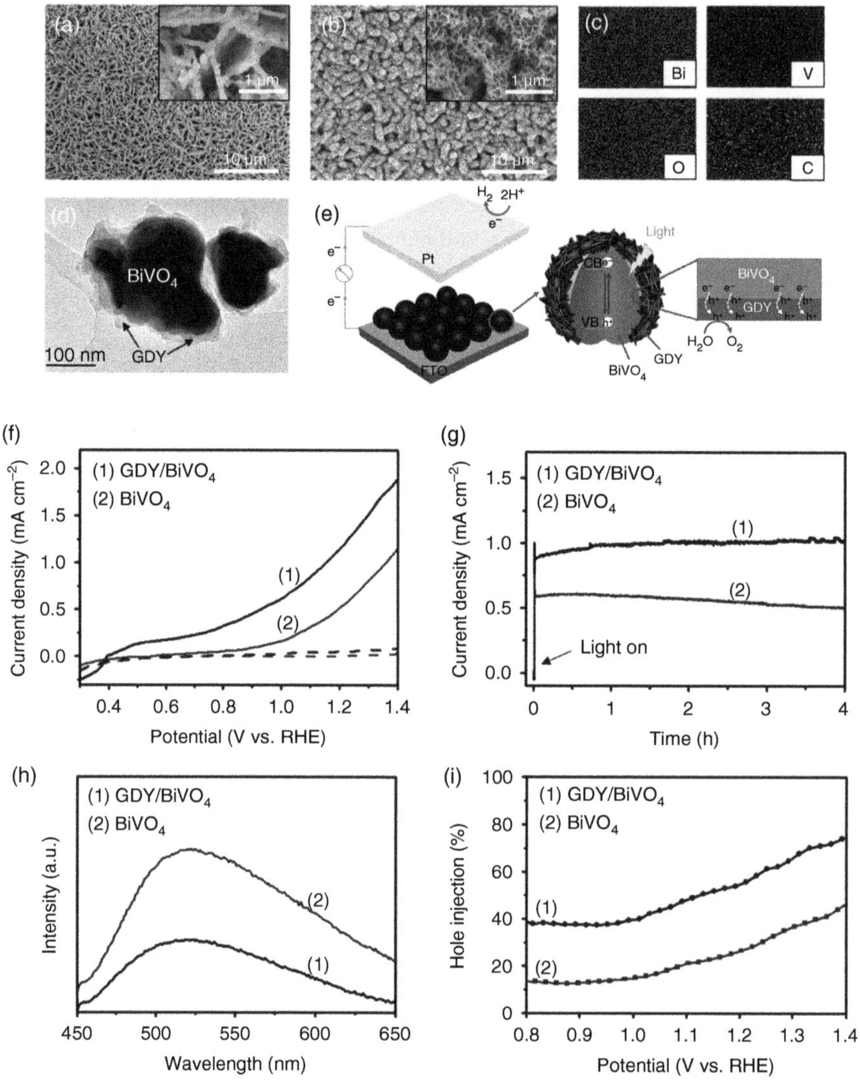

Figure 5.24 SEM images of (a) $BiVO_4$ and (b) the as-prepared GDY/$BiVO_4$. (c) Elemental mapping of bismuth, vanadium, oxygen, and carbon by energy dispersive spectroscopy (EDS). (d) TEM image of GDY/$BiVO_4$. (e) Schematic illustration of GDY/$BiVO_4$ photoanodes in a PEC setup and the migration of the photogenerated excitons at the interface. (f) Current–voltage curves of $BiVO_4$ and GDY/$BiVO_4$ photoanodes under Xe lamp illumination (solid line) and in the dark (dashed line). (g) J–t curves of $BiVO_4$ and GDY/$BiVO_4$ photoanodes in a four hours test. (h) Photoluminescence curves of $BiVO_4$ and GDY/$BiVO_4$ films. (i) Hole injection yield of $BiVO_4$ and GDY/$BiVO_4$ photoanodes. Source: Gao et al. [66]. ©2017, Wiley-VCH Verlag GmbH & Co. KGaA.

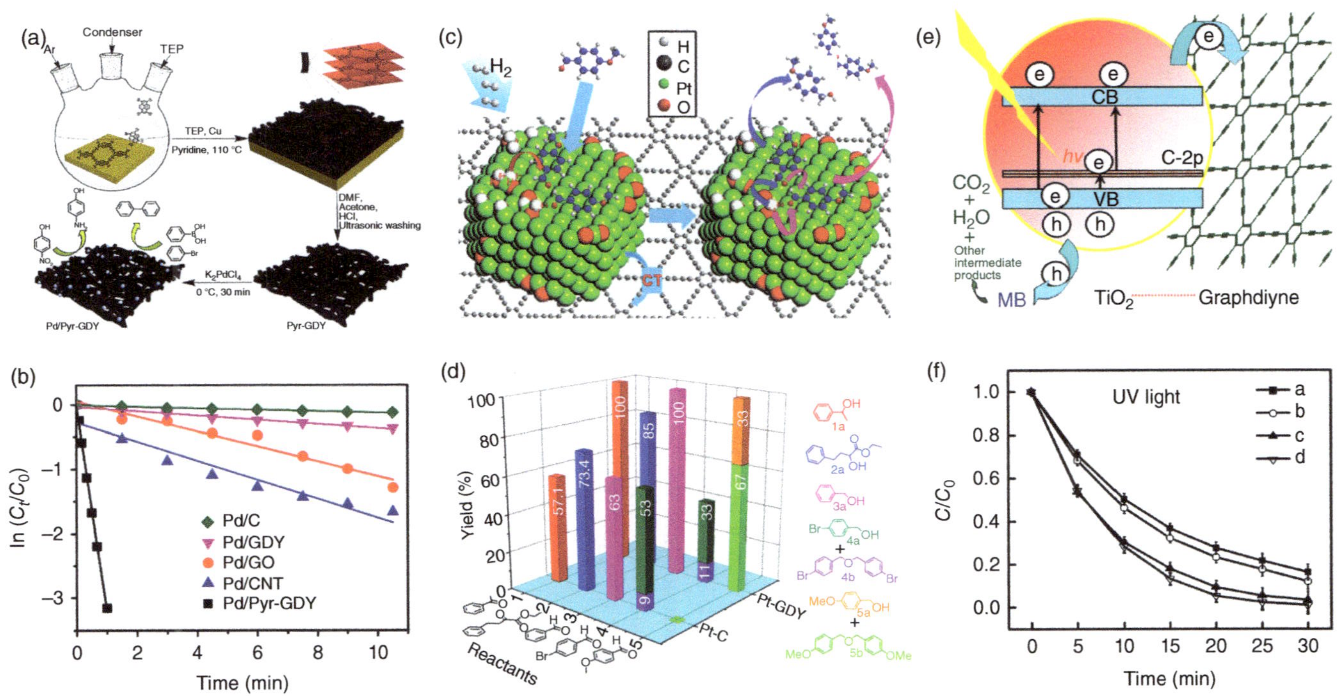

Figure 5.25 (a) Preparation of the Pd/Pyr-GDY composite and its catalytic reactions. (b) Plots of ln(C_t/C_0) as a function of the reaction time for the reduction of 4-nitrophenol catalyzed by Pd/Pyr-GDY, Pd/GO, Pd/CNT, Pd/GDY, and commercial Pd/C, respectively. (c) Proposed catalysis mechanism of the hydrogenation of aldehydes and ketones to the corresponding alcohols or dibenzylether. (d) Hydrogenation of different ketones, aldehydes using Pt–GDY as the catalyst. Conditions: 5 mg of Pt–GDY, 1 mmol of substrate, 20 ml of CH_3CH_2OH at 30 °C for four hours under 1 bar initial hydrogen pressure. The yield was determined by the integration of ^{1}H NMR spectra signals of the starting materials and the targeting materials. (e) Schematic structure of P25–GDY and tentative processes of the photodegradation of methylene blue (MB) over P25–GDY. (f) Photocatalytic degradation of MB under UV irradiation over (a) P25, (b) P25–CNTs, (c) P25–GR, and (d) P25–GDY. Source: (a, b) Yang et al. [71]. ©2019, Royal Society of Chemistry), (c, d) Shen et al. [72]. © 2019, American Chemical Society, (e, f) Wang et al. [73]. © 2012, John Wiley & Sons.

chemically through a facile hydrothermal treatment, forming a novel P25-GDY nanocomposite, which was directly used for photocatalytic degradation of methyl blue (Figure 5.25e,f). P25-GDY (0.6 wt% GDY) exhibited the best catalytic activity over those of P25-carbon nanotubes and P25-graphene. Their later results showed that the combination of GDY and TiO_2(001) crystal plane could contribute to an efficient carrier separation process and longer photogenerated carrier lifetime, and facilitated photocatalytic oxidation ability [75]. Ramakrishnan et al. [76] further showed that the strong interaction between GDY/GDYO and TiO_2 could all effectively optimize hole transport process and increase photocurrent density of the catalyst, resulting in the enhancement of the catalytic performance [64].

GDY-based heterostructures can also be applied in some important biocatalytic reactions. Mao and coworkers [68] designed an electron transport pathway to realize catalytic specificity on a heterostructure fabricated by interfacing redox-active methylene green (MG) and semiconducting GDY to obtain a selective catalyst (MG/GDY) in one single step (Figure 5.26a). Both theoretical and experimental studies clearly indicated the formation of MG dimers within the interlayer space of GDY nanosheets, which consequently played a critical role in modulating both activation overpotential and electron transfer rates. The MG/GDY exhibited high selectivity in bioelectrocatalysis by accelerating oxidation of dihydronicotinamide adenine dinucleotide at 7.06×10^{-2} cm s^{-1} in an electron-hopping pathway or decelerating oxidation of ascorbic acid at 6.60×10^{-5} cm s^{-1} in an electron-tunneling pathway (Figure 5.26b–f). Li's group also employed N-GDY as metal-free photocatalyst for nicotinamide adenine dinucleotide (NADH) regeneration (Figure 5.26g,h). It was reported that the catalytic activity increased by increasing the N content from N1 to N3 and reached 35% for N3-GDY [77].

5.3 Graphdiyne-Based Metal-Free Catalysts

Although metal-based materials have been widely used for catalysis, energy conversion and storage, and many other important industrial processes, their high cost and poor stability have seriously limited their practical uses [34]. Carbon materials possessing rich and adjustable chemical or electronic structure and excellent resistance to acidic/alkaline condition have been demonstrated to be effective for large numbers of catalytic processes. Because of its unique properties, GDY has shown great promising advantages for metal-free catalysis.

5.3.1 Applied for Water Splitting

In Xing's work [78], a GDY precursor substituted with fluorine atoms was used to controllably grow a three-dimensional flexible fluorinated graphdiyne (p-FGDY/CC) electrode material on carbon cloth (p-FGDY/CC). Because of the unique chemical structure, p-FGDY/CC electrode has high HER, OER, and OWS performance and stability. For example, in acidic and alkaline environments, when the current density is 10 mA cm^{-2}, the hydrogen evolution overpotentials are only

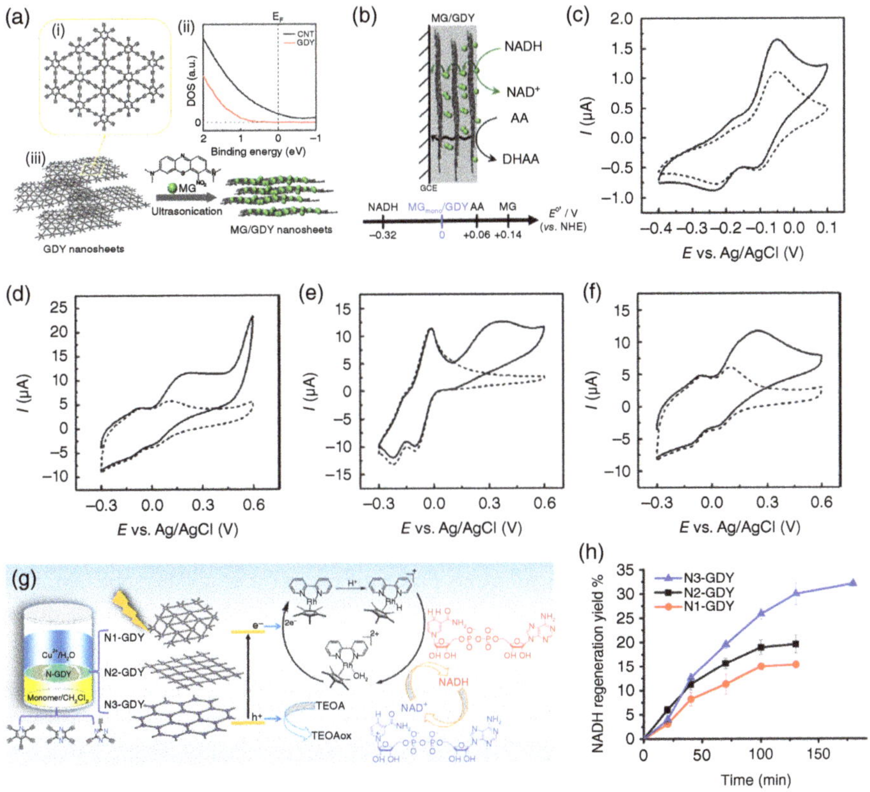

Figure 5.26 (a) Formation of MG/GDY nanosheets: (i) sp–sp^2-hybridized carbon framework of a single-layered GDY; (ii) DOS of pristine GDY as compared to that of CNTs; (iii) Schematic illustration of intercalative modification of GDY nanosheets with MG by ultrasonication. (b) Electron-hopping pathway for NADH oxidation and electron-tunneling pathway for AA oxidation catalyzed by the MG/GDY nanosheets. (c, d) CVs of NADH (1.5 mM in phosphate buffer at pH 7.4) oxidation at MG/GDY-modified glass carbon electrode (GCE) (c) and MG-GDY-modified GCE (d). (e, f) CVs of AA (1.5 mM in phosphate buffer at pH 7.4) oxidation at MG/GDY-modified GCE (f) and MGGDY-modified GCE (g). Solid lines and dashed lines represent the presence and absence of substrates, respectively. Scan rate is 10 mV s^{-1}. (g) Schematic illustration of the liquid/liquid interface preparation of N-GDYs from the corresponding monomers and photocatalytic NADH regeneration in the presence of [Cp*Rh(bpy)(H)]$^+$ in aqueous solution. (h) Comparison of the regeneration rates of the three systems. Source: (a–f) Guo et al. [68];©2020, American Chemical Society; (g, h) Pan et al. [77]. © 2019, American Chemical Society.

92 and 82 mV, the catalytic activity cannot be significantly attenuated after 3000 and 8500 CV cycles, and the performance exceeds the metal-free catalysts and most metal catalysts reported at that time. In addition, p-FGDY/CC also shows great OER performance in acidic and alkaline environments. When the current density is 10 mA cm^{-2}, the oxygen evolution current is only 600 and 475 mV. DFT calculations show that the strong C—F bond leads to the redistribution of p-electron orbitals, which enhances the electron-rich characteristics of the C2 site, thereby improving the electron transfer ability. This ensures that p-FGDY/CC has higher

selectivity for the adsorption/desorption of various O/H intermediates, and ensures that p-FGDY/CC has excellent overall water-splitting performance under full pH conditions. Zhao et al. produced [79] the GDY-based OER catalyst codoped with N and S, by adjusting the ratio of N and S atoms to control the OER activity. When the current density is 10 mA cm^{-2}, the oxygen evolution overpotential is only 299 mV, which is lower than RuO_2 catalyst (305 mV), and the performance is also better than N or S alone doped GDY. This is because of the synergistic effect of double doping and three-dimensional position. Among them, sp-N occupies a dominant position in all N configurations, which can significantly reduce the overpotential, and further introduction of S can increase the current density, which makes it have better catalytic activity and faster kinetics (Figure 5.27).

5.3.2 Applied for Oxygen Reduction Reactions

Heteroatom-doped GDY not only exhibits excellent properties in the field of water electrolysis, but also molecularly designed and orderly doped GDY is of great significance to promote the kinetics of ORRs. Li and coworkers [80] designed an N, F codoped GDY catalyst; in alkaline environment (1.0 M KOH), the catalyst exhibited catalytic activity equivalent to that of the Pt/C electrode and had excellent stability. After 6000 CV cycles, the performance did not change significantly. And a disposable zinc–air battery was self-made with *N*, *F* co-doped graphdiyne (NFGD) as the cathode catalyst. Test results show that the open circuit voltage of the battery with NFGD as the cathode catalyst is 1.18 V, which is close to commercial batteries. Zhao et al. [81] introduced a new form of nitrogen doping-sp-hybridized nitrogen (sp-N) into the chemically defined site of ultra-thin GDY by substituting the acetylene group in the ring, and designed and synthesized an sp-N doped GDY to catalyze ORR (Figure 5.28). The sp-N doping promotes the adsorption and electron transfer of O_2 on the catalyst surface, so that the material has great electrocatalytic ORR performance. In alkaline environment, compared with Pt/C (76 mV dec^{-1}), sp-N doped GDY (60 mV dec^{-1}) has smaller Tafel slope. At the same time, it has a half-wave potential (0.87 V) close to that of Pt/C catalyst (0.86 V), and has better methanol tolerance; under acidic conditions, the performance is slightly lower than Pt/C, but the performance is still better than other metal-free catalysts.

Zhang and coworkers [82] prepared an N-doped GDY. Electrochemical test shows that the limiting current density of N 550-GD at 0.05 V vs. RHE in 0.1 M KOH reaches about 4.5 mA cm^{-2}, which is comparable to 20% Pt/C, and N 550-GD exhibits excellent stability. To illustrate the influence of N doping on the electronic structure of GDY, quantum mechanical calculations have been carried out. The calculations show that the C atoms adjacent to the N dopants have a significantly higher positive charge density to offset the strong electron affinity of N atoms and promote the transfer of the electrons from the anode, thereby promoting ORR. The calculation results also found that imine N has a greater positive effect on the charge density of adjacent C atoms to pyridine N.

Huang and coworkers [83] synthesized pyridine N-doped graphyne on the basis of hydrogen-substituted graphyne (HsGDY) for electrocatalytic ORR. In HsGDY, there

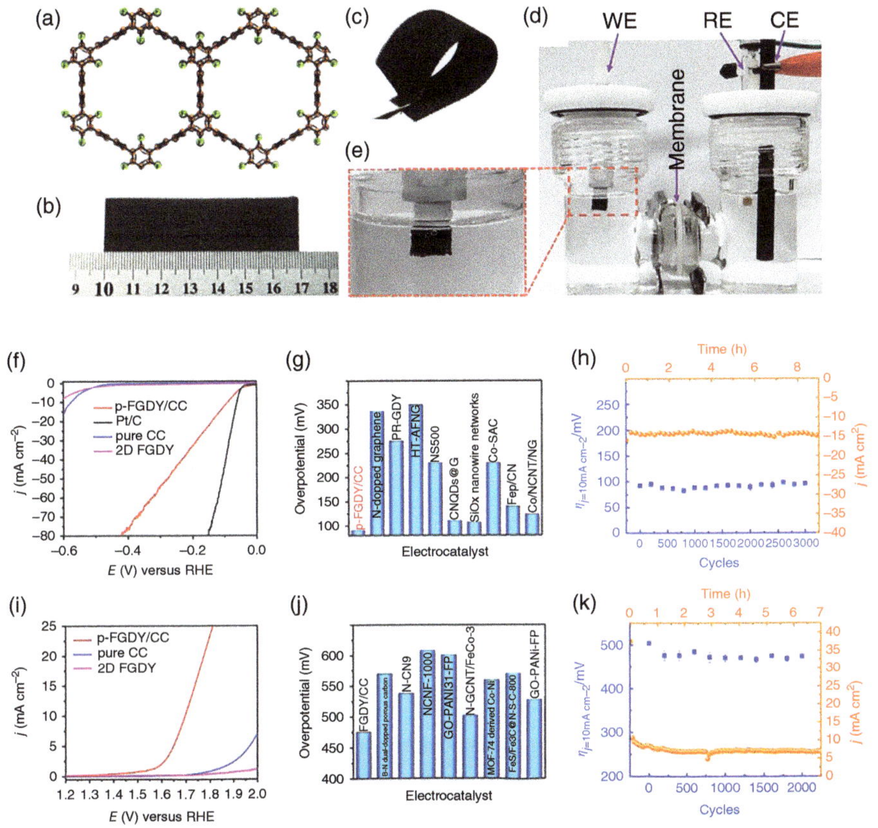

Figure 5.27 (a) Schematic of the structure of FGDY. (b) The photograph of the pFGDY/CC material. (c) Photographs of the p-FGDY/CC electrode with bended morphology exhibiting its flexibility nature. Photographs of (d) the three-electrode system (WE, working electrode; RE, reference electrode; CE, counter electrode) and (e) enlarged image of the WE of the marked in (d). (f) HER polarization curve of the catalyst in 0.5 M H_2SO_4. (g) Comparison of the HER performance of p-FGDY/CC with the reported metal-free and metal-based catalyst. (h) pFGDY/CC in the HER process at 0.5 M H_2SO_4 (i) OER polarization curve of the catalyst in 1.0 M KOH. (j) Comparison of p-FGDY/CC and reported OER performance of metal-free and metal-based catalysts. (k) OER process long-term stability test of pFGDY/CC in 1.0 M KOH. Source: Xing et al. [78]. © 2019, Wiley-VCH Verlag GmbH & Co. KGaA.

are three carbon atoms on the benzene ring combined with hydrogen, which can be used as active sites to selectively dope with pyridine N. The synthesized HsGDY was heated under NH_3 conditions at 700, 800, 900, and 1000 °C to obtain N-HsGDY catalyst. In an oxygen-saturated 0.1 M KOH solution, at half-wave potential of 0.85 V vs. RHE, the current density of N-HsGDY-900 is 3.03 mA cm^{-2}, which shows the highest catalytic activity compared with other nonmetallic catalysts. The onset potential of N-HsGDY-900 is 1.02 V vs. RHE, the half-wave potential is 0.85 V vs. RHE, and the limiting current density is 6.2 mA cm^{-2}, which is 1.6 times higher than commercial platinum-carbon catalysts (Figure 5.29). In addition, N-HsGDY also has better stability and methanol tolerance. Under acidic conditions, it also has the same

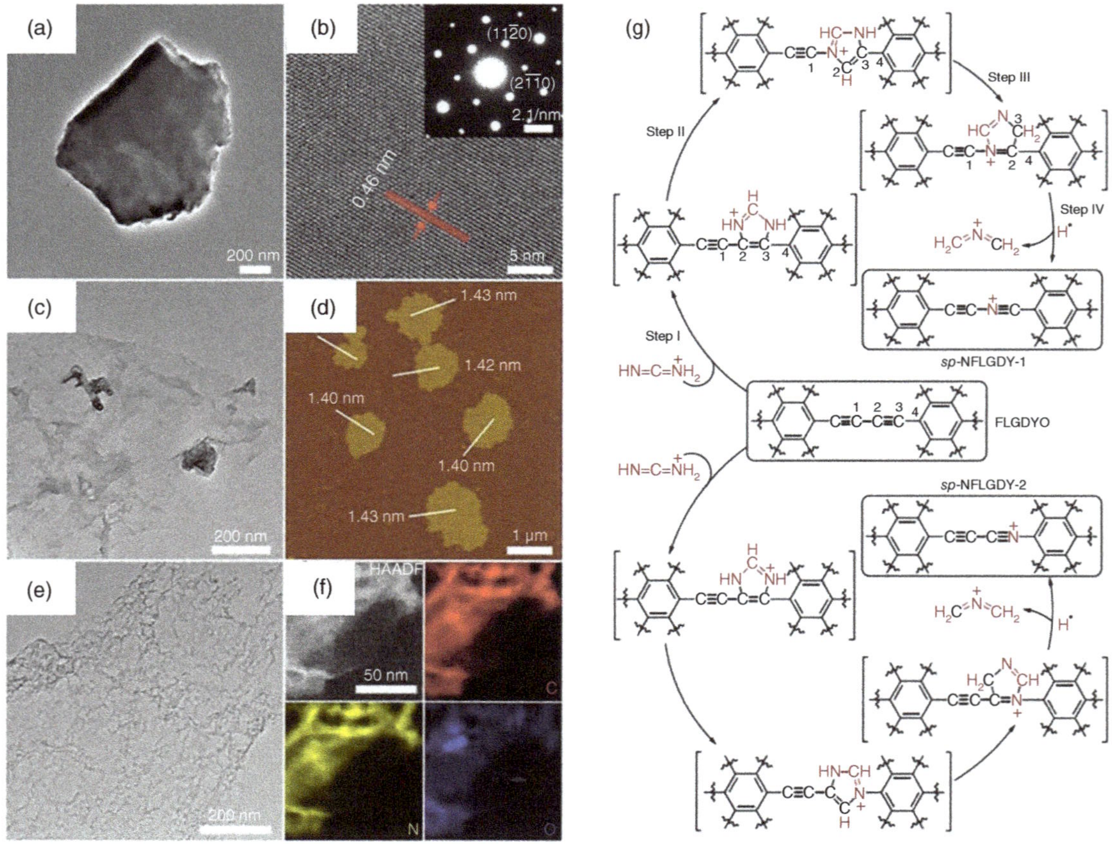

Figure 5.28 (a) Two-dimensional TEM image of boron-doped graphdiyne (BGDY); (b) HRTEM image of BGDY; (c) TEM image of few-layer oxidized graphdiyne (FLGDYO), confirming the morphology of the nanosheet; (d) AFM image of FLGDYO and the thickness of the FLGDYO nanosheet. (e) TEM image of NFLGDY-900c, showing obvious wrinkles and wrinkles. (f) High-angle annular dark field (HAADF) and electron energy loss spectroscopy (EELS) mapping of C, N, O atoms in NFLGDY-900c; (g) Process of sp-hybridized nitrogen atom doping graphyne. Source: Zhao et al. [81]. © 2018, Springer Nature.

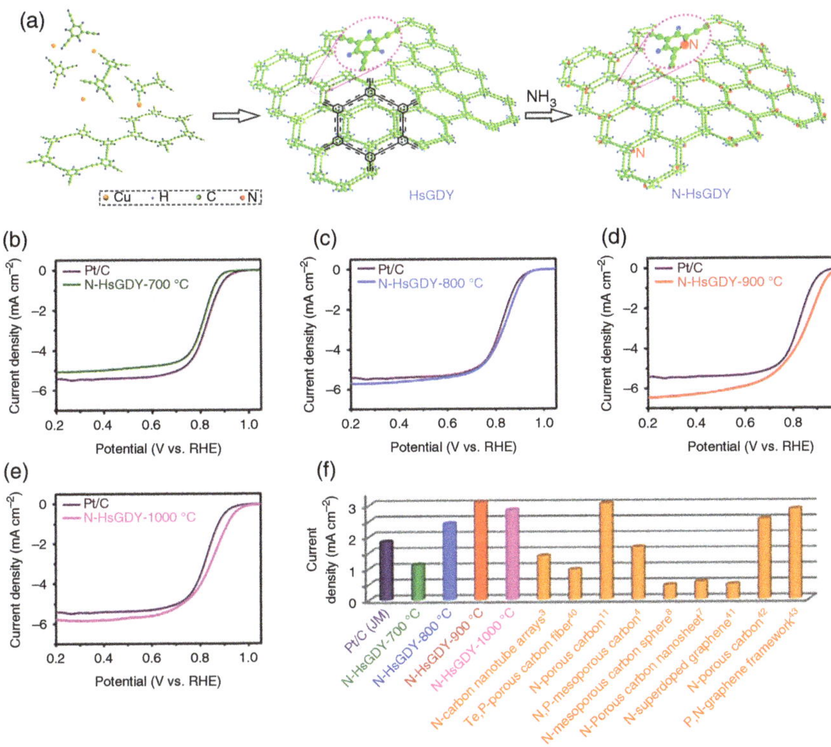

Figure 5.29 (a) Schematic illustration of the preparation process for the hydrogen-substituted graphdiyne (HsGDY) from the monomer triethynylbenzene. Performance of catalysts in alkaline media. (b)-(d) Linear sweep voltammetry curves of pyridinic nitrogen-doped hydrogen-substituted graphdiyne. (N-HsGDY) treated at 700 °C (N-HsGDY-700 °C) (b), N-HsGDY treated at 800 °C (N-HsGDY-800 °C) (c), N-HsGDY treated at 900 °C (N-HsGDY-900 °C) (d), and N-HsGDY treated at 1000 °C (N-HsGDY-1000 °C) (e), compared to commercial carbon-based platinum catalyst (Pt/C (JM)) for the oxygen reduction reaction (ORR). (f) Comparison of current density at 0.85 V vs. reversible hydrogen electrode (RHE) of Pt/C, N-HsGDY catalysts in this work and previously reported metal-free catalysts. Source: Lv et al. [83].

performance as the platinum carbon catalyst. From theoretical calculations, it is known that hydrogen-substituted graphdiyne doped with pyridine nitrogen can reduce oxygen more effectively than that only doped with pyridine nitrogen.

5.3.3 Applied for Photocatalysis

As a direct-gap natural semiconductor, GDY has great electron and hole transport capability, and has been used as an excellent metal-free material for photocatalytic processes. Pan et al. [77] used pyridyl, pyrazinyl, and triazine-based monomers as raw materials to prepare three new nitrogen-doped GDYs through polymerization at the liquid/liquid interface (Figure 5.30). As the proportion of nitrogen increases, the surface of the material gradually changes from wettability to hydrophilicity. Experimental results show that under illumination, N3-GDY with the best

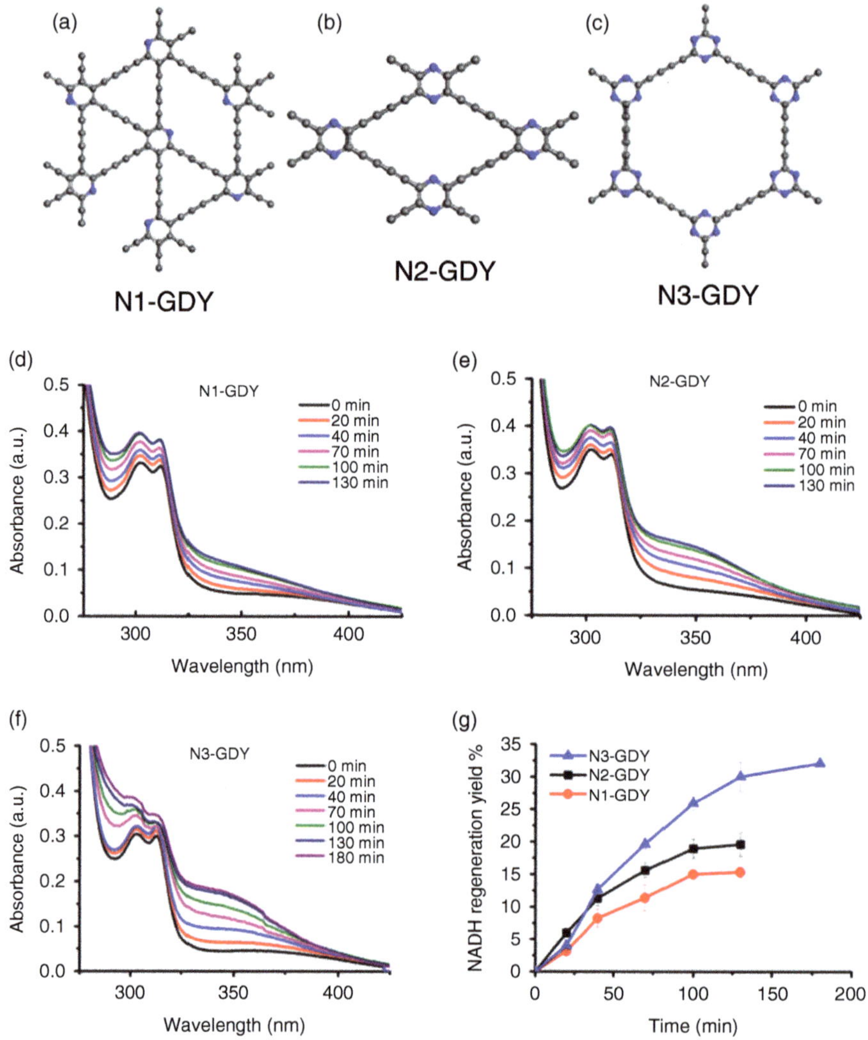

Figure 5.30 Proposed structure of (a) N1-GDY, (b) N2-GDY, and (c) N3-GDY; Experimental results of the NADH regeneration rates in three systems: (d) N1-GDY, (e) N2-GDY, and (f) N3-GDY. (g) Comparison of the regeneration rates of the three systems. Source: Pan et al. [77]. © 2019, American Chemical Society.

hydrophilicity exhibits the highest catalytic activity, for example, the NADH conversion rate can reach 35% within three hours. Compared with N1-GDY and N2-GDY, N3-GDY has a larger negative potential, and at the same time, higher hydrophilicity ensures that the catalyst is in close contact with the solution and promotes the NADH regeneration reaction.

Benefiting from the excellent hole transport capability and strong interaction between g-C_3N_4 and GDY, the life of g-C_3N_4/GDY photogenerated electrons has been increased by seven times. GDY quickly transfers the photogenerated holes

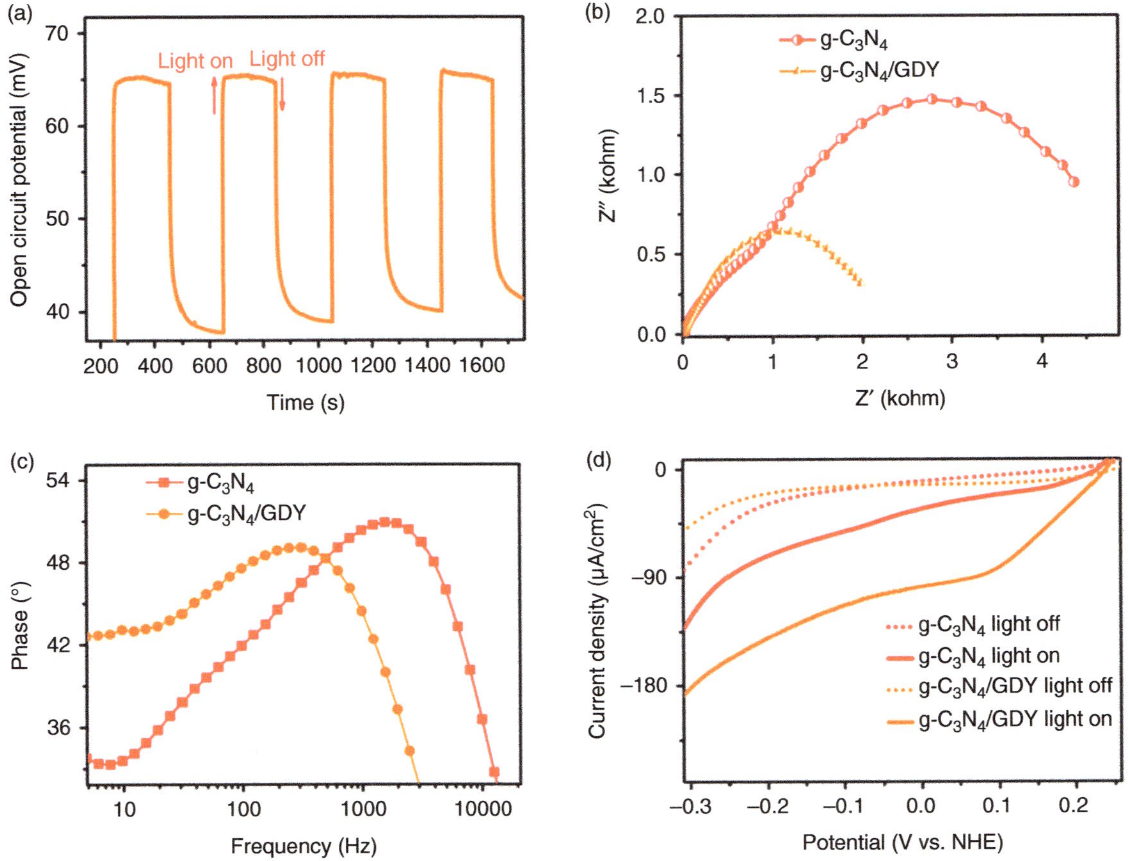

Figure 5.31 (a) Photocurrent response spectrum of g-C_3N_4/GDY photoelectrode; (b) Nyquist spectrum of the material under illumination; (c) Boed phase diagram of the material; (d) Linear scan spectrum of the material under light and dark conditions. Source: Han et al. [49]. © 2018, John Wiley & Sons.

of g-C_3N_4, effectively inhibits the recombination of photogenerated carriers, and improves the photoelectrocatalytic performance of heterostructures (Figure 5.31). Compared to g-C_3N_4 ($-32\,\mu A\,cm^{-2}$), at a potential of 0 V, the photocurrent density of the g-C_3N_4/GDY has been greatly increased ($-98\,\mu A\,cm^{-2}$), showing a more superior photocatalytic water-decomposing ability [49]. Recently, Xu et al. [84] combined graphitized carbon nitride and graphyne to form a heterostructure, which improved the separation efficiency of charge carriers and prolonged the life of charge carriers. The electron flow in the photocatalyst was accelerated and reduced. The overpotential of the hydrogen evolution reaction was analyzed.

References

1 Zuo, Z., Wang, D., Zhang, J. et al. (2019). Synthesis and applications of graphdiyne-based metal-free catalysts. *Advanced Materials* 31: 1803762.
2 Qiao, B., Wang, A., Yang, X. et al. (2011). Single-atom catalysis of CO oxidation using Pt_1/FeO_x. *Nature Chemistry* 3: 634–641.
3 Wang, A., Li, J., and Zhang, T. (2018). Heterogeneous single-atom catalysis. *Nature Reviews Chemistry* 2: 65–81.
4 Yang, X.-F., Wang, A., Qiao, B. et al. (2013). Single-atom catalysts: a new frontier in heterogeneous catalysis. *Accounts of Chemical Research* 46: 1740–1748.
5 Zhang, L., Ren, Y., Liu, W. et al. (2018). Single-atom catalyst: a rising star for green synthesis of fine chemicals. *National Science Review* 5: 653–672.
6 Li, Z., Ji, S., Liu, Y. et al. (2020). Well-defined materials for heterogeneous catalysis: from nanoparticles to isolated single-atom sites. *Chemical Reviews* 120: 623–682.
7 Xiong, Y., Dong, J., Huang, Z.-Q. et al. (2020). Single-atom Rh/N-doped carbon electrocatalyst for formic acid oxidation. *Nature Nanotechnology* 15: 390–397.
8 Xue, Y., Huang, B., Yi, Y. et al. (2018). Anchoring zero valence single atoms of nickel and iron on graphdiyne for hydrogen evolution. *Nature Communications* 9: 1460.
9 Hui, L., Xue, Y., Yu, H. et al. (2019). Highly efficient and selective generation of ammonia and hydrogen on a graphdiyne-based catalyst. *Journal of the American Chemical Society* 141: 10677–10683.
10 Yu, H., Xue, Y., Huang, B. et al. (2018). Ultrathin nanosheet of graphdiyne-supported palladium atom catalyst for efficient hydrogen production. *iScience* 11: 31–41.
11 Yu, H., Xue, Y., Hui, L. et al. (2020). Graphdiyne based metal atomic catalysts for synthesizing ammonia. *National Science Review* https://doi.org/10.1093/nsr/nwaa213.
12 Hui, L., Xue, Y., Yu, H. et al. (2020). Loading copper atoms on graphdiyne for highly efficient hydrogen production. *ChemPhysChem* 21: 2145–2149.
13 Yu, H., Hui, L., Xue, Y. et al. (2020). 2D graphdiyne loading ruthenium atoms for high efficiency water splitting. *Nano Energy* 72: 104667.

14 He, J.J., Ma, S.Y., Zhou, P. et al. (2012). Magnetic properties of single transition-metal atom absorbed graphdiyne and graphyne sheet from DFT+U calculations. *Journal of Physical Chemistry C* 116: 26313–26321.

15 Li, J., Zhong, L., Tong, L. et al. (2019). Atomic Pd on graphdiyne/graphene heterostructure as efficient catalyst for aromatic nitroreduction. *Advanced Functional Materials* 29: 1905423.

16 Huang, B. (2016). 4f fine-structure levels as the dominant error in the electronic structures of binary lanthanide oxides. *Journal of Computational Chemistry* 37: 825–835.

17 Huang, B. (2017). The screened pseudo-charge repulsive potential in perturbed orbitals for band calculations by DFT+U. *Physical Chemistry Chemical Physics* 19: 8008–8025.

18 Yin, X.-P., Wang, H.-J., Tang, S.-F. et al. (2018). Engineering the coordination environment of single-atom platinum anchored on graphdiyne for optimizing electrocatalytic hydrogen evolution. *Angewandte Chemie International Edition* 57: 9382–9386.

19 Yin, X.-P., Tang, S.-F., Zhang, C. et al. (2020). Graphdiyne-based Pd single-atom catalyst for semihydrogenation of alkynes to alkenes with high selectivity and conversion under mild conditions. *Journal of Materials Chemistry A* 8: 20925–20930.

20 Walter, M.G., Warren, E.L., McKone, J.R. et al. (2010). Solar water splitting cells. *Chemical Reviews* 110: 6446–6473.

21 Pham, T.A., Ping, Y., and Galli, G. (2017). Modelling heterogeneous interfaces for solar water splitting. *Nature Materials* 16: 401–408.

22 Popczun, E.J., McKone, J.R., Read, C.G. et al. (2013). Nanostructured nickel phosphide as an electrocatalyst for the hydrogen evolution reaction. *Journal of the American Chemical Society* 135: 9267–9270.

23 Li, H., Tsai, C., Koh, A.L. et al. (2016). Activating and optimizing MoS_2 basal planes for hydrogen evolution through the formation of strained sulphur vacancies. *Nature Materials* 15: 48–53.

24 Yin, H., Zhao, S., Zhao, K. et al. (2015). Ultrathin platinum nanowires grown on single-layered nickel hydroxide with high hydrogen evolution activity. *Nature Communications* 6: 6430.

25 Voiry, D., Yamaguchi, H., Li, J. et al. (2013). Enhanced catalytic activity in strained chemically exfoliated WS_2 nanosheets for hydrogen evolution. *Nature Materials* 12: 850–855.

26 Cheng, N., Stambula, S., Wang, D. et al. (2016). Platinum single-atom and cluster catalysis of the hydrogen evolution reaction. *Nature Communications* 7: 13638.

27 Shi, M.-M., Bao, D., Wulan, B.-R. et al. (2017). Au sub-nanoclusters on TiO_2 toward highly efficient and selective electrocatalyst for N_2 conversion to NH_3 at ambient conditions. *Advanced Materials* 29: 1606550.

28 Lv, C., Yan, C., Chen, G. et al. (2018). An amorphous noble-metal-free electrocatalyst that enables nitrogen fixation under ambient conditions. *Angewandte Chemie International Edition* 57: 6073–6076.

29 Tao, H., Choi, C., Ding, L.-X. et al. (2019). Nitrogen fixation by Ru single-atom electrocatalytic reduction. *Chem* 5: 204–214.

30 Wang, J., Yu, L., Hu, L. et al. (2018). Ambient ammonia synthesis via palladium-catalyzed electrohydrogenation of dinitrogen at low overpotential. *Nature Communications* 9: 1795.

31 Shi, M.-M., Bao, D., Li, S.-J. et al. (2018). Anchoring PdCu amorphous nanocluster on graphene for electrochemical reduction of N_2 to NH_3 under ambient conditions in aqueous solution. *Advanced Energy Materials* 8: 1800124.

32 Gao, Y., Cai, Z., Wu, X. et al. (2018). Graphdiyne-supported single-atom-sized fe catalysts for the oxygen reduction reaction: DFT predictions and experimental validations. *ACS Catalysis* 8: 10364–10374.

33 Gong, K., Du, F., Xia, Z. et al. (2009). Nitrogen-doped carbon nanotube arrays with high electrocatalytic activity for oxygen reduction. *Science* 323: 760–764.

34 Liu, X. and Dai, L. (2016). Carbon-based metal-free catalysts. *Nature Reviews Materials* 1: 16064.

35 Guo, D., Shibuya, R., Akiba, C. et al. (2016). Active sites of nitrogen-doped carbon materials for oxygen reduction reaction clarified using model catalysts. *Science* 351: 361–365.

36 Xue, Y., Guo, Y., Yi, Y. et al. (2016). Self-catalyzed growth of Cu@graphdiyne core shell nanowires array for high efficient hydrogen evolution cathode. *Nano Energy* 30: 858–866.

37 Yu, H., Xue, Y., Hui, L. et al. (2018). Efficient hydrogen production on a 3D flexible heterojunction material. *Advanced Materials* 30: 1707082.

38 Yu, H., Xue, Y., Hui, L. et al. (2018). Controlled growth of MoS_2 nanosheets on 2D N-doped graphdiyne nanolayers for highly associated effects on water reduction. *Advanced Functional Materials* 28: 1707564.

39 Hui, L., Xue, Y., He, F. et al. (2019). Efficient hydrogen generation on graphdiyne-based heterostructure. *Nano Energy* 55: 135–142.

40 Yao, Y., Jin, Z., Chen, Y. et al. (2018). Graphdiyne-WS_2 2D-nanohybrid electrocatalysts for high-performance hydrogen evolution reaction. *Carbon* 129: 228–235.

41 Hui, L., Xue, Y., Huang, B. et al. (2018). Overall water splitting by graphdiyne-exfoliated and -sandwiched layered double-hydroxide nanosheet arrays. *Nature Communications* 9: 5309.

42 Yu, H., Xue, Y., Hui, L. et al. (2019). Graphdiyne-engineered heterostructures for efficient overall water-splitting. *Nano Energy* 64: 103928.

43 Fang, Y., Xue, Y., Hui, L. et al. (2019). In situ growth of graphdiyne based heterostructure: toward efficient overall water splitting. *Nano Energy* 59: 591–597.

44 Hui, L., Jia, D., Yu, H. et al. (2019). Ultrathin graphdiyne-wrapped iron carbonate hydroxide nanosheets toward efficient water splitting. *ACS Applied Materials & Interfaces* 11: 2618–2625.

45 Shi, G., Fan, Z., Du, L. et al. (2019). In situ construction of graphdiyne/CuS heterostructures for efficient hydrogen evolution reaction. *Mater Chem Front* 3: 821–828.

46 Berardi, S., Drouet, S., Francas, L. et al. (2014). Molecular artificial photosynthesis. *Chemical Society Reviews* 43: 7501–7519.

47 Zhang, B. and Sun, L. (2019). Artificial photosynthesis: opportunities and challenges of molecular catalysts. *Chemical Society Reviews* 48: 2216–2264.

48 Li, J., Gao, X., Liu, B. et al. (2016). Graphdiyne: a metal-free material as hole transfer layer to fabricate quantum dot-sensitized photocathodes for hydrogen production. *Journal of the American Chemical Society* 138: 3954–3957.

49 Han, Y.-Y., Lu, X.-L., Tang, S.-F. et al. (2018). Metal-free 2D/2D heterojunction of graphitic carbon nitride/graphdiyne for improving the hole mobility of graphitic carbon nitride. *Advanced Energy Materials* 8: 1702992.

50 Si, H., Deng, Q., Yin, C. et al. (2020). Gas exfoliation of graphitic carbon nitride to improve the photocatalytic hydrogen evolution of metal-free 2D/2D g-C_3N_4/graphdiyne heterojunction. *Journal of Alloys and Compounds* 833: 155054.

51 Lv, J.-X., Zhang, Z.-M., Wang, J. et al. (2019). In situ synthesis of CdS/graphdiyne heterojunction for enhanced photocatalytic activity of hydrogen production. *ACS Applied Materials & Interfaces* 11: 2655–2661.

52 Shen, H., Zhou, W., He, F. et al. (2020). A dehydrobenzoannulene-based three dimensional graphdiyne for photocatalytic hydrogen generation using Pt nanoparticles as a co-catalyst and triethanolamine as a sacrificial electron donor. *Journal of Materials Chemistry A* 8: 4850–4855.

53 Li, Y., Yang, H., Wang, G. et al. (2020). Distinctiveimproved synthesis and application extensions graphdiyne for efficient photocatalytic hydrogen evolution. *ChemCatChem* 12: 1985–1995.

54 Xue, Y., Zuo, Z., Li, Y. et al. (2017). Graphdiyne-supported $NiCo_2S_4$ nanowires: a highly active and stable 3D bifunctional electrode material. *Small* 13: 1700936.

55 Shi, G., Xie, Y., Du, L. et al. (2020). Stabilization of cobalt clusters with graphdiyne enabling efficient overall water splitting. *Nano Energy* 74: 104852.

56 Yin, X.-P., Lu, D., Wang, J.-W. et al. (2019). 2D/2D heterojunction of Ni–Co–P/graphdiyne for optimized electrocatalytic overall water splitting. *Chem Cat Chem* 11: 5407–5411.

57 Si, H.-Y., Deng, Q.-X., Chen, L.-C. et al. (2019). Hierarchical graphdiyne@NiFe layered double hydroxide heterostructures as a bifunctional electrocatalyst for overall water splitting. *Journal of Alloys and Compounds* 794: 261–267.

58 Shi, G., Yu, C., Fan, Z. et al. (2019). Graphdiyne-supported NiFe layered double hydroxide nanosheets as functional electrocatalysts for oxygen evolution. *ACS Applied Materials & Interfaces* 11: 2662–2669.

59 Huang, H., Li, F., Zhang, Y. et al. (2019). Two-dimensional graphdiyne analogue Co-coordinated porphyrin covalent organic framework nanosheets as a stable electrocatalyst for the oxygen evolution reaction. *Journal of Materials Chemistry A* 7: 5575–5582.

60 Li, J., Gao, X., Li, Z. et al. (2019). Superhydrophilic graphdiyne accelerates interfacial mass/electron transportation to boost electrocatalytic and photoelectrocatalytic water oxidation activity. *Advanced Functional Materials* 29: 1808079.

61 Cui, J., Liu, J., Wang, C. et al. (2020). Efficient electrocatalytic water oxidation by using the hierarchical 1D/2D structural nanohybrid of CoCu-based zeolitic imidazolate framework nanosheets and graphdiyne nanowires. *Electrochimica Acta* 334: 135577.

62 Zhang, S., Yin, C., Kang, Z. et al. (2019). Graphdiyne nanowall for enhanced photoelectrochemical performance of Si heterojunction photoanode. *ACS Applied Materials & Interfaces* 11: 2745–2749.

63 Yi, Z., Ye, J., Kikugawa, N. et al. (2010). An orthophosphate semiconductor with photooxidation properties under visible-light irradiation. *Nature Materials* 9: 559–564.

64 Guo, S., Jiang, Y., Wu, F. et al. (2019). Graphdiyne-promoted highly efficient photocatalytic activity of graphdiyne/silver phosphate pickering emulsion under visible-light irradiation. *ACS Applied Materials and Interfaces* 11: 2684–2691.

65 Si, H.-Y., Mao, C.-J., Zhou, J.-Y. et al. (2018). Z-scheme Ag_3PO_4/graphdiyne/g-C_3N_4 composites: Enhanced photocatalytic O_2 generation benefiting from dual roles of graphdiyne. *Carbon* 132: 598–605.

66 Gao, X., Li, J., Du, R. et al. (2017). Direct synthesis of graphdiyne nanowalls on arbitrary substrates and its application for photoelectrochemical water splitting cell. *Advanced Materials* 29: 1605308.

67 Guo, S., Yan, H., Wu, F. et al. (2017). Graphdiyne as electrode material: tuning electronic state and surface chemistry for improved electrode reactivity. *Analytical Chemistry* 89: 13008–13015.

68 Guo, S., Yu, P., Li, W. et al. (2020). Electron hopping by interfacing semiconducting graphdiyne nanosheets and redox molecules for selective electrocatalysis. *Journal of the American Chemical Society* 142: 2074–2082.

69 Huang, C., Li, Y., Wang, N. et al. (2018). Progress in research into 2D graphdiyne-based materials. *Chemical Reviews* 118: 7744–7803.

70 Qi, H., Yu, P., Wang, Y. et al. (2015). Graphdiyne oxides as excellent substrate for electroless deposition of Pd clusters with high catalytic activity. *Journal of the American Chemical Society* 137: 5260–5263.

71 Yang, L.-L., Wang, H.-J., Wang, J. et al. (2019). A graphdiyne-based carbon material for electroless deposition and stabilization of sub-nanometric Pd catalysts with extremely high catalytic activity. *Journal of Materials Chemistry A* 7: 13142–13148.

72 Shen, H., Li, Y., and Shi, Z. (2019). A novel graphdiyne-based catalyst for effective hydrogenation reaction. *ACS Applied Materials & Interfaces* 11: 2563–2570.

73 Wang, S., Yi, L., Halpert, J.E. et al. (2012). A novel and highly efficient photocatalyst based on P25–graphdiyne nanocomposite. *Small* 8: 265–271.

74 Tan, X., Xu, J., Huang, T. et al. (2019). Graphdiyne bearing pillar[5]arene-reduced Au nanoparticles for enhanced catalytic performance towards the reduction of 4-nitrophenol and methylene blue. *RSC Advances* 9: 38372–38380.

75 Yang, N., Liu, Y., Wen, H. et al. (2013). Photocatalytic properties of graphdiyne and graphene modified TiO_2: from theory to experiment. *ACS Nano* 7: 1504–1512.

76 Ramakrishnan, V., Kim, H., and Yang, B. (2019). Improving the photo-cathodic properties of TiO_2 nano-structures with graphdiynes. *New Journal of Chemistry* 43: 12896–12899.

77 Pan, Q., Liu, H., Zhao, Y. et al. (2019). Preparation of N-graphdiyne nanosheets at liquid/liquid interface for photocatalytic NADH regeneration. *ACS Applied Materials & Interfaces* 11: 2740–2744.

78 Xing, C., Xue, Y., Huang, B. et al. (2019). Fluorographdiyne: a metal-free catalyst for applications in water reduction and oxidation. *Angewandte Chemie International Edition* 58: 13897–13903.

79 Zhao, Y., Yang, N., Yao, H. et al. (2019). Stereodefined codoping of sp-N and S atoms in few-layer graphdiyne for oxygen evolution reaction. *Journal of the American Chemical Society* 141: 7240–7244.

80 Zhang, S., Cai, Y., He, H. et al. (2016). Heteroatom doped graphdiyne as efficient metal-free electrocatalyst for oxygen reduction reaction in alkaline medium. *Journal of Materials Chemistry A* 4: 4738–4744.

81 Zhao, Y., Wan, J., Yao, H. et al. (2018). Few-layer graphdiyne doped with sp-hybridized nitrogen atoms at acetylenic sites for oxygen reduction electrocatalysis. *Nature Chemistry* 10: 924–931.

82 Liu, R., Liu, H., Li, Y. et al. (2014). Nitrogen-doped graphdiyne as a metal-free catalyst for high-performance oxygen reduction reactions. *Nanoscale* 6: 11336–11343.

83 Lv, Q., Si, W., He, J. et al. (2018). Selectively nitrogen-doped carbon materials as superior metal-free catalysts for oxygen reduction. *Nature Communications* 9: 3376.

84 Xu, Q., Zhu, B., Cheng, B. et al. (2019). Photocatalytic H_2 evolution on graphdiyne/g-C_3N_4 hybrid nanocomposites. *Applied Catalysis B: Environmental* 255: 117770.

6

Graphdiyne-Based Materials in Rechargeable Batteries Applications

Zicheng Zuo and Yuliang Li

Beijing National Laboratory for Molecular Sciences, Institute of Chemistry, Chinese Academy of Sciences, Beijing, Zhongguancun North First Street 2, 100190, PR China

6.1 Introduction

Sustainable development of human society will be a long-term issue due to the striking growth in global energy demand and serious environment problems. It is urgent to reduce the strong dependence on nonrenewable energy, especially on fossil fuels [1]. The effective harvest, conversion, storage, transportation, and utilization of nonfossil energy has become one of the hot research topics for world-wide scientists in recent years. The technologies related to clean and sustainable energy, such as solar energy, wind energy, and tidal energy, have made many great progresses. The electrochemical power sources play a vital role in the efficient use of these new forms of energy [2, 3]. Lithium-ion batteries (LIBs) are a prominent representative of the electrochemical power sources, and great successes in commercialization have been achieved [4]. The LIBs have played a significant role in the informatization, mobility, artificial intelligence, transportation, and so on [5, 6]. Meanwhile, this has put forward higher requirements in the overall performance of LIBs [7]. Facing some long-term issues in the energy density, lifespan, safety, and environmental adaptability of LIBs, researchers have become increasingly aware that to overcome these challenges requires the innovations in material designs and preparations [8–10].

Carbon element has sp, sp^2, and sp^3 hybrid forms, and can theoretically form a large number of carbon allotropes [11–16]. Different combinations of carbon elements lead to diverse physical and chemical properties, making carbon materials range from soft to hard, insulating to conducting, and dense to loose. The adjustable properties of carbon materials offer them broad application prospects in many fields, such as optics, electronics, magnetism, mechanics, and electrochemistry [11, 17, 18]. Specially, carbon materials played an indispensable role in promoting the performance of the electrochemical energy devices, due to [9, 17, 19] their unique chemical and physical properties.

In recent years, with the help of advanced instruments, researchers have conducted a lot of deep explorations and analyses on the key scientific issues in the

Graphdiyne: Fundamentals and Applications in Renewable Energy and Electronics,
First Edition. Edited by Yuliang Li.

electrochemical energy fields [7, 20]. A large number of fundamental investigations demonstrate a general consensus that new breakthroughs in the preparation and application of carbon materials are needed to build more excellent electrochemical energy devices. Since the preparation of conventional carbon materials is carried out under harsh conditions such as high temperature, high pressure, and inert gas protection, it is difficult to accurately control the molecular structure and aggregation morphologies, which severely restricts their better applications in electrochemical energy fields [14, 21–23]. In this case, the emergence of graphdiyne, a large π-conjugated two-dimensional porous carbon allotrope, broke the traditional way of preparing carbon materials. Graphdiyne was first synthesized from Professor Yuliang Li's group in 2010 [24]. The successful preparation of graphdiyne has attracted widespread attention in the worldwide scientific community [25–28]. Graphdiyne is formed by the periodic connection of acetylene bonds and double bonds in a two-dimensional plane, which is different from the traditional carbon materials made up of sp^2- or sp^3-hybrid carbon atoms. Through the precise design of molecular structure, graphdiyne can have many derivatives [13, 15] (Figure 6.1a). The one-dimensional linear characteristics of the acetylene bond and the two-dimensional planar characteristics of the benzene ring construct the natural porous structure of graphdiyne, and the pore size can be precisely controlled by the number of acetylene bonds. For electrochemical energy, two-dimensional graphdiyne with a large conjugated structure has two inherent advantages - low-temperature preparation and adjustable in-plane pores, both of which perfectly complement conventional carbon materials. Thus, the graphdiyne opens a new chapter for the applications of carbon materials in green electrochemical energy, and provides novel solutions to many key problems of electrochemical energy storage [33–36].

Graphdiyne is formed by the cross-coupling reaction of the precursors with certain structures in present of copper catalysts. The reaction process is mild and the precursor structures are diverse. Therefore, graphdiynes with corresponding structures can be designed and synthesized according to actual requirements. The all-carbon skeleton perfectly inherits the excellent physical and chemical stability of carbon materials, while the acetylene-rich structure also endows graphdiyne with good electron and hole conductivity [37, 38]. The combination of graphdiyne with traditional materials can produce new concepts and ideas, and bring new phenomena and properties [39–42]. The preparation of graphdiyne on a Cu substrate is an original classic synthesis method, and the large-area graphdiyne films can be obtained based on this method [24, 27, 41, 43]. This is also the current mainstream method for exploring and studying graphdiyne in electrochemical applications. Through precise control of the number of acetylene carbons between the benzene rings, the pore size in the graphdiyne plane can be precisely controlled at the atomic level [15, 44]. Via selectively adjusting and designing the reaction catalysts, solvent systems, ligands, templates, etc. in this method, the controllable and efficient growth of graphdiyne can be achieved [45–49]. Up to now, we have seen many reports on the synthesis of graphdiyne family members [24, 50–52]. Except for the solution method of graphdiyne preparation, the chemical vapor deposition (CVD) [53–56] method

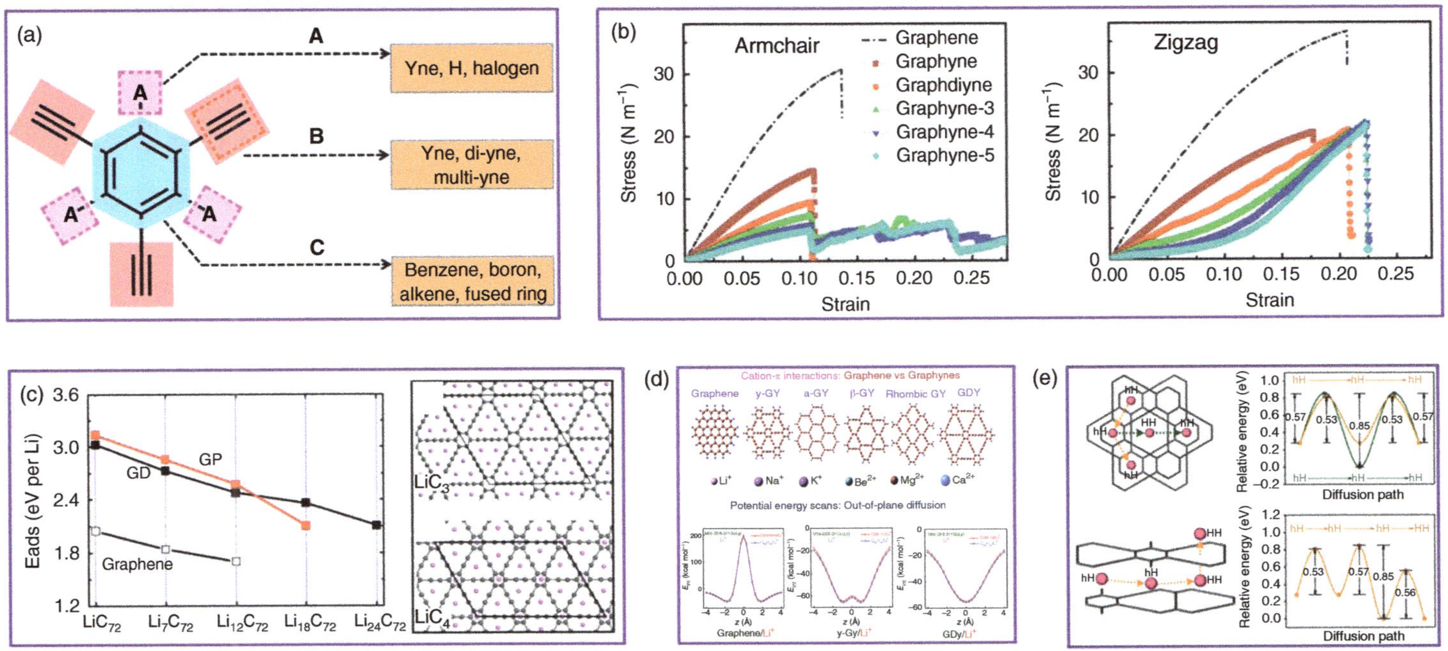

Figure 6.1 (a) Structural character of the precursors for the graphdiyne derivatives; (b) mechanical properties of the graphdiyne derivatives. Source: Modified from Kang et al. [29]. (c) Storage mechanisms of Li ions in the graphdiyne. Source: Sun et al. [30]. © 2012, American Chemical Society. (d) The interaction between the graphdiyne and various alkali metal ions. Source: Shekar et al. [31]. © 2015; American Chemical Society. (e) The energy barriers of the in-plane and cross-plane Li ion transfer in graphdiyne. Source: Zhang et al. [32]. © 2011, American Chemical Society.

and the solid-phase reaction [57–59] were achieved, respectively. Based on these systematical synthetic methods developed, we can efficiently promote the application of graphdiyne in wider ranges, especially for the electrochemical energy fields.

Theoretical simulation technology shows that graphdiyne has good mechanical properties (Figure 6.1b), the in-plane mechanical modulus can reach as high as up to $166\,N\,m^{-1}$ [29, 60], and the design of the number of acetylene bonds is the key to directly control the mechanical strength and strain range of the material [44, 61, 62]. The adjustable stress–strain properties of graphdiyne can meet the requirements of different applications in the mechanical characteristics of carbon materials. In the storage of alkali metal ions (Li^+, Na^+, K^+, etc.), a large number of experiments and theoretical calculations have fully revealed that graphdiyne is a promising negative-electrode material (Figure 6.1c,d). The interlayer distance of graphdiyne planes is 3.6 Å, which is larger than that of graphite (3.4 Å). Compared with graphite, graphdiyne has a higher theoretical storage density (lithium ion storage capacity up to $744\,mAh\,g^{-1}$) and a smaller interlayer transmission energy barrier. It is a new type of carbon material for high-energy density and high-power density batteries [30, 31, 63–66]. The interlayer stacking of graphdiyne is determined to be ABC stacking model [45, 47, 67], and this kind of stacking will not block the ions cross-plane transfer channels in the applications (Figure 6.1e). In this chapter, we will focus on the research progress of graphdiyne in electrochemical energy storage. The above discussion on the comprehensive properties of graphdiyne (mechanics, electronics, chemistry, stability, etc.) fully shows the great application potential of graphdiyne as an emerging energy carbon material.

6.2 Lithium-Ion Battery Anodes

With the rapid development in electric vehicles, the sluggish advancements in energy density of LIBs are becoming one of the main obstacles. Increasing the energy density of LIBs attracts wide attention. The practical capacity of the graphite in commercial LIBs has been close to its theoretical limit ($372\,mAh\,g^{-1}$), and it is difficult to produce remarkable breakthroughs in the energy density based on graphite anode. Exploring new materials for anode applications in LIBs is a hot research topic. Graphdiyne is an emerging two-dimensional carbon material. The precise control in the structure demonstrates that more ideal electrochemically active sites can be added to the graphdiyne. This not only adjusts the intrinsic electronic structure of graphdiyne, but also produces high-performance LIB anode materials. Theoretical studies have shown that the acetylene bonds and porous network of graphdiyne provide abundant positions for storing lithium ions. The theoretical value is more than twice that of conventional graphite ($372\,mAh\,g^{-1}$) [32, 65, 66]. Not only that, the special three-dimensional ion transport channel of graphdiyne also ensures the rapid transfer of lithium ions, thus beneficial for increasing the power performance. The all-carbon skeleton network of graphdiyne guarantees a stable structure in a long lifespan. In recent studies, researchers have tuned the aggregation structures and controlled the doping states of graphdiyne, and the effects of morphologies and doping states on the lithium storage were studied.

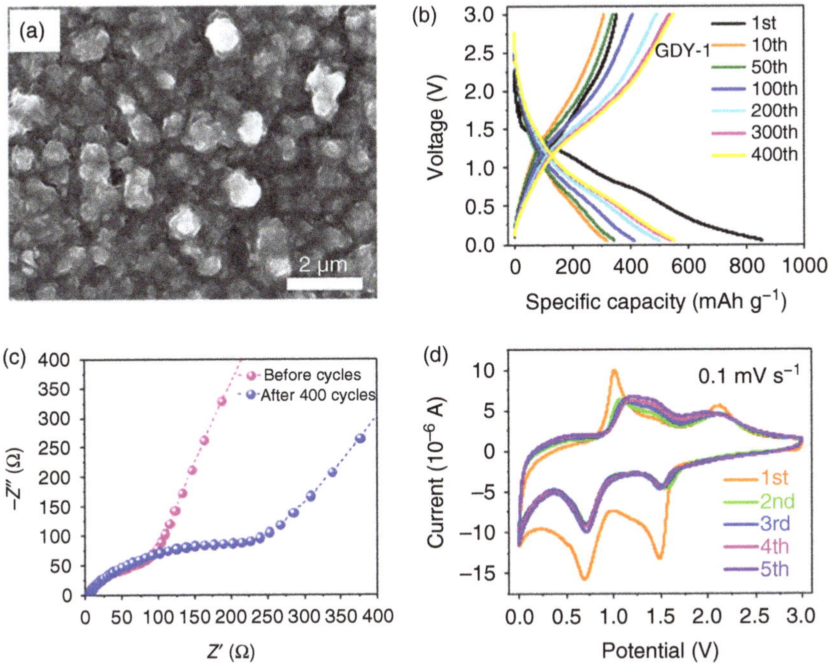

Figure 6.2 Performance of graphdiyne anode in LIB. (a) Morphology of the graphdiyne film after 400 cycles. (b) Charge/discharge profile of the graphdiyne electrode at a current density of 500 mA g^{-1}. (c) Nyquist plots of the graphdiyne electrode before and after 400 cycles at a current density of 500 mA g^{-1}. (d) Cyclic voltammetry curves of the graphdiyne electrode at a scan rate of 0.1 mV s^{-1}. Source: Huang et al. [68]. © 2015, Elsevier.

Professor Huang et al. [68] first explored the performance of graphdiyne film on copper foils in storing the lithium ion. Since the graphdiyne film can be in situ grown on the copper surface, it can be directly used as the electrode for investigating its property in storing the lithium ion without any polymer binders or conductive additives (Figure 6.2). In the test, the battery has a reversible specific capacity of up to 520 mAh g^{-1} after 400 cycles under a current density of 500 mA g^{-1}. At a high current density of 2 A g^{-1}, the graphdiyne electrode still maintains its capacity as high as 420 mAh g^{-1} after 1000 cycles. The graphdiyne films with different thicknesses were also investigated for storing the Li atoms, showing robust performance. Their further theoretical simulation indicates that the mechanism for storing Li ions in such graphdiyne film is mainly via the interlayer intercalation/extraction and surface adsorption/desorption. Through this mechanism, the graphdiyne electrode can realize high specific capacity, good power performance, and stable cycling.

Similar with the method for increasing the capacity of conventional carbon materials, improving the porous structure of graphdiyne is also an effective approach to increase the electrochemical performance. Zhang et al. [69] found the method to control the pores in the graphdiyne film (Figure 6.3). The as-prepared graphdiyne electrode has a hierarchical structure, including the microporous and mesoporous

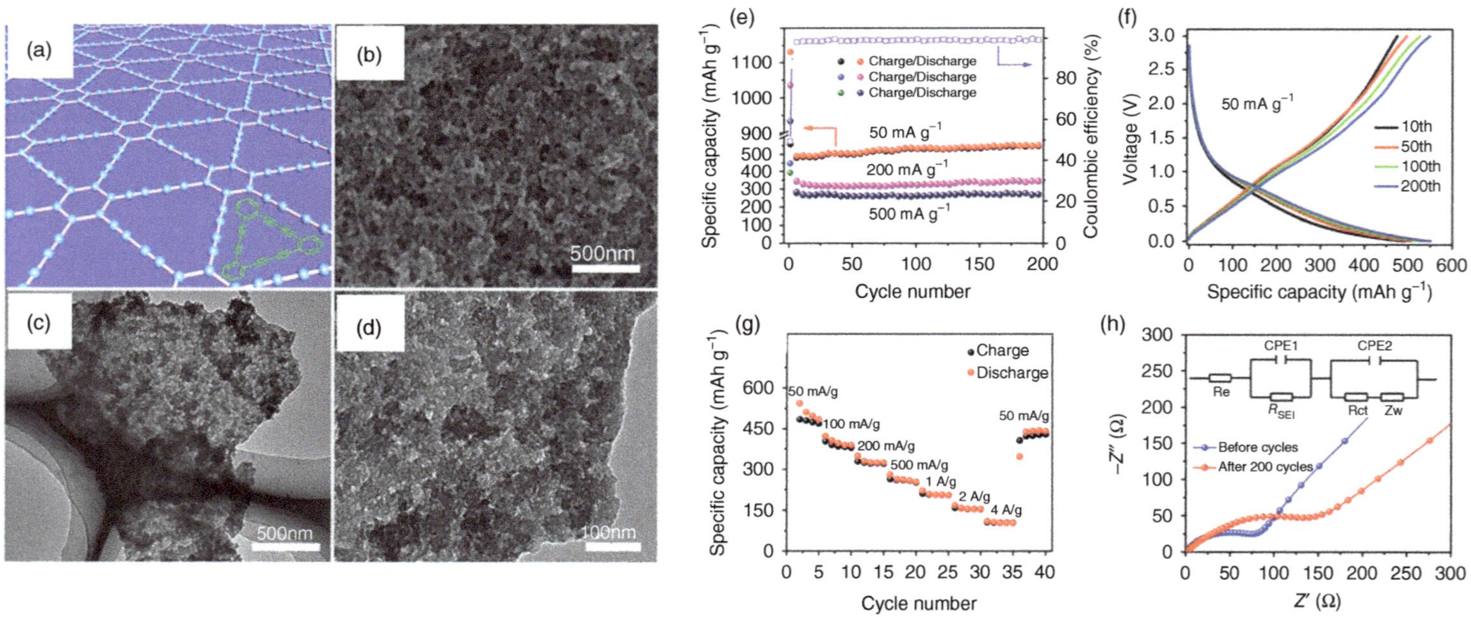

Figure 6.3 (a) Structure of graphdiyne; (b) SEM and (c, d) TEM images of graphdiyne sample; (e) cycle performance and coulombic efficiency; (f) charge–discharge profiles and (g) rate performance of the graphdiyne electrode; (h) Nyquist plots of the anode before and after 200 cycles, and the inset is the equivalent circuit. Source: Zhang et al. [69]. © 2015, Royal Society of Chemistry.

pores (Figure 6.3a–d). The specific area can be increased up to 320.9 $m^2\ g^{-1}$, and the pore size is mainly distributed in 0.81, 1.56, 2.98, and 8.12 nm. Such a porous structure in graphdiyne electrode is beneficial for the high-speed conductivity of electrons and ions and also offers more active sites to store Li^+. Using such a graphdiyne as an anode, the reversible capacity is of 552 $mAh\ g^{-1}$ at a current density of 50 $mA\ g^{-1}$. Meanwhile, it maintains a good stability after 200 cycles (Figure 6.3e,f). Due to the characteristics of hierarchical pore structure, the rate performance of the electrode is excellent (Figure 6.3g,h). When the charge and discharge current density is 1, 2, and 4 $A\ g^{-1}$, the high reversible capacities of 210, 158, and 105 $mAh\ g^{-1}$ are maintained.

The thermal treatment has strong influence on the performance of graphdiyne film as the LIB anode [70]. After a calcination treatment at 200 °C for two hours, the scanning electron microscope (SEM) measurement demonstrates that there are many graphdiyne nanoparticles with a diameter of about 700 nm on the surface of the graphdiyne film (Figure 6.4a). Further magnification shows that the film is

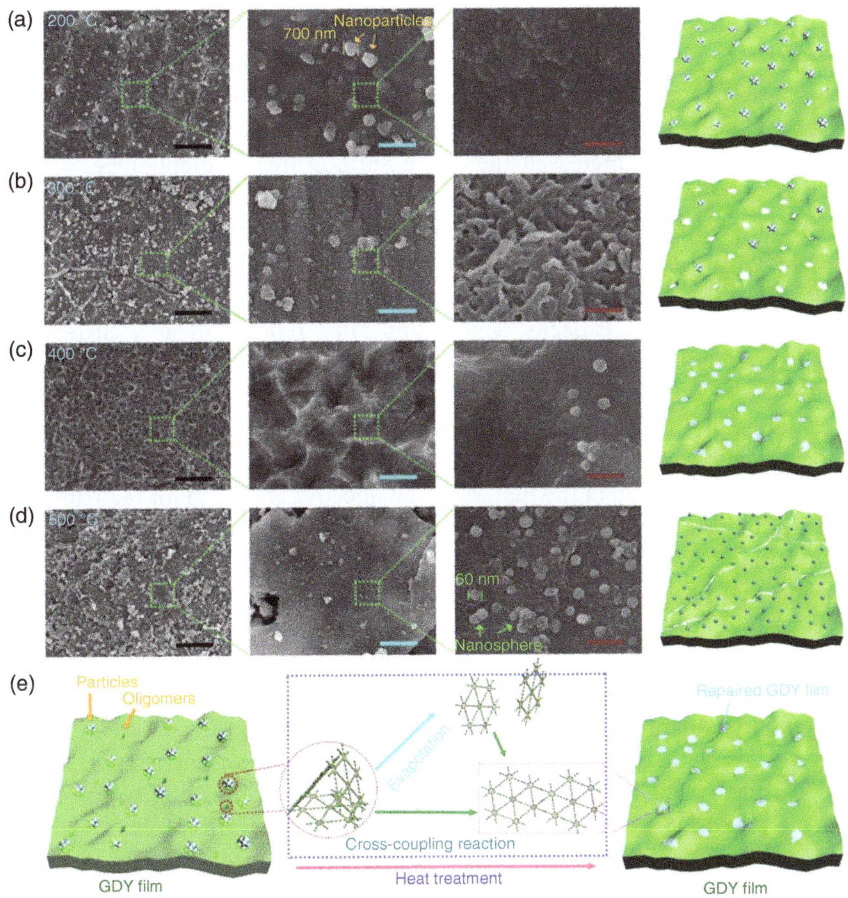

Figure 6.4 SEM images of graphdiyne film treated at different temperatures: (a) 200 °C for 2 hours, (b) 300 °C for 2 hours, (c) 400 °C for 2 hours, and (d) 500 °C for 2 hours. Scale bars: 20 mm (black), 2 mm (green), 200 nm (red). (e) The proposed growth mechanism of uniform graphdiyne film. Source: He et al. [70]. © 2018, Wiley-VCH Verlag GmbH & Co. KGaA.

composed of molten nanoparticles. After calcination at 300 °C, many of the nanoparticles were eliminated, because the oligomers with lower vaporization temperature were evaporated (Figure 6.4b). When the temperature was increased to 400 °C, most of the nanoparticles disappeared, forming a uniform and smooth graphdiyne film (Figure 6.4c). This phenomenon is ascribed to the evaporation of oligomers and the self-repairing process of graphdiyne under such condition. However, as the temperature is increased up to 500 °C, the uniform film was destroyed, and many graphdiyne nanospheres with a diameter of 60 nm were produced (Figure 6.4d,e). The electrochemical performance shows that, except for the poor cycle performance of the sample treated at 500 °C, the thermal treatment makes the graphdiyne electrode give a better cyclability. After 150 cycles at a current density of 100 mA g^{-1}, the reversible capacity of the samples treated at 200, 300, 400, and 500 °C was about 272, 689, 901, and 578 mAh g^{-1}, respectively. The low capacity of the free-standing film treated at 200 °C is due to the reduction of electronic conductivity. However, compared with the untreated sample, the performances of the samples are significantly increased, because the conductivity of graphdiyne film was greatly improved after the thermal treatments.

Different from the membrane structure, the graphdiyne nanowall array [43] can be obtained by modified solution method. Professor Jin Zhang's group adjusted the solution system, and a large-area graphdiyne nanowall structure was obtained. The nanowall structure well exhibited the two-dimensional planar properties of graphdiyne and further increased the active surface area of graphdiyne, which is beneficial for the rapid diffusion and transport of lithium ions. The nanosheet array has a three-dimensional interconnected structure (Figure 6.5a–h). Therefore, such graphdiyne also shows excellent properties in storing the lithium ions as the anode of LIBs. The reversible capacity is 908 mAh g^{-1} at a current of 0.05 A g^{-1}, and there is no obvious capacity degradation after 1000 cycles at a current of 1 A g^{-1} (Figure 6.5i). While cycling at 2 A g^{-1} for 1000 cycles, the specific capacity of the electrode remains as high as 526 mAh g^{-1} (Figure 6.5j). Furthermore, the graphdiyne nanowall can also be used for high-performance sodium ion batteries and capacitor electrodes [71].

Reducing the dimension of the material is an effective approach for realizing the high-performance electrode. Above observations have proved the great potential of graphdiyne for storing the Li^+. For further increasing the performance of graphdiyne, the highly active copper nanowires with a high surface area were used as the catalyst and substrate for the in situ growth of graphdiyne (Figure 6.6a–i) [72]. The reaction was carried out under more mild conditions, and the reaction time was greatly shortened. The highly active catalyst makes the average thickness of the graphdiyne nanosheets about 2 nm. The ultra-thin nanosheets well display the two-dimensional characteristics of graphdiyne. More transport channels and active sites were exposed, which is advantageous for performing potential of graphdiyne in storing the lithium ions. In the electrochemical test, the reversible specific capacity of graphdiyne prepared by this method is as high as 1388 mAh g^{-1} (Figure 6.6j). The rate performance is impressive. When the charge and discharge current density is increased up to 10 and 20 A g^{-1}, the sample still retains a capacity

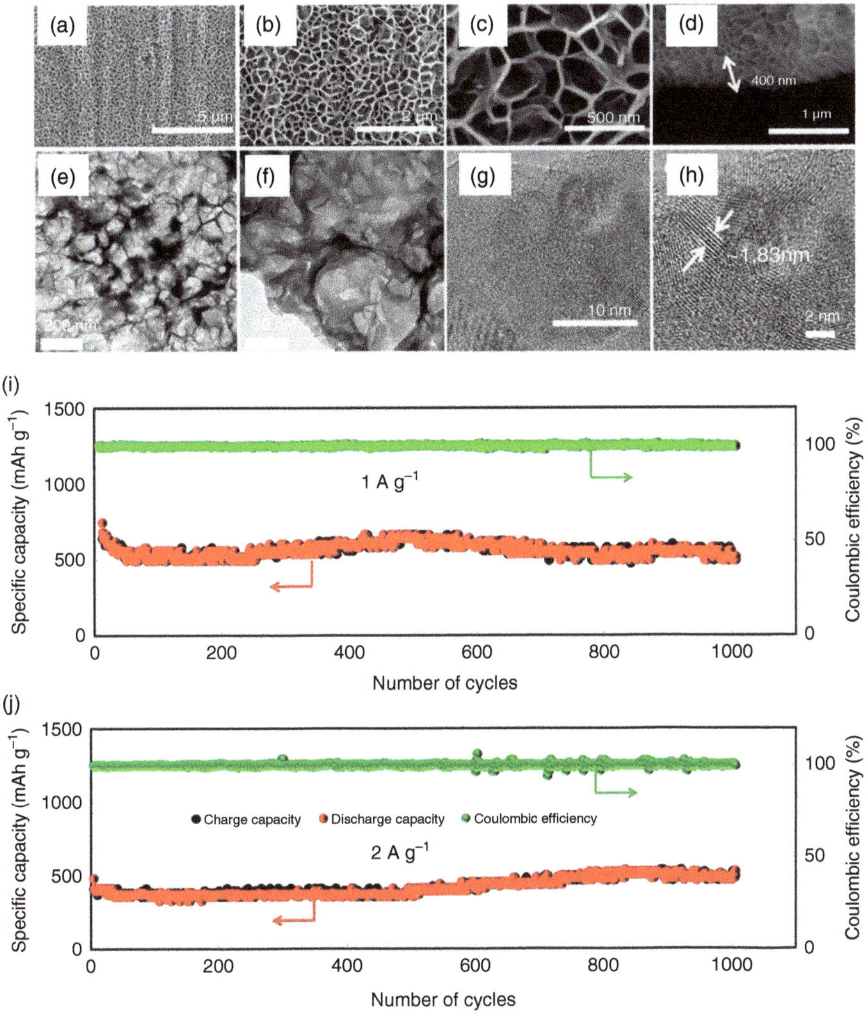

Figure 6.5 Morphologies of graphdiyne nanowall: (a–c) Top view of graphdiyne nanowall in different magnifications; (d) cross section of graphdiyne nanowall; (e–h) TEM image of the samples in different scale; the cycle performance of the graphdiyne nanowall at the current density of (i) 1 A g^{-1} and (j) 2 A g^{-1}. Source: Wang et al. [71]. © 2017, Elsevier.

of 870 and 449.8 mAh g^{-1}, respectively (Figure 6.6k,l). The sample not only has good rate performance, but also has good cycle stability. The results demonstrate that graphdiyne is a promising negative-electrode material for LIBs.

The graphdiyne in situ grown on the Cu substrates can be directly used as the anode for the batteries. However, the high cost and low specific surface area (SSA) of Cu substrates severely prevent the large-scale preparation and application of graphdiyne. It is necessary to evaluate the performance of graphdiyne prepared with low amount of Cu. The researchers prepared three-dimensional porous graphdiyne

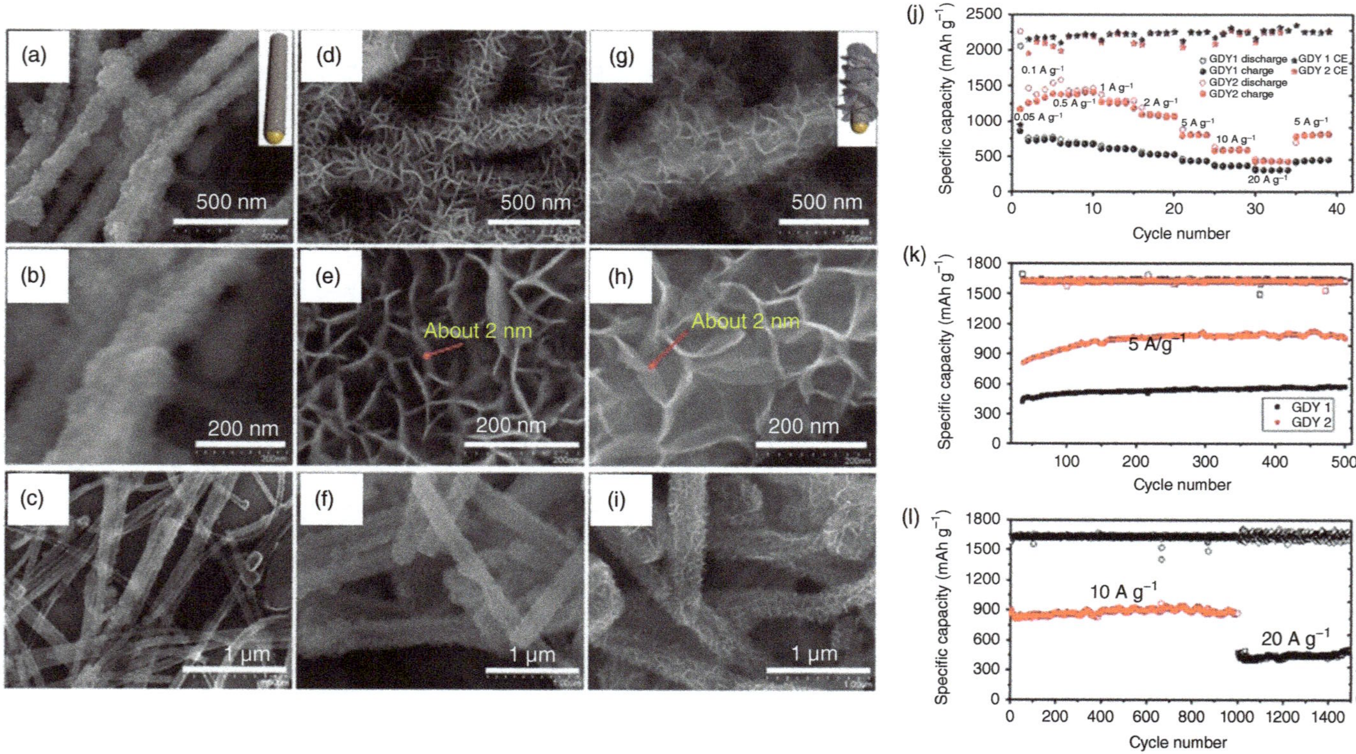

Figure 6.6 Morphologies of the ultrathin graphdiyne catalyzed by the Cu nanowires using different amounts of precursor and their morphologies after removing Cu nanowires: (a–c) 1 mg; (d–f) 2.5 mg; (g–i) 5 mg; (j) variations in specific capacity at different current density; long-term stability of samples at (k) 5 A g^{-1} and (l) at 10 and 20 A g^{-1}. Source: Shang et al. [72]. © 2018, Wiley-VCH Verlag GmbH & Co. KGaA.

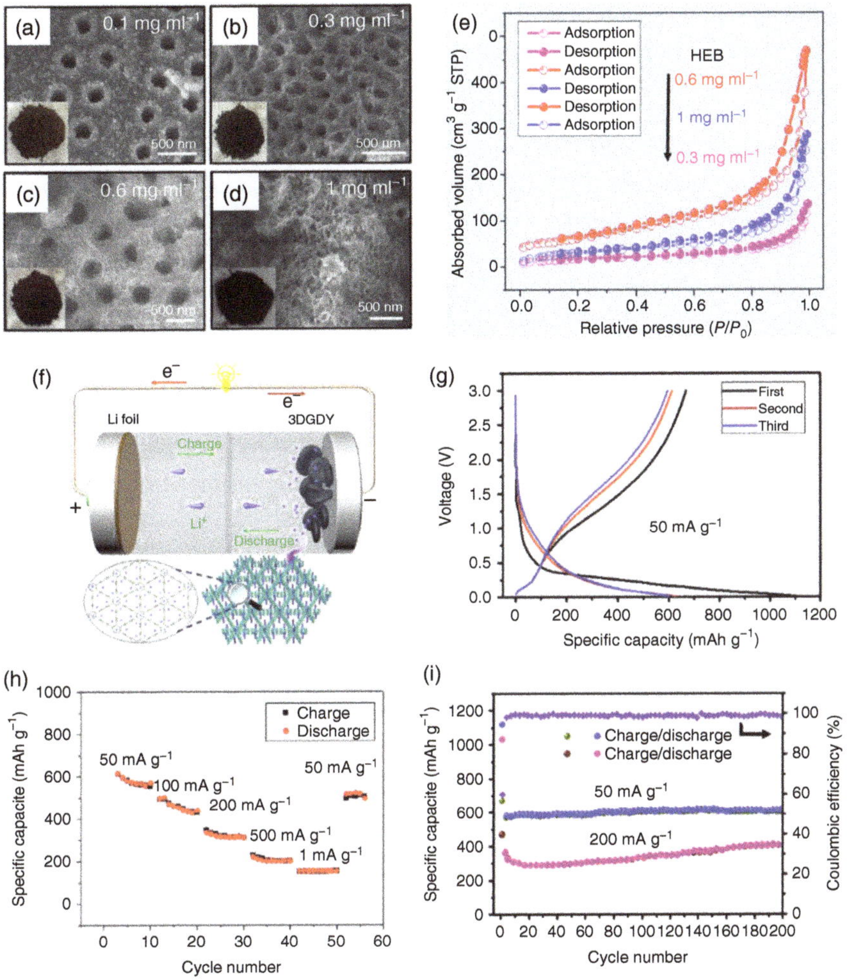

Figure 6.7 (a–d) SEM images and (e) nitrogen adsorption–desorption isotherms of the samples prepared with the precursor of (a) 0.1, (b) 0.3, (c) 0.6, and (d) 1 mg ml^{-1}; (f) schematic illustration of the sample for LIB. (g) charge/discharge curves, (h) rate performance, and (i) cycle performance at different current densities of 50 and 200 mA g^{-1}. Source: Li et al. [73]. © 2018, Wiley-VCH Verlag GmbH & Co. KGaA.

using the cheap diatomaceous earth as a template (Figure 6.7a–d) [73]. First, they used a simple metal displacement reaction to adsorb copper nanoparticles on the surface and pores in the diatomite. The adsorbed copper nanoparticles and diatomaceous earth act as the catalyst and a template, respectively, so that the graphdiyne was grown on the surface of the diatomaceous earth via the cross-coupling reaction. After the reaction, graphdiyne nanosheets were wrapped on the diatomaceous earth to form a unique three-dimensional architecture. After removing residual copper and diatomaceous earth, a three-dimensional graphdiyne with a hierarchical structure is obtained. The SSA of the prepared graphdiyne sample can be controlled at 220–369 $m^2\ g^{-1}$ (Figure 6.7e). In the test, the three-dimensional graphdiyne electrode showed stable cycle performance at a current density of 50 $mA\,g^{-1}$ (Figure 6.7f–h), and after 200 cycles, the reversible capacity was still as high as 610 $mAh\,g^{-1}$ (Figure 6.7i). After the seventh cycle, the electrode was activated, and the Coulombic efficiency of the battery was significantly improved, with the efficiency being more than 98%. This three-dimensional graphdiyne also exhibited good cycle stability at a current density of 500 $mA\,g^{-1}$, and its reversible capacity was 250 $mAh\,g^{-1}$ after 400 cycles.

Graphdiyne has abundant unsaturated acetylene bonds, and the doping behavior of graphdiyne may be more diverse than traditional carbon materials, which can bring new inspirations for electrochemical applications. Similar to traditional carbon materials, doping strategies may be helpful in improving the electrochemical activity of graphdiyne. Heat treatment in the presence of an external nitrogen source (ammonia gas) can achieve nitrogen doping of graphdiyne. Researchers used this method to improve the lithium ion storage performance of graphdiyne (Figure 6.8a–e) [74]. After N doping, the electrochemical performance of the sample has been significantly improved (Figure 6.8e-i). The N-doped graphdiyne electrode performed a high reversible capacity of 785 $mAh\,g^{-1}$ after 200 cycles at a current density of 200 $mA\,g^{-1}$. On the contrary, the reversible capacity of the undoped graphdiyne can only deliver a low capacity of 584 $mAh\,g^{-1}$. At a higher charge–discharge current density of 500 $mA\,g^{-1}$, the reversible capacity of the N-doped graphdiyne electrode was still as high as 761 $mAh\,g^{-1}$ after 400 cycles, which is much higher than the 559 $mAh\,g^{-1}$ of the undoped graphdiyne. Although the capacities of the N-doped and undoped graphdiyne electrodes were found to be slightly increased in the first 200 cycles, the capacities were greatly maintained in the following 200 cycles. This phenomenon was attributed to the gradual activation process of graphdiyne. It can be concluded that the N-doped method provides more active sites for graphdiyne in storing the lithium ion.

The graphdiyne precursor has six terminal acetylene bonds and has high reactivity. The active graphdiyne precursor can react rapidly under the thermal treatment [75]. Under different condition, the reaction can be controlled with the production of diverse nanostructures. Via the gradual thermal treatment from room temperature to 120 °C in nitrogen (10 $°C\,min^{-1}$), the light yellow precursor turns black without volume change. When this treatment process was carried out in air and the temperature reaches 90 °C, a popcorn-like explosion occurs, and the volume of the sample was increased significantly by six times. This means that oxygen catalyzed the dehydrogenation reaction and accelerated the coupling reaction. If the precursor is directly put in an air atmosphere at 120 °C, it immediately causes a more violent explosion, resulting in a 48-fold increase in volume. This method is a large-scale and efficient method for the preparation of graphdiyne, and it is also convenient for more extensive application of graphdiyne.

It is found that graphdiyne prepared through such method has different morphologies, including nanoribbons and nanochains. Because the graphdiyne nanochain has the largest SSA, the researchers tested the electrochemical properties (Figure 6.9). For Li^+ storage, at 50 $mA\,g^{-1}$, it can provide a reversible capacity of

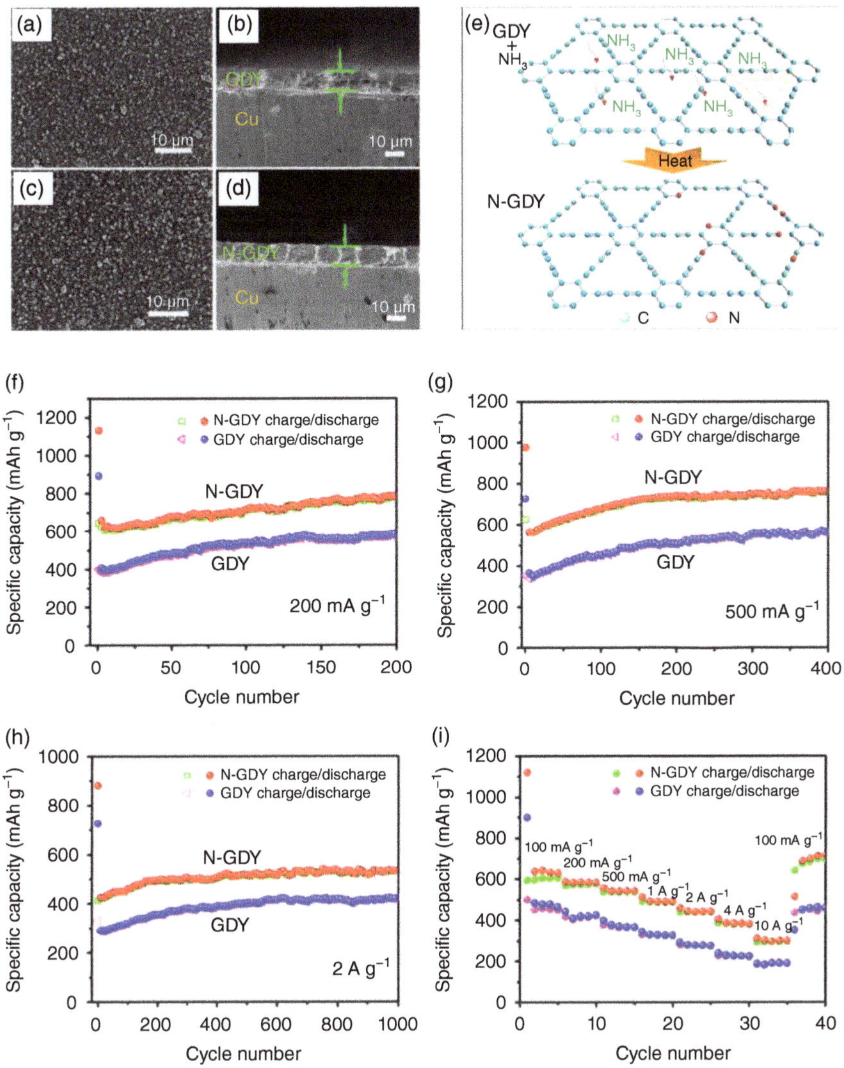

Figure 6.8 (a–d) SEM images of samples; (e) Scheme showing the N-doping of graphdiyne; (f–h) cycle performance of *N*-graphdiyne and graphdiyne electrodes under 200 mA g^{-1}, 500 mA g^{-1}, and 2 A g^{-1}, respectively; (i) different rate performance of *N*-graphdiyne and graphdiyne electrodes at varied current density. Source: Zhang et al. [74]. © 2016, American Chemical Society.

910.8 mAh g^{-1}. The first cycle Coulomb efficiency is 49.34%. The irreversibility is caused by the formation of the solid electrolyte interface in the first cycle. Graphdiyne nanochains show excellent rate performance, maintaining capacities of 410 and 250 mAh g^{-1} at high rates of 1000 and 2500 mA g^{-1}, respectively. At high current densities of 500, 1000, and 2500 mA g^{-1}, even after 500 cycles, it still shows high capacity retention. After 200 cycles at 2500 mA g^{-1}, the battery capacity is 98% of its initial specific capacity. According to the cyclic voltammetry (CV)

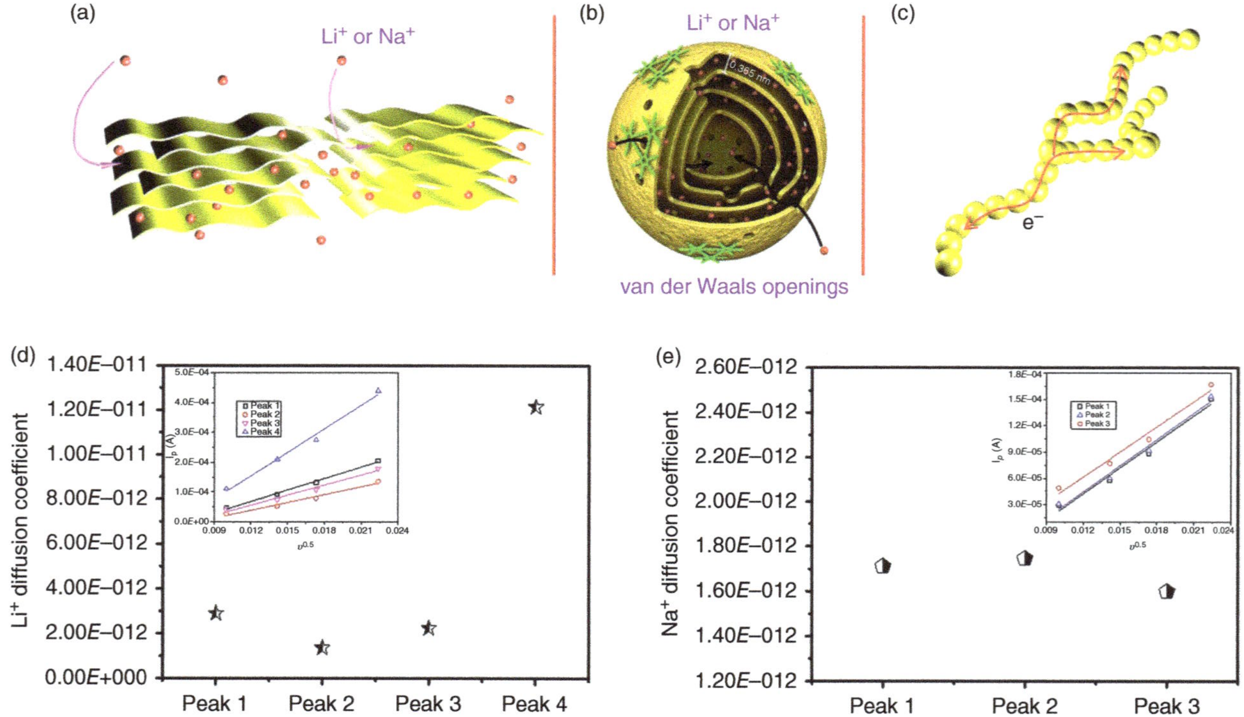

Figure 6.9 Schematic illustrations of the possible metallization/demetallization processes and electron transfer in graphite (a) and the graphdiyne nanochain (b, c), the as-obtained Li^+ (d) and Na^+ (e) diffusion coefficiency in graphdiyne samples. Source: Zuo et al. [75]. © 2017, Royal Society of Chemistry.

curves at several rates, the Li^+ diffusion coefficient in the graphdiyne ranges from 1.36×10^{-12} to 1.21×10^{-11} $cm^2\ s^{-1}$. The sodium ion diffusion coefficient is about 1.7×10^{-12} $cm^2\ s^{-1}$, slightly lower than that of the Li^+ (Figure 6.9d,e).

6.3 Graphdiyne Derivatives for LIB Anodes

One of the natural advantages of graphdiyne is that it can be prepared under mild conditions and the molecular structure of graphdiyne can be precisely adjusted by controlling the precursor. The design and preparation of different graphdiyne derivatives can enrich the understanding of the intrinsic properties of graphdiyne, and can also meet different requirements. Researchers used elements to replace the acetylene bonds on the hexaynylbenzene, and then through the typical copper-catalyzed cross-coupling reaction to synthesize different element-substituted graphdiynes, and discussed the effect of element substitution on the electrochemical behavior in storing the lithium ions.

The meta-substitution is currently the most used in the preparation of graphdiyne derivatives. The meta-position of the hexaethynylbenzene can be substituted with hydrogen atoms [76]. Using the precursor of triethynylbenzene, the H-graphdiyne can be prepared (Figure 6.10). It can be seen from the ideal molecular structure that the reduction in the number of terminal alkyne leads to the reduction of the coupling sites, so such molecular structure has a larger in-plane cavity with the diameter of 16.3 Å. Researchers can easily grow and prepare centimeter-level H-graphdiyne film. It has large-scale continuity and is transparent. Due to the reduction in the number of coplanar-conjugated acetylene bonds of the precursor, the π–π stacking effect between monomers has also been greatly weakened. Therefore, the microscale continuity of the H-graphdiyne film obtained under such preparation conditions becomes worse. This can be clearly drawn from the characterizations by the scanning electron microscopy and transmission electron microscopy. Scanning electron microscopy shows that the H-graphdiyne film has a large number of nanoscale holes, and the continuity and compactness of the H-graphdiyne film are significantly reduced. From the microscopic morphologies and molecular structure of H-graphdiyne, it can be seen that this film has a multilevel pore structure with both the macropores and nanopores. This kind of hierarchical pore structure provides more active sites and accessible nanochannels for active ions, and is also conducive for the rapid adsorption and desorption of active ions. This feature is conducive to obtaining high-power performance electrodes. Researchers used it as a negative electrode for the LIBs and sodium ion batteries, and obtained both good performances in storing these active ions. Among them, the reversible storage capacity of Li^+ is as high as 1050 mAh g^{-1} (Figure 6.10a), and the storage capacity of Na^+ is as high as 650 mAh g^{-1}. Ascribing to the 3D porous structures, the as-prepared film electrodes have high rate performance (Figure 6.10b) and robust electrochemical stability at a high current density of 1 A g^{-1} (Figure 6.10c,d). Ascribed to the continuity, this thin-film electrode can also be used to construct a flexible electrochemical energy storage device (Figure 6.10e,f).

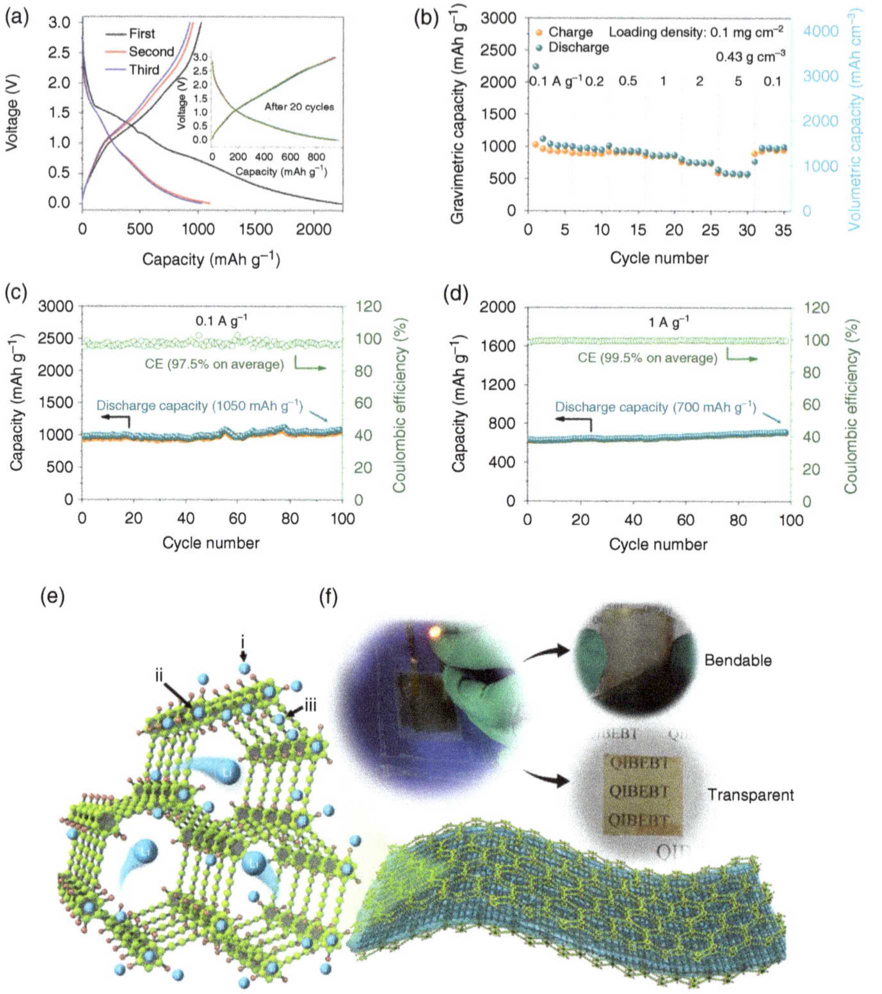

Figure 6.10 The electrochemical performance of H-graphdiyne (H-GDY) electrodes. (a) The charge–discharge profiles of the H-GDY at 0.1 A g^{-1}. (b) The rate performance of the electrode. Cycle performance of H-GDY at (c) 0.1 A g^{-1} and (d) 1 A g^{-1}. (e) The mechanism of Li storage. (f) Photo of the transparent LIB is made up of H-GDY. Source: He et al. [76]. © 2017, the Nature Publishing Group.

The halogen elements are electron-withdrawing groups and are usually used to tune the electronic distribution in the π-conjugated network. The halogen substitution in the graphdiyne is also easy to achieve. According to the molecular structure engineering, the chlorine-substituted triethynylbenzene can be synthesized by taking advantage of the different activities of the chlorine and iodine in the reaction of chlorobenzene and iodobenzene with trimethylsilylacetylene. Using this precursor, the chlorine-substituted graphdiyne (Cl-graphdiyne) was obtained by the corresponding cross-coupling of the precursor on the copper foil [77]. The chlorine element has a strong electronegativity. Through the substitution of chlorine, the electronegativity of the graphdiyne conjugated skeleton is adjusted, and the active sites of the graphdiyne are increased. Through this bottom-up synthesis method, the controllable Cl doping in graphdiyne is easily achieved. On the contrary, the Cl doping of conventional carbon materials can only be produced in a top-down way, and the doping process is low efficiency and uncontrollable for conventional carbon materials. Similar to the structure of H-graphdiyne above, the Cl-substitution reduces the conjugation of the network and weakens the intermolecular stacking. As a result, the Cl-graphdiyne film was not dense with many pores. Combined with the intrinsic in-plane pores, this membrane also has a hierarchically porous structure, offering the capability for the rapid ion transport. After the electrochemical performance test, it can be seen that the anode material has the advantages of high specific capacity (1150 $mAh\,g^{-1}$), good rate performance, and strong cycle stability (Figure 6.11). According to the theoretical calculation, the Cl-groups

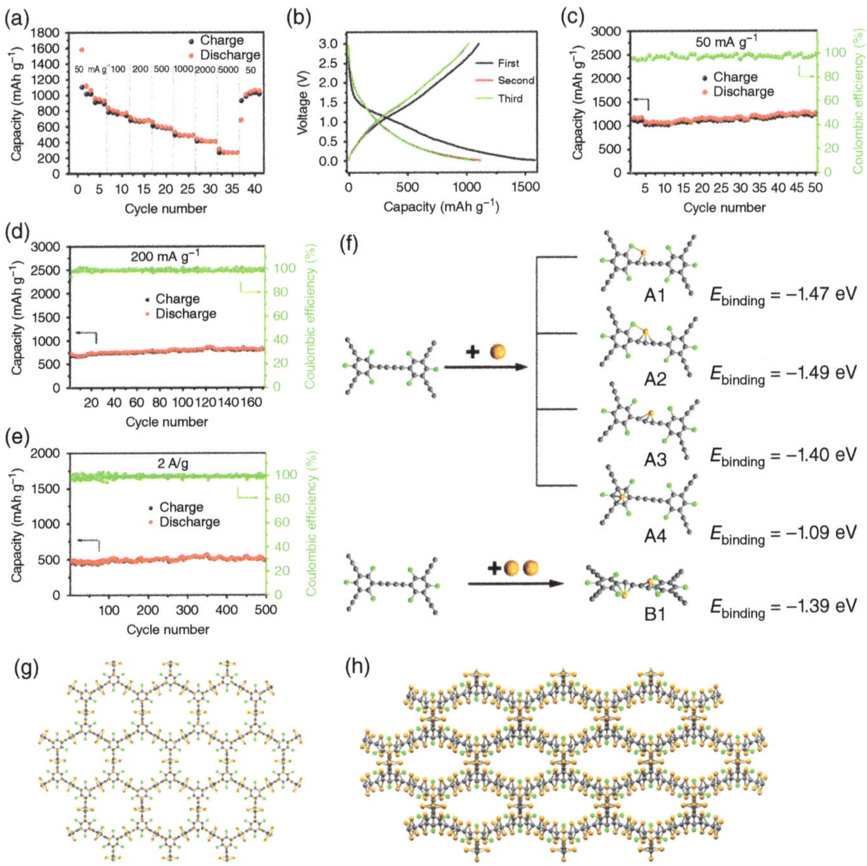

Figure 6.11 Electrochemical performance of Cl-graphdiyne in half-cell testing. (a) Rate performance; (b) the charge–discharge curves of Cl-graphdiyne in first three cycles at 50 $mA\,g^{-1}$; (c–e) The cycle performance of the flexible electrode at 50 $mA\,g^{-1}$ (c), 200 $mA\,g^{-1}$ (d), 2 $A\,g^{-1}$ (e). (f) The geometries and formation energies of four optimized Li^+ $C_{28}Cl_6$ (A1–A4) complex and one optimized $2Li^+$ $C_{28}Cl_6$ (B1). (g, h) The possible Li storage sites in Cl-graphdiyne and the function of chlorine. C: gray, Cl: green, Li: orange. Source: Wang et al. [77]. © 2017, John Wiley & Sons.

tune the binding energy between the Cl-graphdiyne and the Li atom. The Li atom on the butadiyne groups is more stable than that on the benzene ring, ascribing to the synergistic stabilization of Cl-group on the ortho-position of butadiyne group (Figure 6.11f–h). While Li^+ intercalated into the Cl-graphdiyne, the peak of Cl—C bond shifts to lower energy, further confirming the interaction between the Cl and the Li in experiment.

The electronic structure of the graphdiyne can be further adjusted by the introduction of the more electronegative fluorine element to obtain F-graphdiyne. With the help of copper-catalyzed cross-coupling reaction, large-area F-graphdiyne can be obtained [78]. The as-synthesized F-graphdiyne film is intertwined by nanowires. This film is less dense than that of the original graphdiyne. The porous morphology of F-graphdiyne can also be attributed to the reduced interlayer π–π stacking. The film can be used as flexible thin-film anode, and the electrochemical performance of the flexible electrode is tested in the potential range of 0.005–3 V. The lithium storage performance of the F-graphdiyne at a current density of 0.05–5 A g^{-1} is obtained. When the current density is 0.05 A g^{-1}, the reversible capacity is about 1700 mAh g^{-1}, and when the current density is increased to 5 A g^{-1}, it can still maintain a capacity of 300 mAh g^{-1}. Researchers attribute its high rate retention to the superior conductivity of F-graphdiyne. It can also be seen from the electrochemical curve that F-graphdiyne has an obvious peak around 0.9 V, which is different from the unsubstituted graphdiyne. The author attributed this phenomenon to the introduction of fluorine atoms, which successfully adjust the electronic structure of graphdiyne. At the same time, the cyclic voltammetry curve of fluorinated graphdiyne has characteristic peaks corresponding to this redox reaction, indicating that a new binding interaction with Li^+ is formed. This is confirmed by the evolutions in the ex situ X-ray photoelectron spectroscopy (XPS) spectra of F-graphdiyne electrode at different potential (Figure 6.12a). The theoretical simulations demonstrate that the introduction of the F-element increases the sites for storing the Li atoms; thus, the theoretical capacity of F-graphdiyne is increased up to 1867 mAh g^{-1} (Figure 6.12b,c). There is no obvious irreversibility found during the charge/discharge process. The F-graphdiyne electrode has excellent stability and still has high capacity retention after 9000 cycles at 2 A g^{-1} (Figure 6.12d).

Except for the substitution of the acetylene bonds in the network, the benzene ring can also be chemically modified. The substitution of the benzene ring in the graphdiyne network is also an important strategy to expand the graphdiyne derivatives. Via this method, the certain heteroatoms with well-defined configurations can be obtained, and the pore structure of the graphdiyne can be efficiently controlled. Contrary to the traditional carbon materials, the intrinsic electronic structure of graphdiyne can be accurately adjusted, greatly enriching the synthesis, properties, and applications of graphdiyne.

Although the thermal treatment of graphdiyne in the N-containing atmospheres is a simple way to realize N-doped graphdiyne, the N-configuration doped by this treatment shows poor controllability. Based on the unique bottom-up synthetic methodology of graphdiyne, if the structure of the benzene ring is adjusted, graphdiyne doped with well-defined N configurations can be realized. Then,

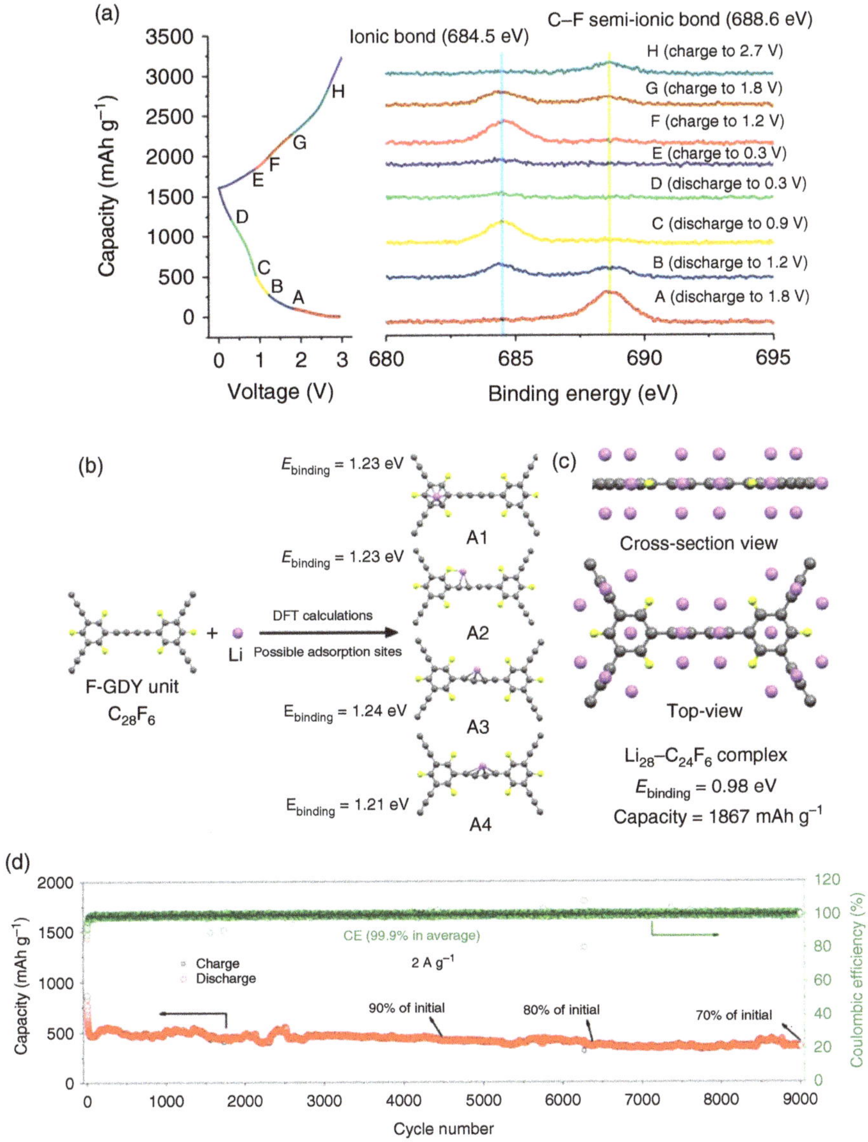

Figure 6.12 The interaction between Li atom and F-graphdiyne. (a) The ex situ XPS spectra of F-graphdiyne at different potential; (b) the binding energies of Li atom at different sites on F-graphdiyne; c top view and cross-section view of $Li_{28}-C_{24}F_6$ complex and its binding energy; (d) The cycle performance of F-graphdiyne at the current density of $2\,A\,g^{-1}$. Source: He et al. [78]. © 2018, Royal Society of Chemistry.

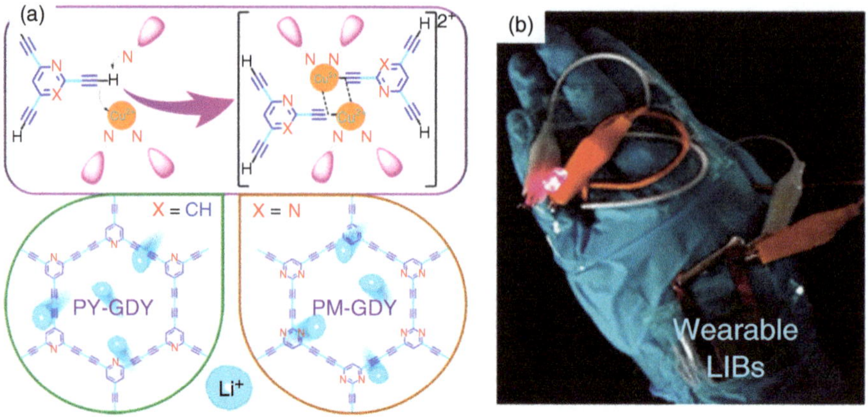

Figure 6.13 (a) Structure of N-doped graphdiyne with well-defined N-configuration for storing the Li^+, (b) photo of the assembled LIB for lighting up the light-emitting diode (LED). Source: Yang et al. [79]. © 2018, American Chemical Society.

researchers designed the precursors with the well-defined nitrogen atom to prepare the N-doped graphdiyne, which has a precisely controlled nitrogen structure and content (Figure 6.13) [47, 79]. The new films are named as pyrimidine graphdiyne and pyridine graphdiyne according to the N-configuration in the precursor. These films are self-standing, uniform, continuous, flexible, and transparent. Theoretical predictions indicate that these N-doped structures are beneficial for the storage of Li^+ because pyridine N can enhance the binding energy with the Li^+. The experimental results show that pyrimidine graphdiyne and pyridine graphdiyne have excellent electrochemical performance as anodes for LIBs. The pyrimidine graphdiyne and pyridine graphdiyne have obtained the high capacities of 1168 and 1165 mAh g^{-1} at a current density of 100 mA g^{-1}, respectively. Meanwhile, both the pyrimidine graphdiyne and pyridine graphdiyne exhibited good stability even after 4000 and 1500 cycles at a high current density of 5000 mA g^{-1}.

Using the ethylene to replace the benzene ring, a precursor of tetraethynylethylene can be designed, and the content of sp-hybridized carbon atom in the precursor is further increased up to 80%. The sp-hybridized carbon atom is one of the main reasons that provide graphdiyne a higher theoretical lithium storage capacity. From the ideal structure, it can be seen that this graphdiyne derivative has the quadrilateral pores with large pore size (Figure 6.14a). This new graphdiyne derivative is named as the carbon ene-yne [80]. The synthetic preparation of the material is also carried out under very mild conditions, and a thin-film structure can be obtained. Theoretical calculations show that the band gap of carbon ene-yne is very small, about 0.05 eV, and the conductivity of the film obtained from the experiment is increased up to 1.4×10^{-2} S m^{-1} (Figure 6.14b). These two characteristics demonstrate that the carbon ene-yne is a good electronic material. When the material is used for lithium ion storage, it can be seen that the electrode has excellent electrochemical performance (Figure 6.14c–g). At a high current density of 748 mA g^{-1}, a reversible specific capacity of 410 mAh g^{-1} can be obtained, showing great potential for LIBs.

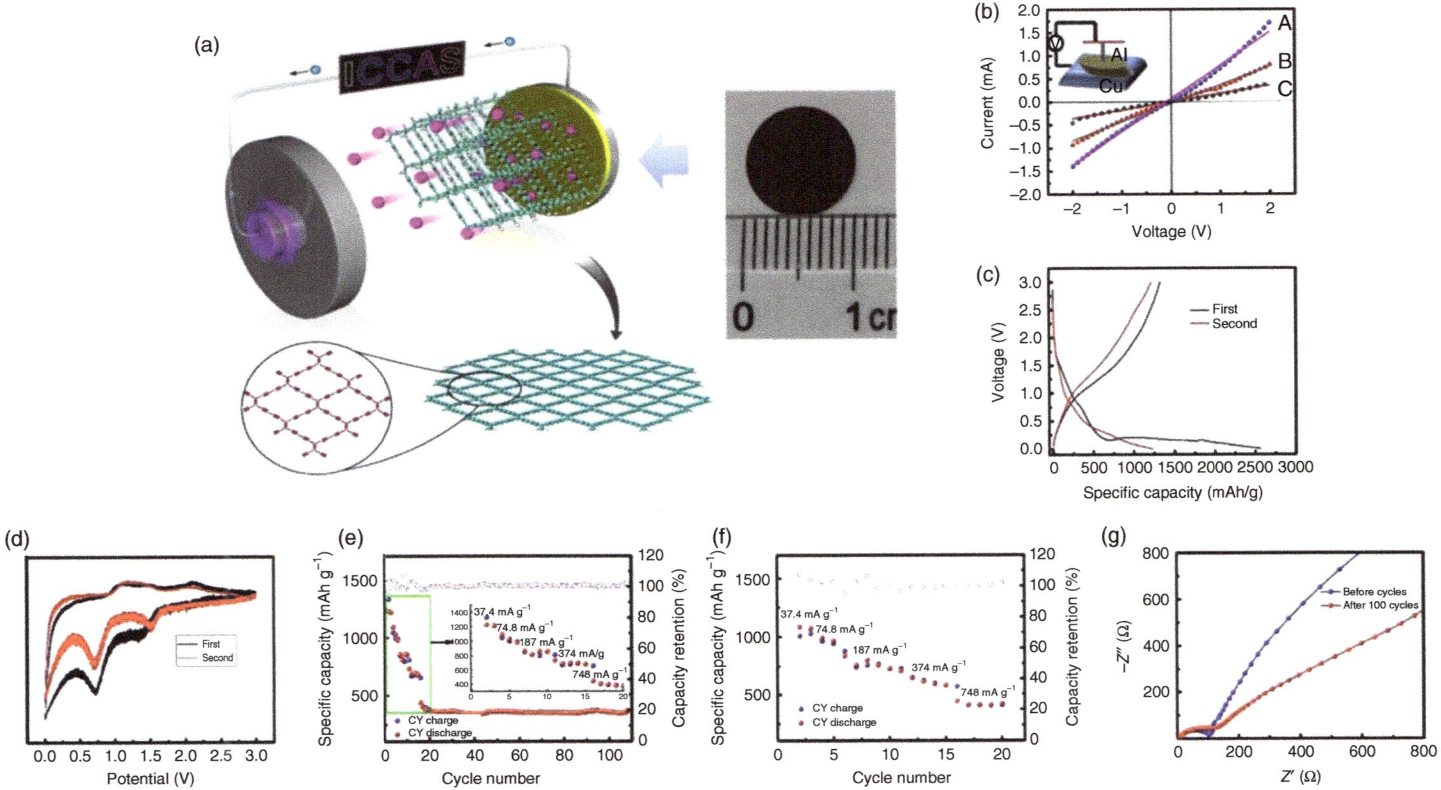

Figure 6.14 (a) Schematic representation and photo of a carbon ene-yne-based battery. (b) *I*–*V* curve of carbon ene-yne film with different thickness; (c) galvanostatic charge/discharge profile of a carbon ene-yne film electrode at 37.4 mA g^{-1}; (d) cyclic voltammetry curves of the carbon ene-yne electrode; (e) rate performance of carbon ene-yne film; (f) rate performance of carbon ene-yne film after 120 cycles; (g) Nyquist plots of this electrode before and after 120 cycles at 748 mA g^{-1}. Source: Jia et al. [80]. © 2017, Elsevier.

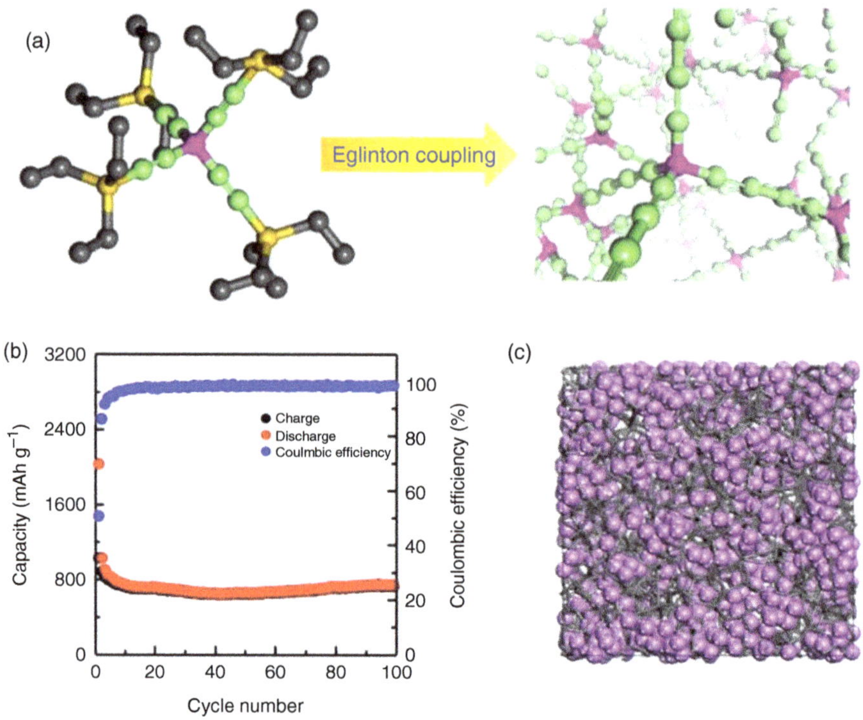

Figure 6.15 (a) Formation of polytetraethynylmethane via the Eglinton coupling reaction; (b) cycling performance of polytetraethynylmethane at a current density of 200 $mA\,g^{-1}$ for 100 cycles; (c) simulated capacity of lithium ions (purple) in polytetraethynylmethane. Source: Zhao et al. [81]. © 2018, John Wiley & Sons.

Furthermore, a tetraethynylmethane precursor can be obtained by organic synthesis. The precursor has a tetrahedral structure containing sp- and sp^3-hybridized carbon atoms. The researchers synthesized a new three-dimensional porous carbon material by the Eglinton coupling method using this precursor, and it was named as polytetraethynylmethane (Figure 6.15a) [81]. The content of the sp-hybridized carbon in this material will be further increased up to 90%, the highest value of the sp-hybridized carbon in the carbon materials. Due to the complicated growth process, this as-prepared carbon material is an amorphous structure. However, according to the experimental measurements, it can be found that this carbon material with the sp- and sp^3-hybridized carbon atoms combines high electronic conductivity, high porosity, and large Li^+ adsorption space. Polytetraethynylmethane is proved to be a disordered wide-bandgap semiconductor, and the electron transport occurs through phase-coherent tunneling. In the electrochemical test of lithium ion storage, a reversible specific capacity of 748 $mAh\,g^{-1}$ is delivered after 100 cycles at the current density of 200 $mA\,g^{-1}$ (Figure 6.15b). Under the same conditions, the reversible capacity of polytetraethynylmethane exceeds that of conventional graphite. No obvious platform is observed in the charge–discharge curves, which is consistent with that in the CV curve. Dynamic calculations and simulations reveal that it is easy for Li^+ to pass through the microporous structure of polytetraethynylmethane (Figure 6.15c), and its diffusion coefficient is about $4\times10^{-4}\,cm^2\,s^{-1}$, which is similar with the measured value in the polymer electrolyte. Such value is much higher than that in the prevailing electrode materials, indicating that the polytetraethynylmethane is a high-power-density material. Interestingly, the cycle stability of polytetraethynylmethane is higher than that of graphite in the overcharge experiment. The scanning electron microscopy found that no lithium metal dendrites were formed on the surface of polytetraethynylmethane electrode. On the contrary, there were obvious lithium metal dendrites formed on the graphite surface. This result reveals that polytetraethynylmethane can inhibit the formation of harmful lithium dendrites, indicating that polytetraethynylmethane is a safe anode material for LIBs. This may be ascribed to the special three-dimensional porous network, beneficial for rapid transfer of the Li^+.

6.4 Sodium Ion Battery Anodes

In recent years, the LIB technology has received more and more successes in commercialization. However, due to the severe shortage in resources for LIBs, the search for key materials for alternative energy sources is becoming one of the next important assignments for scientists. Because of the abundance of the sodium resources, scientists have already carried out a lot of investigations on the availability of the abundant sodium resources, and the preparation of electrode materials for the sodium ion batteries has gradually received more and more attention. To find the low-cost, high-performance anode materials is essential for the development of commercially available sodium ion batteries. Some remarkable advancements in exploring the candidate anodes have already been achieved in the past decades. Among these candidate materials, hard carbon is one of the promising ones due to its potential low cost, scalability, and long cycle performance. Even so, the design and development of anode materials with higher capacity is necessary for assembling the sodium ion battery with energy density close to the commercial LIBs.

The investigations of 2D graphdiyne with many in-plane cavities and active positions in storing the sodium ions are interesting for the development of sodium ion batteries. Not only theoretical calculations revealed the potential of graphdiyne in the storage of sodium ions, but also many experiments have demonstrated the excellent performance of graphdiyne in storing sodium ions [82]. The three-dimensional graphdiyne nanowalls prepared according to the reported method were used as the anode and assembled with metallic sodium for a half cell. It was found that the graphdiyne nanowalls showed excellent sodium ion storage performance and had good cycle performance. The three-dimensional nanostructure and intrinsic in-plane pores of graphdiyne have abundant electrochemically active sites, and such structure can fully optimize the storage behavior, thereby promoting the rapid ion transfer and electrode reaction. Through in situ Raman spectroscopy, the great contribution ascribed to the acetylene bond in graphdiyne to the capacity of sodium ions can be directly observed (Figure 6.16). The electrode based on graphdiyne

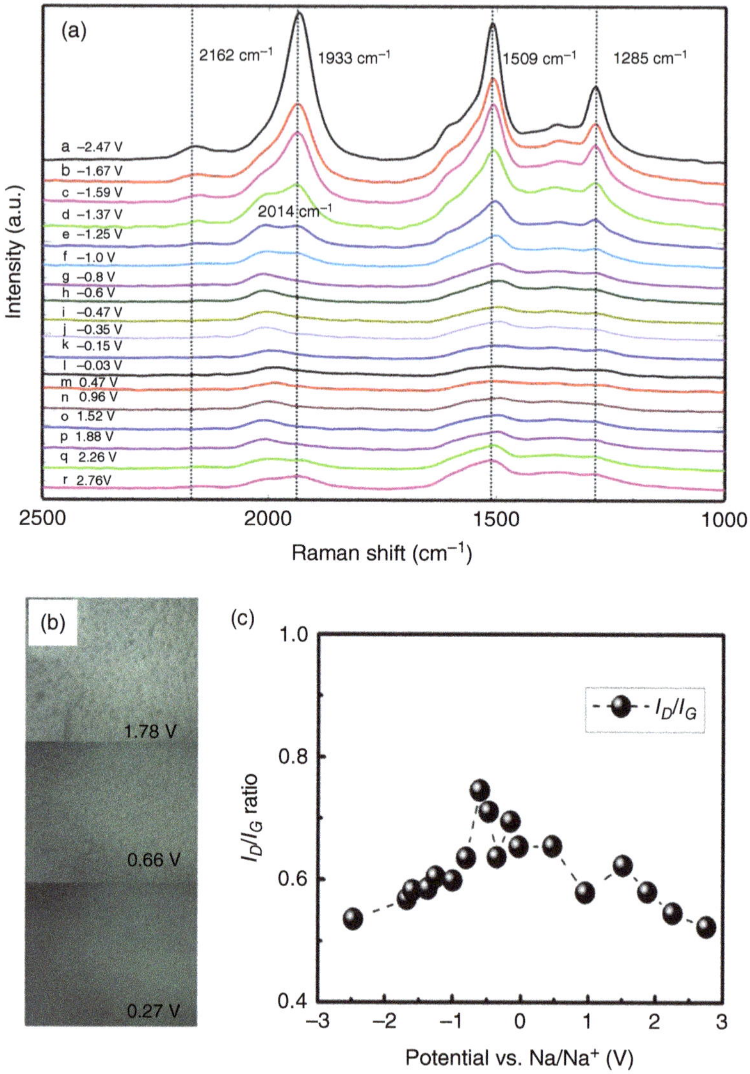

Figure 6.16 (a) In situ Raman spectra during the first discharge process; (b) optical micrograph shows the accompanying changes in the color of the graphdiyne anode during the sodiation at different voltages; (c) evolutions of the intensity ratio between the D and G bonds. Source: Wang et al. [82]. © 2017, American Chemical Society.

nanosheets obtained a reversible capacity of 812 $mAh\,g^{-1}$ of sodium ions at a current density of 0.05 $A\,g^{-1}$. While at a current density of 1 $A\,g^{-1}$, the electrode still remained a capacity of more than 405 $mAh\,g^{-1}$ after 1000 cycles. The results indicate that the graphdiyne is a more promising electrode than many reported materials.

Recently, researchers have done work on the substitution of the basic unit of the benzene ring and studied its sodium storage behavior. The boron is used to substitute the benzene ring to obtain a honeycomb structure of B-doped graphdiyne (Figure 6.17) [83]. The formed honeycomb structure has a smaller pore size than H-graphdiyne but larger than the conventional graphdiyne, which is a much more efficient transmission channel for sodium ion. First, researchers conducted theoretical investigation on the sodium ion storage performance of such new material and concluded that the sodium storage capacity of the material is as high as 1617 $mAh\,g^{-1}$. The sodium atoms are preferred to stay both near the sp-carbon atoms and the B atoms, due to the appropriate electronegativity of the B atoms and butadiyne linkage (Figure 6.17a–e). The storage of Li on the top of B indicates that the B elements increased the active sites for Li storage. In experiments, the characterization of the as-prepared samples shows that the B-doped graphdiyne film is the aggregate of nanoparticles, and the surface structure of the film is not dense and smooth, which may be related to the weakened stacking among the monomers. The conductivity of the B-doped graphdiyne film is $1.33 \times 10^{-2}\,S\,m^{-1}$. The researchers used the material as the negative electrode for sodium ion batteries. The tests showed that the material has a sodium storage capacity of 600 $mAh\,g^{-1}$, and at a high current density of 5 $A\,g^{-1}$, it can maintain a stable capacity output in 4000 cycles (Figure 6.17f–h).

6.5 Electrochemical Interface

6.5.1 Function of Interface

Although the electrode interface is a small proportion of the entire electrode composition, it plays a vital role in the device. It involves interface structure, interface reaction, thermodynamics and kinetic behaviors, etc., affecting all aspects of the performance of the electrochemical energy electrodes, including efficiency, life, power performance, and safety performance [84]. In the field of LIBs, the instability of the electrode interface is the main reason for the increase of the interface charge transfer resistance, the decay of the electrode structure, and the production of gas in the electrode. In the use of anode and cathode materials with high-energy density (silicon, germanium, oxide, high nickel cathode materials, etc.), the intercalation/deintercalation process of a large amount of Li ions will produce greater volume changes and structural phase transitions, thus greatly reducing the interfacial stability. The problems are particularly prominent. In the current research, a large amount of research work is on how to use conventional materials to modify and protect the electrochemical interface of the electrode. In alkaline metal (lithium, sodium, potassium) batteries, unstable electrode interfaces are important reasons for the formation of metal dendrites, low efficiency of electrochemical processes, poor safety, and short service life. The construction of an artificial electrochemical interface to regulate the nucleation of alkali metals, control the ordered growth of alkali metals, and stabilize the ion and electron transport at the interface is very effective in inhibiting metal dendrites, improving electrochemical efficiency, and increasing electrode safety.

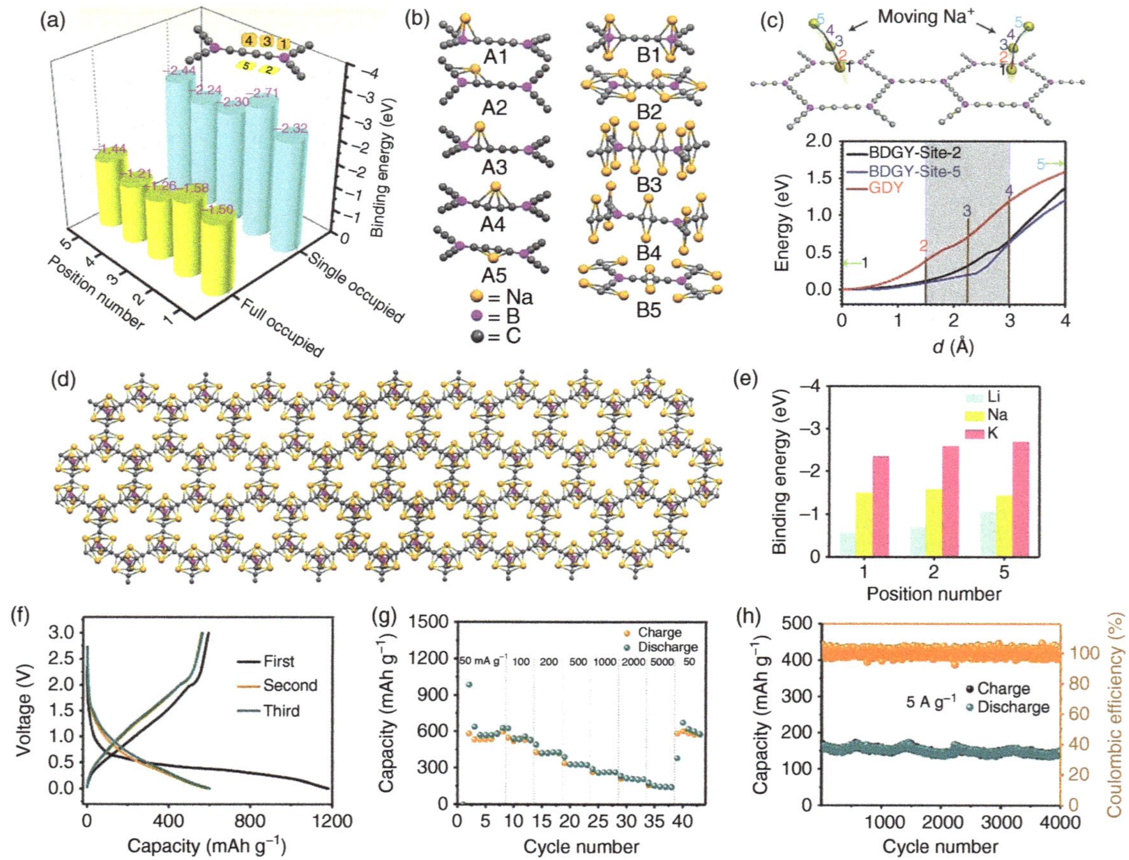

Figure 6.17 (a) Calculated binding energies at five storage sites. (b) The geometries of optimized single occupied mode Na + $C_{16}B_2$ (A1–A5) and fully occupied mode *n*Na + $C_{16}B_2$ (B1–B5). (c) The energy potential curves along the trajectories for the Na atom transfer from the plane of B-doped graphdiyne (B-GDY) through the sites 2 and 5. (d) The possible Na storage sites in B-GDY. (e) The binding energy comparison between the repeat unit ($C_{16}B_2$) and Li, Na, K, respectively. (f) The charge/discharge profiles at 50 mA g^{-1}, (g) rate performance and (h) cycle performance at 5 A g^{-1} of B-GDY electrode for sodium ion battery (NIBs). Source: Wang et al. [83]. © 2018, John Wiley & Sons.

Carbon materials are an essential component in the use of various electrochemical energy electrode materials. The use of carbon materials carries many tasks such as stabilizing the electrode structure, optimizing the conductive network, strengthening the safety of the electrode, enhancing the stability of high and low temperatures, and extending the lifespan of the electrode. Research and practical applications have proved that conventional carbon materials can play a certain role, but there are still many shortcomings, especially the slow progress in the construction and protection of electrode interfaces. This is also the bottleneck of the conventional carbon materials in the field of electrochemical energy. This is caused by the high temperature and harsh preparation conditions of conventional carbon materials. Such conditions for their preparation are difficult to match the structural stability of general electrode materials. Therefore, the construction and protection of electrode interface with the carbon materials progress sluggishly far below than the practical applications. Graphdiyne, as a carbon material with a large conjugated electronic structure, has the unique characteristics of mild preparation and porous structure, which perfectly compensates for the deficiencies of conventional carbon materials. In the construction and protection of electrode interface, these unique properties of graphdiyne may play an important role in its wide applications (Figure 6.18). It provides a lot of new inspirations for solving the problems related to interface stability that are common in electrochemical energy storage, and it is conducive for further solving the lifespan and safety of high-energy density batteries [84]. In this section, we mainly focus on the construction and application of the graphdiyne interface in the battery cathodes and anodes.

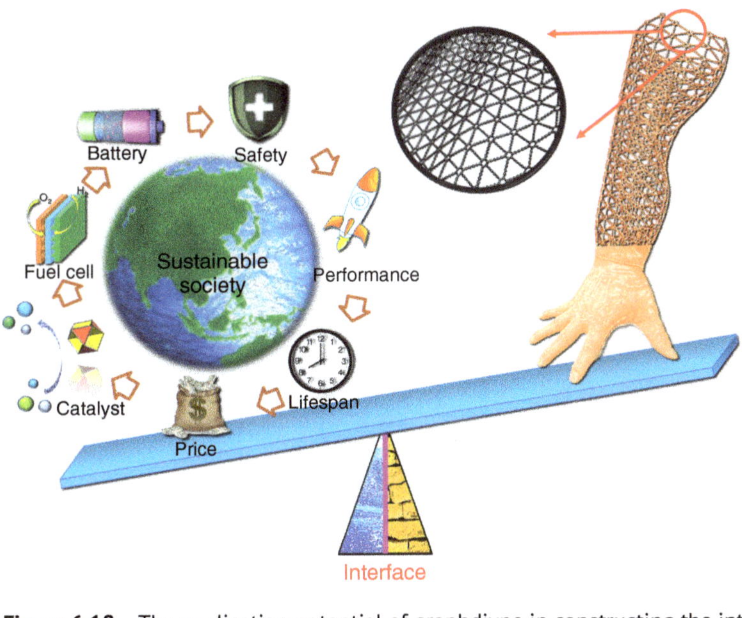

Figure 6.18 The application potential of graphdiyne in constructing the interface of the electrochemical energy. Source: Zuo et al. [84]. © 2019, Elsevier.

6.5.2 Protection for LIBs Anodes

Except for the development of the new materials for the LIBs, how to efficiently deliver the theoretical performance of the current electrode materials is still one of the main directions. Therefore, scientists paid attention to the prevailing electrode materials, including graphite anode, silicon anode, high-energy-density oxide anodes, high-nickel cathodes, high-voltage cathodes, and resource-rich alternative electrodes. For the LIBs, the low capacity is becoming one of the bottlenecks, because the commercial graphite anode has gradually reached its theoretical specific capacity of 372 $mAh\,g^{-1}$. There is no room for the big breakthrough in energy density based on the graphite anode. Since graphdiyne is a new carbon allotrope, its preparation and properties show many unique advantages. The highly efficient applications of graphdiyne bring new possibilities and insights for improving the performance of the prevailing anodes.

The silicon anode has a theoretical capacity more than 4000 $mAh\,g^{-1}$, 10 times more than that of the graphite. The successful use of the silicon anode is of great significance for increasing the energy density of LIBs [85, 86]. However, the use of silicon anode faces severe volume changes larger than 300%, which cause serious damage in the conductive network and the unstable interfacial structure of the electrode [86–88]. These two issues seriously affect the reversibility and safety of the silicon anode. As a result, the commercial applications of high-content silicon anode are still facing huge challenges. In this research field, the coating of carbon materials is the promising strategy. According to the coating of carbon materials, it can effectively apply the excellent properties of carbon materials in terms of mechanical performance, electron conductivity, and chemical stability to resolve the above problems of the silicon anode [89, 90]. However, the coating technologies of the silicon anode with carbon materials need to be carried out at extremely high temperatures, resulting in a large investment in equipment and energy consumption. Importantly, these methods using the conventional carbon materials are difficult to construct porous structures to relieve the remarkable volume changes of silicon particles during charge and discharge.

In the research of protecting the silicon anode, the unique advantages of graphdiyne were well exhibited [91]. The growth of graphdiyne on the copper foil is an important method to prepare the graphdiyne, and the copper foil is the current collector for the anode in the LIBs. This is a great feature for the application of graphdiyne in the LIBs anode. In the fundamental research, the copper nanowires were used to initiate the growth of graphdiyne under ambient temperature and pressure, thereby forming a seamless graphdiyne protective layer on the surface of silicon nanoparticles. The method not only achieves complete protection of silicon nanoparticles and forms a stable interface protection layer, but also in situ constitutes a three-dimensional network with enhanced mechanical and electrical properties. Via such approach, it significantly increases the interaction among the components of the electrode, including the interaction between silicon particles and copper nanowires, copper nanowires and copper nanowires, silicon particles and graphdiyne. All morphological characterizations have confirmed that

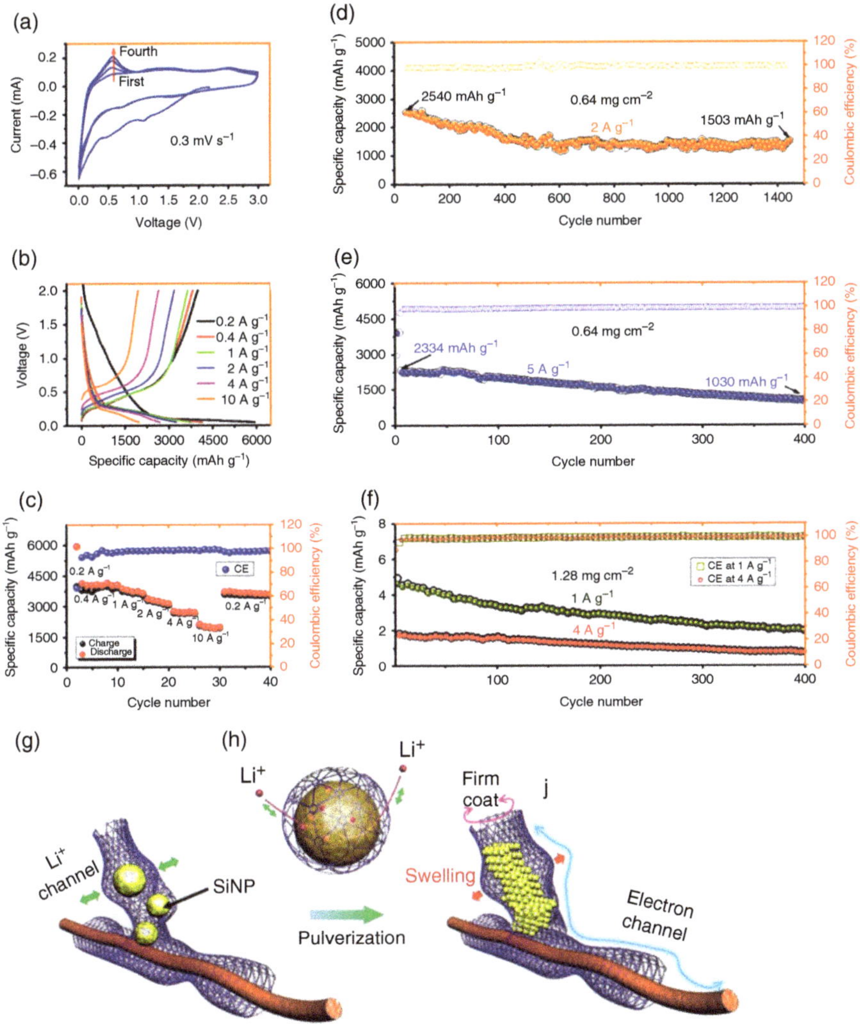

Figure 6.19 Electrochemical performance of the freestanding silicon anode. (a) CV curves of the first four cycles at 0.3 mV s^{-1}; (b) charge/discharge curves and (c) capacities at different current densities; the lifespan of the sample at current densities of (d) 2 and (e) 5 A g^{-1} with an areal mass loading of 0.64 mg c^{-2}; (f) the lifespan of the samples at the high current density of 1 and 4 A g^{-1} with an areal mass loading of 1.28 mg cm^{-2}. Source: Shang et al. [91]. © 2018, John Wiley & Sons.

the graphdiyne effectively protects silicon anode particles. This method greatly improves the mechanical stability of the electrode when subjected to large volume changes, effectively inhibiting the pulverization of silicon particles, and relieves fracture of the conductive network. Therefore, problems such as increased polarization and resistance in electrochemical processes were solved (Figure 6.19). The performance of the as-obtained silicon anode is significantly improved with the capacity as high as 4120 mAh g^{-1}. The capacity retention is robust in the long-term cycling at a high current density of $2\,A\,g^{-1}$, and the capacity is still as high as $1503\,mAh\,g^{-1}$ after 1450 cycles (Figure 6.19b). This method can be used to prepare silicon anode with a high areal capacity. The silicon anode with an areal capacity of up to $4.72\,mA\,h\,cm^{-2}$ also maintains a good stability in the long-time cycling at high rate. The long-term stability and high capacity are obviously ascribed to the enhanced interface stability and conductive network according to the in situ growth of the graphdiyne protection layer (Figure 6.19g,h).

Copper foil is an important substrate for in situ growing graphdiyne and as the current collector in LIBs. Due to this reason, the applications of graphdiyne can well match the large-scale preparation process of LIBs anodes. In the work by Li et al. [92], the slurry of silicon nanoparticles was uniformly loaded on the surface of the copper foil by the doctor blade, and then the electrode, after drying, was immersed in a solution containing the precursor for in situ coating of graphdiyne. Since the copper substrate can initiate the growth of graphdiyne nanosheet, the as-grown graphdiyne nanosheets are conformally and seamlessly coated on the surface of the silicon nanoparticles, thus forming a good all-carbon protective interface (Figure 6.20). Because it is synthesized in situ, it can be seen that the three-dimensional framework of graphdiyne has good continuity and integrity, forming a good conductive and mechanical network. At the same time, the morphological measurement demonstrates that this method is the first to realize the strong binding between the electrode active material and the current collector through the chemical bonds. It can effectively prevent the separation phenomenon of the silicon anode from the current collector due to the high volumetric strain in the charge/discharge reactions. The in situ–grown graphdiyne nanosheet also enhances

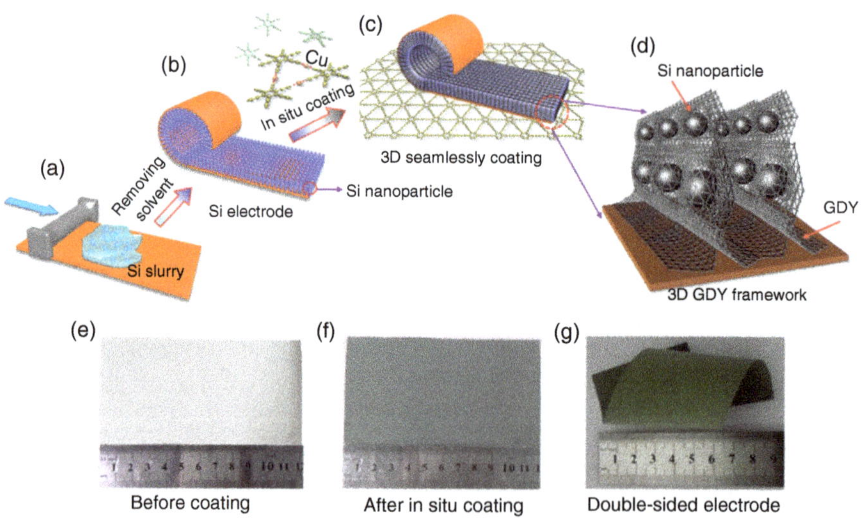

Figure 6.20 (a–c) Schematic illustration of the preparation of large-scale silicon anode coated by GDY nanosheets. (d) Scheme showing the seamless support and strong bind between silicon nanoparticles and the current collector. (e–g) Photographs of a large-scale silicon anode before and after in situ coating of GDY nanosheets. Source: Li et al. [92]. © 2018, Elsevier.

the charge transfer between electrode components. In the electrochemical process, because graphdiyne nanosheet has the selective transmission of Li^+, other components in the electrolyte are difficult to pass through the graphdiyne protective layer. So, the interfacial graphdiyne on the silicon nanoparticle can efficiently suppress the side reactions between the electrolyte and the silicon anode. This is beneficial for providing a stable interface protection layer in the electrochemical process, thus greatly improving the cycle stability of the electrode. Electrochemical performance manifested that the cycle stability of the electrode constructed by this method was significantly improved, and the power performance was significantly enhanced. The silicon anode constructed by this method can be coupled with the high nickel cathode, forming a full LIB with high volumetric energy density up to 1343 Wh l^{-1}. Therefore, the in situ construction of the graphdiyne interface has great potential for the preparation of large-area high-performance silicon anodes.

Metal oxides are the promising candidates for the metal ion batteries anode. A large number of metal oxide anodes have been proved with high theoretical capacities [93–95]. In the electrochemical process of metal oxides, the redox reactions take place with the production of lithium oxide and reduced metal. This storage mechanism of metal oxides leads to serious volume changes and structural pulverization during the charge and discharge process, resulting in low efficiency, poor cycling, and poor rate performance of metal oxides in the electrochemical application. Such problems greatly hinder the practical applications of metal oxide anodes. Due to the significant volume changes, the coating layer constructed by the conventional inorganic materials cannot have good mechanical properties to offer a stable interface protection. The carbon materials have stable electrochemical properties and can offer robust mechanical performance confronting the large volume change. However, using the composition method cannot realize a seamless interfacial protection by prevailing carbon materials. Ascribed to the harsh preparation conditions, the conventional carbon materials cannot grow in situ on the surface of metal oxides. It can be seen that the advantages of conventional carbon materials are difficult to be adopted for solving the problems of the metal oxide anodes. Therefore, the development of a method for in situ growth of carbon materials on the surface of metal oxides has important universal significance for the in-depth application of metal oxides.

As a contrast, the graphdiyne can be prepared in large area and high quality at room temperature. Because it is prepared under mild conditions, the structures of the metal oxides will not be changed during the in situ coating process of graphdiyne. Based on the superiorities of graphdiyne, a universal protection method for metal oxides was obtained [96]. Researchers directly load the metal oxides on the surface of the copper foil. Keeping the samples in the reaction solution for one day, the graphdiyne can be successfully coated on the surface of the oxide in situ. The morphological characterizations by the scanning electron microscopy and transmission electron microscopy proved that graphdiyne nanosheets are continuously grown along the contour of the metal oxide, having a good interfacial contact between the two components. This method shows good universality, and the graphdiyne nanosheets can be readily in situ coated on different oxides with

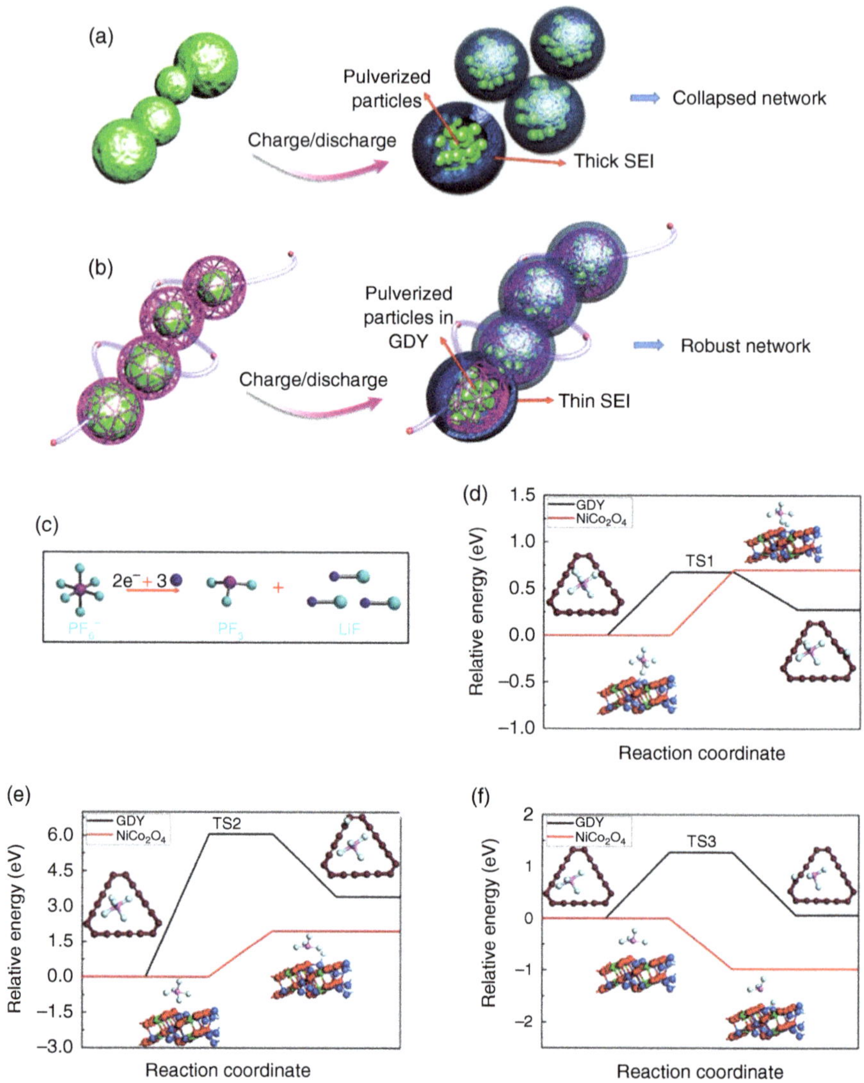

Figure 6.21 Schematic illustrations showing (a) the pulverization of the bare $NiCo_2O_4$ electrode with thick solid electrolyte interphase, and (b) the robust network of $NiCo_2O_4$@GDY with thin solid electrolyte interphase; (c) possible side reaction of $LiPF_6$ on the $NiCo_2O_4$ with and without GDY protection; (d–f) nudged-elastic band calculated energy profiles for the decomposing processes of PF_6^- over the single-layer GDY (black lines) and $NiCo_2O_4$ (311) surface (red lines), respectively. Source: Wang et al. [96]. © 2019, John Wiley & Sons.

different architectures. The metal oxides before and after the graphdiyne coating were tested as the anodes for the LIBs, and it was found that the performance of the metal oxide coated with the graphdiyne was greatly improved, and the lifespan was significantly enhanced. Systematic research on the results showed that graphdiyne-coated layer stabilized the primary and secondary structures of the electrode during the electrochemical process (Figure 6.21a,b). Due to the increased energy barrier of the sider reaction of electrolyte on the graphdiyne nanosheet, it greatly inhibited the decomposition reaction of the electrolyte on the surface of the metal oxides (Figure 6.21c–f). Because metal oxides have wide application prospects in electrochemical catalysis, electrochemical supercapacitors, photocatalysis, and other fields, for stabilizing the metal oxides this method can offer many novel inspirations on the application of metal oxides in these fields.

6.5.3 Protection for LIB Cathodes

The above Section is 6.5.2 summarized the applications of graphdiyne in improving the LIB anodes. For developing next-generation LIBs with higher-energy density, not only should the issues in LIB anodes be solved, but also those in cathodes should be finished. The deep utilization of the high-energy-density cathodes based on existing materials also confronts some issues, such as the long-term stability, safety, and low-temperature performance. Lithium cobalt oxide ($LiCoO_2$) is an earlier material [97, 98] developed and utilized as LIBs cathode. At present, commercial $LiCoO_2$ cathodes have been used in the era of high-voltage 4.45 V vs. Li^+/Li and high capacity (200 mAh g^{-1}). The $LiCoO_2$ has the highest compaction density, good voltage platform, and high specific capacity. All the efforts on $LiCoO_2$ are made for realizing the potentials in assembling the single cell with the highest volumetric energy density up to 700 Wh l^{-1} [99]. Due to such high energy density, the $LiCoO_2$ cathode becomes the first choice for the consumer electronics (mobile phones, computers, mobile devices, etc.) market that pursues extreme energy density. However, it also faces many problems. With the increase of charging voltage, problems such as irreversible phase transition, surface and interface instability, safety issues, and performance degradation are extremely intractable in $LiCoO_2$ cathode, greatly restricting its better application. Stopping the particle pulverization, improving the interface stability of deeply delithiated $LiCoO_2$, and inhibiting the side reaction of the electrolyte at the interface under high voltage are the essential approaches to achieve safer and more stable applications of $LiCoO_2$. The element doping strategies are widely applied in modifying the $LiCoO_2$ cathode. The stability of both interfacial and bulk structure can be improved; therefore, the reversibility of the phase transition during charging and discharging processes is enhanced [99]. Except for that, the coating of $LiCoO_2$ is one another strategy to solve the above problems. Due to the harsh preparation methods of conventional carbon materials, it is difficult to effectively carry out their coating strategies on the surface of $LiCoO_2$.

The preparation process of graphdiyne under mild conditions shows important inspiration for solving this problem. Up to now, although there were no experiments carried out to realize the coating of graphdiyne on $LiCoO_2$, in theoretical investigation, Qian Wang's group made a good theoretical prediction on graphdiyne coating of $LiCoO_2$ [100] (Figure 6.22). First, they calculated the electrochemical window of graphdiyne in the prevailing electrolyte system for $LiCoO_2$ through first principles and investigated the electrochemical stability of graphdiyne. Calculations show that the highest discharge potential of graphdiyne is 1.02 V (Li^+/Li), and the lowest charge potential is 7.85 vs. Li^+/Li. Graphdiyne has a wide electrochemical

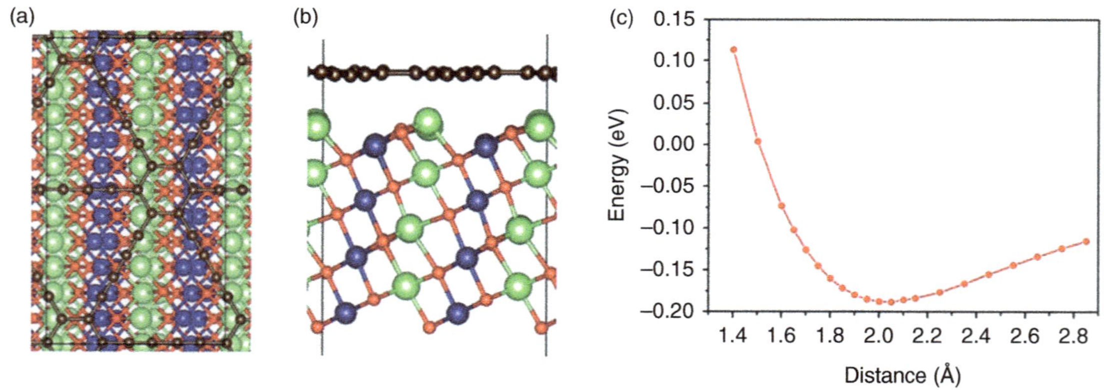

Figure 6.22 (a) Top and (b) side views of the heterostructure between graphdiyne and the $LiCoO_2$ (101) surface. (c) Binding energy variation as a function of the interlayer distance between the graphdiyne and the $LiCoO_2$ surface. Source: Gong et al. [100]. © 2018, Royal Society of Chemistry.

stability window, which satisfies the requirements of prevailing cathode materials for LIBs, indicating that graphdiyne has great potentials in this field. Taking the single-layer graphdiyne as the calculation model, it is found that the single-layer graphdiyne can be well matched with the electrolyte system (Figure 6.22a,b). Lithium ions can quickly pass through the in-plane pores of the graphdiyne, and the other components in the electrolyte are very efficiently blocked. The single-layer graphdiyne and $LiCoO_2$ particles have good compatibility, and their binding energy is strongest when the distance between them is 2 Å (Figure 6.22c). Graphdiyne has a close contact with $LiCoO_2$, which efficiently reduces the interface resistance, improves the electronic conductivity of the system, and helps improve the power performance of the $LiCoO_2$-based battery. Their theoretical simulation also shows the application prospects of graphdiyne as a unique carbon material coating in other conventional cathode materials.

In recent years, with the rapid development of new energy vehicles and smart electronic products, the market demand of LIBs has greatly increased. However, the important components of conventional LIBs, such as the lithium, cobalt, and nickel resources, are limited and unevenly distributed on the earth. These natural resources are hard to support the explosive growth of the requirements. In addition, with the expiration of large amount of LIBs, the pressure in the LIBs recovery will gradually appear. The recycling industry of traditional LIBs is faced with some problems. The recycling processes are high cost and cause heavy environment pollution [101]. Therefore, the development of low-cobalt and -nickel cathodes and new alkaline metal ion batteries is of great significance and can fundamentally alleviate the problems in the resource scarcity and recycling pollution.

Small-molecule organic cathode materials have many outstanding advantages such as high capacity, easy preparation, abundant resources, and good structure controllability, and can store a variety of alkali metal ions [102–106] (Figure 6.23a). Meanwhile, the batteries based on small-molecule organic cathode can be easily processed and effectively recycled [108, 109]. Therefore, the development of small-molecule organic cathode materials is an effective way to alleviate the resource shortage and the pollution problems in traditional inorganic LIBs. However, small-molecule organic cathode materials generally face the serious problems of poor electronic conductivity and easy dissolution during charge and discharge processes (Figure 6.23a). The poor conductivity can be compensated by adding large amount of conductive additives. As a result, the proportion of active materials is usually less than 60% [110–112], which is much lower than 90% of the inorganic electrode, greatly reducing the practical value and the high-capacity advantages of organic cathodes (Figure 6.23b). Moreover, the dissolution and shuttle effect of small organic molecules are another difficult problem to solve. Suppressing the dissolution and shuttle effect by the polymerization will result in great reduction in capacity, molecule stacking, and tap density. Such problems directly cause the performance of the organic battery to be far lower than the practical requirements. It is necessary to explore a stable and effective coating technology to inhibit the dissolution of small-molecule organic cathodes while increasing the electronic conductivity and the active material content to realize the practical requirements.

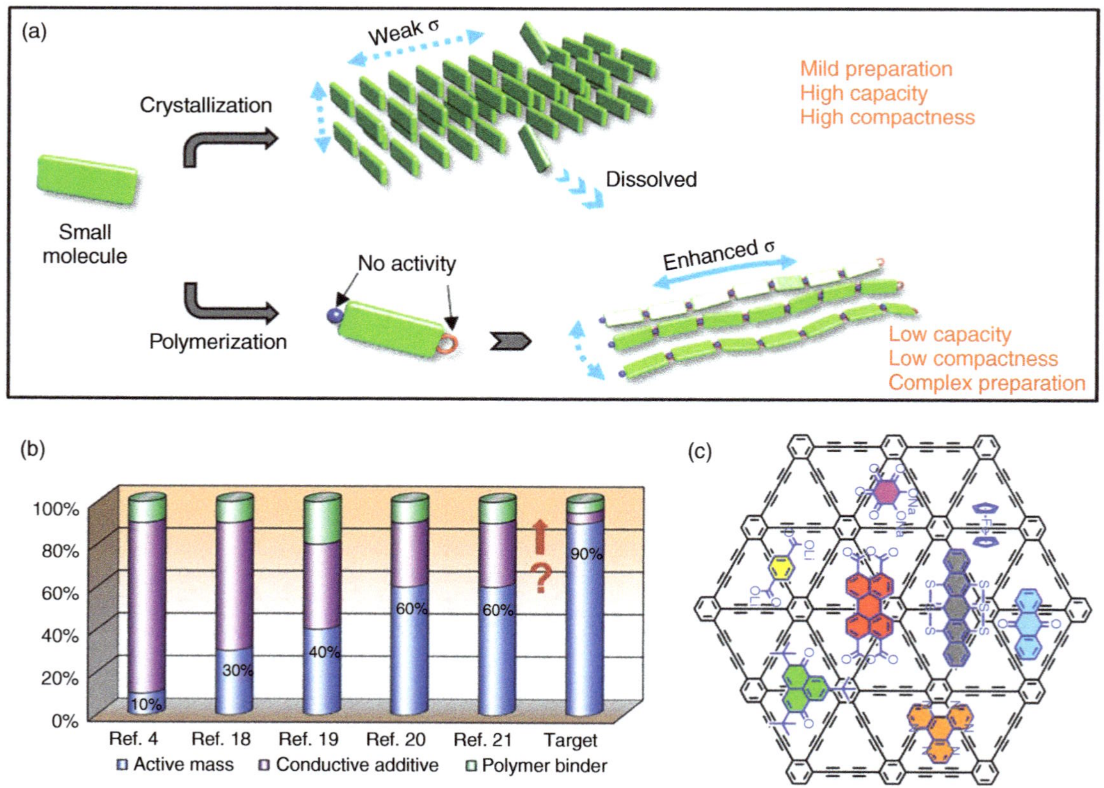

Figure 6.23 (a) Advantages and the disadvantages of organic-based cathodes for the batteries; (b) the composition comparison of the prevailing electrodes based on the organic materials; (c) the structural comparisons between the prevailing organic cathodes and the graphdiyne. Source: Li et al. [107]. © 2020, John Wiley & Sons.

The carbon material–coating technology can not only improve the electronic conductivity but also maintain the long-term cycle stability. However, due to the harsh preparation conditions of conventional carbon materials, the structure of small-molecule organic cathode will be destroyed during coating process, and the carbon-coating strategy for these organic cathodes is hard to develop. Graphdiyne has a large two-dimensional conjugated structure with many in-plane pores, conducive for both the electron transport and alkali metal ion transport. Importantly, the growth of graphdiyne under mild conditions offers the possibility of in situ coating all-carbon materials on the small organic molecules, and it is possible to realize the practical application of organic cathode materials (Figure 6.23c) [107].

In this field, researchers recently realized the in situ construction of a seamless graphdiyne coating layer on the surface of the organic small-molecule organic cathode (Figure 6.24) [107]. In the work, two typical organic small-molecule cathodes were chosen. The 3,4,9,10-perylenetetracarboxylic dianhydride (PTCDA) and rhodizonic acid salt (SR) were fully investigated as the promising organic cathodes. They represent the covalent compounds and the ionic compounds in the organic cathodes. Since the working voltage of perylenetetracarboxylic dianhydride and rhodizonic acid salt is lower than the oxidation potential of copper foil, researchers chose the copper foil as the current collector and loaded the two cathode materials on the copper foil in large scale through a common coating method. According to this method, the active material content is as high as 95%. Then, the electrode is immersed into the reaction solution containing the precursor, and the graphdiyne nanofilm is in situ grown on the surface of the material, forming a seamless all-carbon protective layer. This method requires that the small organic cathodes have poor solubility in the solution where the graphdiyne is grown; otherwise, the organic molecules will dissolve in the solution, resulting in the loss of active materials and ineffectively coating. For greatly reducing the solubility of organic materials, the solvents for the reaction can be effectively tuned. In this research, the two selected organic small molecules have poor solubility in the solution, so the coating process can proceed readily. It is found that the graphdiyne is uniformly coated on small organic molecules via the morphological characterization in the SEM and transmission electron microscopy (TEM) images in Figure 6.24. The all-carbon graphdiyne coating layer has the characteristic selectivity for transporting alkali metal ions. Thus, it can not only effectively improve the conductivity of organic electrodes, but also inhibit their dissolution in the electrochemical process. After coating, the active mass loading of organic electrode is still increased up to 93%, similar with the inorganic electrodes. The as-obtained electrodes have greatly improved power performance and long-term stability. Although the perylenetetracarboxylic dianhydride has a low working potential of 2.5 V and a capacity of 140 $mAh\,g^{-1}$, its practical mass energy density can reach up to 310 $Wh\,kg^{-1}$, much better than many results in reports. This is because the active mass loading of the organic materials has been increased to 93%, much higher than them. It is foreseeable that if an organic material with higher voltage plateau and higher capacity is used as the cathode, the energy density of the organic batteries will be much higher than this value. It is possible to construct the organic batteries with high-energy density close to prevailing inorganic batteries. With this method, the research on

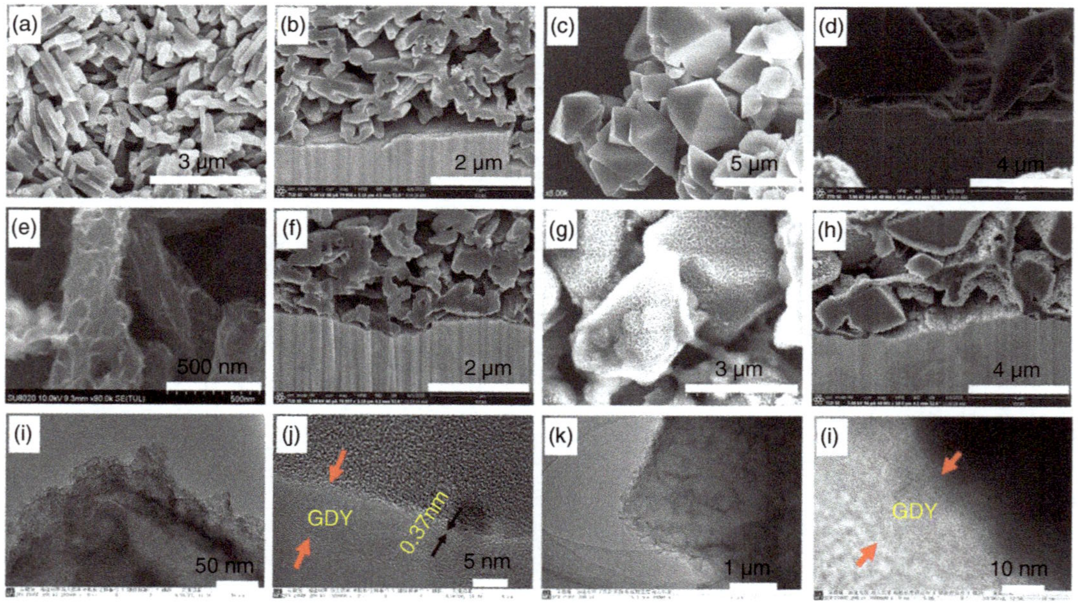

Figure 6.24 Morphological characterization of electrodes before and after in situ coating graphdiyne. SEM images of (a, b) PTCDA; (c, d) SR; (e, f) PTCDA@graphdiyne; (g, h) SR@graphdiyne. (i, j) TEM images of PTCDA@graphdiyne; (k, l) TEM images of SR@graphdiyne. Source: Li et al. [107]. © 2020, Wiley-VCH Verlag GmbH & Co. KGaA.

small-molecule organic cathodes was no longer focused on the design and synthesis of different organic materials, but instead on the in-depth investigation of several organic materials with high potential. Since the organic cathodes can be used to store other alkali metal ions, this method will also help promote the developments in the diversification of batteries and low-cost batteries.

6.6 Lithium-Sulfur Battery

Compared to the commercial LIBs, the lithium-sulfur batteries have remarkable potentials as the candidates of the next-generation energy storage technologies. Similar to the cathode, sulfur has much higher capacity of 1675 A h kg^{-1}, and it is abundant with low price. The theoretical energy density of the lithium-sulfur batteries is increased up to 2500 Wh kg^{-1}, which enables the practical lithium-sulfur batteries to have an energy density close to 400–600 Wh kg^{-1}. Such value is about two or three times higher than that of the commercial LIBs. Although the advantages of S are attractive, the challenges are great. Sulfur is an electronic insulator and will be dissolved during the discharge process. The poor conductivity of sulfur leads to low-active materials in the electrode, and the dissolution and the following shuttling effect cause the rapid performance decay in the cycling. To solve the problems in the sulfur cathode, porous carbon materials with high surface areas and high active sites are usually applied, which not only improve the electronic conductivity of sulfur, but also well suppress the polysulfide shuttle effect.

The polysulfides shuttle effect is the result of interfacial reaction, ion transfer, mass diffusion, phase change, etc. [113–115]. Reducing the shuttle of polysulfides in lithium-sulfur batteries and controlling the reaction process are considered to be effective in enhancing the performance of lithium-sulfur batteries. In recent research, the metal nanoparticles, metal sulfide [116–119], and heteroatoms were introduced into the porous carbon framework [120, 121], to absorb the polysulfides, inhibit the shuttle effect, and catalyze the cathode reactions. Ionic polymers are a class of materials [122–124] that can promote mass transfer and ion migration near the interface. It is widely used to improve the mass transfer outside the primary nanostructures in sulfur cathodes. But, it has never been used to optimize the internal mass transfer in the primary nanostructures because it is difficult to embed in the prevailing nanomaterials, especially in the all-carbon nanostructures. The internal mass transfer behavior in primary nanostructure in lithium-sulfur batteries has been rarely studied, and the influence of the inner mass transfer in the cathode reaction and polysulfide shuttle has never been revealed. The improvement of the mass transfer in the primary structure is beneficial to balance the electronic/ionic transfer, mass migration, and phase transition of the S cathode in the electrochemical reaction. This will provide new insights for preventing polysulfides from shuttling out of the porous nanostructure.

Graphdiyne is a two-dimensional carbon allotrope and has the in-plane selective cavities. The electron distribution on the sp^2 and sp hybrid carbon atoms is uneven. The uneven electron distribution can provide many atomic-level active centers for electrochemical reactions in lithium-sulfur batteries. The growth characteristics of graphdiyne under mild conditions provide a way to combine the advantages of carbon materials and ionic polymers to build new architectures. Such structural features demonstrate the mixed proton/electron conductivity, and are beneficial for storing sulfur, adsorbing polysulfides, and inhibiting polysulfide shuttles.

For realizing this multifunctional nanostructure, the graphdiyne was grown in situ on the surface of Nafion coated on the copper nanoparticles to construct hollow carbon nanostructures (Nafion@graphdiyne) embedded with polyanions [125]. This method is the first to in situ embed the polyanions in the all-carbon nanostructures (Figure 6.25a). This core–shell nanostructure is multifunctional. Nafion is successfully used to adjust the internal mass transfer behavior in the primary nanostructure (Figure 6.25b), thereby improving the phase transition process in the cathode reaction. Graphdiyne, as a host of sulfur, has good conductivity and ability to catalyze the cathode reaction. This core–shell nanostructure plays a new role in balancing charge transfer, mass diffusion, and phase change in the primary

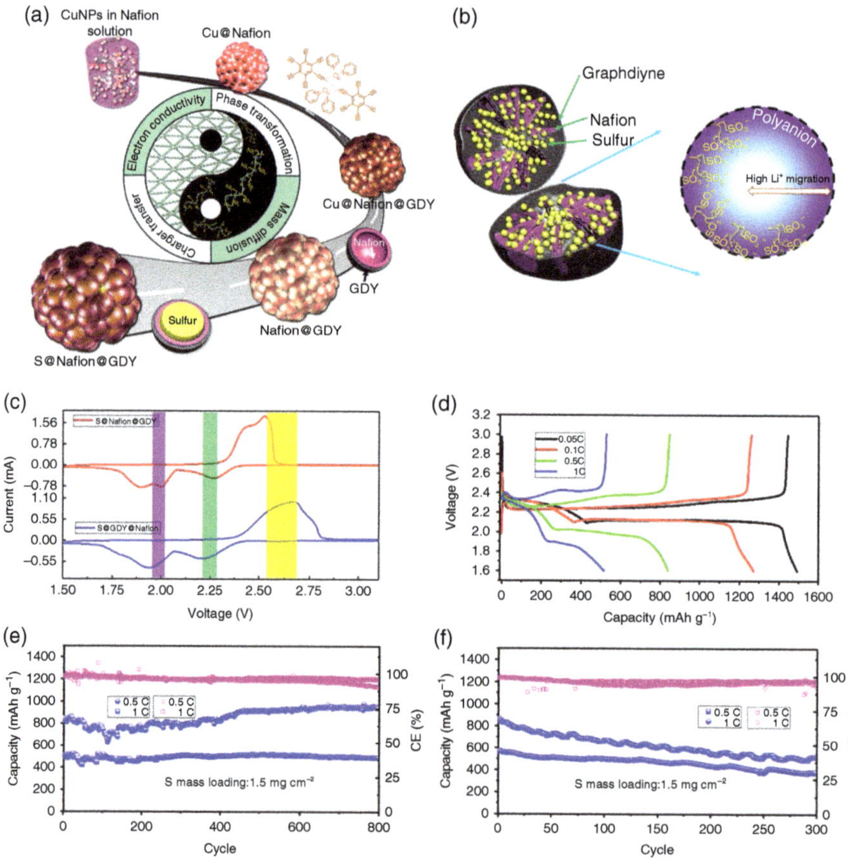

Figure 6.25 (a) Schematic representation of the preparation of Nafion@graphdiyne (Nafion@GDY) core–shell nanostructure and its secondary structure for sulfur host; (b) the schematic illustration of the primary structure of S@Nafion@GDY; (c) the CV curves of Li–S batteries based on the as-prepared electrodes; (d) charge/discharge curves at various rates; (e, f) long-term stability of (e) S@Nafion@GDY and (f) S@GDY@Nafion measured at rates of 0.5C and 1C. Source: Wang et al. [125]. © 2020, Elsevier.

nanostructure, thereby preventing the shuttle effect before the polysulfide shuttle out of the graphdiyne hollow structure. The experimental results show that the lithium-sulfur battery using this core–shell nanostructure shows a high capacity retention rate even if it is cycled for about 800 times at the high current densities of 0.5C and 1C (Figure 6.25c–f). In addition, the author analyzed the electrochemical process in detail through in situ Raman test. Studies have shown that the nanostructure with embedded Nafion polyanion significantly optimizes the electrochemical reaction process inside the nanostructure, accelerates the conversion of internal polysulfides, and inhibits the shuttle of polysulfides to improve cycle performance. Moreover, the electrode has good structural stability, which effectively alleviates the damage of the electrode structure caused by the huge volume change of sulfur in the electrochemical reaction process.

For increasing the conductivity and reducing the polysulfide shuttle, the porous carbon materials are widely used as the host for storing the S in the Li–S batteries. For investigating the potential of graphdiyne as the sulfur host in Li–S battery, the H-graphdiyne with enlarged in-plane pores was used, as shown in Figure 6.26. The

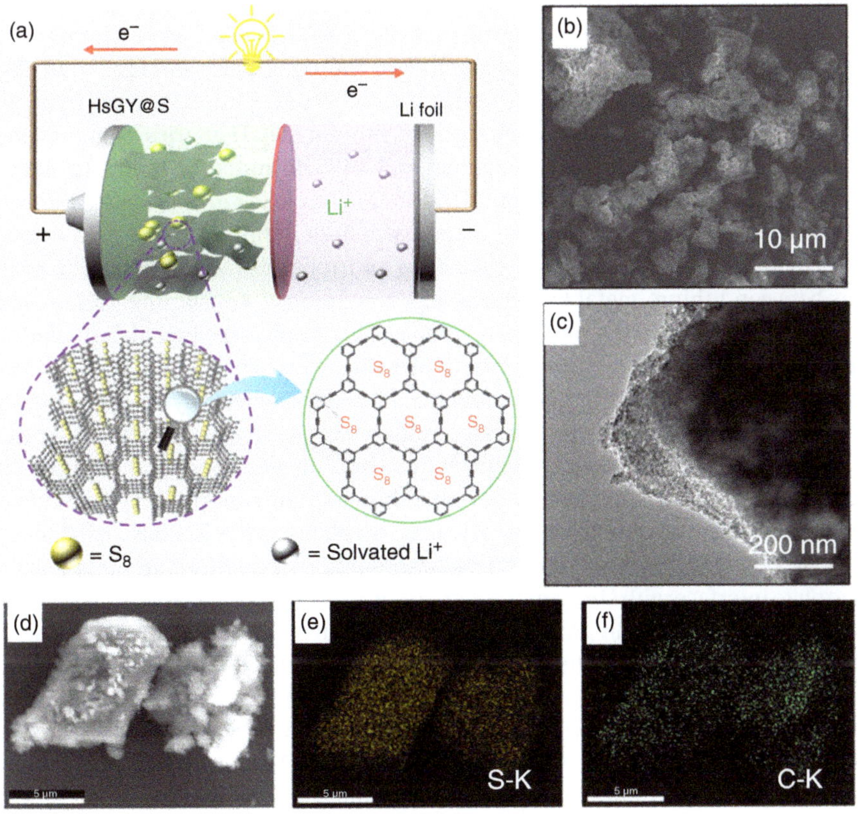

Figure 6.26 (a) Schematic illustration of the H-graphdiyne (HsGY) electrode for the lithium-sulfur battery. (b) SEM image of HsGY@S. (c) TEM image of HsGY@S. (d–f) elemental mapping of HsGY@S composite. Source: Li et al. [126]. © 2019, Wiley-VCH Verlag GmbH & Co. KGaA.

H-graphdiyne was prepared by two-phase solution method and used it as a sulfur storage material. The sulfur is stored according to the thermal treatment, and it is uniformly stored in the form of S_8 molecules in the holes of the H-graphdiyne (Figure 6.26) [126]. The Li–S battery constructed with this electrode shows good performance. Its reversible capacity is as high as 1312 $mAh\,g^{-1}$, and it can achieve a reversible capacity of 799 $mAh\,g^{-1}$ at a high rate of 1C, demonstrating excellent rate performance. Through the cyclic voltammetry and electrochemical reaction impedance spectroscopy tests of the battery at different rates, it is found that the sulfur cathode has good reaction activity, the charge transfer resistance is significantly reduced, and the cycle life of the system is significantly increased. After 200 cycles at 2C, the capacity retention was still as high as 557 $mAh\,g^{-1}$, which was significantly better than the reduced graphene oxide electrode, fully demonstrating the application prospects of H-graphdiyne in Li–S cathodes.

6.7 Lithium Metal Anodes

Among the known negative materials, Li metal is widely regarded as an ideal candidate because of its ultra-high theoretical capacity of 3860 $mAh\,g^{-1}$ and extremely low redox potential of −3.04 V compared to standard hydrogen electrodes [127–129]. For decades, scientists have continued to study the properties of lithium metal, trying to use it in the high-energy density, rechargeable, lithium metal batteries. However, the use of lithium metal in lithium metal batteries still confronts the following issues [130–132]: (i) uncontrollable lithium dendrite growth leads to short cycle life and serious safety problems; (ii) lithium metal has a high Fermi level and high degree of thermodynamic instability, which leads to an irreversible continuous side reactions between lithium metal and electrolyte, continuously consuming lithium and electrolyte, and increasing internal resistance; (iii) in the repeated electro-plating and -stripping process, the lithium metal negative electrode will undergo infinite volume and morphological changes, and the formed solid electrolyte interface layer film is too fragile to support such a large volume change in the lithium metal electrode. The stable and controllable application of lithium metal anodes is essential for the next generation of high-energy density lithium-sulfur batteries and lithium–air batteries. Usually, to solve these key scientific problems, scientists have tried many methods [133–137]. For example: (i) the introduction of additives to build a solid electrolyte interface with stronger mechanical properties; (ii) The porous film is used for the uniform deposition process; (iii) solid electrolyte is used to physically isolate lithium dendrites; (iv) the three-dimensional conductive network is used to store lithium metal and so on.

How to achieve atomic-level uniform deposition and suppress the lithium dendrites are the urgent problems for researchers in the field of lithium metal battery. In recent years, two-dimensional materials research is very active. Two-dimensional materials show the advantages for improving the ions diffusion, nucleation, and growth processes in the batteries. In current two-dimensional materials, the cross-plane diffusion of Li ions mainly depends on the number of defects. But, the formation of these defects is uncontrollable and shows low efficiency using the reported methods. Specially, it is a big challenge for them to achieve accurately selective diffusion of Li ions through these defects. Without the accurate selectivity for Li^+ only transport, the advantages of these two-dimensional materials are difficult to perform in the lithium metal anodes. Graphdiyne is a two-dimensional carbon material with many in-plane cavities (the pore diameter of 5.3 Å). The pore density in graphdiyne nanosheet is much higher any previously prepared porous membrane, in which the barrier-free diffusion of Li^+ among different planes can be realized. Via the graphdiyne film, the ultra-uniform transport of Li^+ can possibly be obtained near the interface of the lithium plating, thus efficiently tuning the plating process. In addition, graphdiyne film is a two-dimensional carbon allotrope, and its higher mechanical strength can also physically prevent the lithium dendrites from penetrating through the separator.

To use the selective transport performance of graphdiyne, a modified surface polymerization method at room temperature was proposed, and the ultrathin graphdiyne nanofilms with thickness of 10 nm was prepared on the copper foils (Figure 6.27) [138]. The size of this film can be scaled up to several centimeters and can be completely transferred on the commercial Celgard separator. Young's modulus of graphdiyne film is increased up to 14 GPa according to the measurements. The graphdiyne film loaded on the Celgard separator was used as the Li^+ ultrafiltration membrane for the first time, and it realizes the uniform diffusion of Li^+ on the electrode interface (Figure 6.28). Via this membrane, it greatly improves the overpotential of the nucleation and growth process of metallic lithium, thereby effectively suppressing lithium dendrites. In the absence of lithium dendrites, the electrode has a high coulombic efficiency and a good cycling life. Through theoretical simulation, it can be found that the two-dimensional graphdiyne film is a promising Li^+ ultrafiltration membrane, through which the Li^+ can pass one by one, achieving ultra-uniform diffusion.

The use of a three-dimensional current collector can optimize the Li metal plating/stripping processes and relieve negative influence of the great volumetric changes. This is because the three-dimensional current collector not only has more abundant nucleation sites for plating lithium metal, but also offers enough space for hosting lithium metal. Many reports reveal that the three-dimensional current collector can significantly increase the cycle life and efficiency. Furthermore, the research results also demonstrate that the initial nucleation process of the lithium metal has a significant impact on the reversibility of Li metal during the Li stripping/plating cycle. When using a three-dimensional current collector, it is also necessary to increase the lithiophilic sites of the current collector for the uniform nucleation and growth of lithium metal. However, it is difficult for traditional methods to achieve uniform and high-areal-density distribution of the lithiophilic sites on these three-dimensional current collectors, especially in atomic level.

Graphdiyne has abundant sp-hybridized carbon atoms. The sp hybrid carbon atoms have a higher electron density than sp^2 carbon materials; thus, the graphdiyne has a hybrid structure with uneven distributed electronic cloud, different from the graphene. The electron-rich sp hybrid carbon atoms possibly make graphdiyne a

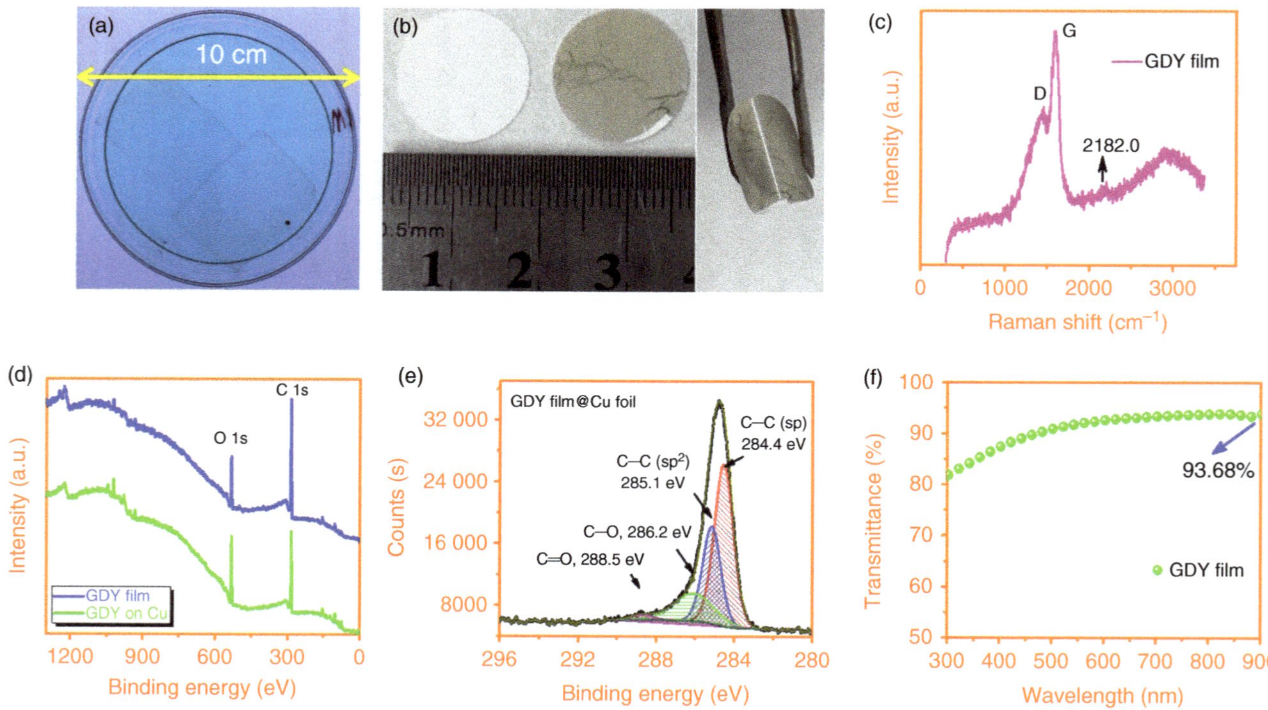

Figure 6.27 Characterization of graphdiyne film. (a) Photograph of large-scale graphdiyne film. (b) Photograph of separator before and after loading thin graphdiyne film and its flexibility. (c) Raman spectrum of graphdiyne film. (d) X-ray photoelectron spectroscopy of graphdiyne film before and after removing Cu foil. (e) The C1s deconvolution of graphdiyne film on Cu substrate. (f) The light transmittance of graphdiyne film loaded on the quartz plate. Source: Shang et al. [138]. © 2018, Elsevier.

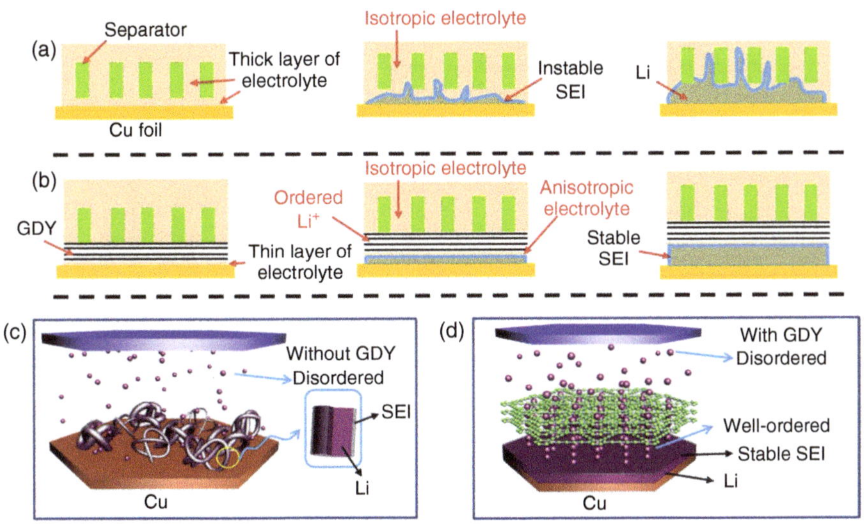

Figure 6.28 Possible mechanism for suppressing lithium dendrites. (a) The disadvantages of the isotropic electrolyte for the lithium dendrite. (b) The mechanism of graphdiyne film in obtaining anisotropic electrolyte. (c, d) Illustrations showing the mechanism of restricting the lithium dendrites via the atomic-level cavities in GDY film. Source: Shang et al. [138]. © 2018, Elsevier.

more lithiophilic interface than that of the graphite. Via the theoretical calculation and comparison, it is found that in the traditional sp^2 hybrid carbon material, the Li atom is located above the center of the benzene ring with a height of 1.754 Å, and the corresponding adsorption energy is −1.18 eV. On graphdiyne, similar with the sp^2 hybrid carbon materials, one position for storing the Li atoms is on the top of the benzene ring with a relatively low binding energy of −1.95 eV. Except for this, the most stable adsorption configuration of Li atoms is located at the triangular pores on the plane of graphdiyne, and the adsorption energy is much lower at −2.60 eV. The results show that the high amount of the sp hybrid carbon in graphdiynes is more lithiophilic than traditional carbon materials, and it is beneficial to the optimization of the lithium nucleation process. The ultra-uniform distribution of the triangular pores with high lithiophilic activity in graphdiyne film provides new inspiration for solving the key problem of lithium metal batteries. The coating of graphdiyne on the current collector can provide a large number of lithiophilic active sites and improve the deposition process of metallic Li on the current collector. To realize this, the ultrathin graphdiyne films are prepared in situ on the copper nanowires. The formation of the three-dimensional seamless graphdiyne coating layer [139] offers the current collector many active sites. Compared with traditional copper foil, copper nanowires have a higher SSA and can provide a large number of reaction sites for catalyzing the in situ growth of graphdiyne. This three-dimensional free-standing current collector can not only provide many lithiophilic active sites, but also provide enough space to accommodate lithium metal (Figure 6.29). As a result, the overpotential of lithium nucleation on the

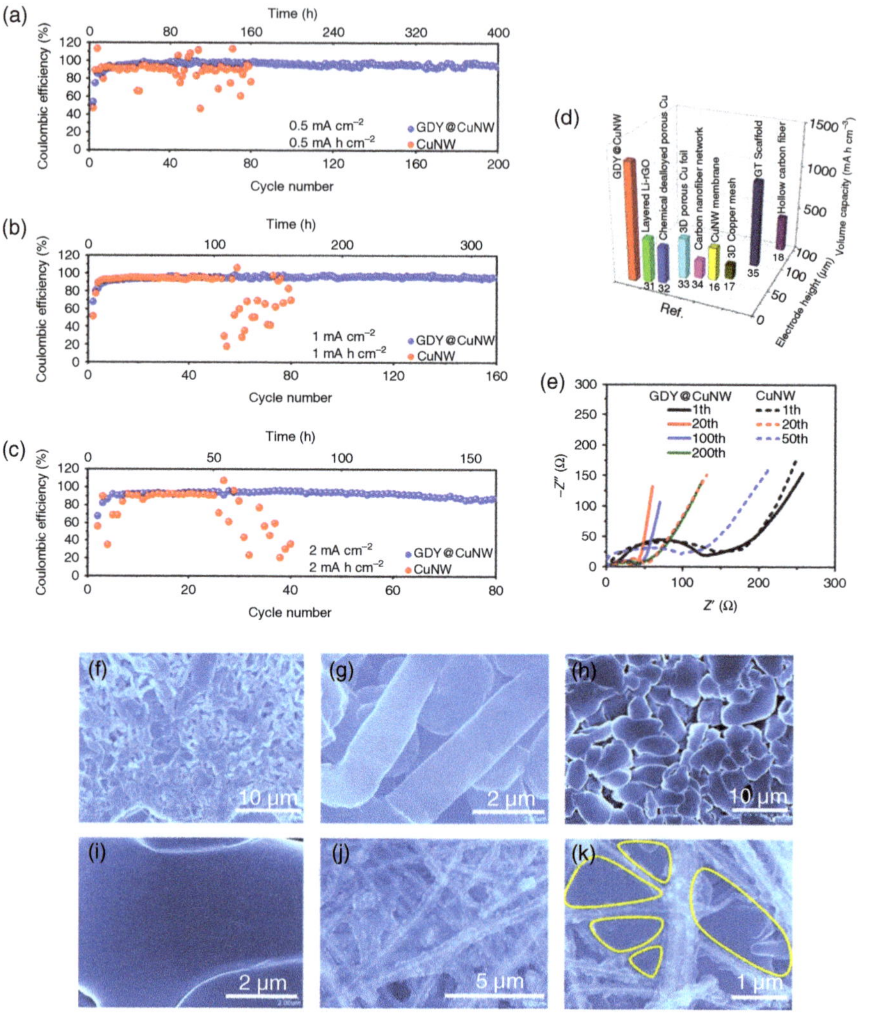

Figure 6.29 (a–c) Coulombic efficiency of copper nanowire (CuNW) electrode with (GDY@CuNW) and without the graphdiyne coating layer at different current densities and capacities; (d) the capacity comparison of the GDY@CuNW electrode with reported current collectors; (e) impedance variations at different cycles; (f, g) SEM images of Li on CuNW; (h–k) SEM images of Li on GDY@CuNW. Source: Shang et al. [139]. © 2019, American Chemical Society.

surface of the graphdiyne-modified copper nanowires is smaller than that on the bare copper nanowires, resulting in the uniform growth of metal lithium. The modification of graphdiyne provides significant improvements in Coulomb efficiency, life span (Figure 6.29a–e), and inhibition of lithium dendrites (Figure 6.29f–k), and inhibition of lithium dendrites. In addition, based on the thin electrode, a volumetric capacity as high as 1333 mAh cm^{-3} can be obtained, which is expected to construct a lithium metal battery with high energy density.

6.8 Supercapacitor Electrodes

Electrochemical supercapacitors, batteries, and fuel cells together constitute the high-efficiency electrochemical energy conversions and storage systems [140, 141]. They have the characteristics of high power density and long lifespan, which can effectively complement the disadvantages of the high-energy-density batteries and fuel cells [142]. In recent years, electrochemical supercapacitors have attracted great attention. The supercapacitor can be used as complementary power system to deliver high power for acceleration in a hybrid electric vehicle and can rapidly store the regenerative braking energy. In addition, electrochemical supercapacitors can also provide backup power to prevent power interruption and supplement the energy storage functions of batteries. Therefore, electrochemical supercapacitors, like batteries, are also very important energy storage devices. Although the theoretical research and practical applications of electrochemical supercapacitors have made significant progresses, electrochemical supercapacitors still have some shortcomings. Low-energy density is one of the main challenges to prevent the further development of electrochemical supercapacitor technology. To solve this problem, the design and development of new structures and synthesis of new materials are the main approaches of current research. Among them, it is prevalent to prepare carbon nanomaterials with ultra-high specific surface. The electrical double-layer capacitance is the storage mechanism in these materials, and a large surface area is conducive to charge storage.

Theoretically, the graphdiyne is a carbon material with a higher SSA than graphene due to the in-plane porous structure. This large π-conjugated two-dimensional material not only has good electronic conductivity, but also has a three-dimensional ion transport channels, ensuring a good accessibility for the electrolyte. Therefore, the graphdiyne structure is an ideal structure for the supercapacitors. Professor Huang's group [143] studied the lithium ion capacitors based on graphdiyne electrodes (Figure 6.30). In this study, they assembled a lithium-ion capacitor using the bulk graphdiyne as anode and activated carbon as cathode. Under the power density of 400.1 W kg^{-1}, the initial specific energy density of the capacitor is as high as 112.2 Wh kg^{-1}, and it can maintain 94.7% of the initial capacitance after 1000 cycles. Working in a voltage range from 2 to 4 V, the capacitor still can deliver an energy density of 95.1 Wh kg^{-1} at a higher power of 1000.4 W kg^{-1}. Results demonstrate that the lithium ion capacitor composed of the graphdiyne anode and activated carbon cathode has both high-energy and high-power density, showing great potential.

Because the graphdiyne precursors have good reactivity, they can quickly react under heat treatment conditions to obtain graphdiyne nanochain with high SSA [75]. This offers the possibility to improve the performance of the capacitor based on the graphdiyne materials. Via tuning the structure of precursors, it was found that the nitrogen-doped graphdiyne can be efficiently prepared from the nitrogen-containing precursor, without any solvents and metal catalysts in the air at 120 °C [57]. The introduction of nitrogen into the graphdiyne skeleton not only adjusts the energy bandwidth of graphdiyne, but also adjusts the size of the resulting

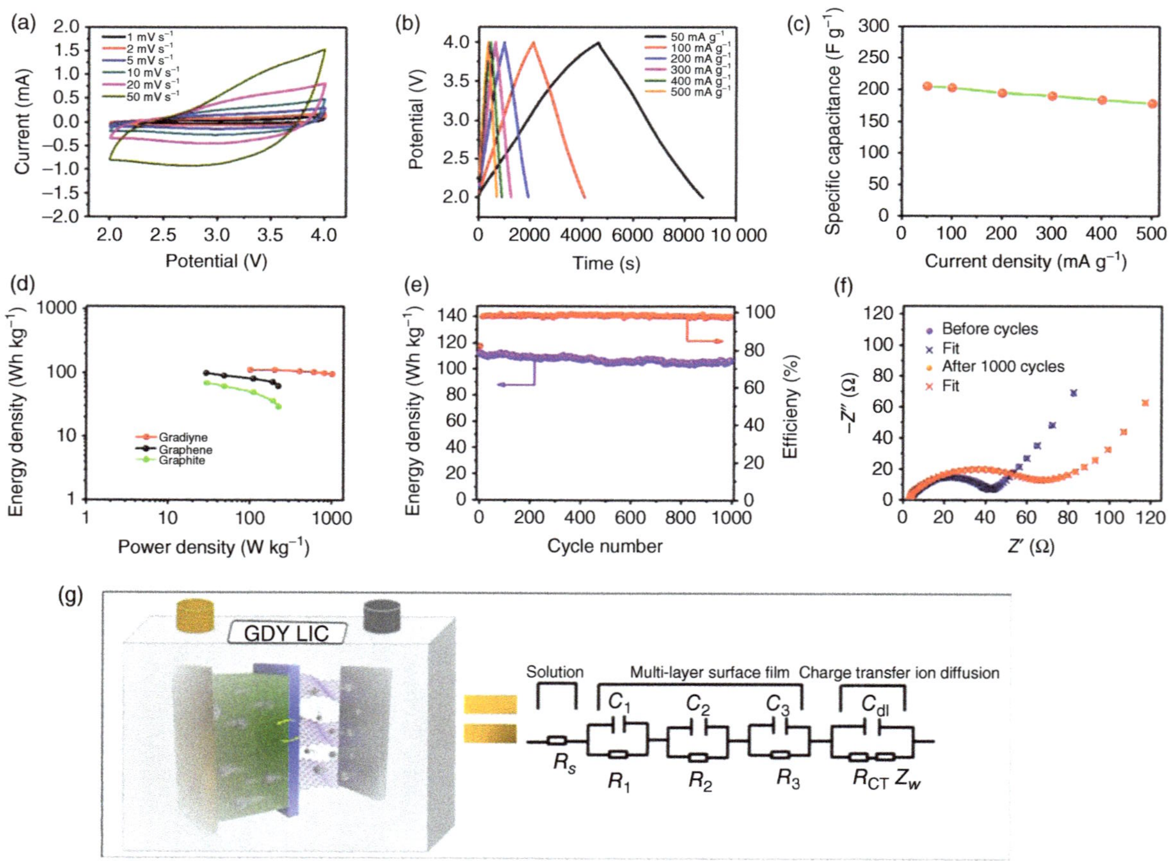

Figure 6.30 Electrochemical performance of capacitor based on graphdiyne. (a) CV curves at various scan rates; (b) Galvanostatic charge/discharge profiles at various current densities; (c) Corresponding specific capacitances at various current densities. (d) Ragone plots of capacitors based on graphdiyne, graphite, and graphene. (e) Cycling stability of at 200 mA g^{-1}. (f) Nyquist plots of the capacitor before and after 1000 cycles. (g) The equivalent circuit of capacitor. Source: Du et al. [143]. © 2016, Elsevier.

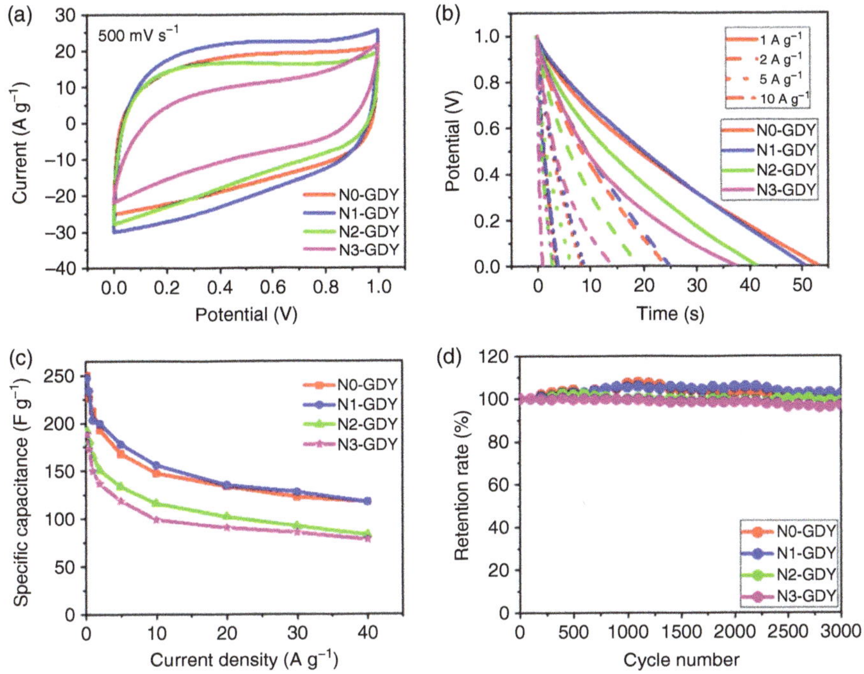

Figure 6.31 Supercapacitor performance based on N-doped graphdiyne electrode in 7.0 M KOH aqueous electrolyte: (a) cycle voltammetry curves at 500 mV s^{-1}; (b) galvanostatic charge/discharge profiles at different current densities; (c) specific capacitance at different current densities; and (d) long-term stability of the devices at 5 A g^{-1}. Source: Shang et al. [57]. © 2018, Elsevier.

nanochains with sizes ranging from 10 to 80 nm. Meanwhile, the N-doping strategy is an efficient way to increase the accessibility of the electrolyte and the capacitance of the electrode (Figure 6.31). The as-prepared graphdiyne nanochain has high SSA and good three-dimensional continuity, and can be used as a high-performance supercapacitor electrode. When the potassium hydroxide is used as the electrolyte, high-energy density (8.66 Wh kg^{-1}) and high-power density (19.3 kW kg) can be realized in a symmetrical two-electrode device.

The above method also shows good scalability, and the graphdiyne can grow rapidly on a variety of substrates [59]. In an air atmosphere, the graphdiyne grown ultra-fast on any substrate, and the as-grown graphdiyne has ultra-fine nanochain structures. It is further confirmed that this is an environment-friendly, metal-free method, which effectively eliminates the impurity pollution that may be faced during the preparation of carbon materials. The growth of seamless graphdiyne on any substrate proves that all-carbon graphdiyne has excellent controllability and processability (Figure 6.32). And this three-dimensional high specific surface structure is exactly what the supercapacitor electrode needs, which can not only ensure that there are abundant specific surface-stored charges, but also ensure that the active ions can reach the electrode surface and realize rapid ion migration. Due to its reasonable porosity and three-dimensional continuity, the prepared sample

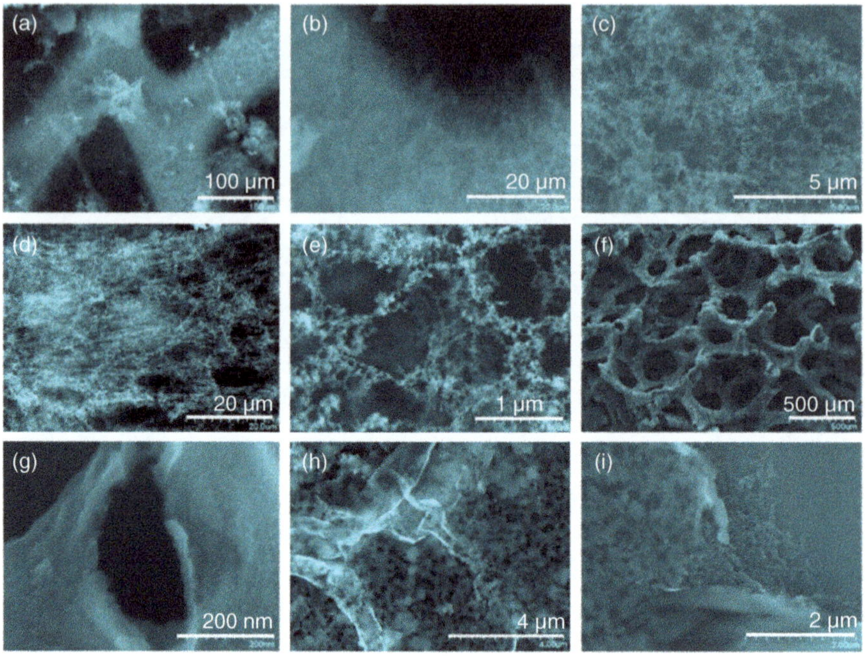

Figure 6.32 SEM images of the graphdiyne nanostructures on the nickel foam: (a–e) graphdiyne nanostructure in different magnification on the nickel foam; (f–i) graphdiyne morphologies after removal of the nickel framework. Source: Wang et al. [59]. © 2019, the American Chemical Society.

can be used as a high-performance binder-free supercapacitor electrode, with a large area capacitance of up to 53.66 $mF\,cm^{-2}$, excellent power performance, and long-term cycle stability (99% after 1300 cycles).

Compositing with metal oxides is an effective way to increase the capacitance of supercapacitors. The study of Jing Xu et al. [144]designed a method for in situ synthesis of manganese dioxide nanospheres and loaded them on graphdiyne oxide, and the resulting composite material was used as an electrode material for supercapacitors (Figure 6.33). Characterization shows that graphdiyne oxide and manganese dioxide have good contact, which is beneficial for the electrochemical performance. Electrochemical tests show that when the current density is 0.2 and 10 $A\,g^{-1}$, this composite electrode can obtain specific capacitances of 301 and 170 $F\,g^{-1}$, respectively. This performance is better than that of the composite electrode of manganese dioxide and graphdiyne, respectively. The cycle stability test shows that composite electrode still has 98% initial capacitance retention after 3000 charge and discharge cycles, and has excellent stability.

6.9 Fuel Cells

Proton-exchange membranes using water as the medium play a vital role in the development of energy storage systems such as fuel cells and flow batteries.

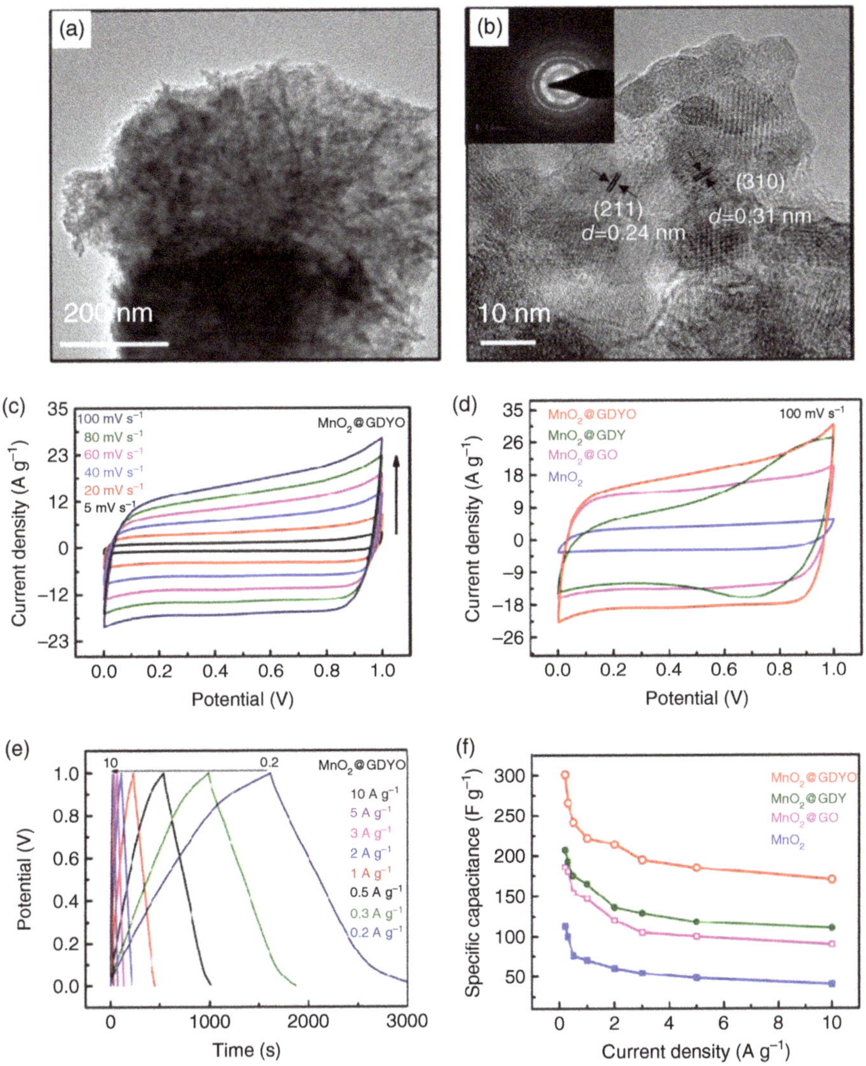

Figure 6.33 (a, b) Morphologies of the MnO_2@graphdiyne oxide (GDYO); (c) CV curves of the MnO_2@GDYO at different scanning rate; (d–f) capacitance comparison among different electrodes. Source: Xu et al. [144]. © 2017, Elsevier.

Compared with hydroxide exchange membranes, proton-exchange membranes have higher conductivity, which is a key factor for assembling the high-power-density fuel cells. There are a large number of commercial proton-exchange membranes based on Nafion, polybenzimidazole, and polyetheretherketone. The proton transport in these membranes is along the ion channels formed in the microphase separation. The ion transport channels formed by the phase separation are about several nanometers, and its size distribution is wide. In these membranes, the selectivity will vary remarkably with the water absorption, and temperature changes. Although high proton conductivity can be obtained through these membranes, serious fuel permeation problems are hard to solve. The penetration of fuel shortens the lifespan of the fuel cell and reduces the efficiency of the fuel. In a direct methanol fuel cell (DMFC), methanol permeability can cause poisoning of the cathode catalyst, which will seriously affect the stability of the catalyst and the power performance of the fuel cell. Although solid proton electrolytes with high selectivity based on metal organic frameworks have been developed in the reports, their proton conductivity at room temperature is still far from requirements. Meanwhile, these membranes also face the problems of the instability when exposed to the acidic conditions for a long time. Developing the membranes that can provide both robust proton conductivity and high selectivity is critical to the practical applications of fuel cells.

Since first discovery of graphene, two-dimensional materials have attracted widespread attention all over the world. In particular, a two-dimensional film will be a new choice for constructing high-selectivity proton exchange membrane for fuel cell. Using present technologies, it is difficult to realize large-area membranes with high-density pores suitable for proton selective transport. In 2014, Geim's group showed that protons can be transported through the uniform pores formed by the electron cloud in the graphene and graphitic boron nitride (hBN), opening up a new way for the design of zero-permeability proton selective membranes. The research work by Holmes et al. showed that when a single layer of graphene was sandwiched in the Nafion membrane, the methanol permeation of Nafion is reduced in the DMFC. However, it is still controversial whether protons can conduct through the pores formed by the electron cloud in the two-dimensional crystal. More theoretical studies have shown that the conduction barrier of hydrated protons in intact graphene or hBN is very high, and it is impossible to achieve cross-plane transmission of protons at room temperature. The results suggest that the proton conductivity observed in the experiment may be mainly due to the atomic defects or test bias.

DMFC has been highly concerned among scientists because of its outstanding advantages in portability, safety, high energy density, and reproducibility. It is a promising power source. In DMFCs, Nafion membrane is still the first choice as proton-exchange membranes due to its high proton conductivity and good electrochemical stability. However, its high proton conductivity is also accompanied by severe methanol permeability. Methanol permeation greatly reduces the utilization rate of methanol, leading to poisoning of the cathode catalyst. Thus, the electrochemical performance is rapidly degraded, which hinders the practical application of DMFCs as the long-term power sources. Methanol permeation is a key problem in this field that scientists urgently need to solve. The phase separation in Nafion constitutes the nanochannels for rapid proton conductivity, and meanwhile leads to the methanol permeation. When the operating temperature increases, the size of the proton transport nanochannel will be expanded, resulting in more serious methanol permeation in the system. For decades, scientists have been committed to solving the problem of methanol crossover. Currently, the use of two-dimensional materials with artificial nanopores is considered to be the most promising method. However, the limitation of the popular two-dimensional materials (graphene, hBN) is that the preparation and perforation technology is uncontrollable, and it is challenging to realize such large-scale two-dimensional materials with uniform artificial nanopores. In addition, without chemical modification, it is difficult to achieve high compatibility between them and the Nafion matrix. Therefore, it is difficult to achieve high selectivity and high stability with proton-exchange membranes for DMFCs based on the existing two-dimensional materials.

The graphdiyne plane is a carbon material with a controllable pore structure at the atomic level. It does not need to use complex and special technology to produce pores for the selective transmission function. It can be seen that the natural pores of all-carbon graphyne provide new possibilities for the selective transmission of protons. In the theoretical calculation [145], the graphynes with the pore size of 0.69 ($n = 1$), 0.95 ($n = 2$), 1.20 ($n = 3$), and 1.45 ($n = 3$), are investigated (Figure 6.34). It is demonstrated that the pore size has a great influence on the selective conduction behavior of protons. Through molecular dynamics simulation, it is found that the selective conduction behavior of protons in a water environment is essentially different from that in a vacuum environment. When $n = 1$, the protons in the hydronium ion phase must dissociate with water and form C—H bonds with the carbon atoms of graphyne; therefore, the corresponding energy barrier is increased up to 2.80 ± 0.03 eV. When $n = 2$, protons can pass through the membrane through a vehicular mechanism in the form of intact H_3O^+, or through the Grotthuss mechanism between two water molecules near both sides of the membrane. This process has a relatively lower-energy barrier. When $n = 3$ and 4, water molecules can penetrate into the pores of graphdiyne to form a continuous water phase. Therefore, protons in the continuous water phase can be conducted through the Grotthuss mechanism. The corresponding low activation energy barriers are 0.27 ± 0.07 eV and 0.19 ± 0.02 eV. At the same time, for graphynes ($n = 3$ and 4), a patterned water/vacuum interface will be formed, which can effectively prevent the penetration of other species such as methanol dissolved in the water phase. According to the potential barrier, the graphyne with $n = 4$ can provide high areal proton conductance of 2.15×10^6 S cm^{-2} and ultra-high proton/methanol selectivity of 1.0×10^{12}. Therefore, in theory, it can be seen that graphynes provide a good structural basis for the development of high-efficiency DMFCs.

Above theoretical calculation reveals the possible advantages of the graphyne with different pore sizes in selective transport. However, based on recent synthetic methodologies, the graphdiyne is the only one which can be readily prepared in large area. The deep investigation in the proton selectivity of graphdiyne was more interesting for practical applications. Wang's group [146]carried the specific calculation in investigating the proton transfer properties via the graphdiyne film (Figure 6.35). According to the calculation, the free-energy barrier of proton transfer across the graphdiyne film via the Grotthuss mechanism is about 2.4 kJ mol^{-1}, which is much closer to that in the water (2.1 kJ mol^{-1}). The low free-energy barrier leads to a high proton conductivity of 0.6 S cm^{-1} through the graphdiyne film, which is about 4 orders of magnitude higher than that in graphene and other 2D material. Due to the small pore size, the cross-plane energy barrier of proton is 0.15 eV, which is much lower than those of methanol (1.77 eV), ethanol (3.66 eV),

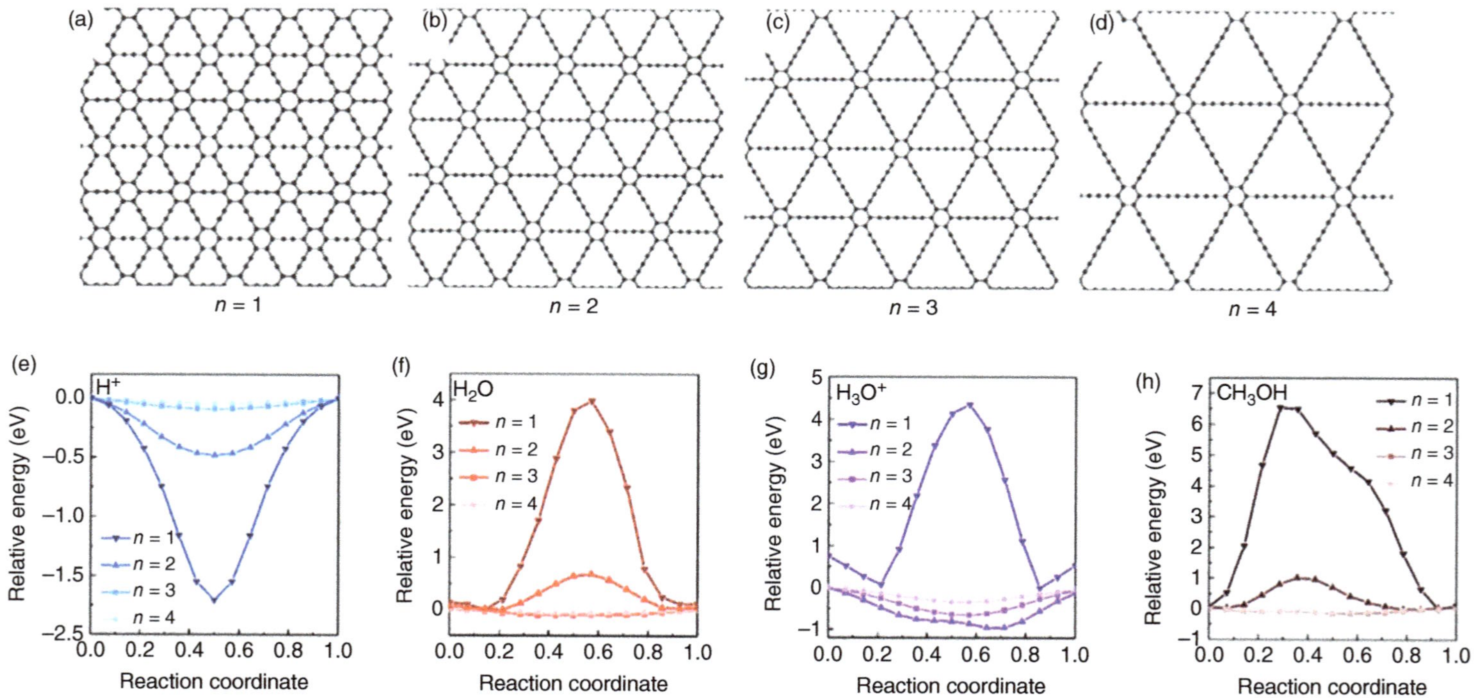

Figure 6.34 (a–d) Molecular structures of the graphyne with different side length; (e, f) penetration energy barriers of different species across graphyne with corresponding side length. Source: Shi et al. [145]. © 2019, the Nature Publishing Group/CC License 4.0.

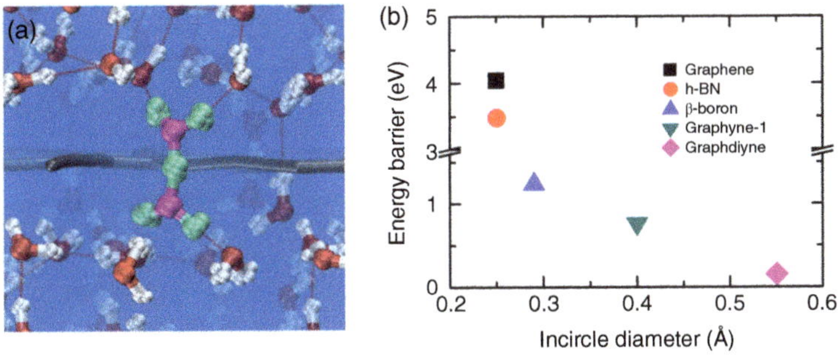

(c)

Table 1 Energy barriers for solutes in aqueous solution passing through nanopore in graphdiyne membrane

Species	Proton	CH_3OH	CH_3CH_2OH	Na^+	Cl^-
Energy barrier (eV)	0.15	1.77	3.66	1.09	1.4

Figure 6.35 (a) Typical quantum configuration of trans-membrane $H_5O_2^+$ complex in path-integral molecular dynamics (PIMD) simulations; (b) energy barriers for proton transport across different two-dimensional (2D) membranes; (c) table showing the energy barriers for different elements cross the graphdiyne in water. Source: Xu et al. [146]. © 2019, the Nature Publishing Group.

Na^+ (1.09 eV), as well as Cl^- (1.4 eV). Therefore, theoretical research demonstrates that the graphdiyne film is a promising nanomesh for transparent proton transport.

In experiments, the ultrathin amino-functionalized graphdiyne was utilized for exploring the selective transport in the DMFC [147]. It is known that the size distribution of proton transfer channels in the Nafion membrane is unstable and wide, and varies greatly with the changes in temperature and water uptake. However, the two-dimensional graphdiyne all-carbon framework has a large number of rigid diacetylene bonds, and in a wide temperature range, the in-plane cavities can still maintain excellent dimensional stability. Therefore, these two materials can greatly complement with each other after combining them together. Up to now, the preparation of single-layer and double-layer graphdiyne films with excellent selective proton transfer function is still a scientific challenge. Additionally, how to improve the compatibility between the graphdiyne and Nafion matrix is also needed to consider. Based on the above reasons, the high-quality amino-functionalized graphdiyne was designed and used in DMFC. The in-plane nanopores of the amino-functionalized graphdiyne are larger than those of the original graphdiyne, which ensures that protons can be transported through even in a thick film. The amino groups are uniformly distributed on the graphdiyne plane, and they have a strong acid–base interaction with the sulfonic acid group in the Nafion molecule (Figure 6.36). Therefore, near the amino-functionalized graphdiyne, the structure of microphase separation of Nafion is small, and the graphdiyne and Nafion matrix show good compatibility. The smaller microphase separation

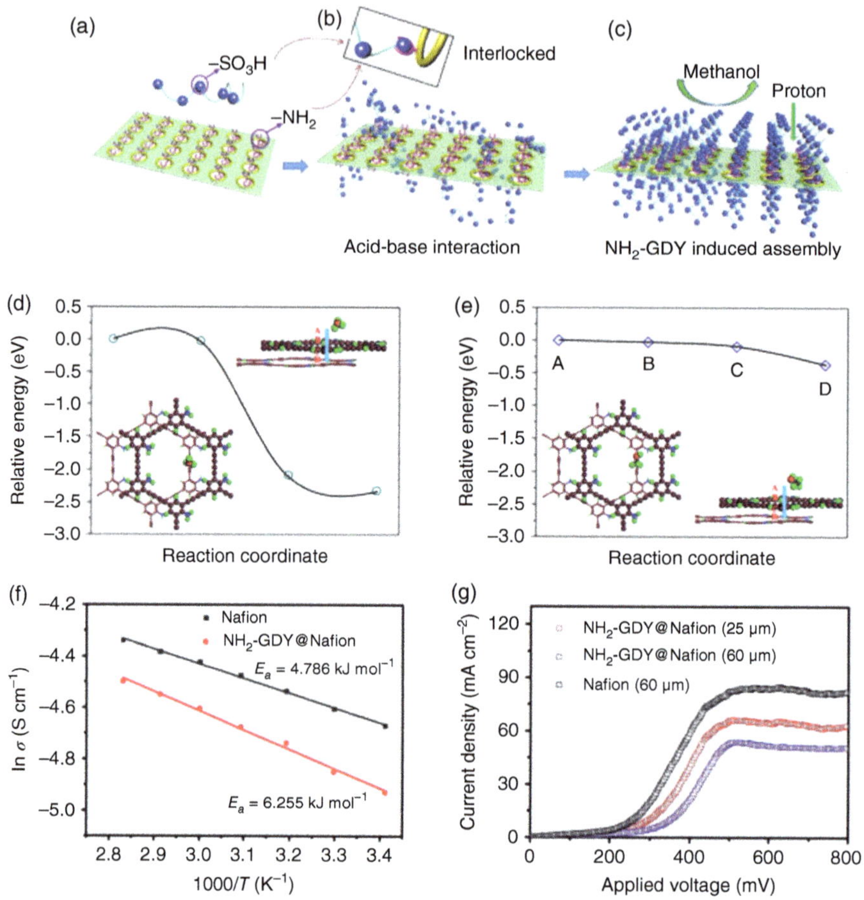

Figure 6.36 (a–c) Possible interaction between Nafion on NH_2-GDY; (d, e) the energy variation when the hydrated proton (d) and the methanol molecule (e) passing through NH_2-GDY; (f) the temperature dependence of the proton conductivity; (g) the methanol crossover current densities at 65 °C in 1 M methanol solution. Source: Wang et al. [147]. © 2019, John Wiley & Sons.

of Nafion and the inherent in-plane pore selectivity of amino-functionalized graphdiyne greatly inhibited the crossover of methanol. Although the addition of the amino-functionalized graphdiyne reduces the proton conductivity of the Nafion membrane in a certain extent, it greatly inhibits the permeation of methanol. The permeation of methanol is reduced by 38%, which not only improves the utilization rate of methanol, but also improves the power performance and stability of fuel cell. This work successfully bypasses the challenges of perforation technology and shows great universality for the construction of two-dimensional porous carbon nanofilms with high selectivity for many potential applications.

References

1 Goodenough, J.B. (2012). Evolution of strategies for modern rechargeable batteries. *Accounts of Chemical Research* 46: 1053–1061.

2 Winter, M. and Brodd, R.J. (2004). What are batteries, fuel cells, and supercapacitors? *Chemical Reviews* 104: 4245–4270.

3 Li, B. and Liu, J. (2017). Progress and directions in low-cost redox flow batteries for large-scale energy storage. *National Science Review* 4: 91–105.

4 Fan, E., Li, L., Wang, Z. et al. (2020). Sustainable recycling technology for li-ion batteries and beyond: challenges and future prospects. *Chemical Reviews* 120: 7020–7063.

5 Massé, R.C., Liu, C., Li, Y. et al. (2017). Energy storage through intercalation reactions: electrodes for rechargeable batteries. *National Science Review* 4: 26–53.

6 Konarov, A., Myung, S.-T., and Sun, Y.-K. (2017). Cathode materials for future electric vehicles and energy storage systems. *ACS Energy Letters* 2: 703–708.

7 Lin, F., Liu, Y., Yu, X. et al. (2017). Synchrotron X-ray analytical techniques for studying materials electrochemistry in rechargeable batteries. *Chemical Reviews* 117: 13123–13186.

8 Cao, G. (2016). Search for better materials for rechargeable electric energy storage. *National Science Review* 4: 16.

9 Xin, S., Guo, Y.G., and Wan, L.J. (2012). Nanocarbon networks for advanced rechargeable lithium batteries. *Accounts of Chemical Research* 45: 1759–1769.

10 Melot, B.C. and Tarascon, J.M. (2013). Design and preparation of materials for advanced electrochemical storage. *Accounts of Chemical Research* 46: 1226–1238.

11 Wassei, J.K. and Kaner, R.B. (2013). Oh, the places you'll go with graphene. *Accounts of Chemical Research* 46: 2244–2253.

12 Yu, L., Shearer, C., and Shapter, J. (2016). Recent development of carbon nanotube transparent conductive films. *Chemical Reviews* 116: 13413–13453.

13 Siemsen, P., Livingston, R.C., and Diederich, F. (2000). Acetylenic coupling: a powerful tool in molecular construction. *Angewandte Chemie International Edition* 39: 2632–2657.

14 Kroto, H.W., Heath, J.R., O'Brien, S.C. et al. (1985). C_{60}: Buckminsterfullerence. *Nature* 318: 162–163.

15 Diederich, F. (1994). Carbon scaffolding: building acetylenic all-carbon and carbon-rich compounds. *Nature* 369: 199–207.

16 Shi, L., Rohringer, P., Suenaga, K. et al. (2016). Confined linear carbon chains as a route to bulk carbyne. *Nature Materials* 15: 634–639.

17 Dai, L. (2013). Functionalization of graphene for efficient energy conversion and storage. *Accounts of Chemical Research* 46: 31–42.

18 Englert, J.M., Dotzer, C., Yang, G.A. et al. (2011). Covalent bulk functionalization of graphene. *Nature Chemistry* 3: 279–286.

19 Biswas, S., Senju, A., Mohr, R. et al. (2017). Minimal architecture zinc–bromine battery for low cost electrochemical energy storage. *Energy & Environmental Science* 10: 114–120.

20 Choi, N.S., Chen, Z., Freunberger, S.A. et al. (2012). Challenges facing lithium batteries and electrical double-layer capacitors. *Angewandte Chemie International Edition* 51: 9994–10024.

21 Li, X.S., Cai, W.W., An, J.H. et al. (2009). Large-area synthesis of high-quality and uniform graphene films on copper foils. *Science* 324: 1312–1314.

22 Hong, G., Zhang, B., Peng, B. et al. (2009). Direct growth of semiconducting single-walled carbon nanotube array. *Journal of the American Chemical Society* 131: 14642–14643.

23 Georgakilas, V., Perman, J.A., Tucek, J. et al. (2015). Broad family of carbon nanoallotropes: classification, chemistry, and applications of fullerenes, carbon dots, nanotubes, graphene, nanodiamonds, and combined superstructures. *Chemical Reviews* 115: 4744–4822.

24 Li, G.X., Li, Y.L., Liu, H.B. et al. (2010). Architecture of graphdiyne nanoscale films. *Chemical Communications* 46: 3256–3258.

25 Huang, C. and Li, Y. (2016). Structure of 2D graphdiyne and its application in energy fields. *Acta Physico-Chimica Sinica* 32: 1314–1329.

26 Li, Y. and Li, Y. (2015). Two dimensional polymers-progress of full carbon graphyne. *Acta Polymerica Sinica* 2: 147–165.

27 Li, Y., Xu, L., Liu, H. et al. (2014). Graphdiyne and graphyne: from theoretical predictions to practical construction. *Chemical Society Reviews* 43: 2572–2586.

28 Lu, C., Yang, Y., Wang, J. et al. (2018). High-performance graphdiyne-based electrochemical actuators. *Nature Communications* 9: 752.

29 Kang, J., Li, J., Wu, F. et al. (2011). Elastic, electronic, and optical properties of two-dimensional graphyne sheet. *Journal of Physical Chemistry C* 115: 20466–20470.

30 Sun, C.H. and Searles, D.J. (2012). Lithium storage on graphdiyne predicted by DFT calculations. *Journal of Physical Chemistry C* 116: 26222–26226.

31 Shekar, S.C. and Swathi, R.S. (2015). Cation–π interactions and rattling motion through two-dimensional carbon networks: graphene vs graphynes. *Journal of Physical Chemistry C* 119: 8912–8923.

32 Zhang, H., Zhao, M., He, X. et al. (2011). High mobility and high storage capacity of lithium in sp–sp^2 hybridized carbon network: the case of graphyne. *Journal of Physical Chemistry C* 115: 8845–8850.

33 Xue, Y., Guo, Y., Yi, Y. et al. (2016). Self-catalyzed growth of Cu@graphdiyne core–shell nanowires array for high efficient hydrogen evolution cathode. *Nano Energy* 30: 858–866.

34 Xue, Y., Zuo, Z., Li, Y. et al. (2017). Graphdiyne-supported $NiCo_2S_4$ nanowires: a highly active and stable 3D bifunctional electrode material. *Small* 13: 1700936.

35 Wang, S., Yi, L., Halpert, J.E. et al. (2012). A novel and highly efficient photocatalyst based on P25-graphdiyne nanocomposite. *Small* 8: 265–271.

36 Xue, Y., Li, J., Xue, Z. et al. (2016). Extraordinarily durable graphdiyne-supported electrocatalyst with high activity for hydrogen production at all values of pH. *ACS Applied Materials & Interfaces* 8: 31083–31091.

37 Long, M.Q., Tang, L., Wang, D. et al. (2011). Electronic structure and carrier mobility in graphdiyne sheet and nanoribbons: theoretical predictions. *ACS Nano* 5: 2593–2600.

38 Chen, J.M., Xi, J.Y., Wang, D. et al. (2013). Carrier mobility in graphyne should be even larger than that in graphene: a theoretical prediction. *Journal of Physical Chemistry Letters* 4: 1443–1448.

39 Yang, N., Liu, Y., Wen, H. et al. (2013). Photocatalytic properties of graphdiyne and graphene modified TiO_2: from theory to experiment. *ACS Nano* 7: 1504–1512.

40 Xiao, J.Y., Shi, J.J., Liu, H.B. et al. (2015). Efficient $CH_3NH_3PbI_3$ perovskite solar cells based on graphdiyne (GD)-modified P3HT hole-transporting material. *Advanced Energy Materials* 5: 1401943.

41 Gao, X., Li, J., Du, R. et al. (2017). Direct synthesis of graphdiyne nanowalls on arbitrary substrates and its application for photoelectrochemical water splitting cell. *Advanced Materials* 29: 1605308.

42 Ren, H., Shao, H., Zhang, L. et al. (2015). A new graphdiyne nanosheet/pt nanoparticle-based counter electrode material with enhanced catalytic activity for dye-sensitized solar cells. *Advanced Energy Materials* 5: 1500296.

43 Zhou, J., Gao, X., Liu, R. et al. (2015). Synthesis of graphdiyne nanowalls using acetylenic coupling reaction. *Journal of the American Chemical Society* 137: 7596–7599.

44 Yue, Q., Chang, S., Kang, J. et al. (2013). Mechanical andelectronic properties of graphyne and its family under elastic strain: theoretical predictions. *Journal of Physical Chemistry C* 117: 14804–14811.

45 Matsuoka, R., Sakamoto, R., Hoshiko, K. et al. (2017). Crystalline graphdiyne nanosheets produced at a gas/liquid or liquid/liquid interface. *Journal of the American Chemical Society* 139: 3145–3152.

46 Matsuoka, R., Toyoda, R., Shiotsuki, R. et al. (2019). Expansion of the graphdiyne family: a triphenylene-cored analogue. *ACS Applied Materials & Interfaces* 11: 2730–2733.

47 Kan, X., Ban, Y., Wu, C. et al. (2018). Interfacial synthesis of conjugated two-dimensional N-graphdiyne. *ACS Applied Materials & Interfaces* 10: 53–58.

48 Li, J., Xiong, Y., Xie, Z. et al. (2019). Template synthesis of an ultrathin beta-graphdiyne-like film using the eglinton coupling reaction. *ACS Applied Materials & Interfaces* 11: 2734–2739.

49 Zhou, J., Xie, Z., Liu, R. et al. (2019). Synthesis of ultrathin graphdiyne film using a surface template. *ACS Applied Materials & Interfaces* 11: 2632–2637.

50 Wu, B., Li, M.R., Xiao, S.N. et al. (2017). A graphyne-like porous carbon-rich network synthesized via alkyne metathesis. *Nanoscale* 9: 11939–11943.

51 Gao, J., Li, J., Chen, Y. et al. (2018). Architecture and properties of a novel two-dimensional carbon material-graphtetrayne. *Nano Energy* 43: 192–199.

52 Li, G., Li, Y., Qian, X. et al. (2011). Construction of tubular molecule aggregations of graphdiyne for highly efficient field emission. *Journal of Physical Chemistry C* 115: 2611–2615.

53 Liu, R., Gao, X., Zhou, J. et al. (2017). Chemical vapor deposition growth of linked carbon monolayers with acetylenic scaffoldings on silver foil. *Advanced Materials* 29: 1604665.

54 Qian, X., Liu, H., Huang, C. et al. (2015). Self-catalyzed growth of large-area nanofilms of two-dimensional carbon. *Scientific Reports* 5: 7756.

55 Zhang, Y.Q., Kepcija, N., Kleinschrodt, M. et al. (2012). Homo-coupling of terminal alkynes on a noble metal surface. *Nature Communications* 3: 1286.

56 Sun, Q., Yu, X., Bao, M. et al. (2018). Direct formation of C-C triple-bonded structural motifs by on-surface dehalogenative homocouplings of tribromomethyl-substituted arenes. *Angewandte Chemie International Edition* 57: 4035–4038.

57 Shang, H., Zuo, Z., Zheng, H. et al. (2018). N-doped graphdiyne for high-performance electrochemical electrodes. *Nano Energy* 44: 144–154.

58 Zuo, T.T., Wu, X.W., Yang, C.P. et al. (2017). Graphitized carbon fibers as multifunctional 3D current collectors for high areal capacity Li anodes. *Advanced Materials* 29: 1700389.

59 Wang, F., Zuo, Z., Shang, H. et al. (2019). Ultrafastly interweaving graphdiyne nanochain on arbitrary substrates and its performance as a supercapacitor electrode. *ACS Applied Materials & Interfaces* 11: 2599–2607.

60 Peng, Q., Ji, W., and De, S. (2012). Mechanical properties of graphyne monolayers: a first-principles study. *Physical Chemistry Chemical Physics* 14: 13385–13391.

61 Degabriele, E.P., Grima-Cornish, J.N., Attard, D. et al. (2017). On the mechanical properties of graphyne, graphdiyne, and other poly (phenylacetylene) networks. *Physica Status Solidi B: Basic Solid State Physics* 254 (9): 1700380.

62 Yang, Y. and Xu, X. (2012). Mechanical properties of graphyne and its family – a molecular dynamics investigation. *Computational Materials Science* 61: 83–88.

63 Xu, Z., Lv, X., Li, J. et al. (2016). A promising anode material for sodium-ion battery with high capacity and high diffusion ability: graphyne and graphdiyne. *RSC Advances* 6: 25594–25600.

64 Farokh Niaei, A.H., Hussain, T., Hankel, M. et al. (2017). Sodium-intercalated bulk graphdiyne as an anode material for rechargeable batteries. *Journal of Power Sources* 343: 354–363.

65 Zhang, H., Xia, Y., Bu, H. et al. (2013). Graphdiyne: a promising anode material for lithium ion batteries with high capacity and rate capability. *Journal of Applied Physics* 113: 044309.

66 Chandra, S.S. and Swathi, R.S. (2013). Rattling motion of alkali metal ions through the cavities of model compounds of graphyne and graphdiyne. *Journal of Physical Chemistry A* 117: 8632–8641.

67 Li, C., Lu, X., Han, Y. et al. (2018). Direct imaging and determination of the crystal structure of six-layered graphdiyne. *Nano Research* 11: 1714–1721.

68 Huang, C., Zhang, S., Liu, H. et al. (2015). Graphdiyne for high capacity and long-life lithium storage. *Nano Energy* 11: 481–489.
69 Zhang, S., Liu, H., Huang, C. et al. (2015). Bulk graphdiyne powder applied for highly efficient lithium storage. *Chemical Communications* 51: 1834–1837.
70 He, J., Bao, K., Cui, W. et al. (2018). Construction of large-area uniform graphdiyne film for high-performance lithium-ion batteries. *Chemistry—a European Journal* 24: 1187–1192.
71 Wang, K., Wang, N., He, J. et al. (2017). Graphdiyne nanowalls as anode for lithium—ion batteries and capacitors exhibit superior cyclic stability. *Electrochimica Acta* 253: 506–516.
72 Shang, H., Zuo, Z., Li, L. et al. (2018). Ultrathin graphdiyne nanosheets grown in situ on copper nanowires and their performance as lithium-ion battery anodes. *Angewandte Chemie International Edition* 57: 774–778.
73 Li, J., Xu, J., Xie, Z. et al. (2018). Diatomite-templated synthesis of freestanding 3D graphdiyne for energy storage and catalysis application. *Advanced Materials* 30: e1800548.
74 Zhang, S., Du, H., He, J. et al. (2016). Nitrogen-doped graphdiyne applied for lithium-ion storage. *ACS Applied Materials & Interfaces* 8: 8467–8473.
75 Zuo, Z., Shang, H., Chen, Y. et al. (2017). A facile approach for graphdiyne preparation under atmosphere for an advanced battery anode. *Chemical Communications* 53: 8074–8077.
76 He, J., Wang, N., Cui, Z. et al. (2017). Hydrogen substituted graphdiyne as carbon-rich flexible electrode for lithium and sodium ion batteries. *Nature Communications* 8: 1172.
77 Wang, N., He, J., Tu, Z. et al. (2017). Synthesis of chlorine-substituted graphdiyne and applications for lithium-ion storage. *Angewandte Chemie International Edition* 56: 10740–10745.
78 He, J., Wang, N., Yang, Z. et al. (2018). Fluoride graphdiyne as a free-standing electrode displaying ultra-stable and extraordinary high Li storage performance. *Energy & Environmental Science* 11: 2893–2903.
79 Yang, Z., Shen, X., Wang, N. et al. (2019). Graphdiyne containing atomically precise N atoms for efficient anchoring of lithium ion. *ACS Applied Materials & Interfaces* 11: 2608–2617.
80 Jia, Z., Zuo, Z., Yi, Y. et al. (2017). Low temperature, atmospheric pressure for synthesis of a new carbon Ene-yne and application in Li storage. *Nano Energy* 33: 343–349.
81 Zhao, Z., Das, S., Xing, G. et al. (2018). A 3D organically synthesized porous carbon material for lithium-ion batteries. *Angewandte Chemie International Edition* 57: 11952–11956.
82 Wang, K., Wang, N., He, J. et al. (2017). Preparation of 3D architecture graphdiyne nanosheets for high-performance sodium-ion batteries and capacitors. *ACS Applied Materials & Interfaces* 9: 40604–40613.
83 Wang, N., Li, X., Tu, Z. et al. (2018). Synthesis and electronic structure of boron-graphdiyne with an sp-hybridized carbon skeleton and its application in sodium storage. *Angewandte Chemie International Edition* 57: 3968–3973.

84 Zuo, Z. and Li, Y. (2019). Emerging electrochemical energy applications of graphdiyne. *Joule* 3: 899–903.

85 Obrovac, M.N. and Chevrier, V.L. (2014). Alloy negative electrodes for Li-ion batteries. *Chemical Reviews* 114: 11444–11502.

86 Xu, Z., Yang, J., Zhang, T. et al. (2018). Silicon microparticle anodes with self-healing multiple network binder. *Joule* 2: 950–961.

87 Choi, S., Kwon, T.-w., Coskun, A. et al. (2017). Highly elastic binders integrating polyrotaxanes for silicon microparticle anodes in lithium ion batteries. *Science* 357: 279–283.

88 Ogata, K., Jeon, S., Ko, D.S. et al. (2018). Evolving affinity between Coulombic reversibility and hysteretic phase transformations in nano-structured silicon-based lithium-ion batteries. *Nature Communications* 9: 479.

89 Son, I.H., Hwan, P.J., Kwon, S. et al. (2015). Silicon carbide-free graphene growth on silicon for lithium-ion battery with high volumetric energy density. *Nature Communications* 6: 7393.

90 Liu, N., Lu, Z., Zhao, J. et al. (2014). A pomegranate-inspired nanoscale design for large-volume-change lithium battery anodes. *Nature Nanotechnology* 9: 187–192.

91 Shang, H., Zuo, Z., Yu, L. et al. (2018). Low-temperature growth of all-carbon graphdiyne on a silicon anode for high-performance lithium-ion batteries. *Advanced Materials* 30: 1801459.

92 Li, L., Zuo, Z., Shang, H. et al. (2018). In-situ constructing 3D graphdiyne as all-carbon binder for high-performance silicon anode. *Nano Energy* 53: 135–143.

93 Tabassum, H., Zou, R., Mahmood, A. et al. (2018). A universal strategy for hollow metal oxide nanoparticles encapsulated into B/N co-doped graphitic nanotubes as high-performance lithium-ion battery anodes. *Advanced Materials* 30: 1705441.

94 Reddy, M.V., Subba Rao, G.V., and Chowdari, B.V. (2013). Metal oxides and oxysalts as anode materials for Li ion batteries. *Chemical Reviews* 113: 5364–5457.

95 Peng, L., Xiong, P., Ma, L. et al. (2017). Holey two-dimensional transition metal oxide nanosheets for efficient energy storage. *Nature Communications* 8: 15139.

96 Wang, F., Zuo, Z., Li, L. et al. (2019). A universal strategy for constructing seamless graphdiyne on metal oxides to stabilize the electrochemical structure and interface. *Advanced Materials* 31: e1806272.

97 Xiao, X., Liu, X., Wang, L. et al. (2012). $LiCoO_2$ nanoplates with exposed (001) planes and high rate capability for lithium-ion batteries. *Nano Research* 5: 395–401.

98 Van der Ven, A., Aydinol, M.K., and Ceder, G. (1998). First-principles investigation of phase stability in Li_xCoO_2. *Physical Review B* 58: 2975–2987.

99 Zhang, J.-N., Li, Q., Ouyang, C. et al. (2019). Trace doping of multiple elements enables stable battery cycling of $LiCoO_2$ at 4.6 V. *Nature Energy* 4: 594–603.

100 Gong, S., Wang, S., Liu, J. et al. (2018). Graphdiyne as an ideal monolayer coating material for lithium-ion battery cathodes with ultralow areal density and ultrafast Li penetration. *Journal of Materials Chemistry A* 6: 12630–12636.

101 Jaffe, S. (2017). Vulnerable links in the lithiumi Ion battery supply chain. *Joule* 1: 225–228.

102 Wang, C., Xu, Y., Fang, Y. et al. (2015). Extended pi-conjugated system for fast-charge and -discharge sodium-ion batteries. *Journal of the American Chemical Society* 137: 3124–3130.

103 Rodríguez-Pérez, I.A., Yuan, Y., Bommier, C. et al. (2017). Mg-ion battery electrode: an organic solid's herringbone structure squeezed upon Mg-ion insertion. *Journal of the American Chemical Society* 139: 13031–13037.

104 Choi, W., Harada, D., Oyaizu, K. et al. (2011). Aqueous electrochemistry of poly(vinylanthraquinone) for anode-active materials in high-density and rechargeable polymer/air batteries. *Journal of the American Chemical Society* 133: 19839–19843.

105 Peng, C., Ning, G.-H., Su, J. et al. (2017). Reversible multi-electron redox chemistry of π-conjugated N-containing heteroaromatic molecule-based organic cathodes. *Nature Energy* 2: 17074.

106 Fan, L., Ma, R., Wang, J. et al. (2018). An ultrafast and highly stable potassium-organic battery. *Advanced Materials* 30: 1805486.

107 Li, L., Zuo, Z., Wang, F. et al. (2020). In situ coating graphdiyne for high-energy-density and stable organic cathodes. *Advanced Materials* 32: 2000140.

108 Lu, Y. and Chen, J. (2020). Prospects of organic electrode materials for practical lithium batteries. *Nature Reviews Chemistry* 4: 127–142.

109 Muench, S., Wild, A., Friebe, C. et al. (2016). Polymer-based organic batteries. *Chemical Reviews* 116: 9438–9484.

110 Jiang, Q., Xiong, P., Liu, J. et al. (2020). A redox-active 2D metal-organic framework for efficient lithium storage with extraordinary high capacity. *Angewandte Chemie International Edition* 59: 5273–5277.

111 Luo, C., Ji, X., Chen, J. et al. (2018). Solid-state electrolyte anchored with a carboxylated Azo compound for all-solid-state lithium batteries. *Angewandte Chemie International Edition* 57: 8567–8571.

112 Tang, M., Zhu, S., Liu, Z. et al. (2018). Tailoring π-conjugated systems: from π-π stacking to high-rate-performance organic cathodes. *Chem* 4: 2600–2614.

113 Yu, X. and Manthiram, A. (2017). Electrode-electrolyte interfaces in lithium-sulfur batteries with liquid or inorganic solid electrolytes. *Accounts of Chemical Research* 50: 2653–2660.

114 Evers, S. and Nazar, L.F. (2013). New approaches for high energy density lithium-sulfur battery cathodes. *Accounts of Chemical Research* 46: 1135–1143.

115 Manthiram, A., Fu, Y., and Su, Y.S. (2013). Challenges and prospects of lithium-sulfur batteries. *Accounts of Chemical Research* 46: 1125–1134.

116 Zhong, Y., Yin, L., He, P. et al. (2018). Surface chemistry in cobalt phosphide-stabilized lithium-sulfur batteries. *Journal of the American Chemical Society* 140: 1455–1459.

117 Li, Z., Guan, B.Y., Zhang, J. et al. (2017). A compact nanoconfined sulfur cathode for high-performance lithium-sulfur batteries. *Joule* 1: 576–587.

118 Fan, L., Li, M., Li, X. et al. (2019). Interlayer material selection for lithium-sulfur batteries. *Joule* 3: 361–386.

119 Chen, T., Zhang, Z., Cheng, B. et al. (2017). Self-templated formation of interlaced carbon nanotubes threaded hollow Co_3S_4 nanoboxes for high-rate and heat-resistant lithium-sulfur batteries. *Journal of the American Chemical Society* 139: 12710–12715.

120 Demir-Cakan, R., Morcrette, M., Nouar, F. et al. (2011). Cathode composites for Li–S batteries via the use of oxygenated porous architectures. *Journal of the American Chemical Society* 133: 16154–16160.

121 Guo, J., Yang, Z., Yu, Y. et al. (2012). Lithium–sulfur battery cathode enabled by lithium–nitrile interaction. *Journal of the American Chemical Society* 135: 763–767.

122 Jiang, S., Lu, Y., Lu, Y. et al. (2018). Nafion/titanium dioxide-coated lithium anode for stable lithium–sulfur batteries. *Chemistry--An Asian Journal* 13: 1379–1385.

123 Yu, X., Joseph, J., and Manthiram, A. (2015). Polymer lithium–sulfur batteries with a Nafion membrane and an advanced sulfur electrode. *Journal of Materials Chemistry A* 3: 15683–15691.

124 Zhang, S.S. (2012). Binder based on polyelectrolyte for high capacity density lithium/sulfur battery. *Journal of the Electrochemical Society* 159: A1226–A1229.

125 Wang, F., Zuo, Z., Li, L. et al. (2020). Graphdiyne nanostructure for high-performance lithium-sulfur batteries. *Nano Energy* 68: 104307.

126 Li, J., Li, S., Liu, Q. et al. (2019). Synthesis of hydrogen-substituted graphyne film for lithium-sulfur battery applications. *Small* 15: 1805344.

127 Liu, Y., Zhou, G., Liu, K. et al. (2017). Design of complex nanomaterials for energy storage: past success and future opportunity. *Accounts of Chemical Research* 50: 2895–2905.

128 Cheng, X.-B., Zhang, R., Zhao, C.-Z. et al. (2017). Toward safe lithium metal anode in rechargeable batteries: a review. *Chemical Reviews* 117: 10403–10473.

129 Kim, H., Boysen, D.A., Newhouse, J.M. et al. (2013). Liquid metal batteries: past, present, and future. *Chemical Reviews* 113: 2075–2099.

130 Sun, Y., Liu, N., and Cui, Y. (2016). Promises and challenges of nanomaterials for lithium-based rechargeable batteries. *Nature Energy* 1: 16071.

131 Liu, Y., Liu, Q., Xin, L. et al. (2017). Making Li-metal electrodes rechargeable by controlling the dendrite growth direction. *Nature Energy* 2: 17083.

132 Bouchet, R., Maria, S., Meziane, R. et al. (2013). Single-ion BAB triblock copolymers as highly efficient electrolytes for lithium-metal batteries. *Nature Materials* 12: 452.

133 Yamada, Y., Wang, J., Ko, S. et al. (2019). Advances and issues in developing salt-concentrated battery electrolytes. *Nature Energy* 4: 269–280.

134 Zheng, J., Engelhard, M.H., Mei, D. et al. (2017). Electrolyte additive enabled fast charging and stable cycling lithium metal batteries. *Nature Energy* 2: 17012.

135 Wu, X., Hong, J.J., Shin, W. et al. (2019). Diffusion-free Grotthuss topochemistry for high-rate and long-life proton batteries. *Nature Energy* 4: 123–130.

136 Liang, Z., Lin, D., Zhao, J. et al. (2016). Composite lithium metal anode by melt infusion of lithium into a 3D conducting scaffold with lithiophilic coating. *Proceedings of the National Academy of Sciences of the United States of America* 113: 2862–2867.

137 Lin, D., Zhao, J., Sun, J. et al. (2017). Three-dimensional stable lithium metal anode with nanoscale lithium islands embedded in ionically conductive solid matrix. *Proceedings of the National Academy of Sciences of the United States of America* 114: 4613–4618.

138 Shang, H., Zuo, Z., Dong, X. et al. (2018). Efficiently suppressing lithium dendrites on atomic level by ultrafiltration membrane of graphdiyne. *Mater Today Energy* 10: 191–199.

139 Shang, H., Zuo, Z., and Li, Y. (2019). Highly lithiophilic graphdiyne nanofilm on 3D free-standing Cu nanowires for high-energy-density electrodes. *ACS Applied Materials & Interfaces* 11: 17678–17685.

140 Salunkhe, R.R., Kaneti, Y.V., Kim, J. et al. (2016). Nanoarchitectures for metal-organic framework-derived nanoporous carbons toward supercapacitor applications. *Accounts of Chemical Research* 49: 2796–2806.

141 Wang, G., Zhang, L., and Zhang, J. (2012). A review of electrode materials for electrochemical supercapacitors. *Chemical Society Reviews* 41: 797–828.

142 Wang, Y., Song, Y., and Xia, Y. (2016). Electrochemical capacitors: mechanism, materials, systems, characterization and applications. *Chemical Society Reviews* 45: 5925–5950.

143 Du, H., Yang, H., Huang, C. et al. (2016). Graphdiyne applied for lithium-ion capacitors displaying high power and energy densities. *Nano Energy* 22: 615–622.

144 Xu, J., Li, J., Yang, Q. et al. (2017). In-situ synthesis of MnO_2@graphdiyne oxides nanocomposite with enhanced performance of supercapacitors. *Electrochimica Acta* 251: 672–680.

145 Shi, L., Xu, A., Pan, D. et al. (2019). Aqueous proton-selective conduction across two-dimensional graphyne. *Nature Communications* 10: 1165.

146 Xu, J., Jiang, H., Shen, Y. et al. (2019). Transparent proton transport through a two-dimensional nanomesh material. *Nature Communications* 10: 3971.

147 Wang, F., Zuo, Z., Li, L. et al. (2019). Large-area aminated-graphdiyne thin films for direct methanol fuel cells. *Angewandte Chemie International Edition* 58: 15010–15015.

7

Graphdiyne-Based Materials in Solar Cells Applications

Tonggang Jiu and Chengjie Zhao

Institute of Frontier and Interdisciplinary Science, Science Center for Material Creation and Energy Conversion, Shandong University, 72 Binhai Road, Jimo, Qingdao 266237, P. R. China

This chapter provides the recent advances in theoretical and experimental studies of graphdiyne (GDY) and other GDY-based materials in solar cells applications. Due to the unique structure and electronic properties, the emergence of GDY sparks new ideas for the revolutionary development of solar cells. According to theoretical predictions proposed by Shuai and coworkers, GDY sheet displays favorable semiconductor characteristics. Moreover, the calculated in-plane intrinsic electron mobility is able to reach the order of 10^5 cm^2 V^{-1} s^{-1} at room temperature, while the hole mobility is about an order of magnitude lower [1]. The most promising application for solar cell devices is the inclusion of GDY-based materials either as functional materials or as dependable parts of devices. Major properties of GDY that are relevant to solar cells applications will be discussed.

Before touching upon the core elaboration of GDY-based solar cells, let us take a look at some basics of solar cells. Solar cells, as one of renewable energy sources, are devices that convert sunlight into electricity by the photovoltaic effect. The operation of solar cells generally involves three steps, the absorption of light producing either electron–hole pairs or excitons, the separation of charge carriers of opposite types, and the separate extraction of those carriers to an external circuit. Their photovoltaic performance can be examined when exposed to light by the measurable parameters of short circuit current density (J_{sc}), open circuit voltage (V_{oc}), fill factor (FF), and power conversion efficiency (PCE). The history of solar cells can be traced back to 1954 when the first solar cell (silicon p–n junction cell) was successfully demonstrated by Bell Laboratories. Till today, various solar cells are being studied such as silicon solar cell, thin film solar cell, organic solar cell (OSC), dye-sensitized solar cell (DSSC), and perovskite solar cell (PSC). Significant progress has been made since the scientific community is always on the pursuit of high-efficient and low-cost solar cells, which are more commercially applicable to day-to-day life. As displayed in Figure 7.1, the recent decade has witnessed a huge leap of researches on solar cell science and technology.

Graphdiyne: Fundamentals and Applications in Renewable Energy and Electronics,
First Edition. Edited by Yuliang Li.

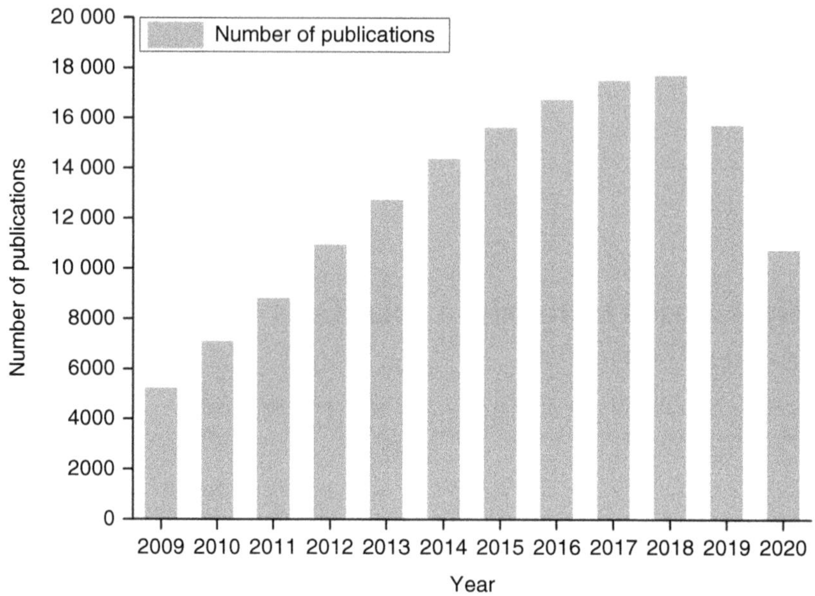

Figure 7.1 Quantitative statistics of research interests in solar cells as of September 2020.

Currently, the mainstream applications of solar energy are dominated by silicon photovoltaic technologies, although their energy-demanding production process remains perplexing. However, emerging photovoltaic technologies featured with low costs, easy fabrication, and promising power conversion efficiencies have attracted considerable attention from scientists worldwide. Except for the commercial single-crystal silicon solar cell, other new-generation photovoltaics cannot be ignored. However, practical demands of stability, materials cost, and toxicity remain a major concern for the technological and social development of humankind in the long run. Note that solar cells mentioned here are emerging photovoltaic devices, including PSCs, OSCs, DSSCs, and quantum dots (QDs) solar cells. Through the evolving history of ongoing efforts for decent efficiencies transcending those of silicon, active and interface materials are main objects under study. Numerous scientific and technological reports have suggested varying material systems in the applications of solar cells, and the presence of GDY-based materials has encouraged further explorations, where the in-depth understanding of work mechanisms and their practical applications for efficient and stable solar cells are revealed. Here the features of GDY-based materials are directly related to the microstructure of active films with varying optoelectronic technologies. In such cases, a thorough study of their applications in solar cells will suggest a visible relationship between a particular sp^2- and sp-hybridized GDY material and concomitant efficacy on photovoltaic performance. In this chapter, we will summarize the representative examples reported in the literature, which are typical of specific influences that GDY-based materials may bring to solar cells at a comprehensive level. The emphasis will be on how GDY affects the performance of solar cells and the understanding of work mechanisms behind this. The following sections of 7.1, 7.2, 7.3 are supported with discussions of recent studies and applications of GDY and GDY-based materials in solar cells, followed by a brief introduction on current research phases of each family.

7.1 Perovskite Solar Cells

PSCs have gained a significant position in the photovoltaic family in terms of efficiency. The soaring rise in power conversion efficiencies has undoubtedly captured widespread attention in industry and academia [2], in addition to their simple and cost-effective manufacture. The fabrication of PSCs based on organic-metal–halide materials was first reported in 2009 with a 3.8% efficiency. Later, photovoltaic technologies have evolved with several structures and different configurations. The most recently reported efficiencies have exceeded 25%, which approaches the efficiency of single-crystalline silicon solar cells. For a better understanding of our discussion in what follows, a general picture of device configuration is given in Figure 7.2. An efficient PSC comprises an intrinsic perovskite absorber (i) sandwiched between a positive (p) and a negative (n) charge-extraction layer [3]. Based on different arrays, device structures can be classified into three categories: mesoporous n–i–p, which is preferred in the early stage until the emergence of planar structured PSCs, planar n–i–p (also known as conventional planar), and planar p–i–n (also known as inverted planar) configuration, where n-type and p-type denote electron transport layers (ETLs) and hole transport layers (HTLs), respectively. For each layer, we will attempt to establish the correlation of key properties in GDY-based PSC devices.

7.1.1 Graphdiyne-Based Materials in Interfacial Layers

Highly efficient PSCs require satisfactory transportation capacities of interface materials and their good contact with perovskite materials. Therefore, seeking superior and appropriate interface materials has been a key research aspect in the field of solar cells, since varying interfacial layers result in different performance parameters. Interfacial engineering is of critical importance for energy-level alignment, electrical conductivity, trap passivation, and subsequently the performance of PSCs. Numerous studies have indicated that employing proper additives is a universal and effective technique for elevating the performance, eliminating the hysteresis as well as improving the stability of PSCs [4–6]. Researchers offer diverse candidates of available additives, including salts, molecules, polymers, and nanoparticles [7]. The acknowledged main efficacy can be one or multiple aspects regarding morphological modification of perovskite films, energy level tuning and nonradiative recombination suppression in PSCs, hysteresis elimination, and operational stability improvement. After understanding the role of interfacial layers, let us take a look at the application of GDY-based materials as additives for interfacial modification of PSCs.

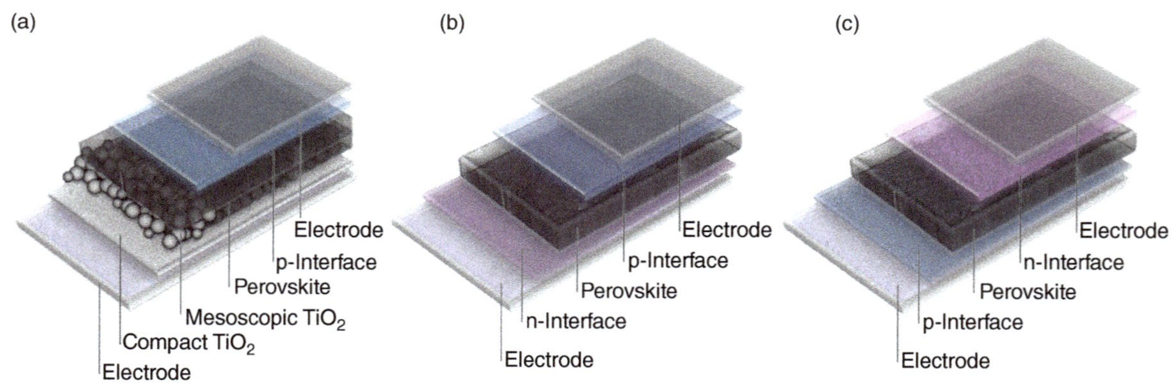

Figure 7.2 (a) Mesoporous n–i–p structure. (b) Planar n–i–p structure. (c) Planar p–i–n structure. Source: Luo et al. [3]. © 2020, Springer Nature BV.

With the incorporation of GDY-based materials, several candidates of interfacial layers have been developed, extending the scope of current interface materials applicable to PSCs system. In PSCs, GDY and its derivatives are introduced to ETLs or HTLs to optimize the device performance, which effectively address major problems deriving from inherent defects of commonly used interface materials or demanding processing requirements of such materials. Taking PCBM, a popular electron transport material in optoelectronics field as an example, the tradeoff between efficient electron transportation and complete film coverage is hard to maneuver. Except optimizing PCBM thickness, doping GDY is also proved effective to improve the film coverage without sacrificing electric properties. Jiu and coworkers constructed an inverted planar heterojunction PSC device with PCBM:GDY ETL and obtained 14.8% efficiency with alleviated hysteresis (Figure 7.3) [8]. A series of characterizations show that GDY is responsible for the improved contact area between ETL and perovskite layer, and the enhanced electron mobility as well as charge extraction. It is speculated that GDY might also be conducive to the passivation of grain boundaries and thus avoid the recombination by decreasing the interface trap states.

Interestingly, in addition to doping GDY into single ETL, an approach of dual doping GDY into electron bilayers of PCBM and ZnO in PSCs is proposed [9], in which electrical properties of GDY have been fully utilized to maximize the merits of current electron transport materials (Figure 7.4). GDY powder is well dispersed

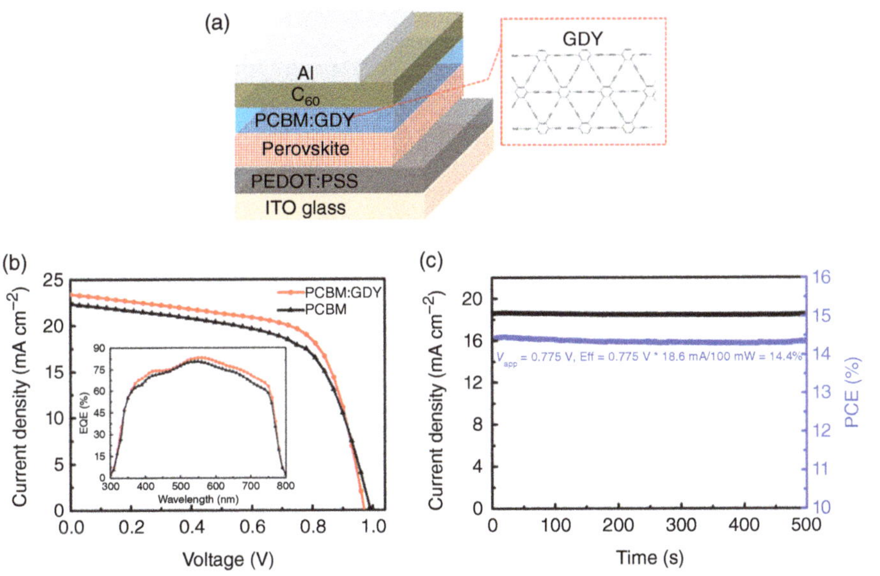

Figure 7.3 (a) Device architecture of perovskite solar cell and chemical structure of graphdiyne (GDY). (b) *J*–*V* characteristic curves of pure PCBM and PCBM:GDY-based perovskite solar cells under AM 1.5G 100 mW cm^{-2} simulated solar light. Inset is corresponding external quantum efficiency (EQE) spectra of the optimal PCBM:GDY- and PCBM-based devices. (c) Steady-state efficiency (blue) with the photocurrent density (black) of the optimal PCBM:GDY-based perovskite solar cell device as a function of time applied at a forward bias of 0.775 V. Source: Kuang et al. [8]. © 2015, American Chemical Society.

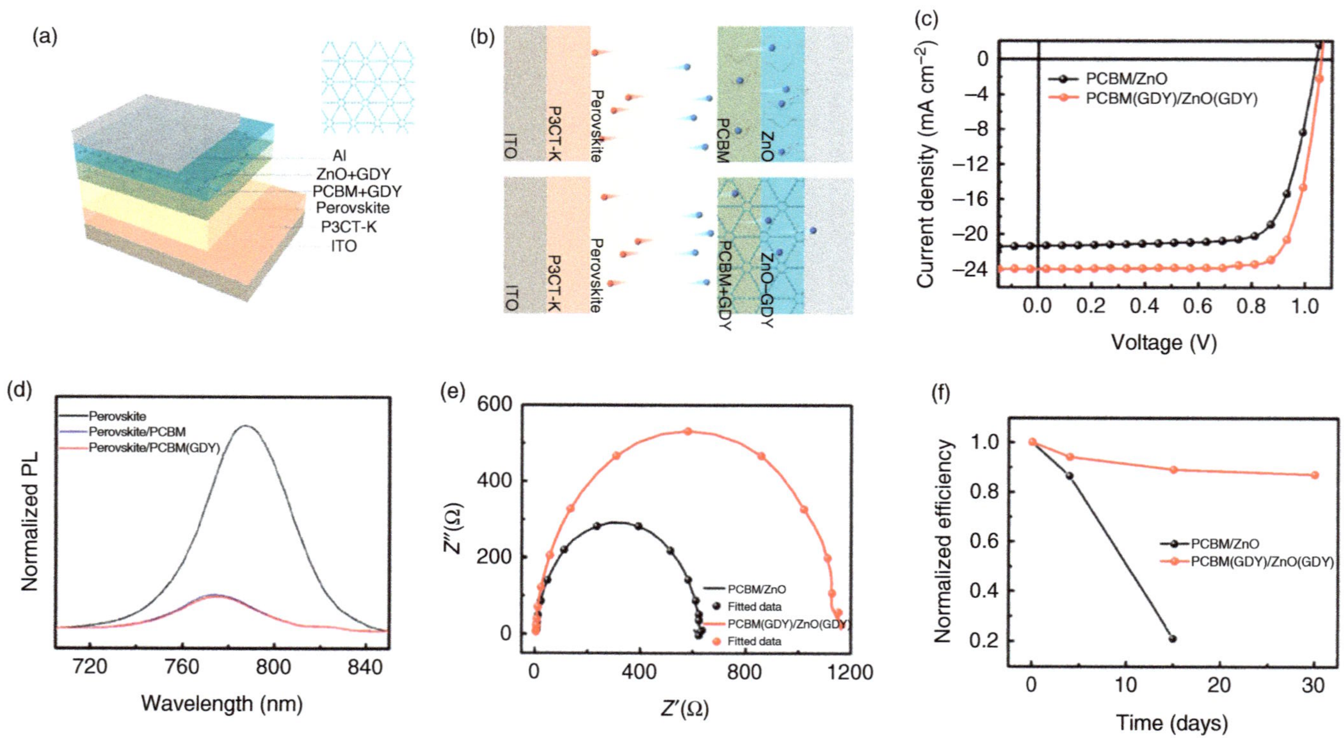

Figure 7.4 (a) Device structure of the photovoltaic device with GDY doped electron layers. (b) Illustration for the process of electron transfer in the devices with different ETLs. (c) Corresponding device performance of the photovoltaic device. Comparison of (d) photoluminescence (PL) quenching and (e) impedance value of perovskite active layer covered with different ETLs. (f) Stability of the PSCs. Source: Li et al. [9]. © 2018, Elsevier.

in chlorobenzene, ensuring the good dispersion of GDY in PCBM and ZnO solution. Morphology studies show the presence of GDY is beneficial to better morphologies of ETLs. After doping GDY into both ETLs, the surface defects are highly passivated, leading to suppressed charge recombination, increased electron transfer rate, and charge transport mobility for each layer. As a result, with an enhanced J_{sc} and FF, $MAPbI_3$-based photovoltaic device gives a best PCE of 20.0% with negligible J–V hysteresis.

In addition, the strategy of incorporating GDY into ETL can also be used to control perovskite crystallization. Li et al. introduced GDY to cross-linkable fullerene [6,6]-phenyl-C61-butyric styryl dendron ester (PCBSD) and fabricated perovskite devices with an optimized 20.19% efficiency and improved stability (Figure 7.5) [10]. It is revealed that a strong π–π stacking interaction between GDY and PCBSD enables to generate a face-on stacked composite film, demonstrating excellent electron mobility, efficient charge extraction, and energy-level tailoring. Therefore, the ordered orientation of GDY-containing composite layer is favorable for the growth and crystallization of the subsequent perovskite film.

Among various low-temperature-processed ETL candidates superior to TiO_2 in fabrication techniques, SnO_2 has been proven to demonstrate good performance

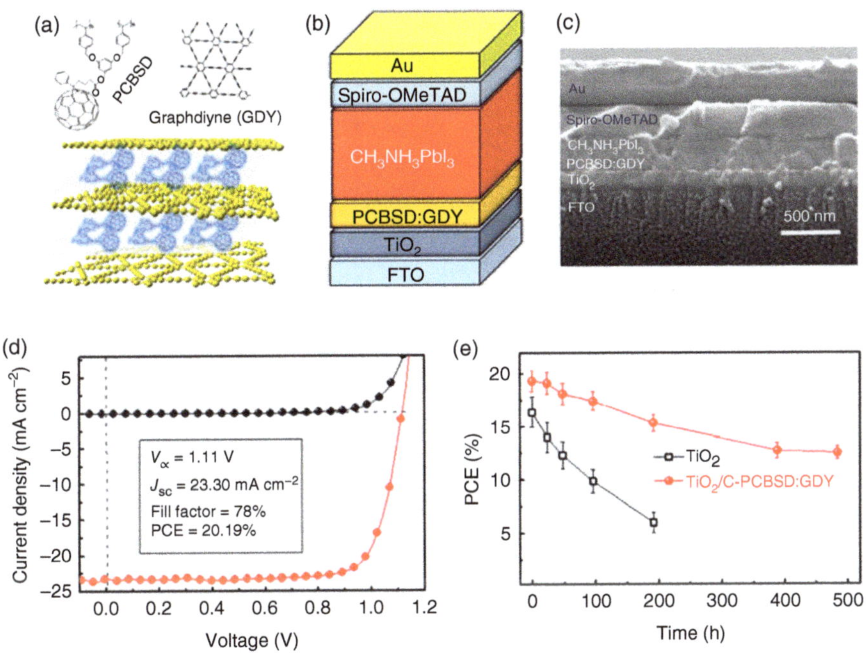

Figure 7.5 (a) Molecule structures of PCBSD and graphdiyne and schematic illustration for the face-on stacked C-PCBSD film owing to the π–π stacking interaction between. (b) Device structure of the perovskite planar heterojunction solar cells. (c) Cross-section SEM image of a typical perovskite device with C-PCBSD: GDY-based ETL. (d) Device performance and (e) stability of PSCs with C-PCBSD: GDY-based ETL. Source: Li et al. [10]. © 2018, Elsevier.

in PSCs owing to its wider band gap (3.6–4.0 eV), higher electron mobility ($240\,cm^{-2}\,V^{-1}\,s^{-1}$), enhanced ultra violet (UV) stability, and more favorable band alignment to perovskite. However, planar PSCs, based on SnO_2 ETL, underperform due to their unmatched photogenerated electron and hole extraction rates. Zhang et al. proposed the strategy of incorporating GDY with multiroles to maximize the collocation between SnO_2 and perovskite aiming to optimize electron extraction rate, induce perovskite growth, and passivate interfacial defect via interface engineering [11]. The electron extraction rate is significantly improved on the basis of elevated electron mobility of SnO_2 and facilitated interfacial energy band alignment (Figure 7.6). Meanwhile, the enhanced hydrophobicity effectively inhibits heterogeneous perovskite nucleation, leading to high-quality film with diminished grain boundaries and lower defect density. Both experimental results and theoretical calculations demonstrate the crucial status of acetylene regarding

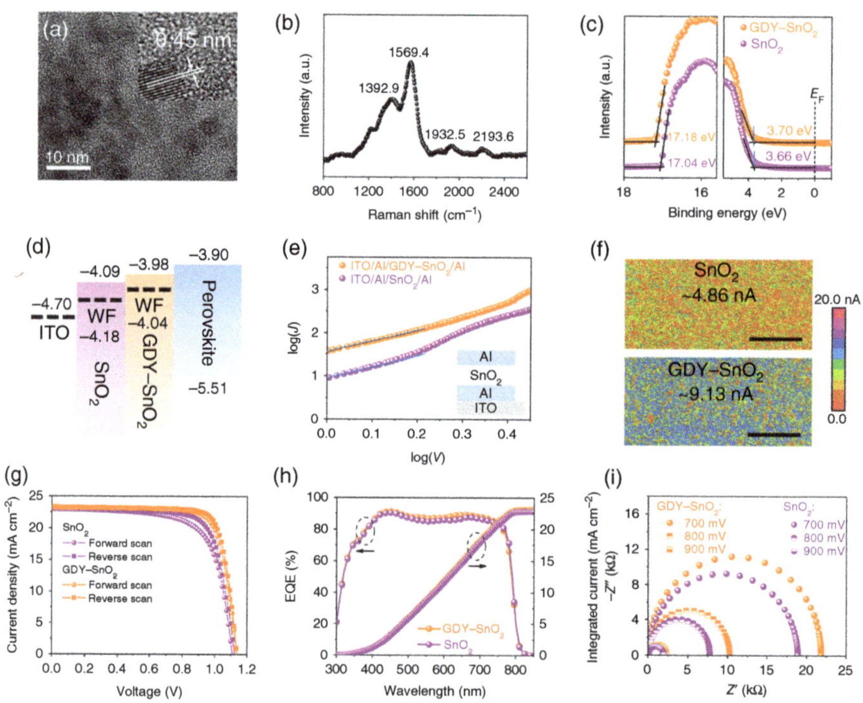

Figure 7.6 Characterization of GDY and GDY–SnO_2 ETLs. (a) TEM image and closeup image of GDY nanoparticles. (b) Raman spectra of GDY powders. (c) Work function (WF) (left) and valence band (right) for SnO_2 and GDY–SnO_2 ETLs measured by UPS. Blue vertical lines represent energy difference of E_V relative to E_F. (d) Schematic of energy band alignment for SnO_2 and GDY–SnO_2 with perovskite. (e) Electron mobility measurement for SnO_2 and GDY–SnO_2 ETLs employing SCLC method; inset depicts SCLC device structure. (f) c-AFM images of SnO_2 and GDY–SnO_2 ETLs deposited on ITO glass substrate. Scale bar: 800 nm. (g) Hysteresis performance of PSCs based on SnO_2 and GDY–SnO_2 ETLs. (h) EQE measurements for PSCs using SnO_2 and GDY–SnO_2. (i) Nyquist plots of PSCs based on SnO_2 and GDY–SnO_2 ETLs under various values of bias in the dark. Source: Zhang et al. [11]. © 2020, Wiley-VCH Verlag GmbH & Co. KGaA.

the interaction with SnO_2, interfacial modification, and passivation for perovskite growth. Consequently, the GDY-containing device yields a 21.11% PCE with negligible hysteresis. This work provides novel insights on the efficacy of multifunctional GDY for precise interface engineering of perovskite devices.

To probe the scenario of other GDY-based materials in PSCs application, chlorine-substituted graphdiyne (ClGD), a derivated GDY with abundant carbon and well-defined features with a two-dimensional GDY-like carbon framework, comprising benzene rings (sp^2 carbon) and butadiyne linkages (sp carbon), comes into sight. Owing to the large pore size in the 2D molecule plane and high theoretical proportion of chlorine atoms, ClGD is expected to afford a promising application in the preparation of photovoltaic materials. Jiu and coworkers developed a novel supramolecular material of chlorine-substituted GDY-PCBM as ETL via a facile method and manufactured $MAPbI_3$-based PSCs with a high efficiency of 20.34% (Figure 7.7) [12]. Moreover, ClGD–PCBM-based devices show suppressed J–V hysteresis. It is experimentally and theoretically demonstrated that the interactions of derivated GDY and PCBM stem from four types of noncovalent bonds, which contribute to the improved device performance.

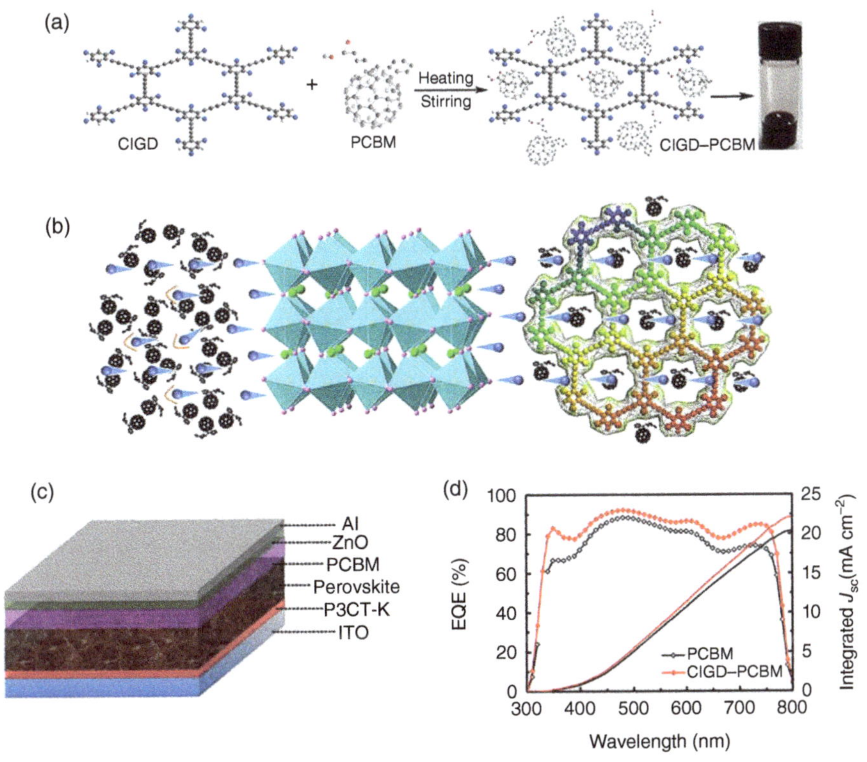

Figure 7.7 (a) Preparation route and photograph of ClGD–PCBM solution. (b) Schematic diagram of the effect of ClGD on charge transfer. (c) Device architecture. (d) EQE curves of devices based on PCBM and GlGD–PCBM as ETLs. Source: Li et al. [12]. © 2019, Wiley-VCH Verlag GmbH & Co. KGaA.

It is concluded that GDY-based materials have been successfully employed into ETLs of PSC via their interaction with organic and inorganic interfacial materials, which further improves the ETL properties. Meanwhile, the addition of GDY-based materials brings multifunctional benefits to ETLs such as the increase of electron mobility, better surface morphology, and traps passivation, leading to the overall improvement of device performance. It is reasonable to infer that similar factors may influence HTLs behaviors as well. In the following, we will discuss the recent progress of GDY-based materials in HTLs of PSCs.

Meng and coworkers employed GDY as a dopant to modify poly(3-hexylthiophene) (P3HT) HTL for the fabrication of PSCs and achieved a high PCE of 14.58% outperforming the pristine P3HT-based devices [13]. The strong π–π stacking interaction of P3HT and GDY nanoparticles leads to the formation of P3HT/GDY composite HTL, which is beneficial to hole transportation, light absorption, and further improvement of device performance (Figure 7.8). In this case, GDY exhibits high charge-transport capability and enables the formation of an efficient percolation pathway in the active layer, which facilitates to accelerate hole extraction in the resultant composite HTL, resulting in enhanced J_{sc} and PCE. Additionally, the scattering of GDY particles induces a lower transmittance and thus increases the harvesting of long-wavelength light in PSC devices. Similar effects can be found when GDY is doped into another hole transport polyelectrolyte of P3CT-K and an elevated 19.5% efficiency is obtained for $MAPbI_3$-based perovskite devices [14].

Aiming to boost the charge extraction and environmental stability of PSCs, Ho and coworkers proposed the strategy of using GDY-containing chlorobenzene as antisolvent onto the upper part of perovskite ($CH_3NH_3PbI_3$) layer during the one-step synthesis of perovskite, in which GDY functions as a hole-transport modifier that benefits hole accumulation and transportation due to its higher Fermi level than perovskite (Figure 7.9) [15]. The oxygen-containing groups of GDY enable a favorable contact with perovskite film. What is more, the Schottky barrier formed on the interface between perovskite and GDY ensures the unidirectional hole transport from perovskite to GDY, thus facilitating further extraction to the HTL. As a result, the GDY-modified device gives a 19.6% PCE with an enhanced J_{sc} of 24.21 mA cm^{-2}. Meanwhile, the device with GDY modification exhibits less than 10% shrinkage after 35 days in ambience owing to the hydrophobic and thermostable properties of GDY.

In the perspective of interface engineering, we can draw the conclusion that using GDY as a dopant into either ETLs or HTLs can be an effective strategy to improve device performance of PSCs owing to efficient charge collection or transport.

7.1.2 Graphdiyne-Based Materials in Active Layers

Perovskite film engineering is one of the indispensable solutions for performance improvement, which is primarily induced by precursor-coordinating molecule interaction and the introduction of antisolvent [16, 17]. Photovoltaic performance is usually correlated with high-quality perovskite films containing few grain boundaries and trap states in the bulk and/or at the interfaces. There are several successful

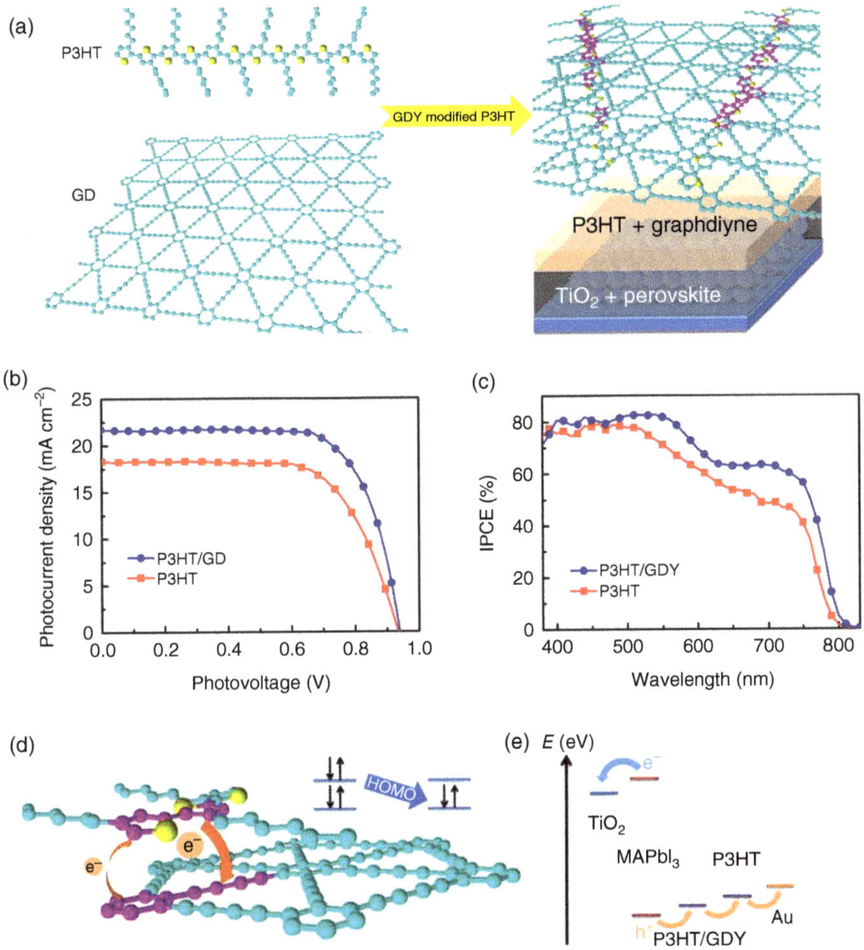

Figure 7.8 (a) Schematic diagram of perovskite solar cells with P3HT hole transporting material modified with graphdiyne (GDY). (b) *J*–*V* curves and (c) incident photo conversion efficiency (IPCE) spectra of P3HT/GDY- and P3HT-based PSCs. (d) Schematic interaction of P3HT and GDY. (e) Energy levels of PSCs with P3HT/GDY. Source: Xiao et al. [13]. © 2015, Wiley-VCH Verlag GmbH & Co. KGaA.

attempts of using GDY in perovskite active layers for further optimization of device performance.

Jiu and coworkers reported that incorporating 25% GDY as a host material into perovskite layer can yield as high as 21.01% efficiency for the as-fabricated device, realizing highly crystalline qualities, large domain sizes, and few grain boundaries [18]. The optimal molar ratio of PbI_3 : MAI : GDY is 1 : 1 : 0.25. The GDY-free perovskite layer shows irregular crystal shapes with scattered pinholes, indicating unfavorable crystal growth orientation. However, due to the introduction of GDY, those pinholes disappear and the crystals in perovskite layer become smoother with larger grain size. Note that iodide vacancies can be easily formed during the heat treatment

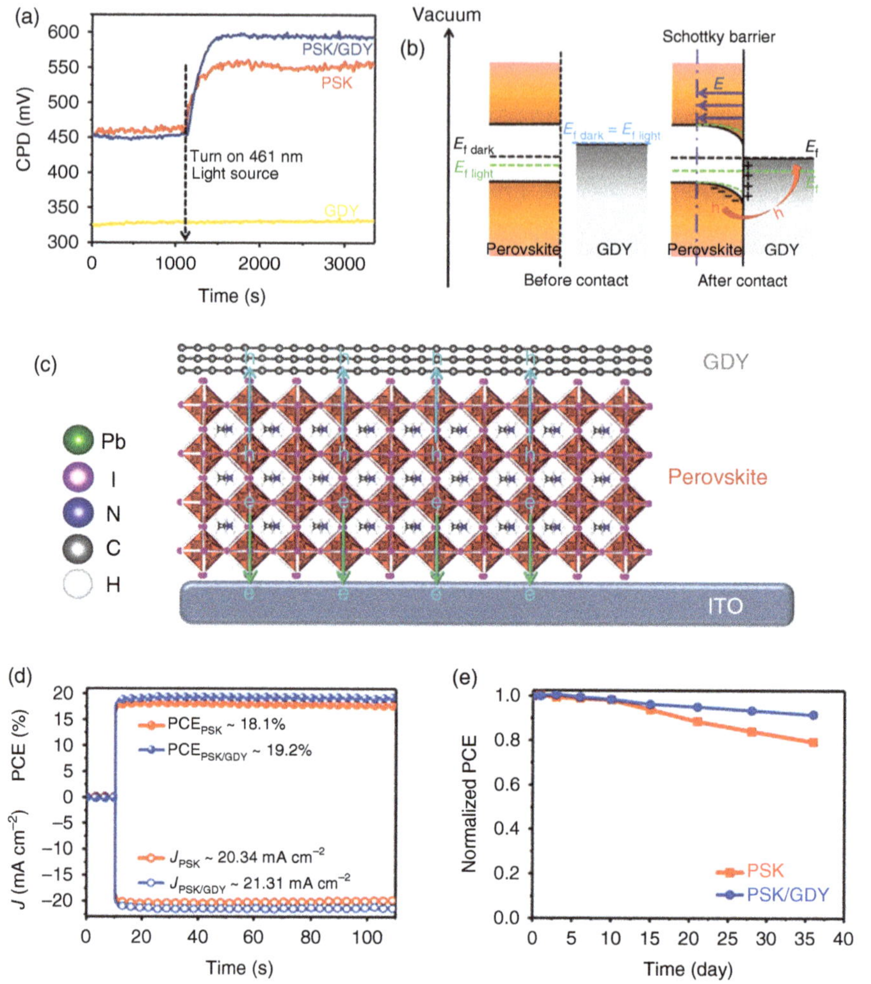

Figure 7.9 (a) In situ contact potential difference (CPD) of perovskite (PSK) and PSK/GDY under dark condition and 461 nm light illumination. (b) Schematic diagrams for the Fermi level of individual perovskite or GDY and combination of perovskite and GDY under dark or 461-nm light illumination. (c) Schematic diagrams for the charge separation in the PSK/GDY. (d) Steady-state output J_{sc} and its corresponding PCE measured at the maximum power point. (e) The long-term stability test of PSCs stored in an air atmosphere with humidity of 50 ± 5%. Source: Zhang et al. [15]. © 2020, Wiley-VCH Verlag GmbH & Co. KGaA.

due to poor crystallinity of perovskite. Through X-ray photoelectron spectroscopy (XPS) examinations, an intermediate adduct of GDY-PbI_2 is formed; hence, Pb atom is able to afford more electrons via coordinate bonding with GDY, which restrains the formation of iodide vacancies and facilitates the enhancement of crystal quality and the stability of $MAPbI_3$ (Figure 7.10). What is more, the current–voltage hysteresis is negligible, and device stability is appreciably improved as well. It is found that GDY, as the host active material, significantly affects the crystallization,

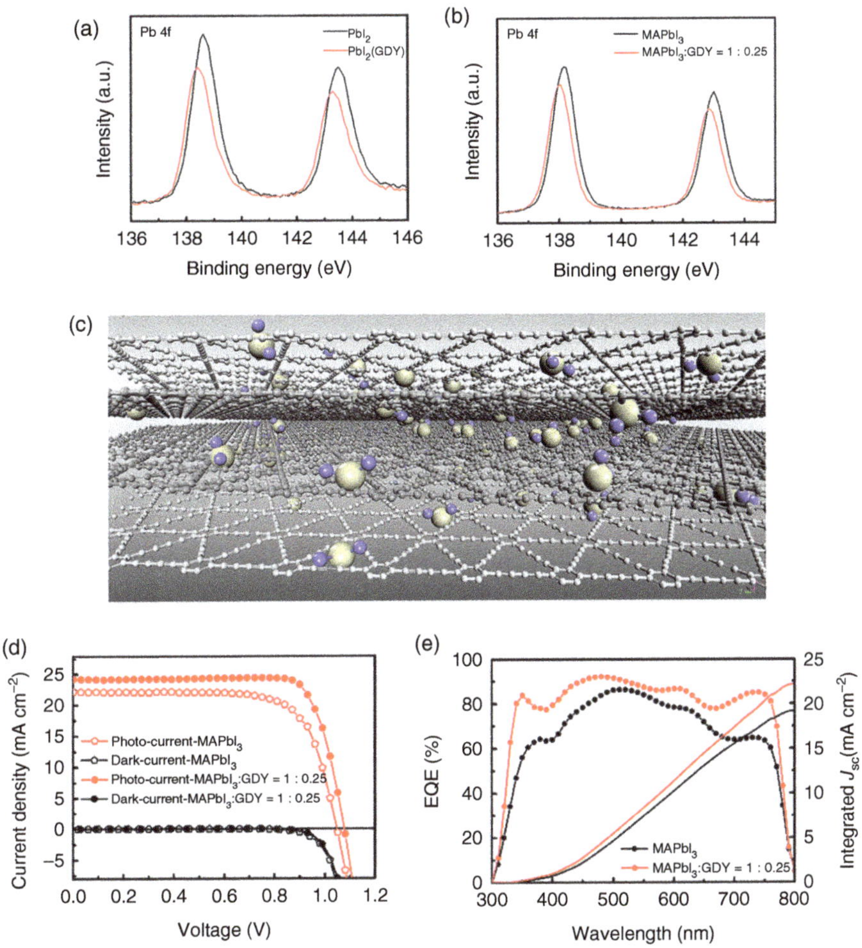

Figure 7.10 XPS spectra of Pb 4f in (a) PbI_2, PbI_2 (GDY) and (b) perovskite. (c) Schematic illustration of the adduct between GDY and PbI_2. (d) Photocurrents and dark currents of the devices with or without GDY and (e) corresponding external quantum efficiency (EQE) spectra. Source: Li et al. [18]. © 2018, American Chemical Society.

film morphology, and a series of optoelectronic properties of perovskite active layer, exhibiting its promising application as a dependable part of photovoltaic devices.

Another promising application of GDY in perovskite is about the construction of bulk heterojunction (BHJ) aiming to accelerate the exciton separation, the extraction, and transportation of photogenerated carriers. Since GDY has higher carrier mobility than perovskite, appropriate band structure, and high chemical stability, the approach of introducing GDY into the perovskite layer by one-step antisolvent method and constructing a graphdiyne/perovskite (GDY/PVSK) BHJ is proposed for efficient and moisture-stable planar PSCs by Li et al. [19]. A 20.54% PCE is obtained with pronounced increase in J_{sc} and FF. The improved J_{sc} is ascribed to an extra channel produced by the GDY/PVSK heterojunction (Figure 7.11), which

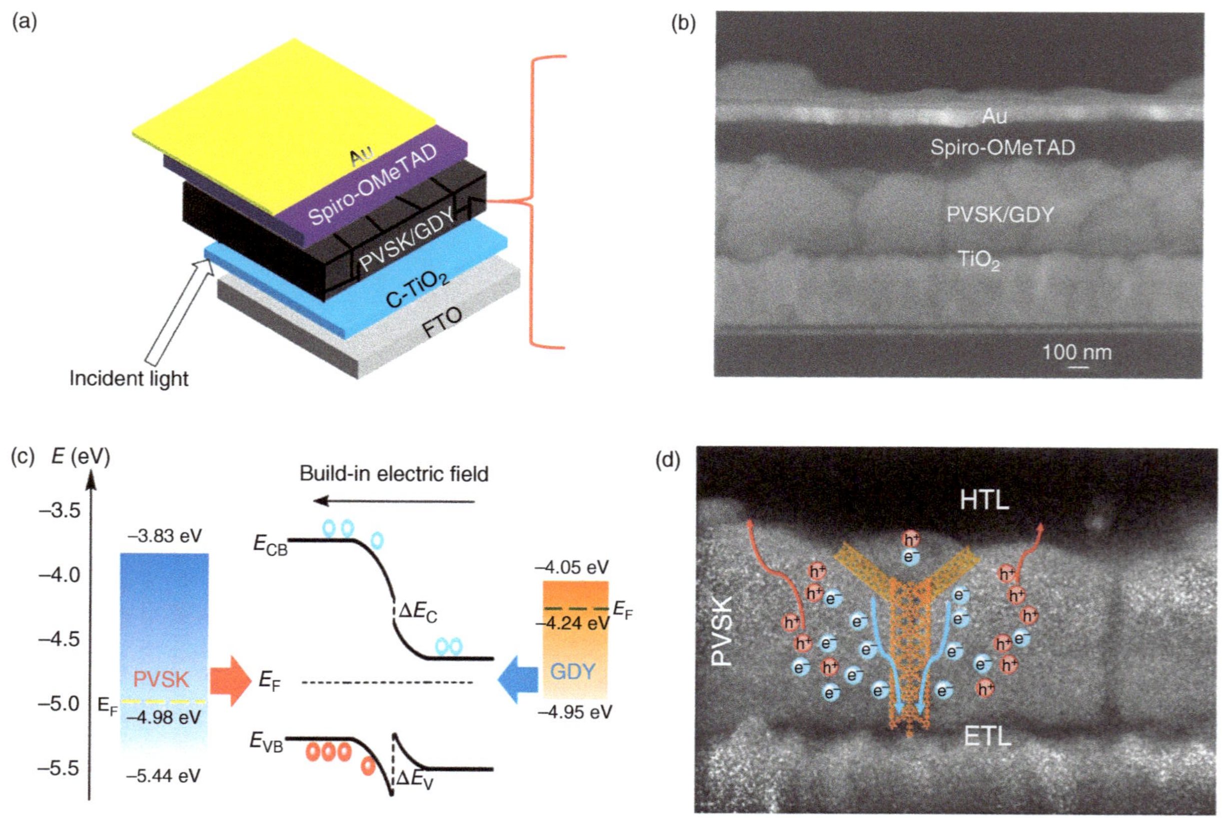

Figure 7.11 (a) Device structure and (b) cross-sectional SEM image of the planar perovskite solar cell with a PVSK/GDY heterojunction. (c) Schematic diagrams of the band structure of the PVSK/GDY bulk heterojunction. (d) Photogenerated carrier transportation processes in the perovskite layer with a PVSK/GDY bulk heterojunction. Source: Li et al. [19]. © 2018, Wiley-VCH Verlag GmbH & Co. KGaA.

can accelerate the photogenerated carrier extraction and transportation by drift and diffusion forces. Meanwhile, a moderate amount of GDY is able to effectively passivate grain boundaries and interface so as to suppress the recombination process, leading to a higher FF. On top of that, the resultant device could retain 95% of initial PCE under ambient condition after 140 days, demonstrating a well-designed GDY-based BHJ in the field of PSCs.

To further understand the underlying interaction of GDY and perovskite, Jiu and coworkers tried to approach this issue by theoretic simulation, where the van der Waals heterostructures of GDY and perovskite, specifically PbI_2- and MAI-terminated systems, are constructed [20]. The first principles calculations unmask the electronic properties of GDY/$CH_3NH_3PbI_3$ perovskite heterostructures. It is revealed that the built-in electrical field generated by the charge transfer inside the heterostructures is advantageous to the separation and the transport of the photogenerated carriers (Figure 7.12). That is, it is the formation of built-in electric field between GDY and perovskite that empowers a high-speed pathway for the transport of carriers.

The polycrystalline structure of perovskite material and its synthesis process inevitably involves the generation of defects. It is assumed that further optimization of PSC performance also relies on defects controlling by which nonradiative charge recombination is suppressed [21]. Particularly on the topic of grain boundaries defects, Jiu and coworkers introduced triazine-graphdiyne (Tra-GD), a nitrogen-rich GDY derivative (Figure 7.13a–e), into perovskite active layer aiming at passivating defects and improving carrier transport [22]. Tra-GD shows good dispersion in the bulk of perovskite. Meanwhile, the two-dimensional highly conjugated

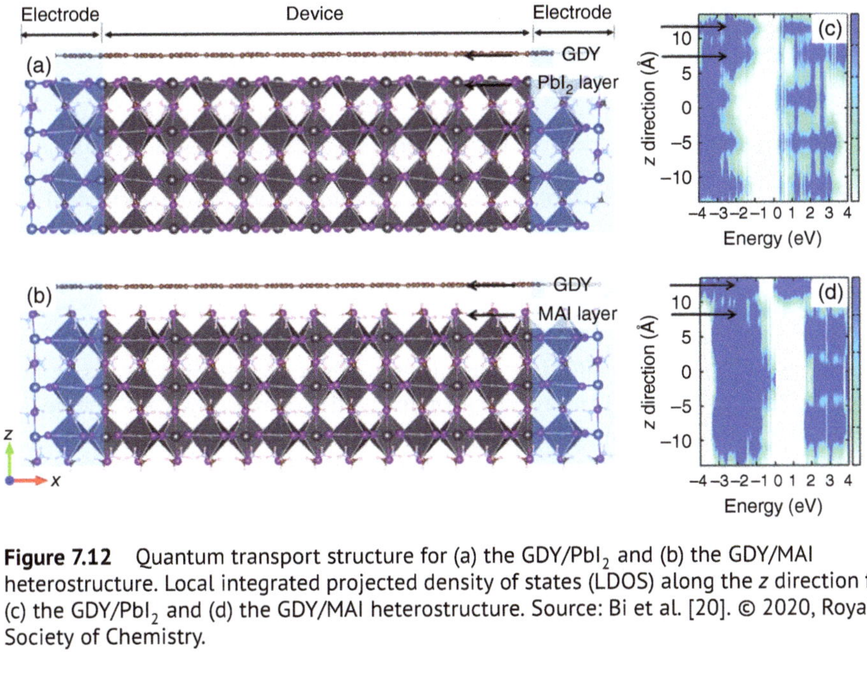

Figure 7.12 Quantum transport structure for (a) the GDY/PbI_2 and (b) the GDY/MAI heterostructure. Local integrated projected density of states (LDOS) along the *z* direction for (c) the GDY/PbI_2 and (d) the GDY/MAI heterostructure. Source: Bi et al. [20]. © 2020, Royal Society of Chemistry.

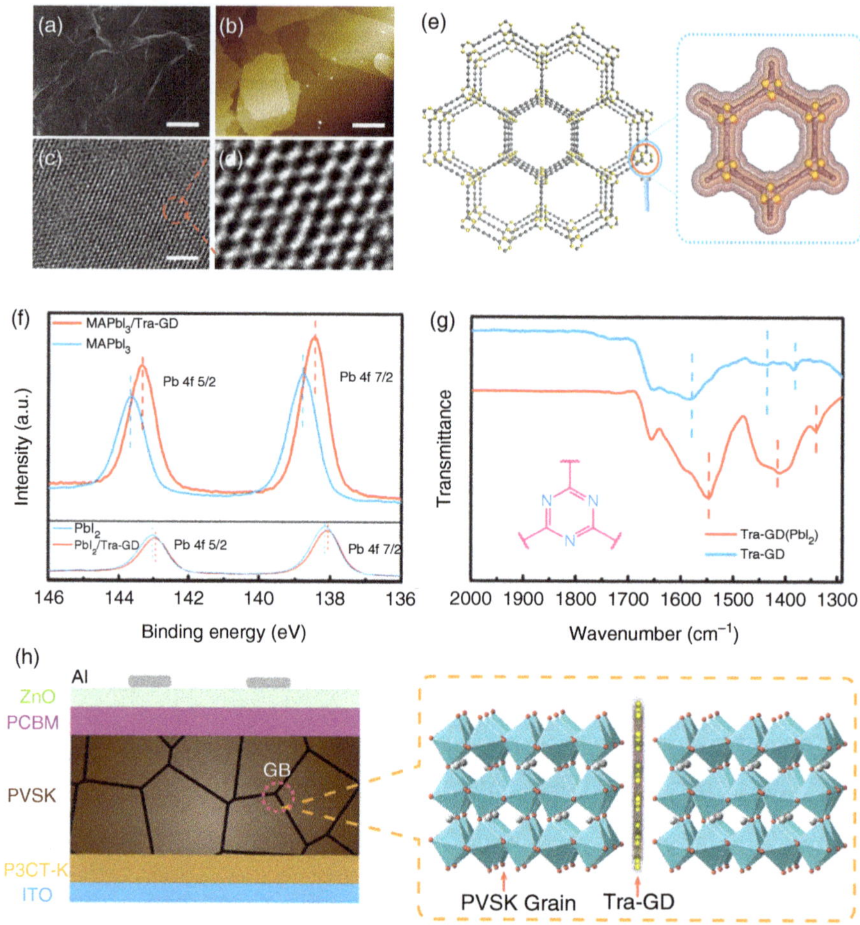

Figure 7.13 (a) SEM, (b) AFM, (c, d) High-resolution transmission electron microscopy (HRTEM) images of Tra-GD synthesized by the interfacial method, the scale bar is 200 nm, 4 μm, and 50 nm, respectively. (e) Periodic structure of Tra-GD. (f) XPS spectra of PbI_2 and $MAPbI_3$ with and without Tra-GD. (g) FTIR spectra of the Tra-GD and PbI_2/Tra-GD hybrid films. (h) Schematic illustration of the structure of the device and Tra-GD on the GB of an active layer. Source: Chen et al. [22]. © 2020, Springer Nature BV.

framework of Tra-GD can afford strong affinity with perovskite surface. As shown in Figure 7.13f,g, XPS and Fourier transform infrared spectroscopy (FTIR) prove the strong interaction between Pb^{2+} and the triazine ring of Tra-GD, enabling Tra-GD effectively passivate perovskite GBs. Furthermore, as illustrated in Figure 7.13h, Tra-GD is proved to demonstrate beneficial effects on the morphology of perovskite film with enlarged grain sizes, resulting in fewer GB defects. More importantly, Tra-GD facilitates the charge extraction and transport at GBs. As a result, the fabricated Tra-GD containing devices yield an optimal efficiency of 20.33% in $MAPbI_3$ system and 21.16% in $FAPbI_3$ counterpart with negligible hysteresis. The successful attempt of Tra-GD passivation presents an effective approach of GB engineering for high-efficiency solar cells.

Wang and coworkers reported the use of novel y-graphdiyne quantum dots (GDY QDs) as a comprehensive dopant for both interfacial layers and the active layer [23, 24]. The overall improvement of device performance has been realized, with enhancement in PCEs exceeding 15% compared to the undoped counterparts. GDY QDs exhibit similar behaviors at each layer of TiO_2, $CH_3NH_3PbI_3$, and Spiro-OMeTAD. Specially, its presence brings higher current density and FF for TiO_2 surface (Figure 7.14), which improves hydrophobicity and eliminates pinholes in the Spiro-OMeTAD film for enhanced device stability. These results indicate that doping GDY QDs is an efficient method to optimize the device performance of perovskite photovoltaic devices.

To conclude, GDY has played a critical role in designing the construction of interfacial layers and the bulk of PSCs, demonstrating its fantastic effects on enhanced carrier mobilities, optimized film qualities of both interfacial layer and bulk layer, and thus improved device performance. For interface engineering, GDY is more inclined to increase the charge transport in interfacial layers; for bulk modification, GDY significantly affects the crystallization, film morphology, and recombination process, contributing to the improvement of device performance in terms of efficiency and stability. GDY-based materials have been developed to extend the

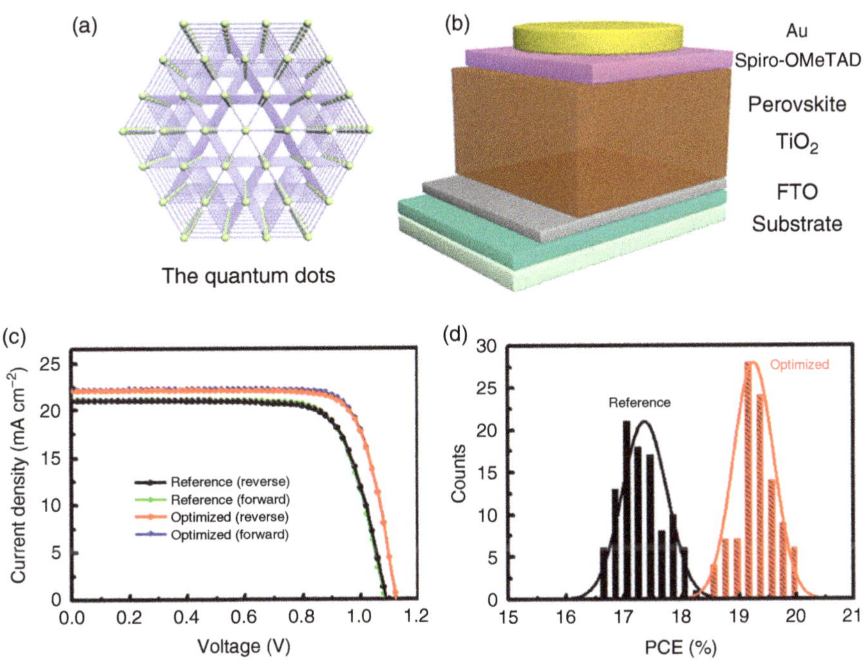

Figure 7.14 (a) Schematic illustration of GDY quantum dots (QDs). (b) Device structure of PSCs. (c) Comparison of champion and (d) average performance of optimized GDY QDs compressive doped devices and those of reference devices. Source: Zhang et al. [24]. © 2017, Wiley-VCH Verlag GmbH & Co. KGaA.

scope of desirable perovskite materials, and further studies on the correlation inside GDY-containing devices are expected to propel the advancement of PSCs.

7.2 Organic Solar Cells

OSCs are usually manufactured from organic chemical compound materials. The merits of light weight, flexibility, and affordability make OSCs become an important branch of the third-generations solar cells. Considerable efforts have been devoted to solution-processed OSCs because of their easy and low-cost fabrication, flexibility and light weight, and the potential of optical transparency [25]. The key concept of BHJ was first reported in the 1990s, where the excitons can be generated from two channels that were later defined as donors and acceptors. The relevant working mechanism is presented in Figure 7.15 [26].

The significant advances of OSCs are often accompanied with the evolution of various material systems, in which many polymers and their composites have been used for BHJ materials. Typical active layer of OSCs usually comprises p-conjugated polymers and small molecules. In 2011, Li and coworkers reported the pioneering work of using GDY in OSCs, in which GDY was introduced into the P3HT:PCBM (poly(3-hexylthiophene):phenyl-C_{61}-butyric acid methyl ester) active layer (Figure 7.16) [27]. The two-dimensional plane of GDY induces a large specific area, leading to the formation of large interfacial contact. Thanks to a high charge transport capability of GDY and the formation of efficient percolation pathways in the active layer, the doping of GDY enables a 56% enhancement in PCE for P3HT/PCBM/GDY based solar cells compared to the counterparts without GDY.

Jiu and coworkers reported a facile approach for preparing graphdiyne-ZnO (GDZO) composite material and introduced GDZO into the matrix of PBDBT:ITIC-M-based OSCs as ETL [28]. ZnO is considered as a good candidate for the application of ETLs in solar cells owing to its relatively high electron mobility, desirable energy level, and compatibility with low-temperature production. However, ZnO-based photovoltaic devices usually deliver underperforming performance, suffering from poor electron mobility and high charge recombination since pristine ZnO possesses surface defects. To address the problem, a novel GDY-containing composite material of GDZO is synthesized, in which Zn and O atoms coordinate bonding with GDY, thus forming C—Zn and C—O bond, respectively. Such adduct affords good dispersion of the composite film, facilitating the formation of high-quality GDZO interlayer. The electronic properties of GDZO are corroborated by first-principles calculations. The density of state (DOS) of GDZO shows the impurity levels are mainly attributed from carbon atoms of GDY. Compared to bulk ZnO, the electronic excitation energy from the occupied isolated impurity levels to conduction bands is smaller in GDZO composite, enabling the electron to be excited more easily (Figure 7.17). Therefore, GDZO film demonstrates better morphology and orderly arrayed distribution of ZnO nanoparticles, leading to the increase of electron transport and the decrease of charge recombination. Later, GDZO is incorporated

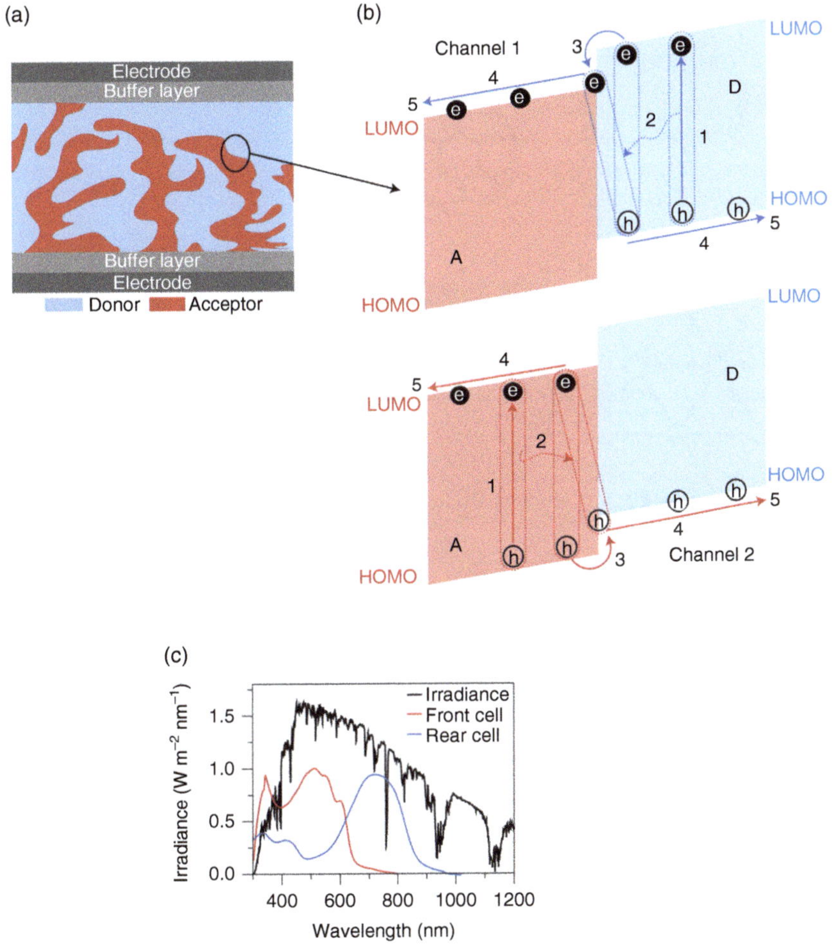

Figure 7.15 Introduction to OSCs. (a) Cross-section of the conceptual device structure of OSCs with the BHJ active layer. (b) The operating mechanism of OSCs: (1) absorption of photons and creations of excitons; (2) diffusion of excitons to D/A interfaces; (3) dissociation of excitons to free charge carriers (holes and electrons) at D/A interfaces; (4) transportation of the charge carriers to electrodes; (5) extraction of the charge carriers at electrodes. (c) Comparison between the AM 1.5G solar spectrum (black curve) and the photon response of different OSCs (red and blue curves). Source: Cheng et al. [26]. © 2018, Springer Nature BV.

in OSCs as ETL, the resulting device efficiency is boosted up to 11.2% with better stability compared to devices with pristine ZnO ETL.

OSCs, using nonfullerene acceptors, display eminent progress owing to their advantages of tunable bandgaps, energy levels, planarity, and crystallinity [26]. The presence of Y6 materials with excellent electronic structure pushes the device efficiency beyond 17% with varying configurations. Highly efficient OSCs based on conventional binary configuration have drawn particular attention for easier manufacture process compared to ternary ones. Morphology tuning of the blend film has been regarded as a powerful approach to the optimization of device performance.

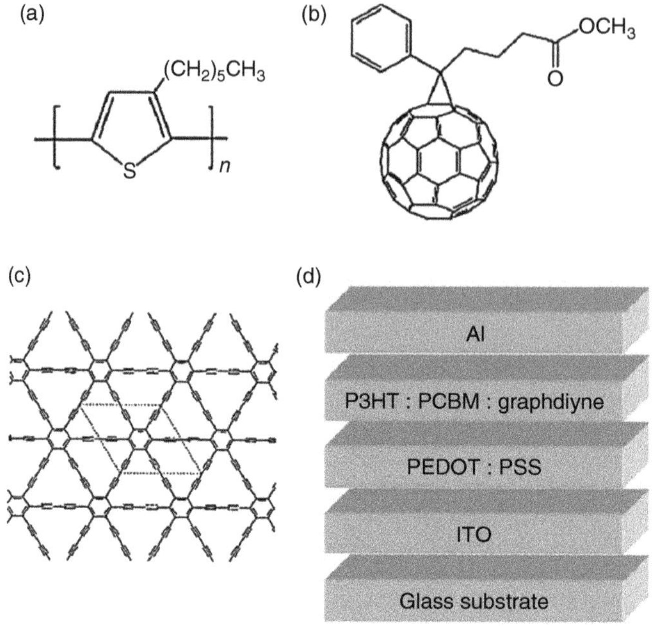

Figure 7.16 Chemical structure of (a) P3HT, (b) PCBM, and (c) GDY, and (d) corresponding device structure of GDY-incorporated organic solar cells. Source: Du et al. [27]. © 2011, Elsevier.

Aiming to further improve the efficiency of binary OSCs with comparable efficiency, Jiu and coworkers proposed the application of a class of carbon material chlorine functionalized-graphdiyne (GCl) as a multifunctional solid additive [29]. The use of GCl solid additive is preferred as compared to the treatment of chloronaphthalene (CN) solvent additive.

Figure 7.18a,b depict the molecular structures of PM6 and Y6, respectively. As shown in Figure 7.18c, GCl consists of repeating hexagonal units of six 1,3,5-trichlorinebenzene rings and linkages of butadiyne, forming a highly π-conjugated 2D carbon planar structure. Its specific structure is able to induce crystallization of conjugated polymer or nonfullerene acceptors. As suggested by Figure 7.18d, no influence on the PM6 films is observed for either CN- or GCl-treatment, due to the high molecular weight of PM6. However, as presented in Figure 7.18e, the main absorption peak of CN-processed Y6 film blueshifts 10 nm, while a 16-nm redshift appears for GCl-processed one. As for PM6:Y6 blend film, the GCl-processed one shows a redshift from 812 to 819 nm, broader than those processed with CN (Figure 7.18f). Therefore, it is concluded that GCl mainly affects Y6, leading to an enhanced ordering of Y6 molecules, which consequently enhances electron mobility and other electrical properties of the blend film of PM6:Y6. After a series of electrical measurements, the addition of GCl is proved to account for the crystallization optimization and enhanced charge transfer efficiency, resulting in the increase of J_{sc} and FF. Meanwhile, GCl-processed blend film displays higher

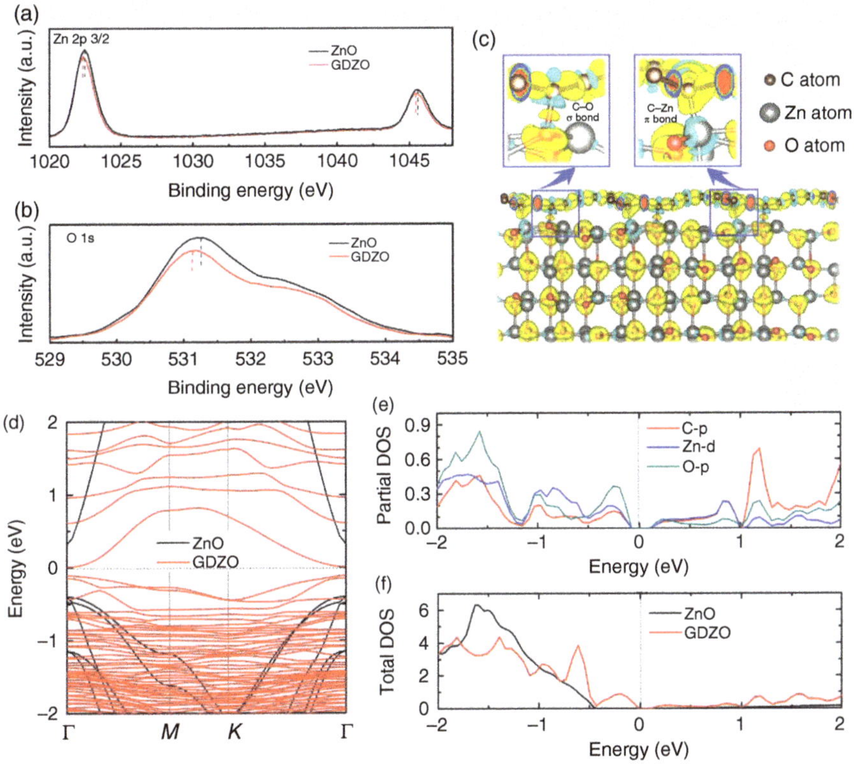

Figure 7.17 (a) XPS spectra of Zn $2p_{3/2}$ and (b) O 1s in ZnO and GDZO. (c) The optimized structure and isosurface of the charge density difference for GDZO surface. (d) Calculated band structure and (e) total density of state (DOS) for bulk ZnO and GDZO system. (f) The partial DOS for GDZO system. The Fermi level is at 0 eV. Source: Li et al. [28]. © 2018, Wiley-VCH Verlag GmbH & Co. KGaA.

electron mobility. According to atomic force microscopy (AFM), transmission electron microscopy (TEM), and grazing-incidence wide-angle X-ray scattering (GIWAXS) characterizations (Figure 7.19), GCl-processed blend films show more pronounced phase separation, which is conducive to the exciton dissociation and charge transport thus benefiting the enhanced J_{sc} and FF. Moreover, GCl-processing method could improve the crystalline quality of the blend film. As a consequence, GCl-processed device (Figure 7.20) delivers a much higher efficiency of 17.3% with a certified PCE of 17.1% along with the simultaneous increase of short-circuit current (J_{sc}) and FF, compared with 15.6% efficiency for CN-processed ones. In addition, devices processed with GCl demonstrate better reproducibility with reduced batch-to-batch variations.

In summary, the introduction of GDY-based materials may result in the improvement of crystallinity, phase separation, and better reproductivity in OSCs. GDY-based materials are promising for further application in other material systems and large area devices.

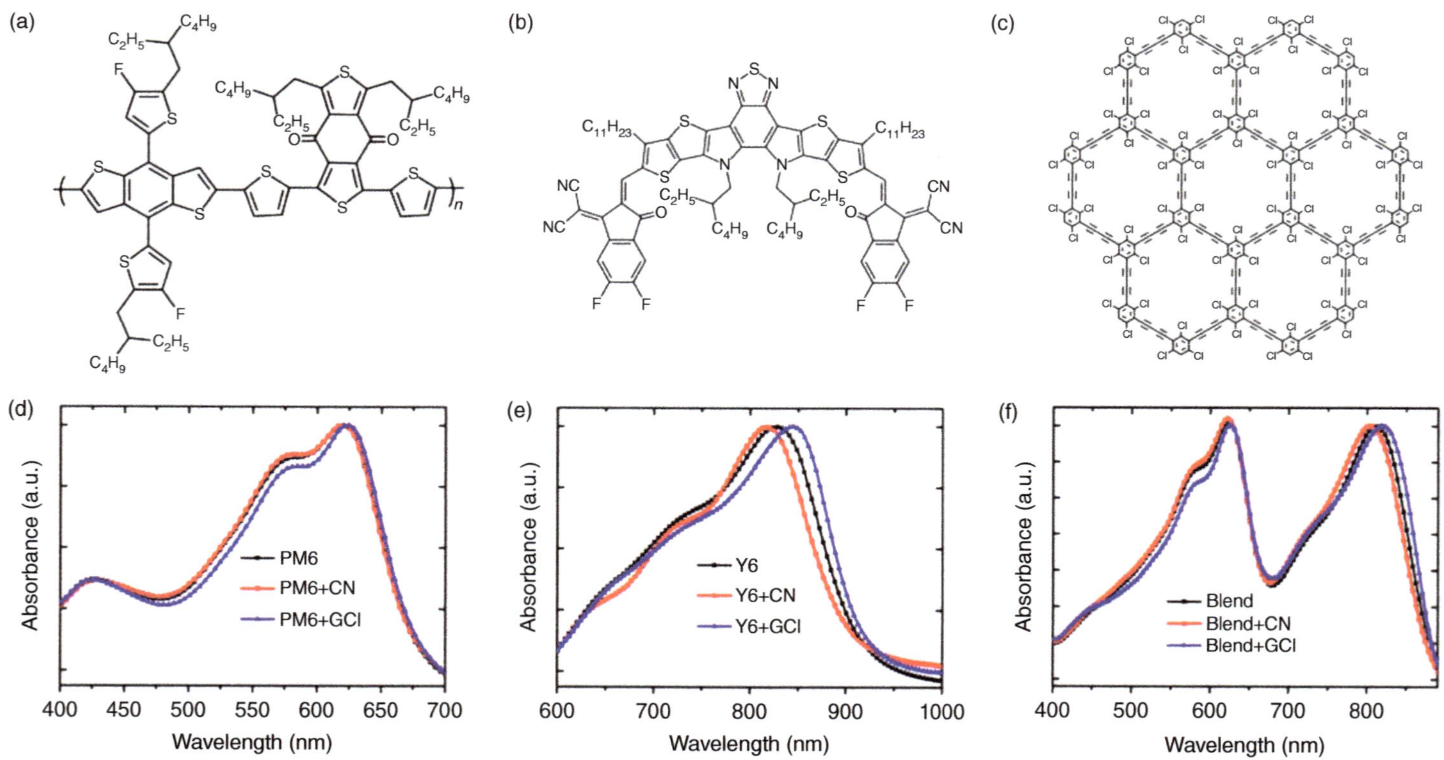

Figure 7.18 Chemical structures of (a) PM6, (b) Y6, (c) GCL. UV–Vis spectra of (d) PM6, (e) Y6, and (f) blend films processed with CN or GCL. Source: Liu et al. [29]. © 2020, Wiley-VCH Verlag GmbH & Co. KGaA.

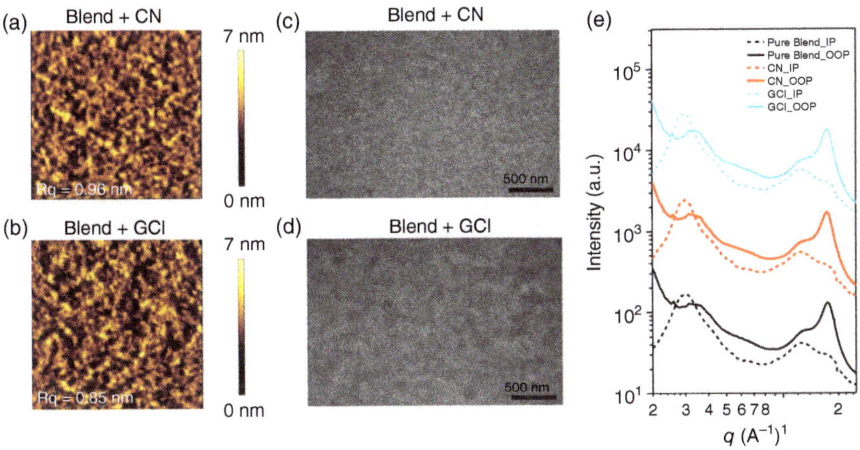

Figure 7.19 AFM height images of PM6:Y6 blend films with (a) CN and (b) GCl, in a scan size of $5 \times 5\,\mu m^2$, respectively. TEM images of PM6:Y6 blend films with (c) CN and (d) GCl, respectively. (e) In-plane and out-of-plane cuts of GIWAXS images of PM6:Y6 blend films with CN and GCl. Source: Liu et al. [29]. © 2020, Wiley-VCH Verlag GmbH & Co. KGaA.

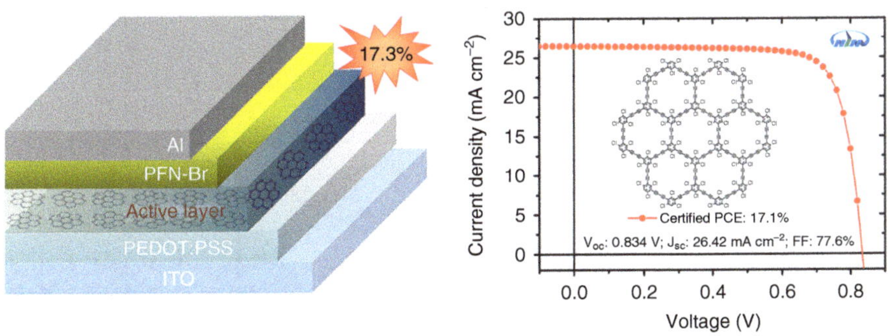

Figure 7.20 Device architecture of GCl-containing organic solar cells and certified *J*–*V* curve. Source: Liu et al. [29]. © 2020, Wiley-VCH Verlag GmbH & Co. KGaA.

7.3 Others

7.3.1 Quantum Dots Solar Cells

In the field of quantum dots solar cells, Wang and coworkers developed and deployed GDY as a hole transfer material, achieving an improved device performance with 10.64% PCE [30]. Because the theoretical calculated hole mobility value of GDY is estimated to be $-104\,cm^2\,V^{-1}\,s^{-1}$, GDY can be applied directly as a hole transfer material. As shown in Figure 7.21, the GDY-containing quantum dots solar cells yields an efficiency of 10.64% with 12% enhancement compared to the reference device. With the introduction of GDY, the work function of PbS CQD is adjusted, which is favorable for hole transfer from the active layer to the anode. Besides, the

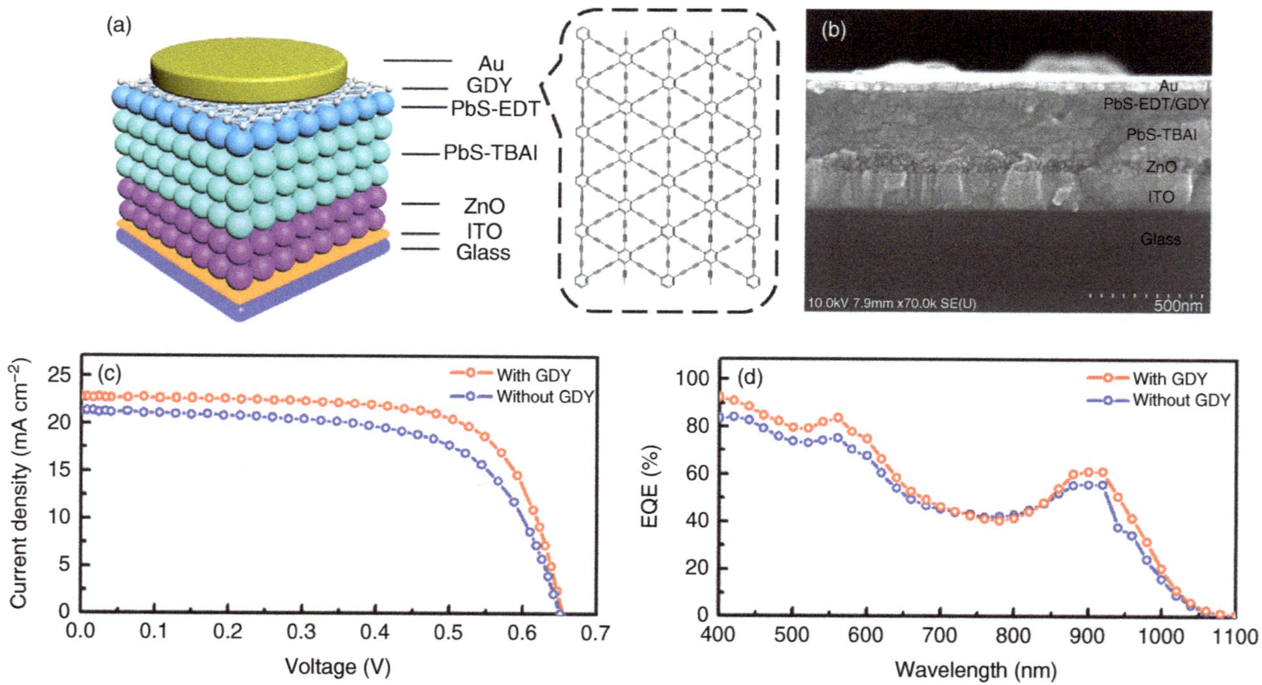

Figure 7.21 (a) Schematic illustration of the PbS CQD solar cells with GDY anode buffer layer. (b) The cross-sectional SEM image of the device. (c) *J*–*V* characteristics under simulated AM 1.5G irradiation. (d) EQE spectra. Source: Jin et al. [30]. © 2016, Wiley-VCH Verlag GmbH & Co. KGaA.

insertion of GDY between PbS-EDT layer and Au anode appreciably improves the interface in between, which is accountable for those phenomenal effects on enhancing carrier lifetime and long-term stability while reducing carrier recombination. The unique structure of the sp^2- and sp-hybridized GDY material is highlighted for promising applications in optoelectronics.

7.3.2 Dye-Sensitized Solar Cells

As an alternative to silicon-based solar cells, DSSCs have drawn considerable attention because of their simple fabrication, low-cost production, and high energy-conversion efficiency. Typical DSSCs primarily consist of a dye-sensitized anode, electrolyte, and a counter electrode. Yu and coworkers reported a facile ion-beam sputtering method to prepare a graphdiyne nanosheet (GDNS) structure modified by platinum nanoparticles (PtNP-GDNS) as the counter electrode of DSSCs (Figure 7.22) [31]. The as-fabricated DSSCs with PtNP-GDNS counter electrodes demonstrate significantly improved catalytic behavior and electron transfer capability, which results from the special "p–n" junction-like structure.

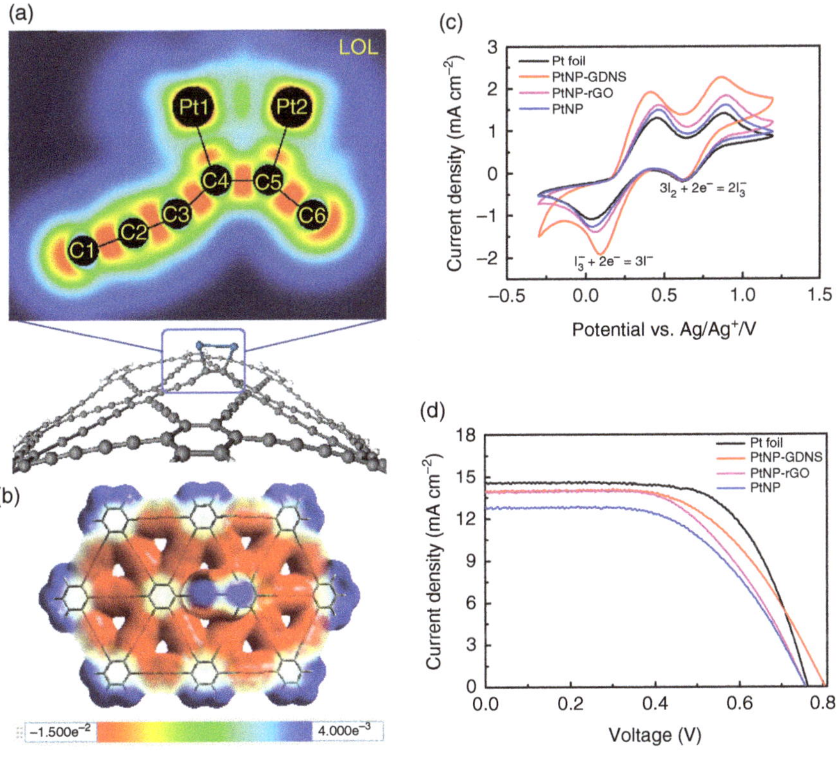

Figure 7.22 (a) Mayer bond-order analysis and maps of localized orbital locator. (b) Electrostatic potential (ESP) surfaces for the Pt_2-GDY. (c) Cyclic voltammetry characteristics and (d) DSSCs' $J-V$ curves with various counter electrodes. Source: Ren et al. [31]. © 2015, Wiley-VCH Verlag GmbH & Co. KGaA.

Density functional theory (DFT) calculations reveal the interaction between Pt atoms and carbon atoms in GDY is formed via the $d_z{}^2$ orbital of Pt and the p_z orbital of C, which is advantageous to the enhancement of charge transfer from GDY to PtNP. Besides, experimental results also confirm that DSSCs with PtNP-GDNS show comparable performance with that of precious Pt foil, and better than that of Pt nanoparticles and rGO/Pt nanoparticle composites. This work demonstrates a new route to GDY nanocomposites that serve as high-performance counter electrodes for photoelectric conversions.

7.4 Future Perspectives

GDY is a newly invented, attractive material that presents multiple excellent properties in different photovoltaic device systems, which can be applicable to future electronic and optoelectronic devices. From its unique properties to possible applications, there is much more to be fully explored. The research area of applying functional GDY-based materials into the architecture of solar cells has been actively studied and keeps growing. Therefore, it is worth emphasizing that several process routes for GDY-based materials have been established for precise control with regard to the dispersion and well-defined bandgap as well as other properties. Further studies on utilizing GDY family in solar cells applications may lead to a new era of the construction of carbon-based solar cells, where the expectations of low-cost and excellent optoelectronic properties may be exceeded.

References

1 Long, M.Q., Tang, L., Wang, D. et al. (2011). Electronic structure and carrier mobility in graphdiyne sheet and nanoribbons: theoretical predictions. *ACS Nano* 5: 2593–2600.

2 Chen, J. and Park, N.G. (2019). Causes and solutions of recombination in perovskite solar cells. *Advanced Materials* 31: e1803019. https://doi.org/10.1002/adma.201803019

3 Luo, D.Y., Su, R., Zhang, W. et al. (2020). Minimizing non-radiative recombination losses in perovskite solar cells. *Nature Reviews Materials* 5: 44–60.

4 Shao, Y.H., Xiao, Z.G., Bi, C. et al. (2014). Origin and elimination of photocurrent hysteresis by fullerene passivation in $CH_3NH_3PbI_3$ planar heterojunction solar cells. *Nature Communications* 5: 5784.

5 Tang, Z.G., Bessho, T., Awai, F. et al. (2017). Hysteresis-free perovskite solar cells made of potassium-doped organometal halide perovskite. *Scientific Reports* 7: 12183.

6 Son, D.Y., Kim, S.G., Seo, J.Y. et al. (2018). Universal approach toward hysteresis-free perovskite solar cell via defect engineering. *Journal of the American Chemical Society* 140: 1358–1364.

7 Liu, S., Guan, Y., Sheng, Y. et al. (2020). A review on additives for halide perovskite solar cells. *Advanced Energy Materials* 10: 1902492.
8 Kuang, C.Y., Tang, G., Jiu, T.G. et al. (2015). Highly efficient electron transport obtained by doping PCBM with graphdiyne in planar-heterojunction perovskite solar cells. *Nano Letters* 15: 2756–2762.
9 Li, J.S., Jiu, T.G., Duan, C.H. et al. (2018). Improved electron transport in MAPbI3 perovskite solar cells based on dual doping graphdiyne. *Nano Energy* 46: 331–337.
10 Li, M., Wang, Z.K., Kang, T. et al. (2018). Graphdiyne-modified cross-linkable fullerene as an efficient electron-transporting layer in organometal halide perovskite solar cells. *Nano Energy* 43: 47–54.
11 Zhang, S.C., Si, H.N., Fan, W.Q. et al. (2020). Graphdiyne: bridging SnO_2 and perovskite in planar solar cells. *Angewandte Chemie International Edition* 59: 11573–11582.
12 Li, J.S., Wang, N., Bi, F.Z. et al. (2019). Inverted $MAPbI_3$ perovskite solar cells with graphdiyne derivative-incorporated electron transport layers exceeding 20% efficiency. *Solar RRL* 9: 3, 1900241.
13 Xiao, J.Y., Shi, J.J., Liu, H.B. et al. (2015). Efficient $CH_3NH_3PbI_3$ perovskite solar cells based on graphdiyne (GD)-modified P3HT hole-transporting material. *Advanced Energy Materials* 5: 1401943.
14 Li, J.S., Zhao, M., Zhao, C.J. et al. (2019). Graphdiyne-doped P3CT-K as an efficient hole-transport layer for $MAPbI_3$ perovskite solar cells. *ACS Applied Materials & Interfaces* 11: 2626–2631.
15 Zhang, J.J., Tian, J.W., Fan, J.J. et al. (2020). Graphdiyne: a brilliant hole accumulator for stable and efficient planar perovskite solar cells. *Small* 16: 1907290.
16 Jeon, N.J., Noh, J.H., Kim, Y.C. et al. (2014). Solvent engineering for high-performance inorganic-organic hybrid perovskite solar cells. *Nature Materials* 13: 897–903.
17 Jung, M., Ji, S.G., Kim, G. et al. (2019). Perovskite precursor solution chemistry: from fundamentals to photovoltaic applications. *Chemical Society Reviews* 48: 2011–2038.
18 Li, J.S., Jiu, T.G., Chen, S.Q. et al. (2018). Graphdiyne as a host active material for perovskite solar cell application. *Nano Letters* 18: 6941–6947.
19 Li, H.S., Zhang, R., Li, Y.S. et al. (2018). Graphdiyne-based bulk heterojunction for efficient and moisture-stable planar perovskite solar cells. *Advanced Energy Materials* 8: 1802012.
20 Bi, F.Z., Yam, C.Y., Zhao, C.J. et al. (2020). Enhanced photocurrent in heterostructures formed between $CH_3NH_3PbI_3$ perovskite films and graphdiyne. *Physical Chemistry Chemical Physics* 22: 6239–6246.
21 Kim, J.Y., Lee, J.W., Jung, H.S. et al. (2020). High-efficiency perovskite solar cells. *Chemical Reviews* 120: 7867–7918.
22 Chen, S.Q., Pan, Q.Y., Li, J.S. et al. (2020). Grain boundary passivation with triazine-graphdiyne to improve perovskite solar cell performance. *Science China Materials* 63: 2465–2476.

23 Zhang, M., Wang, X., Sun, H. et al. (2017). Enhanced paramagnetism of mesoscopic graphdiyne by doping with nitrogen. *Scientific Reports* 7: 11535.

24 Zhang, X., Wang, Q., Jin, Z. et al. (2017). Graphdiyne quantum dots for much improved stability and efficiency of perovskite solar cells. *Advanced Materials Interfaces* 5: 1701117.

25 Li, G., Zhu, R., and Yang, Y. (2012). Polymer solar cells. *Nature Photonics* 6: 153–161.

26 Cheng, P., Li, G., Zhan, X.W. et al. (2018). Next-generation organic photovoltaics based on non-fullerene acceptors. *Nature Photonics* 12: 131–142.

27 Du, H.L., Deng, Z.B., Lu, Z.Y. et al. (2011). The effect of graphdiyne doping on the performance of polymer solar cells. *Synthetic Metals* 161: 2055–2057.

28 Li, J.S., Jian, H.M., Chen, Y.H. et al. (2018). Studies of graphdiyne-ZnO nanocomposite material and application in polymer solar cells. *Solar RRL* 2: 1800211.

29 Liu, L., Kan, Y.Y., Gao, K. et al. (2020). Graphdiyne derivative as multifunctional solid additive in binary organic solar cells with 17.3% efficiency and high reproductivity. *Advanced Materials* 32: e1907604.

30 Jin, Z.W., Yuan, M.J., Li, H. et al. (2016). Graphdiyne: an efficient hole transporter for stable high-performance colloidal quantum dot solar cells. *Advanced Functional Materials* 26: 5284–5289.

31 Ren, H., Shao, H., Zhang, L. et al. (2015). A new graphdiyne nanosheet/Pt nanoparticle-based counter electrode material with enhanced catalytic activity for dye-sensitized solar cells. *Advanced Energy Materials* 5: 1500296.

8

Graphdiyne: Electronics, Thermoelectrics, and Magnetism Applications

Jialiang Xu and Xiaodong Qian

National Institute for Advanced Materials, School of Materials Science and Engineering, Nankai University, Tongyan Road 38, Tianjin 300350, P. R. China

Graphdiyne (GDY) has emerged as the new generation 2D nanostructured carbon material containing sp^2- and sp-hybridized carbon atoms, featuring high degree of π-conjugation, natural uniform porous structure, superior semiconducting properties, along with remarkable chemical and mechanical stability [1–3]. As a novel carbon allotrope, the adjacent benzene rings (sp^2-hybridized carbon) in the GDY structure are linked to each other through butadiyne linkages (sp-hybridized carbon). The fascinating and unique chemical and electronic structures of GDY endow it with extraordinary performances for applications in the fields of electronics, optics, thermoelectrics, and magnetism devices.

8.1 Electronic Devices

One of the most promising applications of GDY arises from its semiconducting characteristics with the intrinsic appropriate band gap and high carrier mobilities [2]. The low effective electron mass and the band gap in the semiconductor range suggest that GDY could work as an ideal semiconductor [4], which promises the application of GDY in electronic devices. The theoretical investigation of GDY nanoribbons (NRs) using the self-consistent field crystal orbital (SCF-CO) method under the periodical boundary condition reveals that the band gap of the GDY strips decreases monotonically as the widths increase [5]. The calculated carrier mobilities of these GDY strips are in the range of 10^2 to 10^6 cm^2 V^{-1} s^{-1} at room temperature. For electron and hole charge carriers, the relationship between mobilities and NR widths is different and the mobilities of electrons are always larger than those of holes. Hence, the GDY NRs studied are possibly more favorable for electron transportation.

Using density functional theory (DFT), the band gap value of GDY was calculated to be 0.44 eV at the local density approximation (LDA) level, while at the theory of GW many-body level, this value was increased to 1.10 eV (Figure 8.1a,b) [6, 7]. The excitons of GDY have both Wannier–Mott and Frenkel excitonic properties. When

Graphdiyne: Fundamentals and Applications in Renewable Energy and Electronics, First Edition. Edited by Yuliang Li.

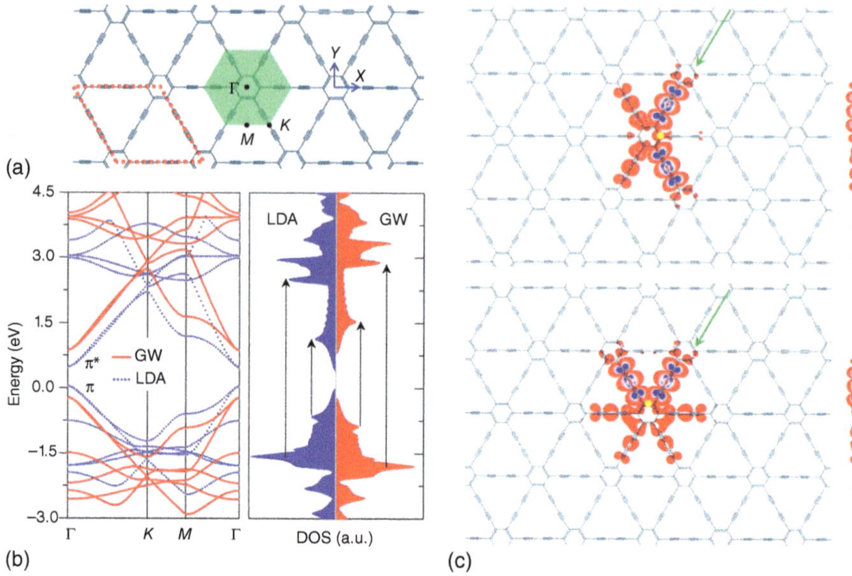

Figure 8.1 (a) Structure of GDY and its first Brillouin zone. (b) Band structures and DOS of GDY. (c) Electron densities of the lowest doubly degenerate bright excitons (the upper figure is top view and the lower one is side view). Source: Luo et al. [6]/American Physical Society.

the hole (yellow dot) is fixed at the density of the largest π_z orbital, the electron density of the lowest double bright denatured excitons A_1 and A_2 (Figure 8.1c). GDY's unique exciton characteristics and remarkable electrical properties endow it with good applicability in field-effect transistor (FET) devices. In systematic investigation of the interfacial properties of GDY in contact with a series of metals (Al, Ag, Cu, Au, Ir, Pt, Ni, and Pd) directed by Pan et al., quantum transport calculations were carried out to study the GDY FET device with Al electrodes [8]. According to the adsorption strength and electronic structures, the contact of GDY with Al, Ag, Cu, and Au surfaces is by physisorption, in comparison with the weak chemisorption with Ir and Pt surfaces and the strong chemisorption with Ni and Pd surfaces. Al, Ag, and Cu electrodes form ohmic or quasi-ohmic contact with GDY, while Pd, Pt, Ni, Ir, and Au electrodes form Schottky contact. The 10-nm channel-length GDY FET reveals an on–off ratio up to 10^4 and a very large on-state current of 1.3×10^4 mA mm^{-1}, indicating the great potential of GDY as the channel of a high-performance nanoscale FET. Dong et al. [9] combined DFT and nonequilibrium Green's function method to systematically study the contact and transport properties of single-layer 4-hT^2 GDY FETs with various electrodes, including metals (Cu, Au, Ni, Al, Ag) and MXenes (Cr_2C, Ta_2C, and V_2C). The different coupling degrees between the electrode materials and the GDY monolayer have been found to affect the electronic structure and transport properties of the contact system, and further exert an influence on the carrier injection. The Au, Al, and Ag electrodes show weak interface coupling with 4-hT^2 GDY, while the Cu, Ni, and MXenes electrodes form covalent bonds with GDY and therefore have higher carrier injection efficiency. The source–drain current of the FET with Au electrode shows linear correction with the applied bias, and the weak Fermi level pinning effect indicates that 4-hT^2 GDY is an appropriate candidate for channel material of FETs.

The first experimental demonstration of the field-effect mobility and semiconductor properties of the GDY material was achieved by fabricating a bottom-gate thin-film transistor (TFT) on OTS/SiO_2/Si, and evaporating Au electrodes directly on top of the GDY film (Figure 8.2) [10]. Compared with the effective field effect mobility of the GDY film ($1\times10^2\,cm^2\ V^{-1}\ s^{-1}$), the GDY nanowires [11] constructed by the vapor–liquid–solid (VLS) growth process have a higher mobility ($7.1\times10^2\,cm^2\ V^{-1}\ s^{-1}$). GDY nanowires with no defects on the surface are excellent semiconductors with the conductivity of $1.9\times10^3\,S\,m^{-1}$. In another study, Li et al. [12] fabricated the GDY-based field effect thin-film transistors (GTFTs) on SiO_2/Si substrate, and the performance of GTFTs can be adjusted continuously and orderly through the lossless adjustment strategy by photoelectric and thermal modulation (Figure 8.3). Under light and heat, the conductivity and switching ratio of GTFT are significantly improved. The higher the power density is, the better the device performs. Remarkably, when the temperature increases from 323 to 393 K, the thermal effect increases the switching ratio from 1.7×10^3 to 7.7×10^3, which is

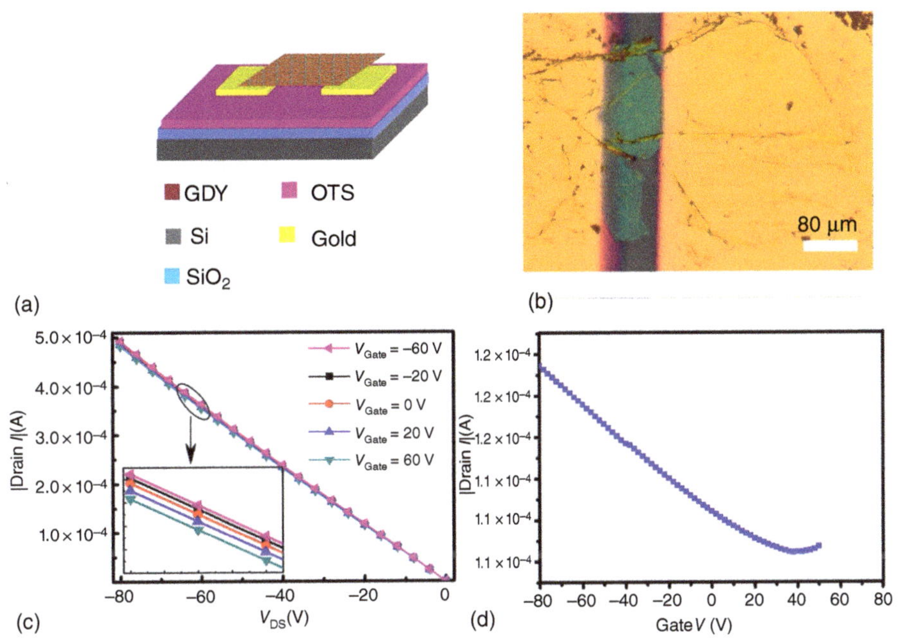

Figure 8.2 (a) Schematic representation of the GDY transistor. (b) Optical microscopy image of a GDY FET device. The typical GDY TFT has a channel length of 80 μm. (c) $I_{ds}-V_{ds}$ characteristics of the device, with the applied gate voltage (V_g) in the range from −60 V (top) to 60 V (bottom). (d) Transfer characteristics, source–drain voltage held at a value of V_{ds} of −20 V. Source: Qian et al. [10]; © 2015, the Springer Nature.

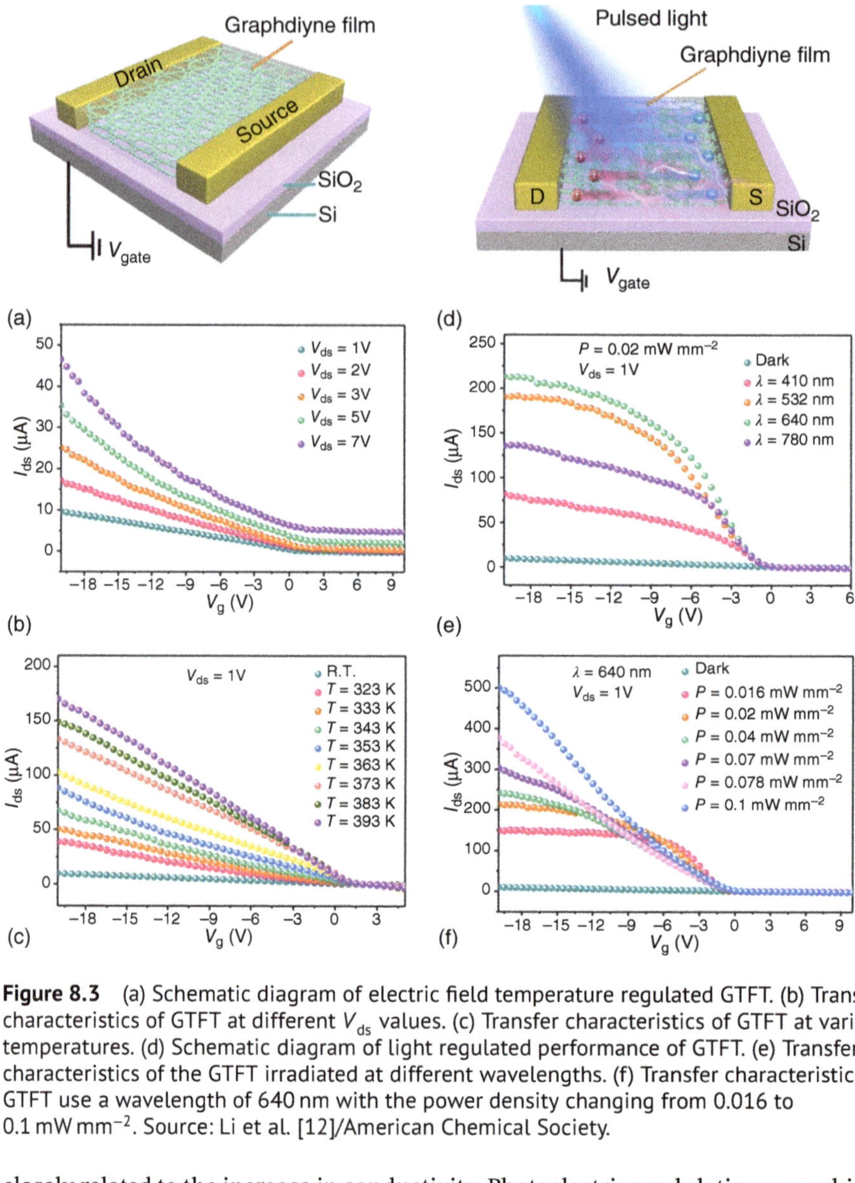

Figure 8.3 (a) Schematic diagram of electric field temperature regulated GTFT. (b) Transfer characteristics of GTFT at different V_{ds} values. (c) Transfer characteristics of GTFT at varied temperatures. (d) Schematic diagram of light regulated performance of GTFT. (e) Transfer characteristics of the GTFT irradiated at different wavelengths. (f) Transfer characteristics of GTFT use a wavelength of 640 nm with the power density changing from 0.016 to 0.1 mW mm^{-2}. Source: Li et al. [12]/American Chemical Society.

closely related to the increase in conductivity. Photoelectric modulation can achieve superior performance optimization. Especially at the wavelength of 640 nm, the conductivity of GDY can be increased to 1.5×10^4 S m^{-1}, and the on–off ratio of GTFT reaches -10^4. It is foreseeable that GTFT devices driven by light and heat will help to achieve multifunctional integration, and promote the application of such thin films composed of GDY materials in new carbon electronic devices.

GDY nanosheets prepared by the liquid-phase exfoliation (LPE) method have been developed as a conductive ink to fabricate a fully printed mesoscale FET device on a silicon substrate (Figure 8.4a) [13]. High-performance GDY-based FET devices with ionic liquid as the gate maintain a stable on–off ratio under different

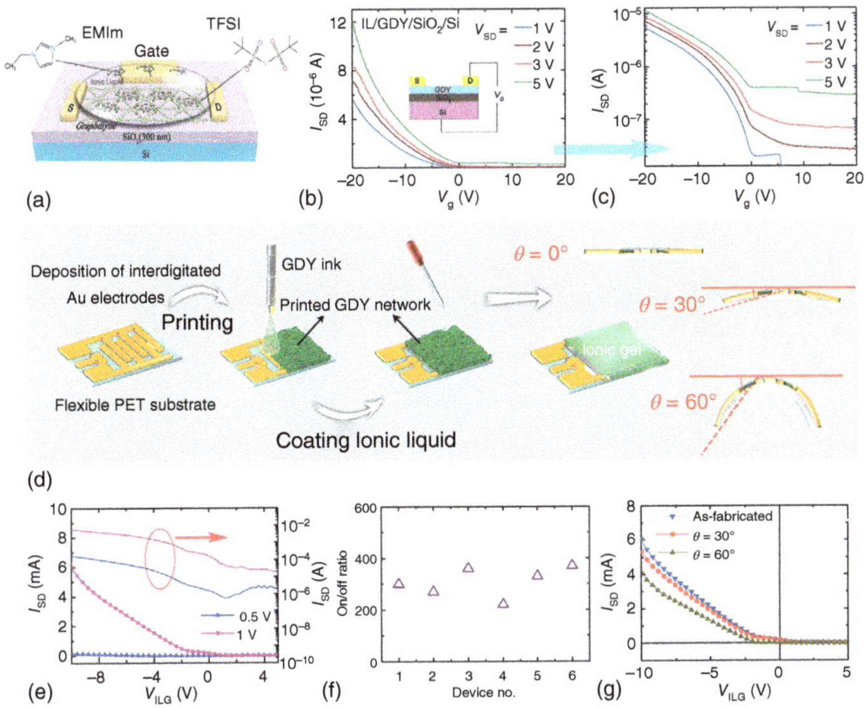

Figure 8.4 (a) Schematic of a GDY-based FET device with an ionic liquid dielectric layer as the gate. (b,c) Transfer curves of GDY FET device with SiO_2 as the gate dielectric at linear coordinates and logarithmic coordinates, respectively. The inset is the illustration of the device configuration. (d) Schematic diagram of the fabrication process of a flexible FET device based on GDY ink and ionic liquid gate dielectric. (e) Transfer curves of the flexible GDY FET device with different V_{SD}. (f) The on–off ratios measured for different numbered devices at V_{SD} = 0.05 V. (g) Transfer curves of the printed FET as a function of bending angles [13]. Source: Zhang et al. [13]/John Wiley & Sons.

V_{SD}, and the off-state current I_{off} gradually increases with the increase of V_{SD}, achieving a significant increase in carrier concentration of GDY to about 10^{14} cm^{-3} (Figure 8.4b,c). When the substrate is changed to polyethylene terephthalate (PET), the stable and flexible GDY-based transistor device obtained still exhibits an on–off ratio of more than 10^2 with an acceptable repeatability (Figure 8.4d). By bending at different angles, the degradation of the device is not obvious, which proves the advantages of the printing methods for preparing flexible GDY-based electronic devices (Figure 8.4e–g). These results reveal the possibility to design wearable or large-area carbon-based electronic products involving GDY semiconductors, which indicates promising new carbon materials for novel electronic devices.

Not only GDY itself can play a direct role in the field of electronic transistor devices, but the developed GDY-based composite material can also be used in transistors. Cui et al. [14] successfully fabricated a FET device based on GDY and a conjugated polymer (*o*-fluoro-*p*-alkoxy-phenyl-substituted benzo[1,2-*b*:4,5-*b*′] dithiophene [PFC]) (Figure 8.5), which can be well dispersed or soluble in organic solvents such as chlorobenzene. The FET device with the structure of top-contact

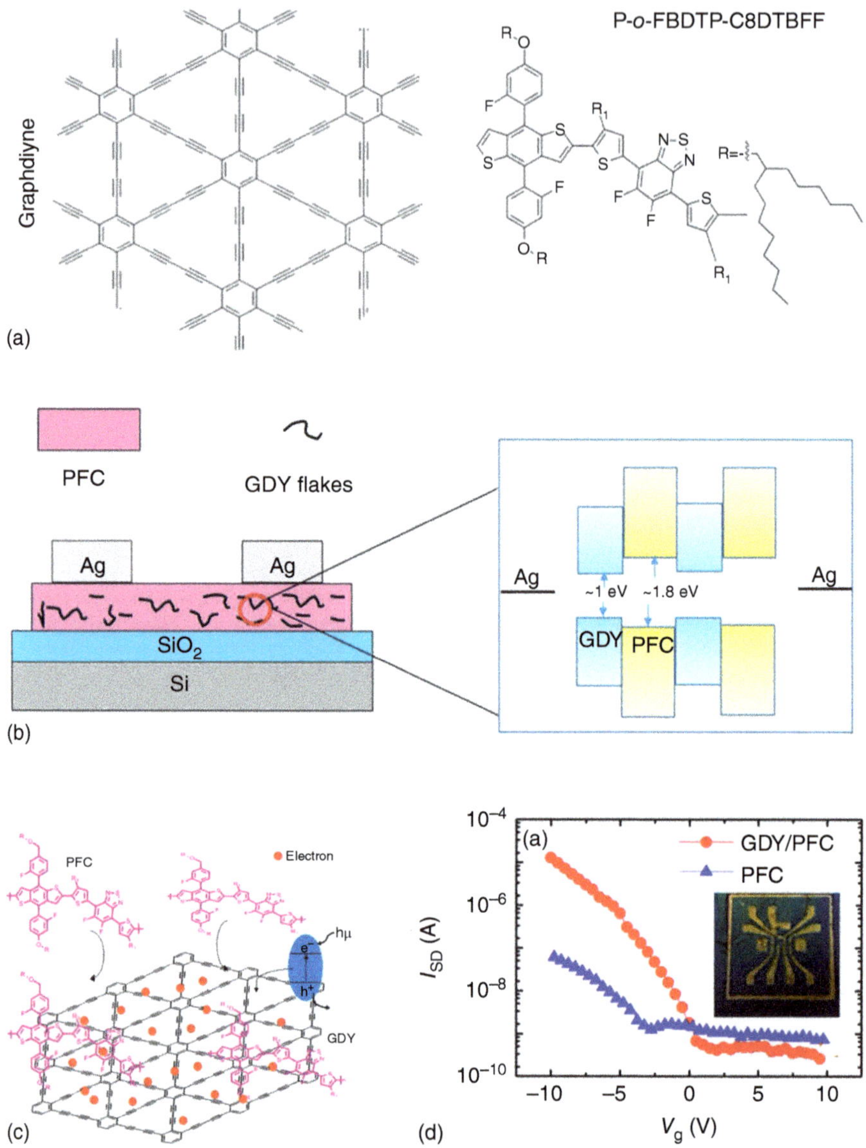

Figure 8.5 (a) Chemical structures of GDY and PFC. (b) Schematic illustration of the FET device and the related energy levels. (c) The photon-generated carrier transport process of GDY/PFC composites under visible light irradiation. (d) Transfer I_{SD}-V_g curves for PFC and GDY/PFC films. The inset shows the photograph of photoetching device. Source: Cui et al. [14]; © 2017, American Chemical Society.

bottom-gate is carried out by spin-coating GDY/PFC on SiO_2-covered silicon wafer and silver-evaporated source/drain electrodes (10 μm for channel length and 5 μm for channel width). The interfacial energy-level matching and improved charge migration channels contribute to the enhanced carrier transport properties of the GDY/PFC film, and its mobility can be significantly increased from 0.002 to 0.69 $cm^2 V^{-1} s^{-1}$. Compared with pure PFC polymer film, the threshold voltage and on/off radio rate of GDY/PFC film are also significantly improved. These results may not only provide a deep insight into the regulation of carrier transport property in organic polymer semiconductors, but also promote the related photoelectric device design by doping with two-dimensional carbon materials.

GDY also shows certain application potential in terms of field emission devices. Zhou et al. adjusted the catalytically active sites by adjusting the ratio of the added organic alkali along with the amount of monomer and grew GDY vertically at these sites first. With more copper ions dissolved, uniform GDY nanowalls formed on the surface of the copper substrate (Figure 8.6a,b) [15]. The E_{to} (turn-on field) is demonstrated to be about 6.6 V μm^{-1}, and E_{thr} (threshold field) is 10.7 V μm^{-1} (Figure 8.6c). The value of E_{to} is higher than that of the vertical graphene nanosheets [16], but the

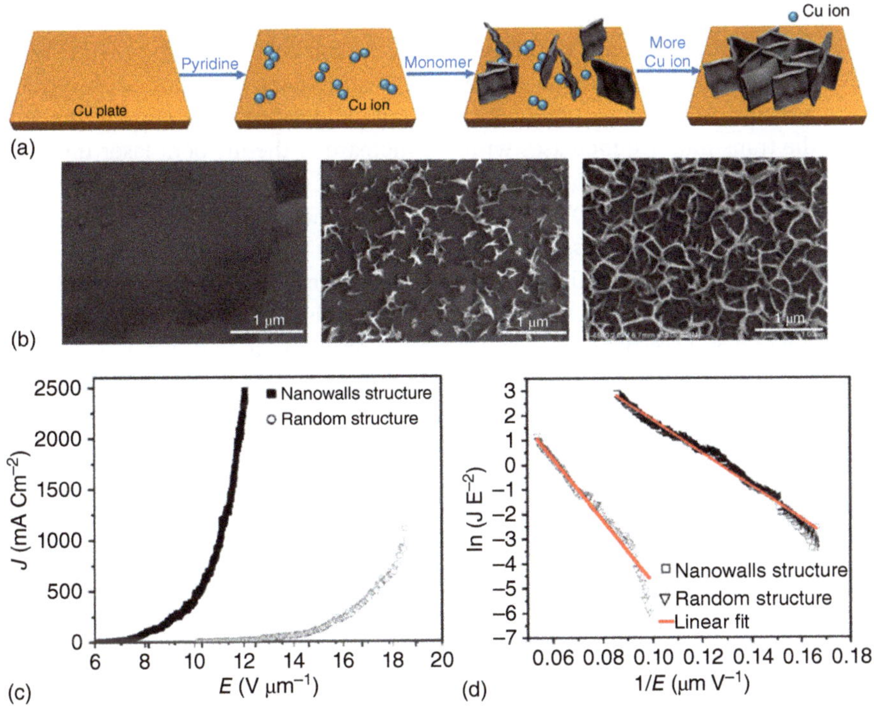

Figure 8.6 (a) Illustration of the preparation process of GDY nanowalls. (b) Scanning electron microscopy (SEM) images showing the formation process of GDY nanowalls in time series: bare Cu plate before reaction and 8 and 10 hours after reaction, respectively (from left to right). (c) Typical plots of the electron-emission current density (*J*) as a function of applied electric field (*E*). (d) Corresponding F–N plots and the linear fittings. Source: Zhou et al. [15]; © 2015, American Chemical Society.

carrier mobility of GDY is similar to that of graphene [17]. The corresponding fitted curve (Figure 8.6d) is in accordance with the Fowler–Nordheim (F–N) mechanism, indicating that the electron emission from the samples is a result of electron tunneling [18]. The GDY nanowalls with highly conjugated structures and uniformly distributed sharp walls show extraordinary and stable field emission properties and therefore could be suitable candidate for use in all-carbon electronic devices.

8.2 Optic Devices

As a rising star of all-carbon nanomaterial, GDY has a direct natural bandgap and features strong light–matter interaction, large optical absorption, superior chemical, and optical stability, promising its broad prospects in the field of photonics and optoelectronics. In particular, the nonlinear optical (NLO) properties of GDY materials have been explored. A series of open-aperture Z-scan measurements have been performed with fs pulses excitation at different wavelengths, revealing the broadband and strong NLO response of GDY [19, 20]. It has been demonstrated that GDY shows a relatively high NLO refraction in the fs regime, surpassing most of known 2D materials. This enhancement is most likely related to the presence of more extended delocalized π-conjugation and the intrinsic bandgap of GDY. With the excitation strong enough, the prepared GDY sample shows saturated absorption characteristics. At the same time, when the sample position (z) is close to the lens focus, the transmittance increases with the increase of the incident laser intensity (Figure 8.7a). It is worth noting that the energy of the excited photon is much greater than the prepared GDY band gap, and the conduction band of GDY is filled under such high photon energy, which leads to Pauli blocking and the subsequent saturable absorption. The fitted-scan curves are in good agreement with the experimental data obtained under 475 to 1800 nm pulse excitations, which indicate the broadband NLO response of 2D GDY from visible light to mid-infrared (mid-IR) band. Meanwhile, a larger absolute value of β (>0.89 cm GW^{-1}) is fitted (Table 8.1), which signifies that 2D GDY is a stronger NLO material than the mainstream NLO materials, such as graphene ($-(0.66 \pm 0.30)$ cm GW^{-1}) [21], black phosphorus (BP) ($-(6.17 \pm 0.19) \times 10^{-3}$ cm GW^{-1}) [22], MoS_2 ($-(4.60 \pm 0.27) \times 10^{-3}$ cm GW^{-1}) [23], Mxene (-0.297 cm GW^{-1}) [24], and metal-organic frameworks (MOFs) ($\approx -3 \times 10^{-2}$ cm GW^{-1}) [25]. Moreover, by using the most commonly used single-photon absorption model [26] to fit the experimental data, the smaller I_s (<10 cm GW^{-1}) of 2D GDY from 475 to 1800 nm is obtained, which indicates that GDY has a strong saturable absorption response (Figure 8.7b). The strong and broadband saturable absorption under the fs pulse in infrared (IR) region suggests that 2D GDY has great research value as a passive mode-locked device of ultrafast lasers.

In subsequent studies, the feasibility of the application of GDY-based materials in the saturable absorber (SA) of ultrashort pulsed lasers has been explored [27]. A uniform GDY–PVP nanocomposite is prepared by mixing high-quality 2D GDY with polyvinylpyrrolidone (PVP). The encapsulation of PVP contributes to the separation of GDY and oxygen in the air, which greatly improves the stability of the composite

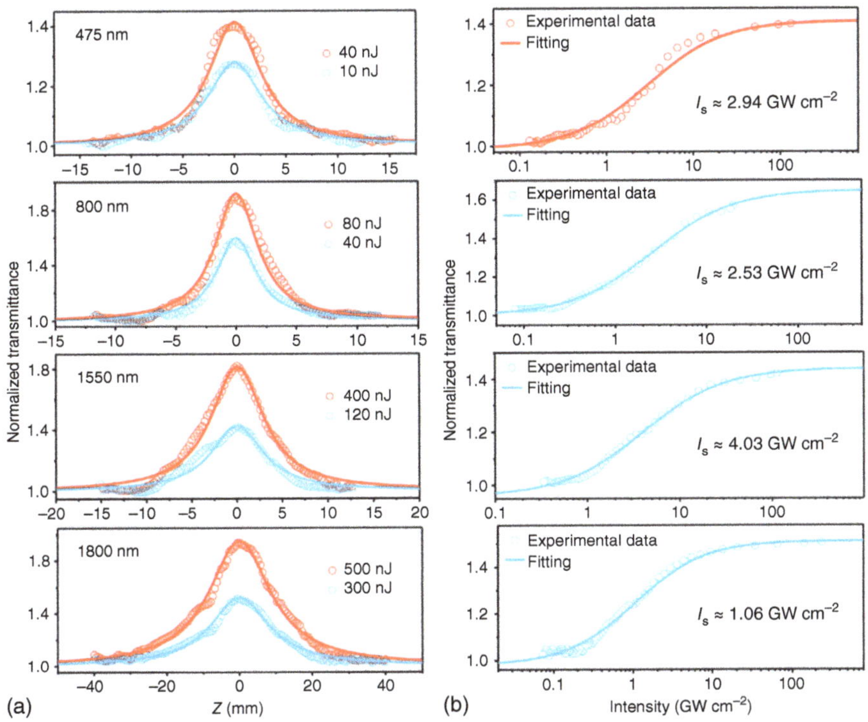

Figure 8.7 Nonlinear optical responses of 2D GDY to fs pulses. (a) Open-aperture Z-scan results of 2D GDY at 475, 800, 1550, and 1800 nm. (b) Changes of normalized transmittance vs. input peak intensity of the fs laser. Source: Guo et al. [20]/John Wiley & Sons.

Table 8.1 NLO parameters of 2D GDY at different wavelengths.

λ (nm)	α_0 (cm^{-1})	β (cm GW^{-1})	I_s (GW cm^{-2})	ΔT (%)	T_{ns} (%)
457	2.96×10^2	−0.89	2.94	41.69	40.84
800	2.19×10^2	−14.37	2.53	64.62	64.88
1550	1.65×10^2	−1.58	4.03	48.13	44.07
1800	1.54×10^2	−1.4	1.06	52.98	51.19

structure. The determined large NLO absorption coefficient (β, >−1 cm GW^{-1}), low saturation intensity (I_s, <13 GW cm^{-2}), and ultra-fast relaxation time (<30 ps) are superior to the reported values of most of the benchmark NLO materials, which indicates the outstanding potential of GDY in photonics among the emerging novel NLO materials. I_s, modulation depth (T_s), and nonsaturable loss (T_{ns}), which are considered as the three common criteria for evaluating the suitability of NLO materials as SA of mode-locked lasers, have been found to be 1.89 GW cm^{-2}, 63.93%, and 63.24% for GDY, respectively. The I_s of GDY at 800 nm is about 1% of BP [22], indicating a stronger saturated absorption response of GDY. SA based on GDY–PVP is integrated into ytterbium-doped fiber (YDF) and erbium-doped fiber (EDF) ring resonators of

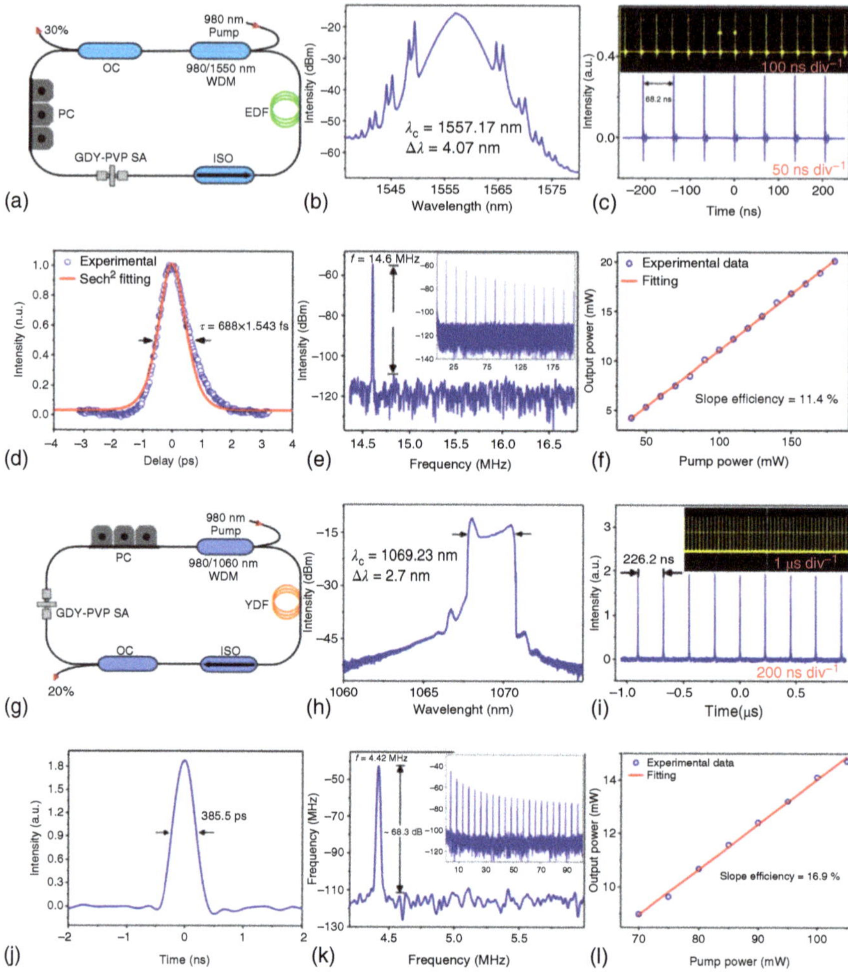

Figure 8.8 (a–f) Mode-locked pulse output performance of the GDY–PVP nanocomposite-based EDF laser. (a) Schematic of the EDF laser resonator. Here, WDM, ISO, PC, and OC denote a wavelength division multiplexer, a polarization-independent isolator, a polarization controller, and an output coupler, respectively. The measured (b) optical spectrum, (c) oscilloscope trace, (d) autocorrelation trace, and (e) radio frequency (RF) spectrum. (f) Changes of output power vs. pump power. (g–l) Mode-locked pulse output performance of the GDY–PVP nanocomposite-based YDF laser. (g) Schematic of the YDF laser resonator. The measured (h) optical spectrum, (i) oscilloscope trace, (j) individual pulse profile, and (k) RF spectrum. (l) Changes of output power vs. pump power. Source: Guo et al. [27]; © 2020, Wiley-VCH Verlag GmbH & Co. KGaA.

the ultrashort pulsed laser to obtain 385.5 ps and 688 fs pulse outputs at 1 and 1.5 μm, respectively (Figure 8.8).

Zhao et al. also realized SA by depositing GDY-based ethanol dispersion on the tapered fiber to obtain stable mode-locked pulsed fiber laser in the optical communication region by interaction with the evanescent field of light [28]. Benefiting from GDY as SA, the proposed fiber laser can obtain a mode-locked laser pulse with a center wavelength of 1564.70 nm, which has a repetition rate of 12.05 MHz and a pulse width of 734 fs. The thermal stability of GDY contributes to the long-term stable pulse generation of the laser at high pump power. When the GDY SA is taken out of the cavity, it is found that the mode-locking cannot be regained by adjusting the polarization controllers (PCs) or changing the pump power, which proved the necessity of GDY SA in the laser model-locking. This proves the broad prospects of GDY in fiber lasers, the great potential in the field of nonlinear optics and ultrafast photonics, and opens up a new path for the application of GDY.

2D GDY has been further demonstrated to exhibit a strong light–matter interaction with high stability to achieve a strong broadband NLO Kerr response [29]. Taking advantage of the large Kerr nonlinearity in GDY, an interesting configuration of GDY/SnS_2 can be used as a photonic device to realize the unidirectional diffraction ring excitation, and this phenomenon of the nonreciprocal light propagation indicates that the configuration possesses good advance of application in passive photonic diode. The spatial self-phase modulation (SSPM) method has been employed to study this novel nonlinear photonic diode with the hybrid structure of the 2D GDY/SnS_2 (Figure 8.9). It is found that: (i) If the laser beam passes from the forward (GDY/SnS_2) direction, the diffraction rings are excited by the first part of

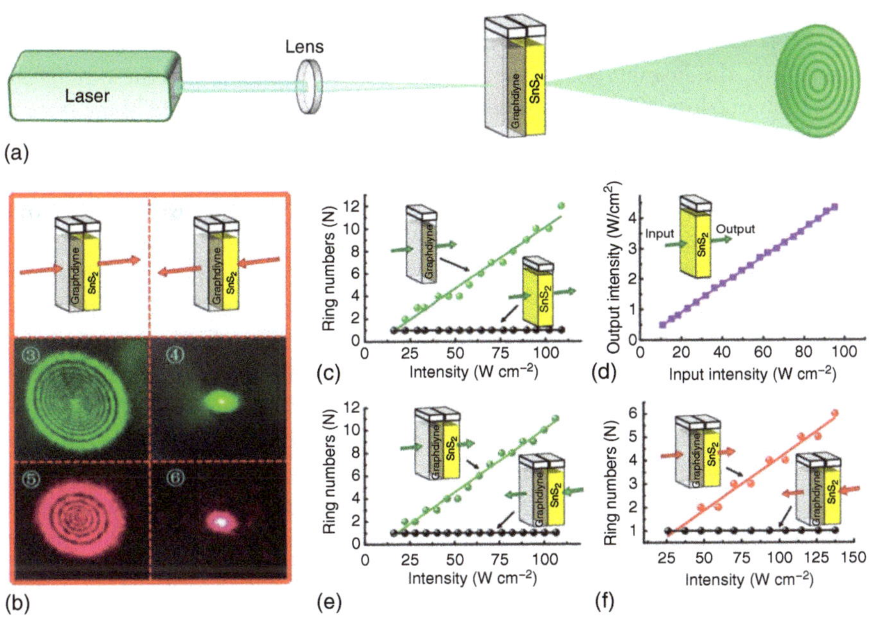

Figure 8.9 (a) The experimental setup for the proposed graphdiyne/SnS_2-based nonlinear photonic diode. (b) The phenomenon of the nonreciprocal light propagation observed from the proposed nonlinear photonic diode. (c) The number of diffraction rings excited from the independent graphdiyne dispersions and SnS_2 dispersions at $\lambda = 532$ nm. (d) The input and output intensity of the 532-nm laser beam before and after passing through the SnS_2 dispersions. (e,f) The results of the nonreciprocal light propagation for the proposed nonlinear photonic diode at $\lambda = 532$ and 671 nm, respectively. Source: Wu et al. [29]; © 2019, Wiley-VCH Verlag GmbH & Co. KGaA.

GDY dispersions. After that the excited diffraction rings continue to pass through the second part of SnS_2 dispersions, and the ring numbers cannot be changed with the intensity being weakened. (ii) If the laser beam passes from the reverse (SnS_2/GDY) direction, the incident laser beam will be weakened by the first part of SnS_2 dispersions. The weakened laser beam, which has a relatively low intensity below the threshold to excite the diffraction rings of GDY Ns, continues to pass through the second part of GDY dispersion. The larger value of N/I indicates that the proposed nonlinear photonic diode can have the more obvious phenomenon of the nonreciprocal light propagation ("N" is the number of diffraction rings and "I" is the intensity of incident laser beam). Therefore, the performance of the nonlinear photonic diode based on GDY/SnS_2 at 532-nm wavelength is better than that at the wavelength of 671 nm.

The GDY materials have been also demonstrated for lasing applications. Using different excitation wavelengths and intensities, multicolor random lasers from GDY nanosheets solution have been unambiguously demonstrated [30]. The spectra of the GDY-based random laser cover most of the visible region from 450 to 700 nm. As the pump intensity increases, sharp emission spikes emerged on the broad emission band, which indicates that the random laser is successfully generated (Figure 8.10a–c). Figure 8.10d–f shows the integrated intensity as a function of the pump intensity at the three pump wavelengths (390, 400, and 410 nm), and the random laser thresholds for the three pump wavelengths have been determined to be 7.3, 16.4, and 23.0 $kW\,cm^{-2}$, respectively. Moreover, the Brown motions of the gain medium, the instabilities of the generated bubbles changes as the time goes, and the high freedom of modes competition due to the multiscattering in the system give the spectral diversity.

Compared with graphene, which is constructed only by benzene rings, GDY with the special structures with rings in combination with chains have many superior physical and chemical properties [31]. GDY has been proven to have a natural band gap with high carrier mobility (-10^4 cm^2 V^{-1} s^{-1}), while the combination of electron-conducting and 2D planar structures make GDY a promising platform for effective detection of ultraviolet and IR light, providing great advantages for GDY to be used as a photodetector (PD) [6, 7]. Single-walled carbon nanotubes (SWNTs), especially semiconductive single-walled carbon nanotubes (s-SWNTs), are regarded as important candidates for IR optical detection because of their tunable light–matter interaction in the IR region caused by chirality-dependent band gap. However, the as-prepared SWNTs are a mixture of SWNTs with semiconducting and metallic properties, and the excitons can be easily quenched by the metallic SWNTs. Besides, the strong binding energy displayed by s-SWNTs obstructs the separation of photoinduced excitons. Among many materials, GDY has been found to promote exciton dissociation while maintaining the on-state current and carrier mobility of s-SWNTs. A novel high-purity s-SWNTs/GDY IR photodetector with a high on–off ratio exceeding 10^5 and a mobility near 25 cm^2 V^{-1} s^{-1} has been reported (Figure 8.11) [32]. In comparison to device fabricated only with s-SWNTs, the electrical transport property of SWNTs/GDY photodetector did not decrease. The s-SWNTs/GDY device shows uniform response in the whole channel area,

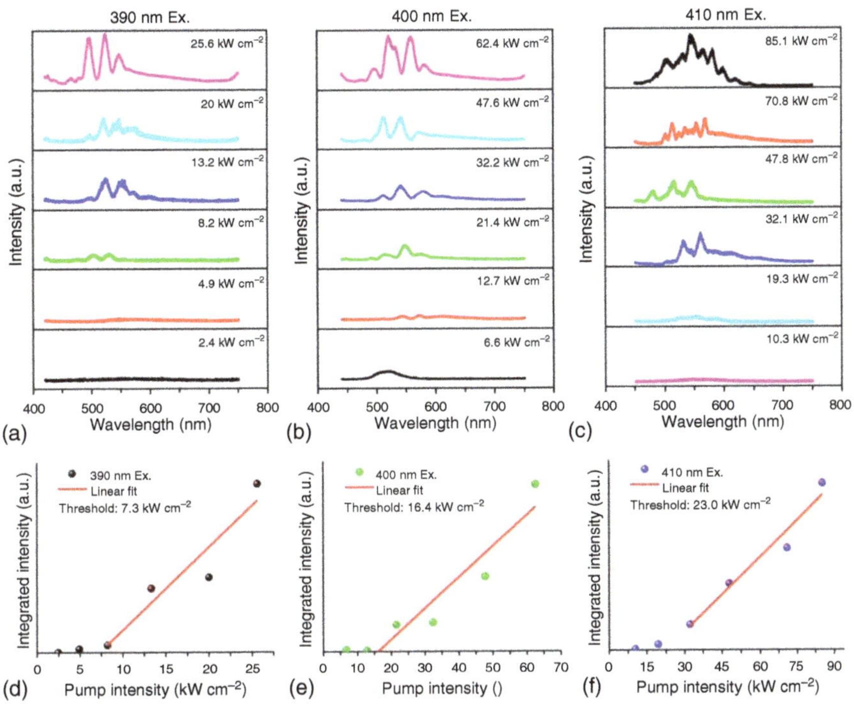

Figure 8.10 (a–c) Random laser performance of GDY nanosheets. The typical emission spectra of GDY nanosheets solution under the excitation of (a) 390, (b) 400, and (c) 410 nm. The pump intensity is shown in the insets of the figures. The random laser emerged when the pump power up to 8.2, 12.7, and 32.1 kW cm^{-2} for the three excitation wavelengths, respectively. (d–f) The integrated emission intensity as a function of pump wavelength at (d) 390 nm, (e) 400 nm, and (f) 410 nm, respectively. The fitted random laser thresholds are shown in the figures. Source: Jiang et al. [30]/American Chemical Society.

and the introduction of GDY greatly improves the efficiency of the dissociation of excitons. The IR detector has been proven to demonstrate a fast response time below 1 ms, with an optimal responsivity of 0.4 mA W^{-1} and detectivity of 5×10^6 cm Hz$^{1/2}$ W^{-1}. It has been proposed that the photoinduced electrons are no longer transferred to the high-energy level of s-SWNTs, but transferred to the lowest unoccupied molecular orbital (LUMO) of the introduced GDY, benefiting the excitons' dissociation and carrier density.

Adopting the LPE method, GDY nanosheets with a height of 3.3–4.0 nm were successfully exfoliated (Figure 8.12a–f) [13]. From the transmission spectrum of GDY nanosheets in the range of 200–1300 nm at room temperature recorded by the UV spectrophotometer (Figure 8.12g), it can be seen that when the wavelength decreased to about 295 nm (T_0 = 69.5%), the transmittance decreases rapidly. The transmittance T_0 is 19.8% at 221 nm, and the minimum transmittance is 0.8% at 200 nm. High transmittance in the ultraviolet (UV) waveband and short ultraviolet (UV) cut-off wavelength implies the probable application of GDY in UV optics.

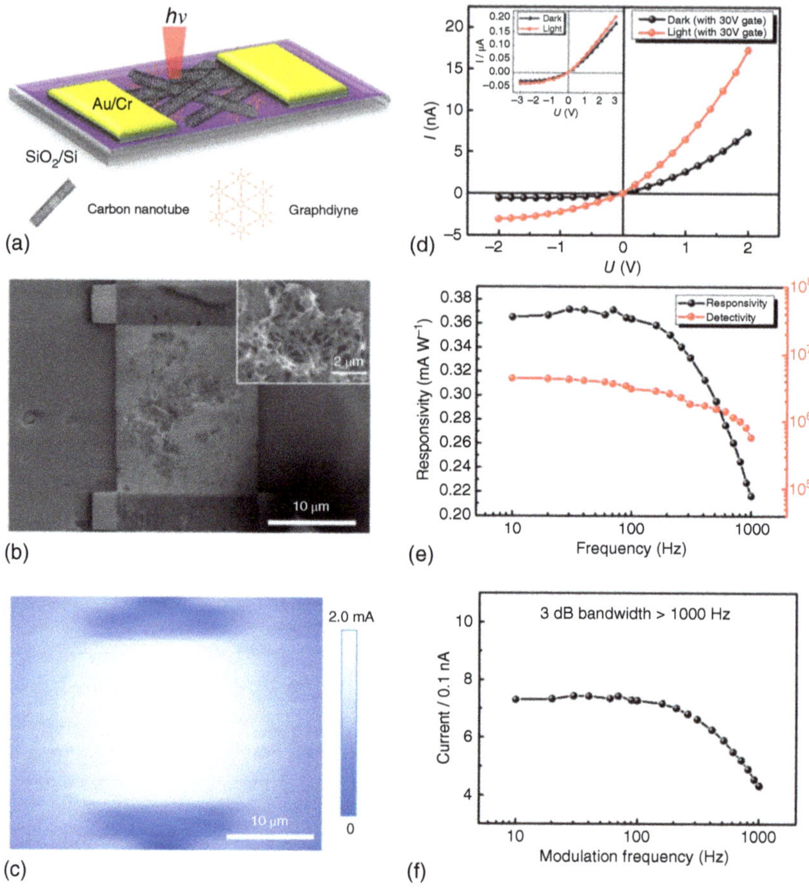

Figure 8.11 (a) Schematic diagram of the device based on s-SWNTs and GDY. (b) SEM image of GDY and s-SWNTs film. The inset shows the magnifying image. (c) IR mapping of the GDY and SWNTs-based IR detector. (d) The *I*–*V* curve of the detector with light on and off at 30 V gate. The inset is the *I*–*V* curve without gate voltage. (e) Responsivity and detectivity of the device at different frequencies. (f) The photocurrent vs. switching frequency. Source: Zheng et al. [32]/John Wiley & Sons/CC BY 4.0.

In this context, GDY nanocomposites have been applied in ultraviolet (UV) photodetectors (PDs) [33–35], which require materials of high sensitivity with UV light. For example, GDY : ZnO nanocomposites prepared by self-assembly of GDY nanoparticles onto the surface of propylamine-modified ZnO nanoparticles (NPs) have been used to fabricate UV photodetectors (Figure 8.13) [35]. Compared to the conventional reference devices, which have relatively low *R* value of 174 $A\,W^{-1}$ and long rise/decay time of 32.1/28.7 seconds, the device based on GDY nanocomposite has a higher responsivity of 1260 $A\,W^{-1}$ and shorter rise/decay time of 6.1/2.1 seconds. This PD device with GDY : ZnO nanocomposite as the active layer exhibits a 7.2 times increase in the value of responsivity and a significant decrease in both rise and decay time. The PN junction formed between GDY NPs and ZnO

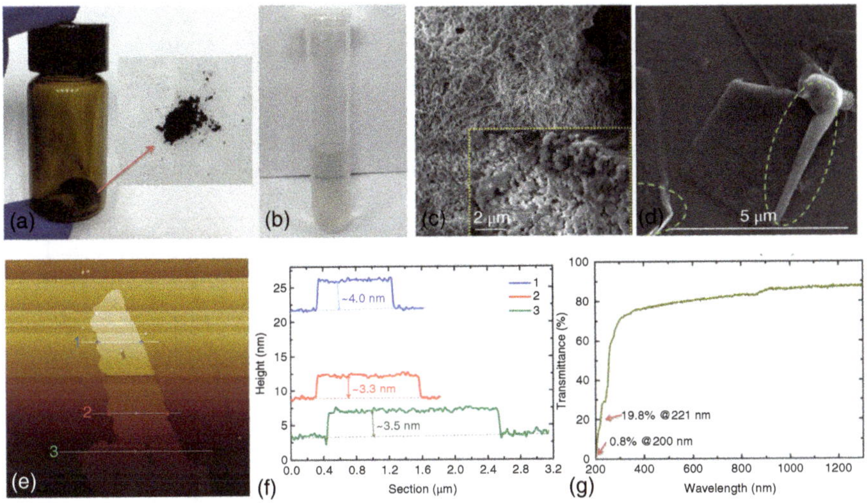

Figure 8.12 (a) GDY powders and (b) the as-prepared GDY nanosheets. SEM images of (c) GDY powder (insert: enlarged view) and (d) GDY nanosheets. (e) Atomic force microscope (AFM) image of a GDY nanosheet and (f) the corresponding heights if traces 1, 2, and 3. (g) The transmission spectrum of GDY nanosheets [13]. Source: Zhang et al. [13].

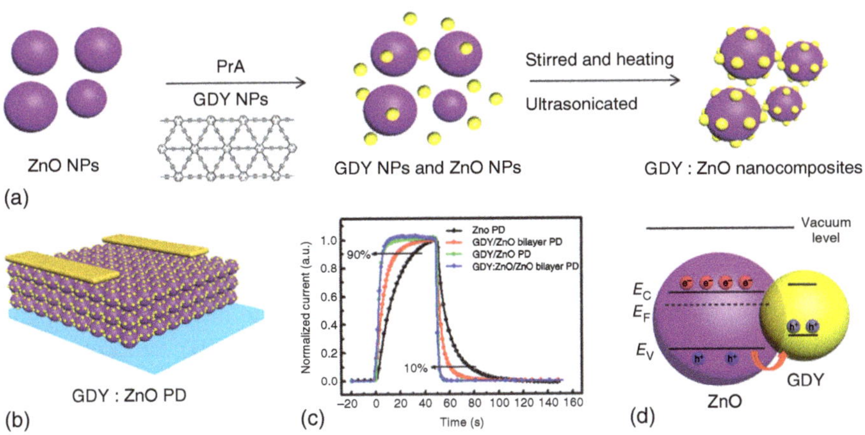

Figure 8.13 (a) Schematic illustration of the preparation route of nanocomposites composing GDY and ZnO. (b) Device structure of PDs based on GDY:ZnO. (c) Rise and decay times of PD with different active layers. (d) Illustration of energy levels of the as-prepared ZnO NPs and GDY NPs. The transfer direction of hole from ZnO NP to GDY NP is also presented by the red arrows. Source: Jin et al. [35]/John Wiley & Sons.

NP can adjust the energy level of the active layer and broadens the depletion region near the surface of ZnO, which greatly improves the carrier-exchange process between the two NPs, and hence significantly enhances the photoresponse.

In the subsequent studies, titanium dioxide (TiO_2) nanocrystals (NCs) with a rational phase of anatase have been encapsulated on GDY particles and intercalated

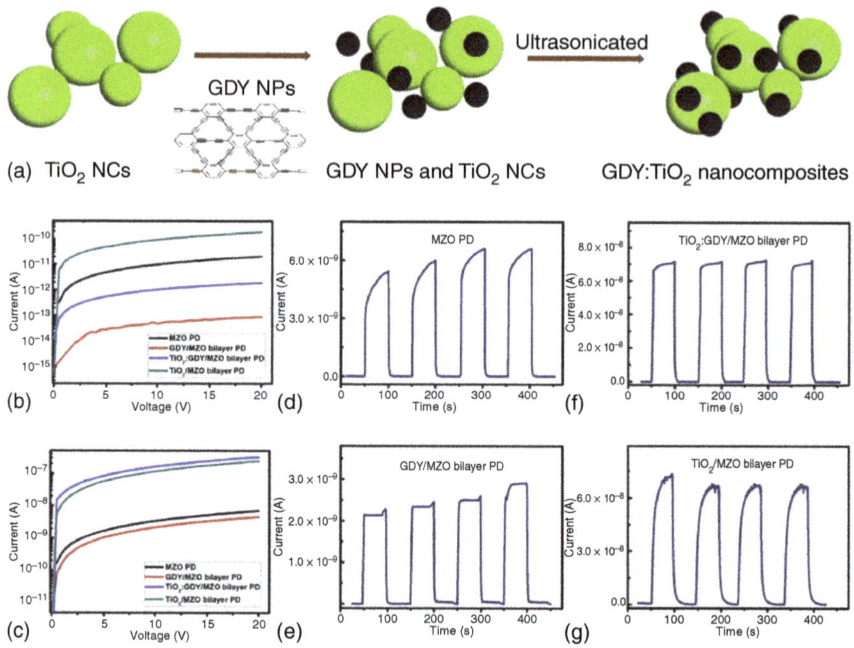

Figure 8.14 (a) Fabrication process for the TiO_2:GDY nanocomposites. (b) *I*-*V* curves of the fabricated PDs in dark and (c) under 254-nm UV light illumination with an intensity of 0.28 mW cm^{-2} ($L = 80\,\mu m$, $W = 8800\,\mu m$). On–off switching properties of: (d) the MZO PD, (e) the GDY/MZO bilayer PD, (f) the TiO_2:GDY/MZO bilayer PD, and (g) the TiO_2/MZO bilayer PD. Source: Li et al. [33]/Elsevier.

into $Mg_{0.3}Zn_{0.7}O$ (MZO) to fabricate a lateral bilayer ultraviolet photodetector (Figure 8.14) [33]. GDY with high specific surface area can optimize the properties of TiO_2 NCs as a proper sustaining material, while TiO_2 NCs can conduct electrons and provide gain for their relatively superior conductivity [36] instead of acting as an electron trap. At the same time, when the optical band gap of MZO is as high as 3.8 eV, the ultraviolet photodetector is highly sensitive to deep ultraviolet light due to the formation of the heterojunction between the TiO_2 : GDY nanocomposite and MZO. The low dark current of TiO_2 : GDY/MZO bilayer PD indicates that GDY NPs effectively restrained the sintering of adjacent TiO_2 NCs in the film during the annealing process, and this dispersed surface morphology can significantly suppress electron mobility. The responsivity of the TiO_2 : GDY/MZO bilayer PD is enhanced by nearly two orders of magnitude under a 10-V bias, and the responsivity at 254 nm is 76.1 mA W^{-1}. The high signal-to-noise ratio (SNR) of 1.5×10^5 is presented in TiO_2 : GDY/MZO bilayer PDs, and the photocurrent rejection of the TiO_2:GDY/MZO bilayer PDs is about three orders of magnitude from 254 to 365 nm. The TiO_2 : GDY/MZO bilayer PD exhibits a good on–off switching property with the rise time and falling time of 3.5 and 2.7 seconds, respectively. This research reveals the potential prospects of GDY in photonic devices applications, and the integration of adjustable GDY NPs with various nanomaterials will help to create photonic devices that can be used commercially.

8.3 Thermoelectric Materials

The unique 2D structure of GDY is predicted to be the most stable diacetylene carbon allotrope with high thermal resistance, high electrical conductivity, and extremely high hardness, which promises its application as a thermoelectric material that can directly convert electricity into heat or heat into electricity [37–40]. The value of lattice thermal conductivity of GDY is 7.30 W m^{-1} K^{-1} as calculated from equilibrium molecular dynamics (EMDs) simulation, which is much lower than that of other carbon allotropes such as graphene and carbon nanotubes, indicating that GDY has the intrinsic advantage in thermoelectric applications [41]. Around the Fermi level ($n = 0$), the Seebeck coefficient S exhibits two obvious peaks, namely −754 μV K^{-1} at $n = 1.13 \times 10^9$ cm^{-2}, and 756 μV K^{-1} at $n = -1.16 \times 10^9$ cm^{-2}, respectively (Figure 8.15a), which are much larger than those of most of the conventional thermoelectric materials. Meanwhile, at the small carrier concentrations, the electrical conductivity is actually very small (Figure 8.15b). For p-type carriers, the power factor can reach a maximum value of 0.21 W m^{-1} K^{-2} when the carrier concentration is tuned to -1.5×10^{12} m^{-1} K^{-2}. In the case of n-type carriers, an even higher power factor of 0.52 Wm^{-1} K^{-2} can be achieved (Figure 8.15c). Additionally, the room-temperature ZT value of GDY for p-type and n-type carriers can be optimized to 3.0 and 4.8 (Figure 8.15d). Such

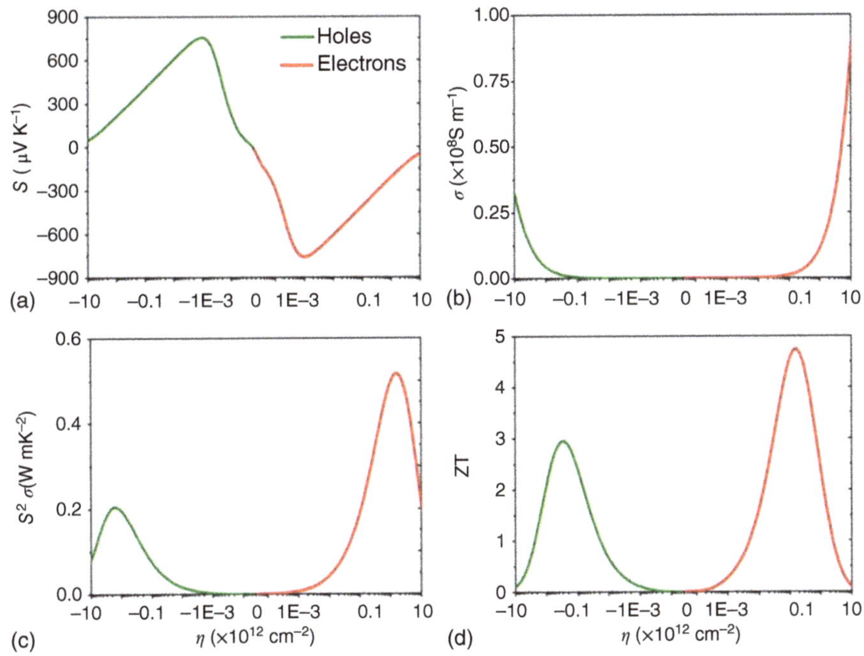

Figure 8.15 The room temperature (a) Seebeck coefficient S, (b) electrical conductivity σ, (c) power factor $S^2\sigma$, and (d) ZT value of GDY as a function of carrier concentration n (using logarithmic coordinates), where the green and red lines correspond to the holes and electrons, respectively. Source: Sun et al. [41]/Elsevier.

high ZT values significantly exceed most laboratory results reported so far, and are either equal to or larger than the target value ($ZT = 3.0$) for the commercial applications of thermoelectric materials, making GDY a reasonable candidate for high-performance thermoelectric materials.

8.4 Magnetism

The combination of extraordinary carrier mobility and long spin diffusion promise the applications of low-dimensional carbon materials for applications in spintronics [42]. The intrinsic magnetic properties of as-prepared and annealed GDY have been investigated (Figure 8.16a–f) [43]. Both the as-prepared GDY (GDY-as) and the GDY samples (GDY-300 and GDY-400) annealed at 300 or 400 °C show spin-half paramagnetism. However, after annealing at 600 °C, the GDY sample (GDY-600) exhibits clear antiferromagnetism. The hydroxyl proportion increases with the increase in the annealing temperature and the hydroxyl groups on the GDY sheet chain have been regarded as the main magnetic source. Similar to the case of hydrogen absorbed on the base plane of the graphene sheet [45], the spin charge density distribution of the hydroxyl groups absorbed on the chain and ring of the GDY sheet exhibits oscillating behaviors. Furthermore, the dangling C atom on the ring forms the covalent bond between the nearest C atom, which can induce structural reconstruction. More importantly, these hydroxyl groups have the high barrier energy of 1.73 eV that migrates from the ring site to the chain site, which is much higher than the value of 0.4 eV for hydroxyl on the basal plane of the graphene sheet [46]. Apparently, the high barrier energy can prevent hydroxyl groups' clustering and may be beneficial to the antiferromagnetism in the annealed GDY. Pristine graphdiyne (pGDY) with high quality was prepared by annealing the as-prepared GDY (as-GDY) in hydrogen at 600 °C (Figure 8.16g–i) [44]. The experimental results show that pGDY is ABC-stacked and has a direct bandgap of around 0.64 eV as measured by the optical absorption spectrum. Furthermore, the pGDY shows spin-half paramagnetism at 2 K, which may be induced by residual hydroxyl groups on the chains of pGDY sheets.

Doping is the practice of adding atoms of impurities to the semiconductor lattice to modify the electrical structure and properties. The advantages of GDY for spintronics in doping are tunable bandgap and conductivity as well as the easily realizable atomic substitution or modification among its special structure [2]. In terms of magnetism, it has been theoretically confirmed that nitrogen (N) doping is an effective method to introduce local magnetic moments [47]. The appearance of special sp-hybridized carbon atoms gives pure GDY material remarkable paramagnetic characteristic [48]. The introduction of nitrogen with N/C ratio of 5.29% into GDY achieves a twofold enhancement of the saturation moment at 2 K. Meanwhile, the conductivity of GDY is also increased to $6.25 \times 10^{-4}\,S\,m^{-1}$ from $3.60 \times 10^{-4}\,S\,m^{-1}$, suggesting that GDY is easier to realize the control of electromagnetic property by

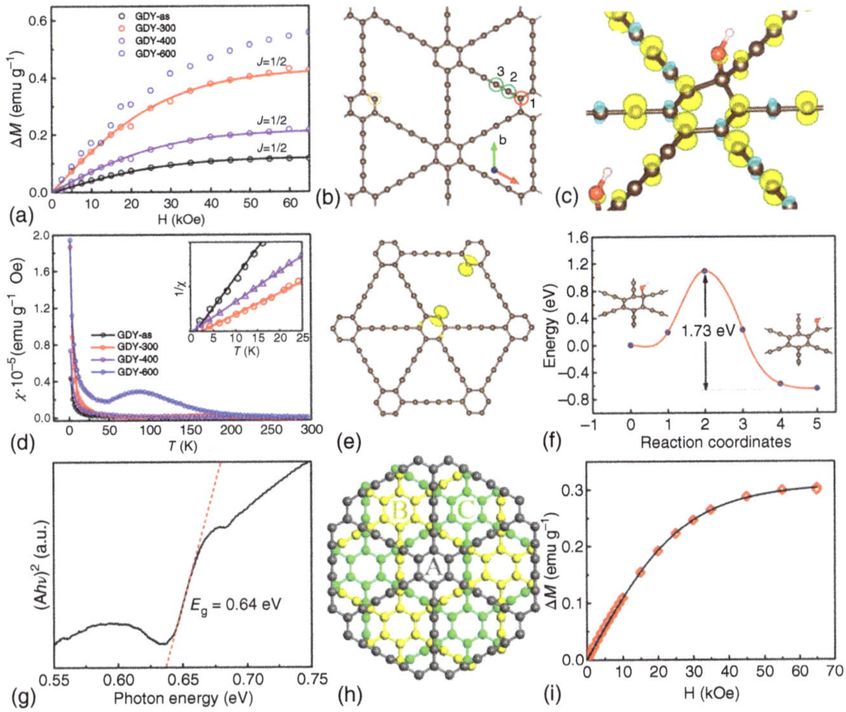

Figure 8.16 (a) Magnetic moment ΔM (after subtracting linear diamagnetic background) as a function of the applied field H measured at 2 K. Symbols are the measurements, and the solid curves are fitted to the Brillouin function. (b) Scheme of functional groups of hydroxyl (OH), carbonyl (C=O) absorbed on the top of the ring site (red as site 1), and chain site (green as sites 2 and 3), epoxide (C–O–C) on the middle of different color sites, and single vacancy model of the GDY sheet. As for C–N chemical environments model, the N element substitutes C element on the color circled site and the pyridine (orange) site. (c) Spin charge density distribution for hydroxyl absorption. (d) Typical χ^{-1} (T) curves under the applied field $H = 1$ kOe. Insets are the corresponding χ^{-1} (T) curves. Symbols are the measurements, and the solid lines are fitted by the Curie–Weiss law. (e) Spin charge density distribution for vacancy. (f) The calculated migration energy barrier of hydroxyl on the GDY sheet migrating from the ring site to the chain site. Source: Zheng et al. [43]. (g) E_g of pGDY, plots of $(Ah\nu)^2$ vs. photon energy ($h\nu$). (h) GDY multilayers with ABC stacking rom the top view. (i) ΔM–H curve of pGDY measured at 2 K. Symbols are the measurements, and the solid curve is fitted to the Brillouin function with $S = 1/2$ and $g = 2$. Source: Lin et al. [44]/Elsevier.

element doping (Figure 8.17). Based on DFT calculations, in the case of substitution with asymmetric pyridinium nitrogen, a larger local magnetic moment value of 0.98 μ_B might be obtained. According to the relative change of the measured magnetic moment between N-GDY and raw GDY, it can be concluded that the form of Py-1N may account for a proportion of 13.6% among the possible five kinds of N-doping positions, further revealing the important role of pyridinic nitrogen in improving magnetic properties of chemically modified GDY. In subsequent studies, the hybrid composite system (Fe/GDY) of GDY and transition metal (Fe)

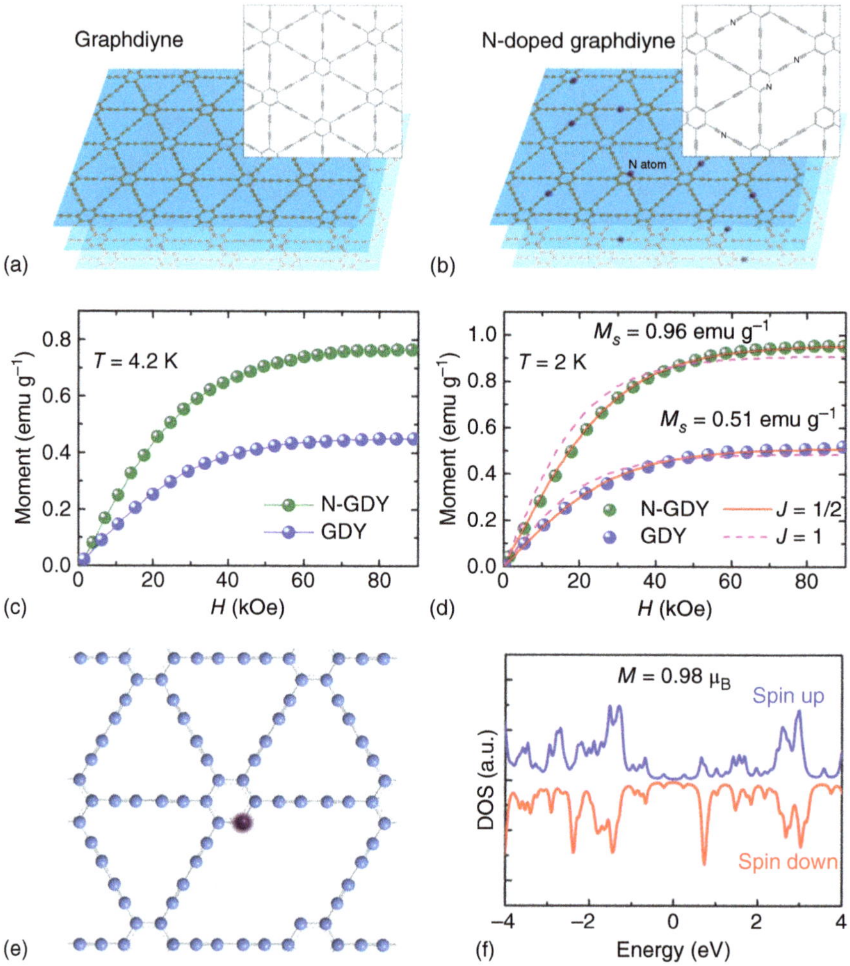

Figure 8.17 (a) Magnetic moment ΔM (after subtracting linear diamagnetic background) as a function of the applied field H measured at 2 K. Symbols are the measurements, and the solid curves are fitted to the Brillouin function. (b) Typical χ^{-1} (T) curves under the applied field $H = 1$ kOe. Insets are the corresponding χ^{-1} (T) curves. Symbols are the measurements, and the solid lines are fitted by the Curie–Weiss law. (c) Scheme of functional groups of hydroxyl (OH), carbonyl (C=O) absorbed on the top of the ring site (red as site 1) and chain site (green as sites 2 and 3), epoxide (C–O–C) on the middle of different color sites, and single vacancy model of the GDY sheet. As for C–N chemical environments model, the N element substitutes C element on the color circled site and the pyridine (orange) site. (d) Spin charge density distribution for vacancy. (e) Spin charge density distribution for hydroxyl absorption. (f) The calculated migration energy barrier of hydroxyl on the GDY sheet migrating from the ring site to the chain site. Source: Zhang et al. [48]/Springer Nature/CC BY 4.0.

exhibited robust ferromagnetic characteristics, and the observed ferromagnetism originated from the remarkable local magnetic moment as well as the electron transfer between carbon atoms and ferric ions [49]. The magnetic data indicate the appearance of ferromagnetism at low temperature for the raw Fe/GDY, as well as the significant ferromagnetic ordering with Curie temperature well above 300 K for the annealed Fe/GDY. The coexistence of ferromagnetism and semiconductivity can not only promote the exploration of new physical properties of GDY, but may also open up new ways for fabricating spintronic devices such as magnetic switching or information storage.

Chemical modification [50] with rational material design [51] has long been considered to be effective approaches to introduce ferromagnetic order. For carbon semiconductors, it is equally crucial to achieve ferromagnetism through chemical modification while maintaining its structure and energy band characteristics. An atomic-level uniformly doping/substitution method using a monomer design strategy [52–54] has been employed to accurately introduce different light elements (H, F, Cl) into the benzene ring of GDY to obtain three carbon-based nanomaterials (HsGDY, FsGDY, and ClsGDY) [55] (Figure 8.18). Both experiments and

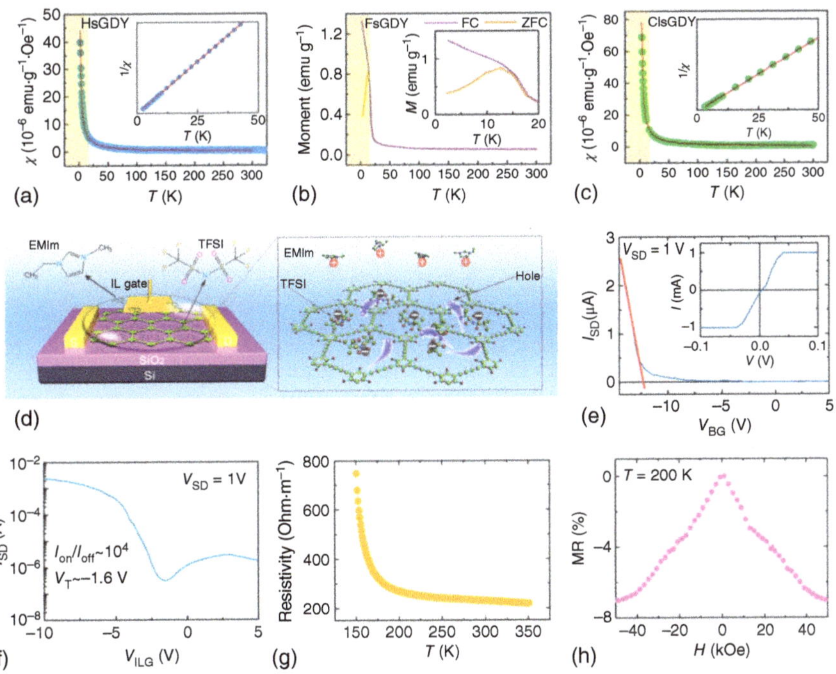

Figure 8.18 Typical χ^{-1} (T) curves of (a) HsGDY, (b) FsGDY, and (c) ClsGDY with the applied magnetic field $H = 2000$ Oe. Red lines correspond to the fitting by Curie law. The M–T curves of FsGDY are measured by zero-field cooling (ZFC) and field cooling (FC), respectively. (d) Schematic diagram of the flexible FET device based on FsGDY film using ionic liquid as the gate. Transfer curves of FsGDY FET fabricated on SiO_2/Si substrate with (e) SiO_2 as gate and (f) ionic liquid as gate, respectively. (g) Temperature-dependent resistivity curve of flexible FsGDY. (h) Magnetoresistance of flexible FsGDY measured at 200 K [55]. Source: Zhang et al. [55]/American Chemical Society.

spin-polarized theory calculations have confirmed that considerable magnetic ordering and enhanced carrier transport properties are closely correlated with electronegativity. Different from the typical paramagnetism in HsGDY and ClsGDY, a robust ferromagnetic order in FsGDY below 16 K has been observed due to the unique electronegativity of fluorine, exhibiting a distinct influence of heteroatom modification. FsGDY exhibits robust ferromagnetism and a high mobility of up to $320\,cm^2\,V^{-1}\,s^{-1}$ to achieve the coexistence of ferromagnetism and semiconductivity. The flexible prototype device manufactured by combining the exceptional film-forming properties and electromagnetic properties of FsGDY film achieves a negative magnetoresistance effect of 8%. All of these carbon matrix materials retain their excellent intrinsic semiconductor properties, and the unique ferromagnetic coupling can be attributed to the most striking charge transfer between carbon and fluorine atoms, demonstrating the advantages of controllable fabrication. The future applications of these 2D carbon materials will be in spintronics such as magnetoresistive random access memory and magnetic switching.

As a 2D full carbon nanomaterial, GDY has many unique advantages that make it a promising 2D carbon material, which has valuable application prospects in electronics, optical devices, thermoelectric, and magnetic equipment. Certainly, there are still some obstacles to the development of GDY materials for their applications in this field. At present, the key challenge we deemed is to realize the standardized synthesis and large-scale industrial production, which would enable GDY to be investigated more widely. Besides, synthesizing high-quality GDY films to obtain theoretically predicted excellent electronic properties is the key to their efficient utilization. Moreover, constructing various high-quality and high-stability GDY nanostructures to meet actual needs is a guarantee for their wide application.

References

1 Li, G.X., Li, Y.L., Liu, H.B. et al. (2010). Architecture of graphdiyne nanoscale films. *Chemical Communications* 46: 3256–3258.

2 Li, Y., Xu, L., Liu, H. et al. (2014). Graphdiyne and graphyne: from theoretical predictions to practical construction. *Chemical Society Reviews* 43: 2572–2586.

3 Matsuoka, R., Sakamoto, R., Hoshiko, K. et al. (2017). Crystalline graphdiyne nanosheets produced at a gas/liquid or liquid/liquid interface. *Journal of the American Chemical Society* 139: 3145–3152.

4 Peng, Q., Dearden, A.K., Crean, J. et al. (2014). New materials graphyne, graphdiyne, graphone, and graphane: review of properties, synthesis, and application in nanotechnology. *Nanotechnology, Science and Applications* 7: 1–29.

5 Bai, H.C., Zhu, Y., Qiao, W.y. et al. (2011). Structures, stabilities and electronic properties of graphdiyne nanoribbons. *RSC Advances* 1: 768–775.

6 Luo, G., Qian, X., Liu, H. et al. (2011). Quasiparticle energies and excitonic effects of the two-dimensional carbon allotrope graphdiyne: theory and experiment. *Physical Review B* 84: 075439.

7 Hybertsen, M.S. and Louie, S.G. (1986). Electron correlation in semiconductors and insulators: band gaps and quasiparticle energies. *Physical Review B: Condensed Matter* 34: 5390–5413.

8 Pan, Y., Wang, Y., Wang, L. et al. (2015). Graphdiyne-metal contacts and graphdiyne transistors. *Nanoscale* 7: 2116–21127.

9 Dong, M.M., Zhang, G.P., Wang, Z.Q. et al. (2020). Pervasive Ohmic contacts of monolayer 4-hT^2 graphdiyne transistors. *Nanotechnology* 31: 225705.

10 Qian, X., Liu, H., Huang, C. et al. (2015). Self-catalyzed growth of large-area nanofilms of two-dimensional carbon. *Scientific Reports* 5: 7756.

11 Qian, X., Ning, Z., Li, Y. et al. (2012). Construction of graphdiyne nanowires with high-conductivity and mobility. *Dalton Transactions* 41: 730–733.

12 Li, Y., Zhang, M., Hu, X. et al. (2020). Light and heat triggering modulation of the electronic performance of a graphdiyne-based thin film transistor. *Journal of Physical Chemistry Letters* 11: 1998–2005.

13 Zhang, M.J., Li, Y., Li, X.D. et al. (2020). Graphdiyne ink for ionic liquid gated printed transistor. *Advanced Electronic Materials* 6: 2000157. https://doi.org/10.1002/aelm.202000157.

14 Cui, W., Zhang, M., Wang, N. et al. (2017). High-performance field-effect transistor based on novel conjugated P-o-fluoro-P-alkoxyphenyl-substituted polymers by graphdiyne doping. *Journal of Physical Chemistry C* 121: 23300–23306.

15 Zhou, J., Gao, X., Liu, R. et al. (2015). Synthesis of graphdiyne nanowalls using acetylenic coupling reaction. *Journal of the American Chemical Society* 137: 7596–7599.

16 Wu, J.H., Luk'yanchuk, B., Chen, H.L. et al. (2009). Light-driven acoustic band gap based on metal nanospheres. *Advanced Materials Research* 74: 17–20.

17 Long, M.Q., Tang, L., Wang, D. et al. (2011). Electronic structure and carrier mobility in graphdiyne sheet and nanoribbons: theoretical predictions. *ACS Nano* 5: 2593–2600.

18 Yu, T., Zhu, Y.W., Xu, X.J. et al. (2005). Controlled growth and field-emission properties of cobalt oxide nanowalls. *Advanced Materials* 17: 1595–1599.

19 Dong, Y., Semin, S., Feng, Y. et al. (2021). Solvent induced enhancement of nonlinear optical response of graphdiyne. *Chinese Chemical Letters* 32 (1): 525–528.

20 Guo, J., Wang, Z., Shi, R. et al. (2020). Graphdiyne as a promising mid-infrared nonlinear optical material for ultrafast photonics. *Advanced Optical Materials* 8: 2000067.

21 Feng, Y., Dong, N., Wang, G. et al. (2015). Saturable absorption behavior of free-standing graphene polymer composite films over broad wavelength and time ranges. *Optics Express* 23: 559–569.

22 Lu, S.B., Miao, L.L., Guo, Z.N. et al. (2015). Broadband nonlinear optical response in multi-layer black phosphorus: an emerging infrared and mid-infrared optical material. *Optics Express* 23: 11183–11194.

23 Wang, K., Wang, J., Fan, J. et al. (2013). Ultrafast saturable absorption of two-dimensional MoS_2 nanosheets. *ACS Nano* 7: 9260–9267.

24 Jiang, X., Liu, S., Liang, W. et al. (2018). Broad band nonlinear photonics in few-layer MXene Ti_3C_2Tx (T = F, O, or OH). *Laser & Photonics Reviews* 12: 1700229.

25 Jiang, X., Zhang, L., Liu, S. et al. (2018). Ultrathin metal-organic framework: an emerging broadband nonlinear optical material for ultrafast photonics. *Advanced Optical Materials* 6: 1800561.

26 Ge, Y., Zhu, Z., Xu, Y. et al. (2018). Broadband nonlinear photoresponse of 2D TiS_2 for ultrashort pulse generation and all-optical thresholding devices. *Advanced Optical Materials* 6: 1701166.

27 Guo, J., Shi, R., Wang, R. et al. (2020). Graphdiyne-polymer nanocomposite as a broadband and robust saturable absorber for ultrafast photonics. *Laser & Photonics Reviews* 14: 1900367.

28 Zhao, Y., Guo, P., Li, X. et al. (2019). Ultrafast photonics application of graphdiyne in the optical communication region. *Carbon* 149: 336–341.

29 Wu, L., Dong, Y., Zhao, J. et al. (2019). Kerr nonlinearity in 2D graphdiyne for passive photonic diodes. *Advanced Materials* 31: 1807981.

30 Jiang, X., Zhao, X., Bao, W. et al. (2020). Graphdiyne nanosheets for multicolor random lasers. *ACS Applied Nano Materials* 3: 4990–4996.

31 Ge, C., Chen, J., Tang, S. et al. (2019). Review of the electronic, optical, and magnetic properties of graphdiyne: from theories to experiments. *ACS Applied Materials & Interfaces* 11: 2707–2716.

32 Zheng, Z., Fang, H., Liu, D. et al. (2017). Nonlocal response in infrared detector with semiconducting carbon nanotubes and graphdiyne. *Advanced Science* 4: 1700472.

33 Li, Y., Kuang, D., Gao, Y. et al. (2020). Titania: graphdiyne nanocomposites for high-performance deep ultraviolet photodetectors based on mixed-phase MgZnO. *Journal of Alloys and Compounds* 825: 153882.

34 Huang, Z., Yu, Z., Li, Y. et al. (2018). ZnO ultraviolet photodetector modified with gaphdiyne. *Acta Physico-Chimica Sinica* 34: 1088–1094.

35 Jin, Z., Zhou, Q., Chen, Y. et al. (2016). Graphdiyne: ZnO nanocomposites for high-performance UV photodetectors. *Advanced Materials* 28: 3697–3702.

36 Yi, X., Ren, Z., Chen, N. et al. (2017). TiO_2 nanocrystal/perovskite bilayer for high-performance photodetectors. *Advanced Electronic Materials* 3: 1700251.

37 Jiang, P.H., Liu, H.J., Cheng, L. et al. (2017). Thermoelectric properties of γ-graphyne from first-principles calculations. *Carbon* 113: 108–113.

38 Ouyang, T. and Hu, M. (2014). Thermal transport and thermoelectric properties of beta-graphyne nanostructures. *Nanotechnology* 25: 245401.

39 Mortazavi, B., Makaremi, M., Shahrokhi, M. et al. (2018). N-graphdiyne two-dimensional nanomaterials: semiconductors with low thermal conductivity and high stretchability. *Carbon* 137: 57–67.

40 Tan, X., Shao, H., Hu, T. et al. (2015). High thermoelectric performance in two-dimensional graphyne sheets predicted by first-principles calculations. *Physical Chemistry Chemical Physics* 17: 22872–22881.

41 Sun, L., Jiang, P.H., Liu, H.J. et al. (2015). Graphdiyne: a two-dimensional thermoelectric material with high figure of merit. *Carbon* 90: 255–259.

42 Kamalakar, M.V., Groenveld, C., Dankert, A. et al. (2015). Long distance spin communication in chemical vapour deposited graphene. *Nature Communications* 6: 6766.

43 Zheng, J., Engelhard, M.H., Mei, D. et al. (2017). Electrolyte additive enabled fast charging and stable cycling lithium metal batteries. *Nature Energy* 2: 17012.

44 Lin, L., Pan, H., Chen, Y. et al. (2019). Identifying the stacking style, intrinsic bandgap and magnetism of pristine graphdyine. *Carbon* 143: 8–13.

45 Yazyev, O.V. and Helm, L. (2007). Defect-induced magnetism in graphene. *Physical Review B* 75: 125408.

46 Boukhvalov, D.W. and Katsnelson, M.I. (2011). sp-electron magnetic clusters with a large spin in graphene. *ACS Nano* 5: 2440–2446.

47 Ma, Y.C., Foster, A.S., Krasheninnikov, A.V. et al. (2005). Nitrogen in graphite and carbon nanotubes: magnetism and mobility. *Physical Review B* 72: 205416.

48 Zhang, M., Wang, X., Sun, H. et al. (2017). Enhanced paramagnetism of mesoscopic graphdiyne by doping with nitrogen. *Scientific Reports* 7: 11535.

49 Zhang, M., Wang, X., Sun, H. et al. (2018). Preparation of room-temperature ferromagnetic semiconductor based on graphdiyne-transition metal hybrid. *2D Materials* 5: 035039.

50 Allouche, F., Lapadula, G., Siddiqi, G. et al. (2017). Magnetic memory from site isolated Dy(III) on silica materials. *ACS Central Science* 3: 244–249.

51 Gui, X., Pletikosic, I., Cao, H. et al. (2019). A new magnetic topological quantum material candidate by design. *ACS Central Science* 5: 900–910.

52 Jia, Z., Zuo, Z., Yi, Y. et al. (2017). Low temperature, atmospheric pressure for synthesis of a new carbon Ene-yne and application in Li storage. *Nano Energy* 33: 343–349.

53 Wang, N., He, J., Tu, Z. et al. (2017). Synthesis of chlorine-substituted graphdiyne and applications for lithium-ion storage. *Angewandte Chemie International Edition* 56: 10740–10745.

54 He, J., Wang, N., Yang, Z. et al. (2018). Fluoride graphdiyne as a free-standing electrode displaying ultra-stable and extraordinary high Li storage performance. *Energy & Environmental Science* 11: 2893–2903.

55 Zhang, M., Wang, X., Sun, H. et al. (2020). Induced ferromagnetic order of graphdiyne semiconductors by introducing a heteroatom. *ACS Central Science* 6: 950–958.

9

Graphdiyne-Based Materials in Sensors and Separation Applications

Yanbing Guo, Chuanqi Pan, and Yuhua Zhu

College of Chemistry, Institute of Environmental and Applied Chemistry, Key Laboratory of Pesticide & Chemical Biology of Ministry of Education, Central China Normal University,152 Luoyu Road, Hongshan District, Wuhan 430079, P. R. China

Graphdiyne (GDY), ever since its discovery, has quickly become a material under spotlight and been applied in various fields. Owing to the unique structural properties and outstanding conductivity, GDY and GDY-based materials demonstrate prominent specificity and superhigh sensitivity in sensor applications [1]. In addition, GDY-based materials show high permeability and selectivity in water desalination, oil water separation, and gas separation because of their thickness of an atomic layer, good mechanical properties, and well-defined nanosized pores [2]. This chapter focuses on introducing the applications of GDY-based materials in sensors and separation.

9.1 Sensors

Sensor is a device that can sense a specified measurement and further convert into a usable output signal according to a certain rule, which is widely equipped in many fields. GDY, as a new two-dimensional (2D) carbon material, has been promising candidate in the construction of sensing devices due to the large specific surface area, good electrical conductivity, and strong adsorption capacities [3, 4]. Because of these unique properties, GDY-based materials exhibit better performance and higher stability than other carbon-based materials for improving the detection capability of sensing devices. However, the research of GDY-based materials in sensing is still in the primary stage, which is mainly reported to detect biomolecules, gas/vapors, and small molecules.

9.1.1 Biomolecules Sensor

9.1.1.1 DNA Detection

Biosensor is an analytical device that converts the biological response into the optical/electrical signal [3]. Biomolecules play indispensable roles in all biological

Graphdiyne: Fundamentals and Applications in Renewable Energy and Electronics,
First Edition. Edited by Yuliang Li.

processes and clinical analyses, so it is critical to accurately identify and detect of biomolecules for disease diagnosis, medical treatment, and food safety [5]. 2D materials have been emerging as promising platforms for biosensing owing to planar structure and ultra-large surface. GDY is a new kind of 2D carbon material with sp- and sp^2-hybridized carbon atoms and a large π-conjugated surface [6], which makes biomolecules easily self-assemble onto the surface of GDY-based materials via π–π stacking and/or hydrophobic interactions [3, 4]. In addition, the biosensor via fluorescence resonance energy transfer and photo-induced electron transfer mechanisms makes these 2D materials feasible for biomolecules detection [7–9]. The good light absorption and/or electron conduction abilities of GDY make GDY-based materials efficient fluorescence quenchers of dye-labeled biomolecules. Therefore, unique structural and optical properties enable GDY to be widely used as fluorescence sensors of biomolecules such as DNA and proteins.

GDY-based materials were first used as fluorescence sensor of biomolecules by Mao's and Li's groups in 2016 [10]. As shown in Figure 9.1a, GDY and GDY oxide were prepared by acid treatment method and were applied as the nanoquenchers for the fluorescent sensing, such as DNA and proteins. The research showed GDY and its oxide could quench the fluorescence of organic dyes-labeled single-stranded DNA (ssDNA) probe via Van der Waals force and π–π stacking interaction between nucleobases and GDY. 5-carboxyfluorescein (FAM)-labeled ssDNA was designed to be complementary to hepatitis C virus target (T_{HCV}), and the FAM-labeled ssDNA was used as the probe (probe of hepatitis C virus sequence [P_{HCV}], 5′-FAM-TAA ACC CGC TCA ATG CCT GGA-3′). When GDY and its oxide were added into the solution, the fluorescence intensity of FAM was effectively quenched (by about 97%), as depicted in Figure 9.1b. Adding the target ssDNA to the mixture created a double-stranded DNA, the probe ssDNA was hybridized with the target DNA, which weakens its interaction with GDY and enabled the fluorescence restoration. Besides, GDY oxide exhibited a higher quenching efficiency toward P_{HCV} than GDY, which can probably be attributed to the fact that the GDY oxide had the high content of hydroxyl and carboxyl groups, which can enable physisorption of nucleobases on the surface of GDY oxide and improve the dispersity of GDY oxide in water. Since the publication of this first paper in biosensor, GDY-based materials have been widely investigated in this field. Density functional theory (DFT) has been carried out to estimate interaction of nucleobases (NBs) and base pairs (BPs) by Swathi's group [12]. The calculation results suggested that NBs and BPs had a strong binding with GDY. The interaction energies of $C_{90}H_{18}$ (a superstructure of GDY) with NBs and BPs can reach −15.7 and −23.4 kcal mol^{-1}, respectively, which demonstrated that GDY can serve as promising templates for the self-assembly of DNA and RNA.

Furthermore, it is crucial to perform the real-time analysis of multiplexed DNA in the practical detection process. Parvin et al. firstly synthesized novel few-layered 2D graphdiyne nanosheets (GDY NSs) and applied them as fluorescent sensing platform for the real-time detection of DNA [13]. Compared with graphene oxide and MoS_2 nanomaterials-based sensors, the GDY NSs-based biosensor showed the best performance for detection of multiplexed DNA (limit of detection as low as 25 pM).

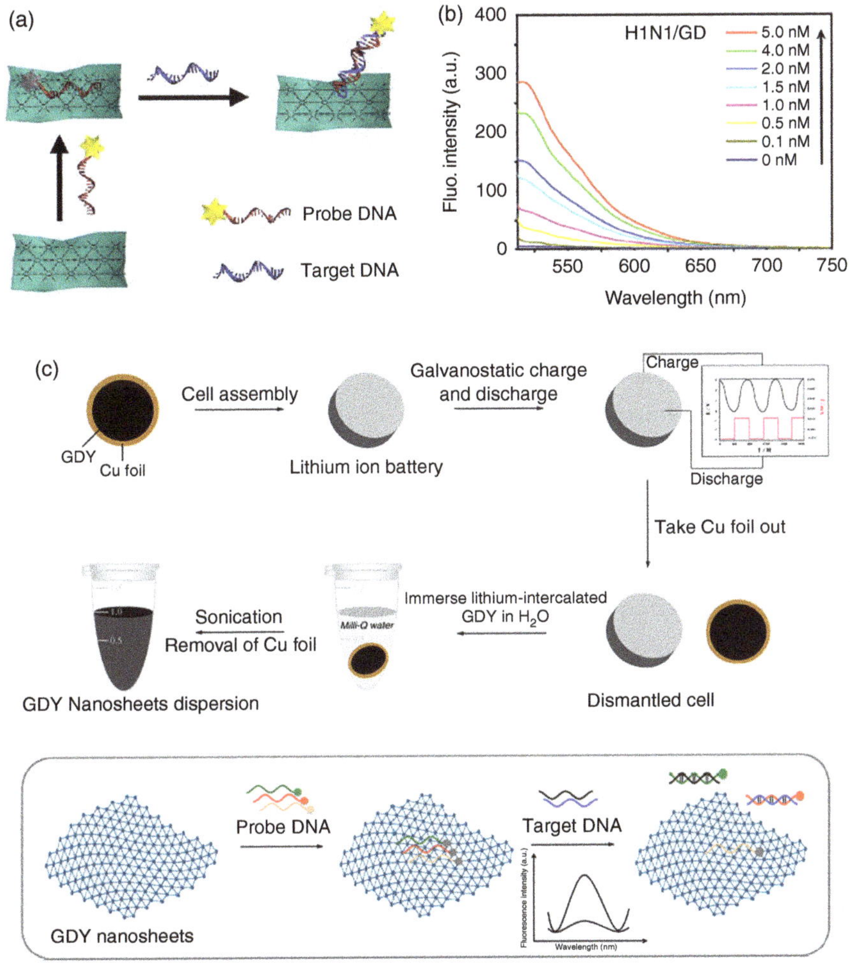

Figure 9.1 (a) Schematic illustration of the GDY-based fluorometric DNA assay. (b) Fluorescence spectra of P_{HCV} (20 nM, black curve), P_{HCV} (20 nM) with addition of GDY oxide (10 μg ml^{-1}, red curve), and P_{HCV} with addition of GD oxide and then hybridized with T_{HCV} (40 nM, blue curve). (c) Scheme for the preparation of GDY NS and GDY-based multiplex DNA analysis. Source: (a, b) Wang et al. [10]. © 2016, Royal Society of Chemistry. (c) Chang et al. [11]. © 2019, American Chemical Society.

Moreover, DFT simulations were carried out to better understand the interactions between dye molecule and GDY. The adsorption energies of FAM with GDY (−344.48 kJ mol^{-1}) were much stronger than those of graphene (−192.98 kJ mol^{-1}), and the optimized interaction distance between FAM and GDY and graphene was found to be 3.583 and 3.594 Å, respectively. The stronger interaction of GDY with FAM was mainly due to the presence of ethyne that consisted of a pair of π bonds in the triple bond of GDY. The research provided a new approach for rapid, facile, and cost-effective multiplexed detection of biological molecules.

Recently, a few-layered GDY NSs have been further applied for the detection of Mycobacterium tuberculosis (Mtb) and its drug-resistant genes [11]. The researchers used an electrochemical lithium intercalation strategy, the thickness of prepared few-layered GDY NSs was only 0.9 nm, which was the thinnest GDY NSs reported in the literature, Figure 9.1c shows the scheme for the preparation of GDY NS and GDY-based multiplex DNA analysis. This GDY NSs sensing platform demonstrated prominent specificity and superhigh sensitivity (limit of detection 25 pM), and they can be engaged to multiplex and quantitative DNA detections. Additionally, it was also allowed to identify DNA mismatch bases and effectively avoided nonspecific interactions. The GDY-based DNA fluorescent sensing platform with a low background and a high signal-to-noise ratio was simple, economical, and time saving, which suggested its potential applications in pathogen diagnosis, clinical drug guidance, and drug resistance monitoring.

9.1.1.2 RNA and Amino Acids Detection

MicroRNAs are endogenous and noncoding RNAs that play a crucial role in cancer diagnosis and treatment. Many researchers have put great efforts on realizing ultrasensitive detection of microRNAs [14]. Photoelectrochemical (PEC) methods are the most broadly used method for constructing assays due to high sensitivity and a low background. GDY has a natural band gap and high electron conductivity, which endows GDY a great application prospect in PEC assay. To the best of our knowledge, the PEC biosensor based on GDY for microRNAs detection was firstly reported by Li et al. [15]. A PEC platform based on AuNPs-loaded GDY was constructed for microRNA let-7a detection as seen in Figure 9.2a. Due to the natural band-gap structure of GDY, the fabricated material could produce hole-electron pairs and make for the plasmon resonance effect. Under optimal condition (potential 300 mV and pH 8.0), the PEC biosensor had a detection limit of 3.3×10^{-19} M, and a good linearity with microRNA let-7a concentration ranged from 1.0×10^{-18} to 1.0×10^{-10} M. In the same year, the group synthesized molybdenum disulfide-graphdiyne (MoS_2-GDY) and successfully applied to the determination of microRNA-141 [16]. The combination of GDY and MoS_2 reduced the electron-hole recombination rate so as to obtain stronger PEC signal. The PEC properties of MoS_2-GDY were characterized by chronoamperometry using three-electrode system. The 1-μM signal amplification amino acid (AA) was selected as electron donor. GDY-modified gold electrode (GE) showed PEC response of 120.3 nA, which was 3.7-fold of the bare GE (32.6 nA), illustrated that GDY had photoactivity. As shown in Figure 9.2b, once the GE was modified with MoS_2-GDY, the PEC signal sharply increased to 3500.9 nA, which was 29.1-fold of that of the GDY-modified GE. The possible mechanism of MoS_2-GDY PEC signal increment is depicted in Figure 9.2c under the light excitation, the photoelectrons produced by GDY transferred to conduction band (CB) of MoS_2. The photogenerated holes produced by MoS_2 migrated to the valence band of GDY, which resulted in the enhancement of the separation of electrons and holes, and the decrease of the recombination rate. Additionally, the highly π-conjugated structure of GDY possessed electrical conductivity and increased the trend of electron flow from GDY to MoS_2, so that

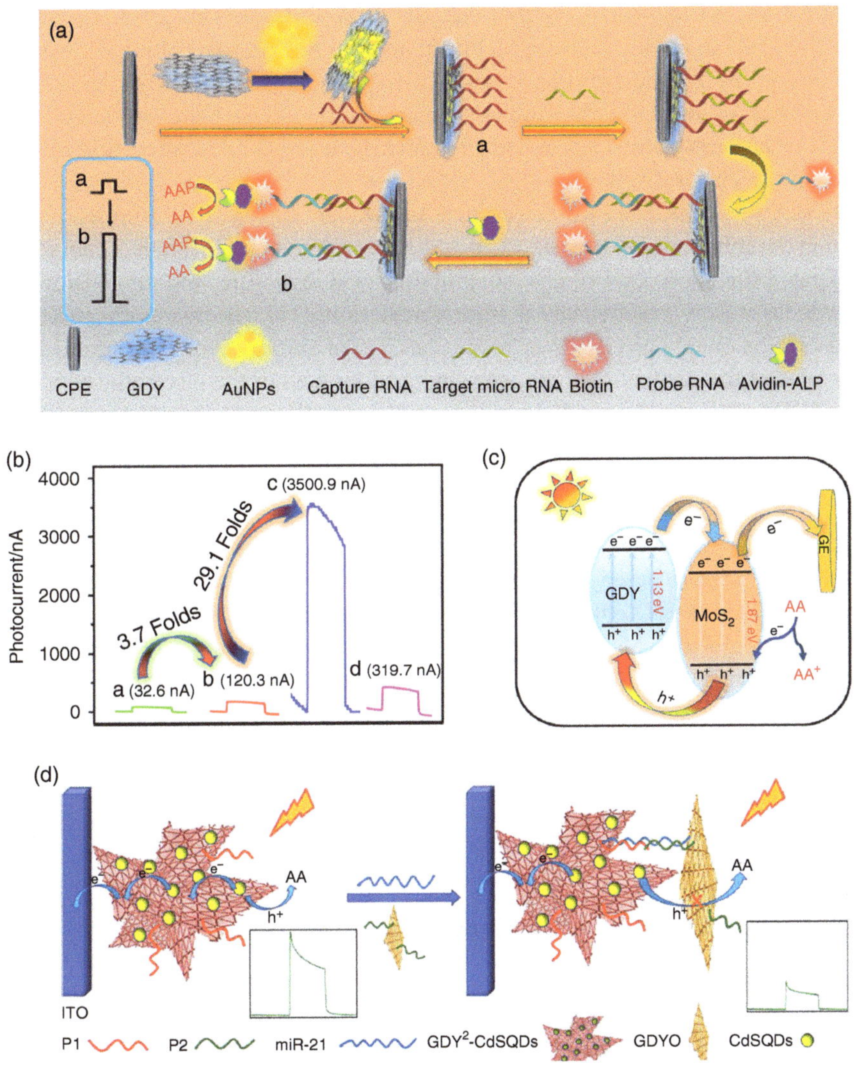

Figure 9.2 (A) Schematic illustration of PEC biosensor based on AuNPs-GDY modified electrode for let-7a microRNA detection. (B) The PEC responses of gold electrode (GE) include GDY/GE; MoS_2-GDY/GE; MoS_2/GE. The PEC test was carried out in pH 7.4 phosphate buffer containing 10 μM AA. (C) Photophysical processes after loading MoS_2 onto GDY. (D) Illustration of the principle for the determination of miR-21. Source: (A) Li et al. [15]. © 2019, Elsevier. (B,C) Li et al. [16]. © 2019, Elsevier. (D) Wang et al. [17]. © 2020, Elsevier.

the recombination rate of electrons and holes was further decreased. Besides, Wang et al. reported acidified tremella-like GDY nanotubes with high dispersibility and synthesized a composite of graphdiyne and cadmium sulfide quantum dots (GDY-CdSQDs) by one-step hydrothermal method [17]. GDY-CdSQDs served as the PEC platform and graphdiyne oxide (GDYO) nanosheets served as the signal inhibitor. The sandwich-type PEC sensor was employed to quantitatively detect microRNA-21. As shown in Figure 9.2d, the GDY-CdSQDs and DNA-capturing probe 1 (P1) acted as PEC sensing platform, and the target object of microRNA-21 was captured by P1 and hybridized with P2-GDYO. The GDY-CdSQDs with high photocurrent response to light irradiation made it a good PEC platform. Therefore, the PEC biosensors opened a promising platform by using GDY to detect microRNA at ultralow levels for diagnoses.

GDY can also be utilized for detecting AAs. AAs are the basic substances of life. Studying the interactions between GDY and AAs is of great importance. Chen et al. investigated the adsorption of glycine, glutamic acid, histidine, and phenylalanine on single-layer GDY/graphene by ab initio calculations [18]. The results showed that the adsorption energy on GDY was larger than that of graphene for each amino acid molecule and the distinct changes were induced by AAs in the electronic conductivity of GDY. This study provided a guidance to future theoretical and experimental study, suggested that GDY was a promising two-dimensional material for sensitively detecting AAs.

In addition, GDY was explored as robust biosensing platform for enzyme immobilization and biosensor fabrication. Wu et al. established an electrochemical tyrosinase biosensor based on GDY for ultrasensitive detection of bisphenol A (BPA). The response of GDY-based tyrosinase biosensor was linear over the range of 1.0×10^{-7}–3.5×10^{-6} M with a high sensitivity of 2990.8 mA cm^{-2} M^{-1} and a low detection limit (24 nM) [19]. The excellent performance of the biosensor was mainly due to the strong π–π interactions between GDY and BPA, which could enrich available BPA concentration on the electrode to react with tyrosinase. Therefore, GDY played an important role for immobilizing tyrosinase and improving the electrochemical performance of fabricated biosensor, which proves that GDY is a potential electrochemical biosensing platform for enzyme-based biosensors.

9.1.2 Small-Molecule Detection Sensor

9.1.2.1 Gas Sensor

Gas-sensing technology receives increasing attention because of its widespread applications such as industrial production and safety, indoor air quality supervision, environmental monitoring, and medical diagnosis. Sensing materials play a significant role in gas sensors [20, 21]. GDY-based materials are predicted to be excellent small-molecule sensing materials due to their high specific surface area and unique electrical properties such as high mobility and low electrical noise. Up to now, the published articles on studying this exciting field of GDY materials-based gas sensors are mainly focus on theoretical calculations. Nagarajan and Chandiramouli studied the adsorption behavior of ammonia molecules on GDY nanosheets and nanotubes by DFT [22]. The highest occupied molecular orbital (HOMO) and lowest unoccupied molecular orbital (LUMO) gap varies upon interaction with ammonia gas molecules on GDY nanostructure, and the interaction of ammonia on GDY nanosheet was observed to be more prominent than GDY nanotube. The findings supported the use of GDY nanosheets for the detection of ammonia molecules. Ebadi and Reisi-Vanani investigated the adsorption behavior of methanol and CO molecules on the pristine GDY and Ca-decorated GDY (Ca-GDY) by modified density functional theory (DFT–D2) calculations [23]. It was seen that methanol and CO molecules were physisorbed on the pristine GDY. Methanol (CH_3OH) and CO capture capacities remarkably increased over Ca-GDY, which reached about 29.81 and 27.10 wt%, respectively. These researches show that the Ca-GDY is a prominent candidate for capture and storage of methanol and CO molecules in practical applications.

GDY-based materials were also predicted for detecting hazardous gases. Nagarajan et al. investigated the adsorption behavior of dimethyl amine (DMA) and trimethyl amine (TMA) molecules on GDY nanosheet by DFT modeling. The adsorption of the DMA and TMA molecules on triangular pores of GDY nanosheet was observed to be optimum. Comparing the energy band gap of DMA/TMA vapor molecules on GDY nanosheet (0.14 eV) with that of GDY (0.5 eV), the energy band gap of DMA/TMA vapor molecules on GDY nanosheet dramatically decreased. The results clearly revealed that the transfer of electron took place between DMA/TMA vapor molecules and GDY nanosheet. The changes in adsorption energy and energy band gap upon adsorption of DMA/TMA vapor molecules on GDY nanosheet suggested that GDY can be effectively applied for the detection of both DMA and TMA vapors [24]. The detection of GDY for formaldehyde (HCHO) and formic acid has also been studied in depth [25]. The two target vapors, HCHO and formic acid, were adsorbed on GDY nanosheet to form three configurations (HCHO and formic acid adsorption at the ring, top, and bridge sites). The adsorption and electronic characteristics for the target vapors adsorbed on GDY nanosheet were computed and explored by adsorption energy, Bader charge transfer, and band structure. For all the configurations, the adsorption energies were negative, and the Bader charge transfer was positive. It is worth noting that the energy band gap values for HCHO and formic acid were 0 and 0.141 eV at ring sites. The results indicated that the GDY nanosheet can be employed as a preferable chief component for gas sensor in detecting HCHO and formic acid. To further improve the detection ability on gas molecules, Chen et al. proposed Sc- and Ti-decorated GDY as the promising materials for high-efficient molecular detection [26]. Figure 9.3 shows the top and side views of HCHO adsorbed on GDY, GDY-Sc, and GDY-Ti. The adsorption of HCHO on Sc- or Ti-decorated GDY (2–3 eV) was found stronger than that on pristine GDY (0.4 eV). Furthermore, for more HCHO molecules on the Sc/Ti adatom, the interactions between the original HCHO and extra HCHOs were found repulsive. In the view of energetics, such property provided a prerequisite for individual molecular sensing. In addition, the adsorption of HCHO on Sc or Ti adatom was robust to high temperature or strain on GDY sheet. The above results indicated that GDY-Sc and GDY-Ti were promising materials for HCHO molecule detection. Bhuvaneswari et al. utilized GDY nanoribbon to detect the existence of explosive vapors like hexogen or cyclonite, hexamethylene triperoxide diamine, and 2,4,6-trinitrotoluene using DFT [27]. The electronic attributes (projected density of states spectrum, band structure, and electron density) and adsorption attributes (adsorption energy, Bader charge transfer, and average

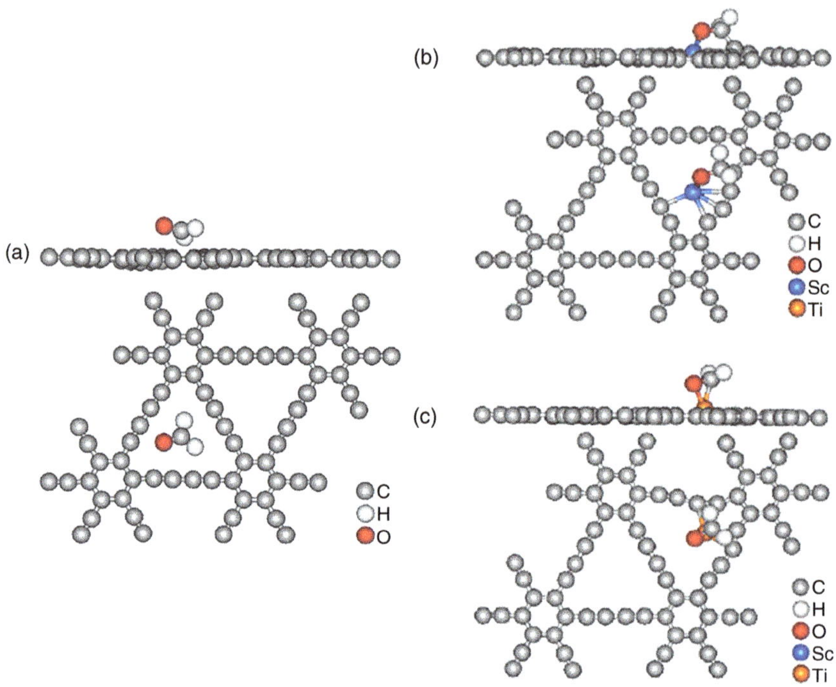

Figure 9.3 (a) Top and side views of HCHO adsorbed on GDY sheet. (b) Top and side views of HCHO adsorbed on GDY-Sc and GDY-Ti. Source: Chen et al. [26]. © 2017, Elsevier.

energy gap variation) indicated that explosive vapors were granted to interact with GDY nanoribbon. Therefore, the utilization of GDY nanoribbon in the gas sensor to detect the existence of the aforementioned explosive vapors was efficient.

Later, Hussain et al. studied the gas-sensing properties of nitrogenated holey graphene (C_2N), GDY and their van der Waals heterostructure (C_2N...GDY) toward particular volatile organic compounds (such as acetone, ethanol, propanal, and toluene) by means of spin-polarized, dispersion-corrected DFT calculations [28]. The results showed that volatile organic compounds interacted weakly with the GDY monolayer, while the bindings can be significantly enhanced with the C_2N monolayer and the hybrid C_2N...GDY heterostructure. The electronic properties of C_2N...GDY heterostructure were changed due to van der Waals formed between the volatile organic compounds (VOCs) and C_2N...GDY by electron localization function analysis. Besides, thermodynamic analysis of each of the VOCs on C_2N, GDY, and C_2N...GDY heterostructure was carried out to estimate sensing characteristics of VOCs under varied conditions of pressure and temperature. As shown in Figure 9.4a–c, C_2N...GDY heterostructure had high selectivity for toluene in comparison with C_2N or GDY only. In more diluted gas, the occupancies were shifted to lower temperature than those of saturated one owing to higher chemical potentials (Figure 9.4d–f). It means lower temperatures were needed so that the diluted VOCs molecules were captured on adsorbent. These results suggested that the C_2N...GDY

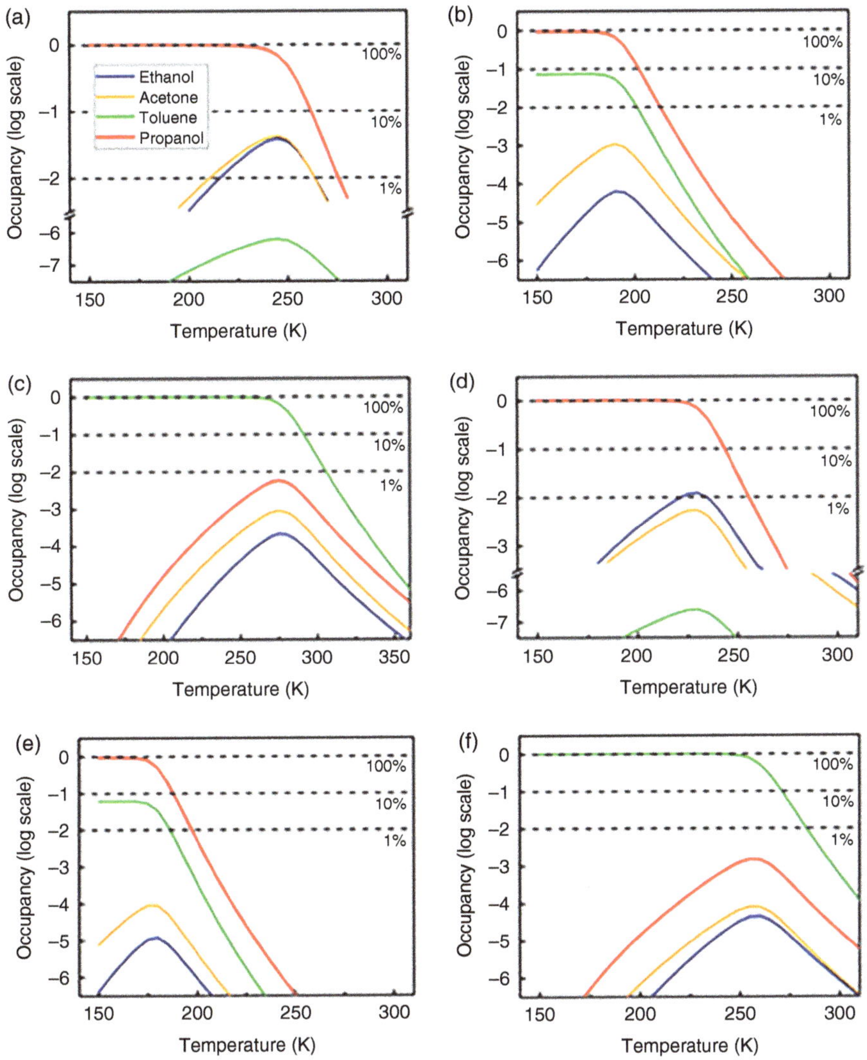

Figure 9.4 Adsorption process of VOCs mixtures on (a) C_2N monolayer, (b) GDY monolayer and (c) C_2N...GDY heterostructure layered structure. Desorption process of VOCs mixtures on (d) C_2N monolayer, (e) GDY monolayer and (f) C_2N...GDY heterostructure layered structure. Source: Hussain et al. [28]. © 2020, Elsevier.

heterostructure was a promising material for sensing of certain volatile organic compounds, especially toluene. Further insights on theoretical studies are also needed to better understand the interaction mechanism between various gases and GDY-based materials, which could facilitate the experimental/device fabrication process in advent of production of enhanced sensors. Meanwhile, it is better to unfold their intrinsic properties and pave a way for their practical application as soon as possible.

9.1.2.2 Humidity Detection

Humidity measurement is more and more essential in many fields such as agriculture, biomedical, and climatology [29, 30]. Various humidity sensors have been invented and developed to improve sensing performance. Among the materials applied in humidity sensing, carbon-based materials are particularly attractive due to their rich surface chemistry with hydrophobic skeleton and water-adsorbing capacity of extended hydrophilic functional groups [31, 32]. For example, graphene oxide (GO) humidity sensor with unique nature of 2D flakes displayed a fast response speed to water molecules because of the ultrathin porous film with superpermeability and high sensitivity toward water molecules [33]. GDY, with sp- and sp^2-hybridized carbon atoms, is forecasted to be quite competitive with sp^2-hybridized carbon systems in meeting all kinds of demands. Yan et al. demonstrated that GDYO possessed an ultrafast humidity response with a response speed of 7 ms for the first time [34]. The ultrafast humidity response of GDYO was mainly due to the stronger electron-withdrawing property of acetylenic bond than that of ethylenic bond in GO and consequently exhibited a faster binding rate with water. The humidity sensor could also be used as a real-time monitor of the respiration rate change of animals and even humans and as a noncontact sensor. This research not only proved the unprecedented humidity-sensing capability of GDYO but also provided a new application prospect for 2D carbon materials.

9.1.2.3 Hydrogen Peroxide Detection

Hydrogen peroxide (H_2O_2), as an important analyte, is widely used in various fields such as food production, biological processes, medical diagnostics, and environmental analysis [35]. Rapid, accurate, sensitive, low-cost detection of H_2O_2 is of great importance [8]. Zhuang et al. reported an in situ synthesis of Prussian blue nanoparticles (PB) on GDYO substrate, and applied them in the electrochemical detection of H_2O_2 [36]. As shown in Figure 9.5a,b, the PB/GDYO nanohybrids, as electrocatalysts for the detection of H_2O_2 in a neutral medium, demonstrated to be a reliable H_2O_2 sensor with a low detection limit (0.5 μM). The PB/GDYO nanohybrids exhibited a much higher stability than PB on GO, and the sensitivity of PB/GDYO toward H_2O_2 was about 3.3 times that of the PB/GO. The negative nature of GDYO and effective electrostatic interaction between GDYO and PB may be beneficial to improving stability.

9.1.2.4 Glucose Detection

Glucose is one of the essential nutrients and a major health indicator of the human body, especially in the surveillance of diabetes mellitus. In addition, glucose is also widely used in the food and pharmaceutical industries [38]. Therefore, glucose detection is significant for human health monitoring, industrial quality control, and processing applications. The typical methods for glucose detection include enzymatic method [39], electrochemical detection [40], and fluorescence method [41]. The enzymatic methods are widely used in glucose detection due to its high sensitivity, low detection limit, and easy operation. Iron as one of the transition metals can form Fe^{3+} and Fe^{2+}, which is an important metal ion in many cofactors of natural

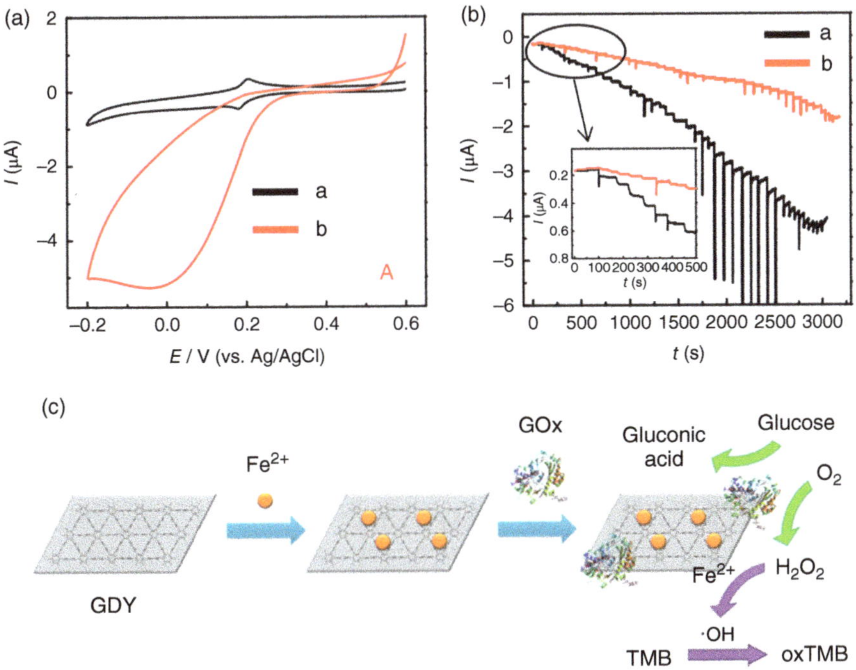

Figure 9.5 (a) Cyclic voltammetry (CVs) of the PB/GDYO-modified glassy carbon electrode (GCE) in 0.1 M phosphate buffer (pH 7.4) a and b, respectively, represent in the absence and in the presence of 10 mM H_2O_2. Scan rate: 50 mV s^{-1}. (b) Chronoamperometric responses of the PB/GDYO(black line) and PB/GO(red line) modified GCE to successive additions of H_2O_2 in 0.1 M phosphate buffer (pH 7.4) at −0.05 V. Inset: amplified calibration plot corresponding to the lower concentration range. (c) Schematic illustration to show the preparation of an Fe^{2+} and glucose oxidase-immobilized GDY composite and the mechanism for glucose detection. Source: (a–b) Zhuang et al. [36]. © 2017, Elsevier. (c) Liu et al. [37]. © 2018, American Chemical Society.

enzymes and plays a key role in the enzymatic catalytic reaction. However, Fe^{2+} was instable and easily oxidized into Fe^{3+} by oxygen in the air. To improve the activity and stability of Fe^{2+}, some researchers prepared iron-based composites. Recently, Liu et al. prepared an Fe^{2+} and glucose oxidase (GO_x)-immobilized GDY composite used for one-step glucose detection [37]. Figure 9.5c displays the schematic illustration of the preparation of an Fe^{2+} and glucose oxidase-immobilized GDY composite and the mechanism for glucose detection. The glucose detection was carried out by adding Fe-GDY/GO_x, 3,3′,5,5′-tetramethylbenzidine and different concentration of glucose solution into pH 4.5 sodium acetate buffer solution. With the increase of glucose concentration, the absorbance of reaction solution at 450 nm was increasing accordingly. The curve exerted great linearity from 5 to 160 μM of the glucose concentration with a high R-square (0.9985), which demonstrated that the as-prepared Fe-GDY/GO_x could be used as a sensitive one-step glucose detection agent. Such high peroxidase activity of Fe-GDY composites could be mainly attributed to two reasons. First, the GDY had greater adsorption capacity to

Fe^{2+} than graphene, and GDY could protect Fe^{2+} from being oxidized, which was mainly due to the heterogeneous surface energy distribution and acetylenic bonds. Second, GDY displayed the better ability of adsorbing glucose oxidase and did not obviously influence the secondary structure of the adsorbed glucose oxidase. The unique structure of GDY was probably beneficial to glucose oxidase adsorbing on its surface.

9.1.3 Other Sensors

GDY-based materials have also been employed to serve as other sensors. Wang et al. reported a patterned GDY stripe array dominated by the superlyophilic grooved templates [42]. The layer thickness of patterned GDY stripes can be manipulated from just several nanometers to hundreds of nanometers by varying the primary concentration of hexaethynylbenzene monomer to meet device requirements. This study showed a proof-of-principle demonstration by using the patterned GDY stripe arrays as a stretchable sensor to monitor the human finger motion. Following the bending motion of the finger and recovery, the sensor can be switched and can generate responsive current signal waves. Hence, the finger movement process could be detected by monitoring the change of the current values. This wettability-facilitated strategy and precisely patterned GDY will presents new insights for flexible electronics and other optoelectronic applications.

9.2 Separation

9.2.1 Gas Separation

Common techniques employed for gas separation include pressure swing adsorption [43], temperature swing adsorption [44], cryogenic distillation [45], complexation adsorption [46], and membrane separation [47–49]. Among these gas separation technologies, membrane separation technology is considered to be the most promising approach, mainly due to its advantages of facile operation, consequently high efficiency and low cost compared with traditional separation technology. GDY is a potential gas separation membrane with one-atom thickness, excellent mechanical properties, and natural uniform pores to endow the desired properties of gas permeation. Notably, the size of the evenly distributed holes in GDY is larger than that of other dimensional nanoporous materials, such as graphyne, polyphenylene, rhombic-graphyne, and square-octagon layer [50] (Figure 9.6), which is appropriate for fabricating efficient membranes for gas separation.

9.2.1.1 Hydrogen Separation

The framework of GDY is constructed by hexagonal carbon rings cross-linked by diacetylene. And the size of van der Waals pores defined by the framework is in between hydrogen and methane/carbon monoxide, which makes GDY possible to be utilized as a separation membrane for hydrogen purification from syngas-a

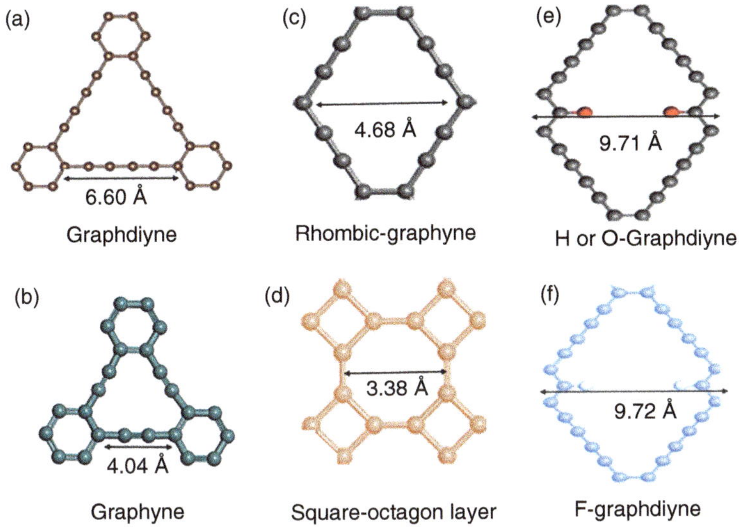

Figure 9.6 Geometry structure of (a) graphdiyne, (b) graphyne, (c) rhombic-graphyne, (d) square-octagon layer; (e) H or O-graphdiyne, (f) F-graphdiyne and a schematic diagram of the holes of the corresponding structure. Source: (a–d) Based on Zhang et al. [51]. © 2012, American Chemical Society, (e–f) Based on Zhao et al.[52]. © 2017, Elsevier.

product of steam-methane reforming by partial oxidation of methane. Jiao et al. provided a precise description of the selectivity of H_2 over CH_4 and CO based on DFT and transition state theory [53]. For gas diffusion, the energy barrier of H_2 passing through the pores on GDY was 0.10 eV, while the barriers for CH_4 and CO passing through the pore were computed to be 0.72 and 0.33 eV (Figure 9.7a), respectively. The values suggested that in comparison with H_2, it is much more difficult for these molecules to diffuse across a GDY-based nanomesh. Additionally, their calculation results showed that the selectivity of H_2 toward CH_4 and CO was about 10^{10} and 10^3 at room temperature (Figure 9.7b), respectively, which was much higher than those of silica and carbon membranes. Zhang et al. performed a first-principle study on the H_2 separation characteristics of these porous sp-sp^2 hybridized carbon allotropes (graphyne, GDY, and rhombic-graphyne) with different pore size [51]. They found that the pore size of graphyne was too small for H_2 separation, while GDY showed a high H_2 selectivity (10^9) over CH_4 and a relatively low H_2 selectivity (10^3) over CO and N_2. And the large difference in molecule diffusion barriers indicated a high H_2 selectivity ($>10^{16}$) over CH_4, CO, and N_2 for rhombic-graphyne (Figure 9.7c,d). Zhao et al. carried out molecular dynamic (MD) simulation to investigate the permselectivity of H_2/O_2, H_2/N_2, H_2/CO, and H_2/CH_4 mixtures through GDY [54]. The MD simulation results show that a large number of H_2 molecules can quickly enter the correct reservoir, while other molecules (N_2, CO, and CH_4) were completely enclosed by GDY. In addition, at pressure range of 0.047–4.5 GPa, H_2 can pass the GDY membrane quickly, while all the O_2, N_2, CO, and CH_4 molecules were blocked. Specifically, at pressure of 47 kPa, the hydrogen flow was 7 mol m^{-2} s^{-1}. With increase of pressure, the hydrogen flow goes up and

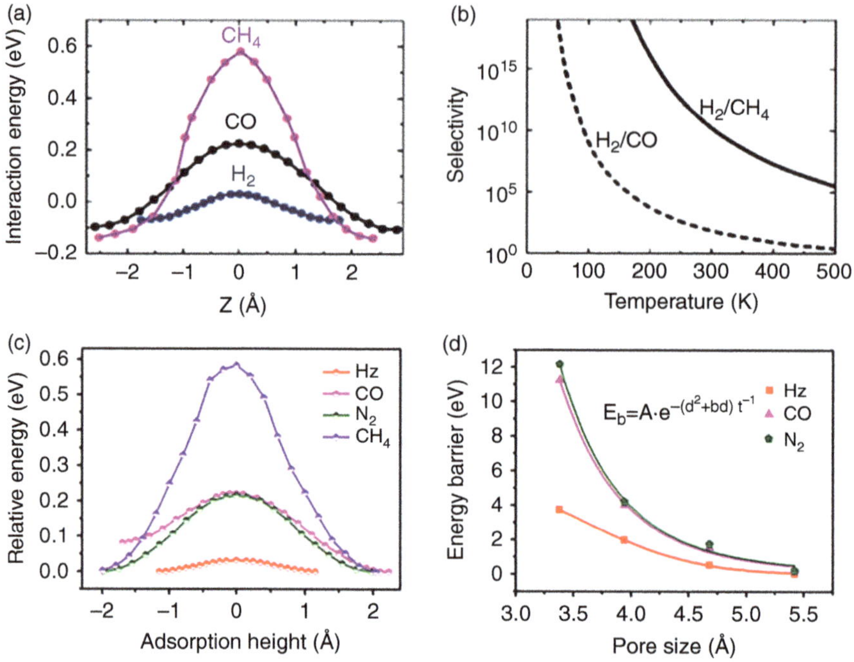

Figure 9.7 (a) Minimum energy pathway for H_2, CO, and CH_4 diffusion through graphdiyne. (b) Selectivity of gas molecules diffusing through the pores on graphdiyne as functions of temperature. (c) Energy profiles for H_2, CO, N_2, and CH_4 passing through the pores of graphdiyne as a function of adsorption height. (d) Diffusion energy barrier as a function of pore size. Source: (a–b) Jiao et al. [53]. © 2011, Royal Society of Chemistry, (c–d) Zhang et al. [51]. © 2012, American Chemical Society.

reaches the maximum of 6×10^5 mol m^{-2} s^{-1} at 1.5 GPa. Compared to other known membranes, GDY can be employed for means of hydrogen purification with the best balance of high selectivity and high permeance.

9.2.1.2 Oxygen Separation

Oxygen has important application value in many fields, such as medical treatment, industrial, and firefighting fields. Therefore, it is essential to find a suitable material that can separate and purify oxygen effectively. For the first time, Liu and coworkers studied the adsorption and diffusion performance of GDY for oxygen and several harmful gases through theoretical calculations [55] (Figure 9.8a). According to the first-principles calculations, the oxidation of the acetylenic bond in GDY needs to surmount an energy barrier of c. 1.97 eV (Figure 9.8b), implying that GDY remained unaffected under oxygen-rich conditions. Electron densities were studied to verify whether gas molecules (O_2, Cl_2, HCl, HCN, CNCl, SO_2, H_2S, NH_3, and CH_2O) can pass through the GDY film smoothly (Figure 9.8c). Obviously, compared with other gases, the electron density overlaps between O_2 and GDY pores are less pronounced. It qualitatively explains the minimum energy barrier of O_2 in all the studied gases, which indicates that oxygen molecules can quickly pass through

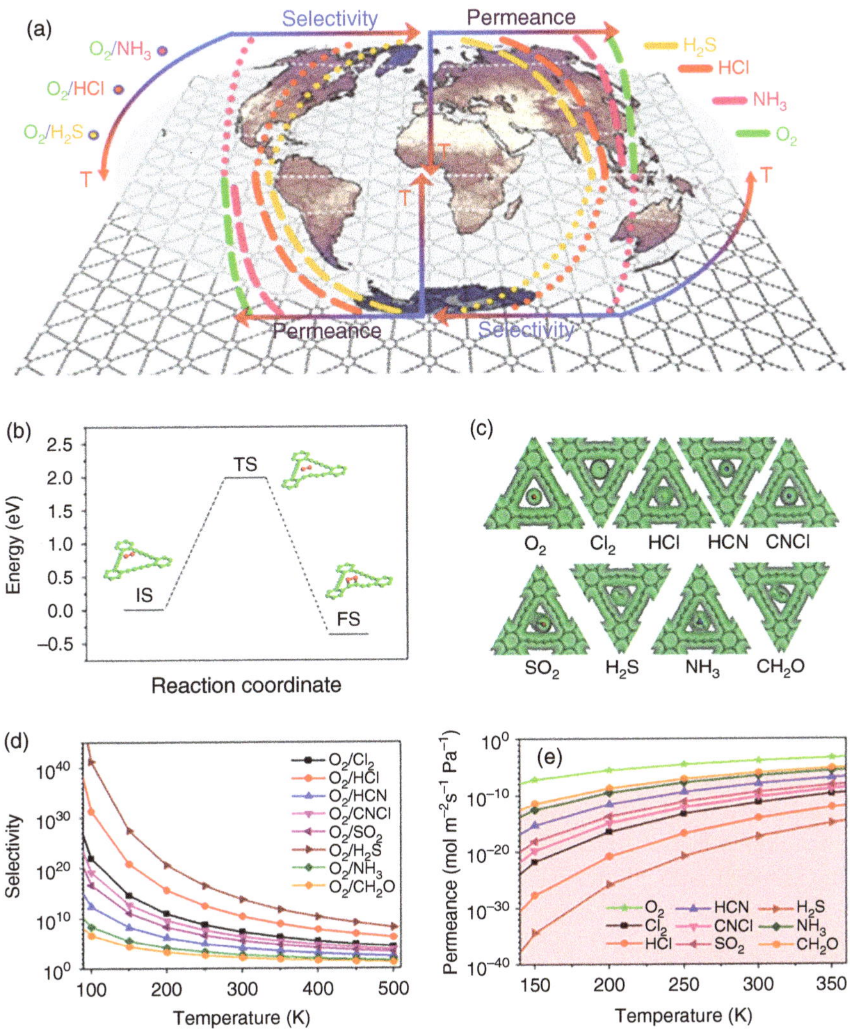

Figure 9.8 (a) Schematic illustration of graphdiyne for separating oxygen from harmful gases. (b) Relative energies of initial state (IS), transition state (TS), and final state (FS) structures in the oxidation process of the acetylenic bond in graphdiyne (C, green balls; O, red balls). (c) Electron densities of the TS structures for O_2, Cl_2, HCl, HCN, CNCl, SO_2, H_2S, NH_3, and CH_2O penetrating the graphdiyne membrane. The isovalue is 0.075 au. (d) Permeances vs. temperature for O_2 over Cl_2, HCl, HCN, CNCl, SO_2, H_2S, NH_3, and CH_2O passing through graphdiyne. (e) Selectivities vs. temperature for O_2 over Cl_2, HCl, HCN, CNCl, SO_2, H_2S, NH_3, and CH_2O passing through graphdiyne. Source: Meng et al. [55]. © 2016, American Chemical Society.

the GDY membrane. If other gas molecules pass through the GDY membrane, external factors may be required, such as increasing temperature and pressure. Besides, in a broad temperature range, GDY with well-defined nanosized pores exhibited a perfect performance for O_2 separation from typical noxious gases (Cl_2, HCl, HCN, CNCl, SO_2, H_2S, NH_3, and CH_2O). For selectivity, the performance of

O_2 separation over H_2S is the best. Even for the O_2/CH_2O separation, the selectivity reaches a considerably high value of 2×10^2 at room temperature (Figure 9.8d). As for permeance, the O_2 permeance of 1×10^{-4} mol m^{-2} s^{-1} Pa^{-1} at 300 K is 10^3 times larger than all reported values at higher temperatures of the mixed ionic-electronic conducting ceramic-based membranes. In terms of selectivity and permeability, GDY can be utilized as an outstanding membrane material to separate oxygen, which will have bright prospects in medical and industrial applications.

9.2.1.3 Carbon Dioxide Separation

Proper pore size plays a decisive role in the efficiency of gas separation. It can enhance the performance of gas separation and even achieve good separation effect for some large molecules (CO_2, CH_4, etc.) by adjusting the size of the holes of GDY. Zhao et al. innovatively proposed three GDY-like monolayer membranes (GDY_X, X = H, F, and O) and confirmed their stabilities through the cohesive energy and phonon dispersion spectra calculations [52]. The dispersion-corrected DFT and MD simulations have been performed to study the separation performance of the three GDY_X membranes for $CO_2(N_2)/CH_4$ binary mixtures. The DFT and MD calculations showed that the GDY_H membrane was unsuitable for the separation of the $CO_2(N_2)/CH_4$ mixtures because of the surmountable energy barriers of 0.06–0.16 eV, but both the GDY_F and GDY_O membranes had a significantly high selectivity of $CO_2(N_2)/CH_4$ (The energy barriers for the CH_4 molecule [0.40 and 0.39 eV] and the energy barriers for CO_2 and N_2 molecules [0.03–0.18 eV]). Especially, the actual selectivity of $CO_2(N_2)/CH_4$ for GDY_F and GDY_O was still as high as 8.9–57 even at 600 K, larger than the industrial standard of 6. In addition, the GDY_O membrane also exhibited an effective selectivity for CO_2/N_2 at temperatures below 300 K. More importantly, the permeances of CO_2 and N_2 passing through GDY_F and GDY_O were about 10^2 times higher than those of previous nanoporous graphene membranes and exceeded the industrial standard by about 10^5 to 10^7 times (Figure 9.9). Furthermore, the relative concentration of binary mixtures was found to have an important influence on the ideal selectivity based on the DFT calculations but showed almost no effect on the actual selectivity based on the MD simulations.

9.2.1.4 Helium Separation

Helium is an irreplaceable natural resource. With the increasing demand and insufficient production of various industrial and scientific applications, this element is extremely scarce. Especially the lighter isotope ^3He is crucial for large neutron scattering. Therefore, it is desirable to use advanced and efficient technology to separate He effectively. The permeation of helium across GDY was studied by Bartolomei et al., and they found that GDY pores permitted an almost unimpeded helium transport while it was much more difficult through the 2D polyphenylene openings (Figure 9.10). Besides, the He/CH_4 selectivity was exceptionally favorable in a wide temperature range, while the $^3He/^4He$ selectivity reaches the acceptable value of 6 at about 20 K [56]. Similarly, Xu et al. explored the He separation performance of the N-modified GDY monolayer (N-GDY) through the first-principles

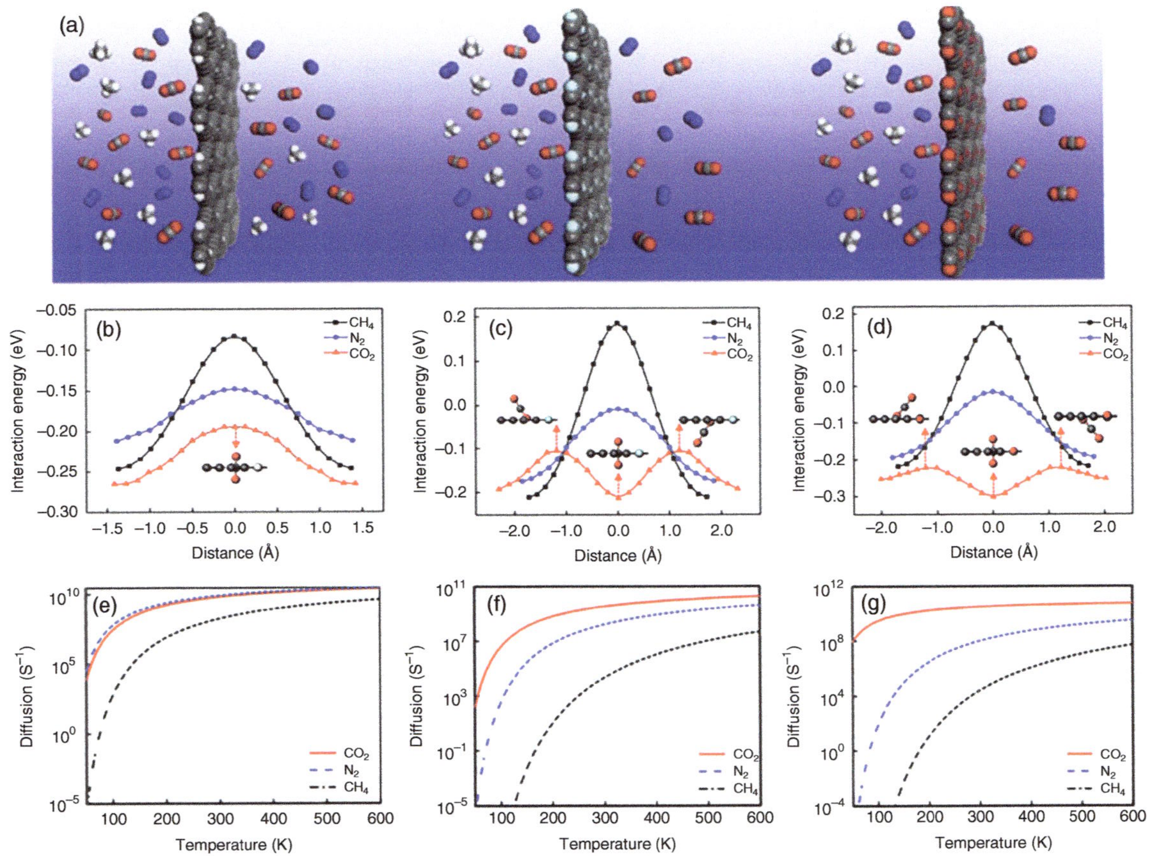

Figure 9.9 (a) MD-simulated configurations of the equimolar $CO_2/N_2/CH_4$ gas mixtures permeating through the membranes of GDY_H, GDY_F, and GDY_O, respectively. (b–d) Minimum energy pathways for CO_2, N_2, and CH_4 passing through (a) GDY_H, (b) GDY_F, and (c) GDY_O. (e–g) Diffusion rate constants of CO_2, N_2, and CH_4 permeating through GDY_H, GDY_F, and GDY_O membranes, respectively. Source: Zhao et al. [52]. © 2017, Elsevier.

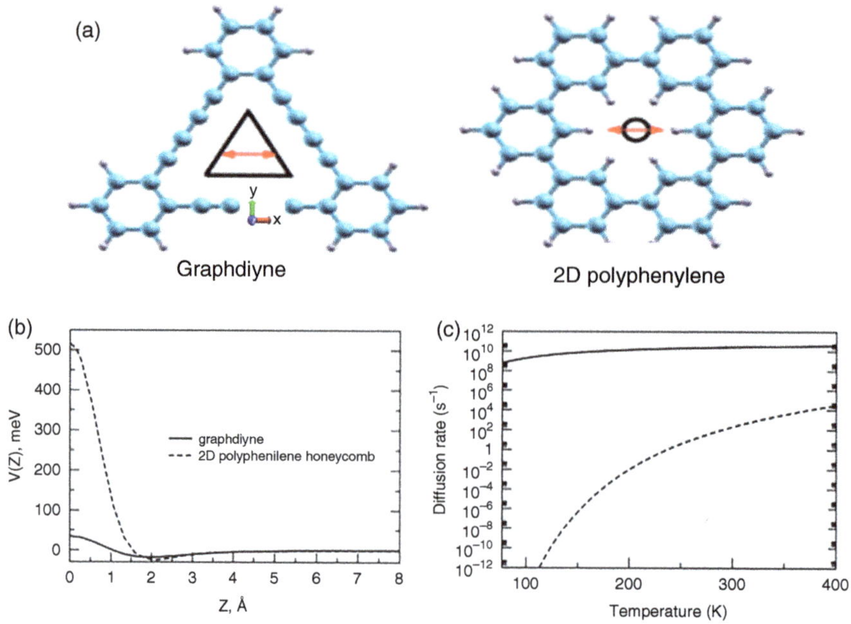

Figure 9.10 (a) Molecular structures used to study the nanopores of graphdiyne and 2D polyphenylene honeycomb. (b) Energy profiles obtained at the "coupled" supermolecular second-order Møller-Plessetperturbation theory level of theory for He perpendicularly approaching the geometric center of graphdiyne and 2D polyphenylene honeycomb pores. (c) Rates for He diffusion through the pores as functions of temperature. Selectivities for different molecular combinations as functions of temperature. Source: Bartolomei et al. [56]. © 2014, American Chemical Society.

DFT and MD simulations [57]. The N-GDY membrane can be penetrated by He molecules with an extremely low-energy barrier of 0.05 eV and high permeance of 4.8×10^{-3} mol m^{-2} s Pa at 100 K. Furthermore, the N-GDY membrane exhibited high selectivity of He over Ar, N_2, H_2O, CO, CO_2, and CH_4 (10^2 to 10^{12} at 300 K).

9.2.2 Oil/Water Separation

The rapid industrial development has led to the existence of a large amount of oil dyes and organics in the water resources we rely on, which greatly threatens the ecological environment and human healthy. Therefore, the purification of water is urgent. The previously DFT results have demonstrated that graphyne, including GDY, can serve as molecular filters for purifying water [58, 59], indicating that the potential application prospect of GDY-based materials in water remediation.

For example, Liu et al. described a successful decoration of GDY on a skeleton of melamine sponge (MS) substrate through a facile in situ Glaser-Hay coupling. The sponge not only maintained its intact porous structure but also became significantly rougher after being completely modified by GDY (Figure 9.11a). Such surfaces are essential for hydrophobic materials. The prepared GDY-coated melamine sponge (GDYMS-15) composite exhibited excellent adsorption ability to oil and organic

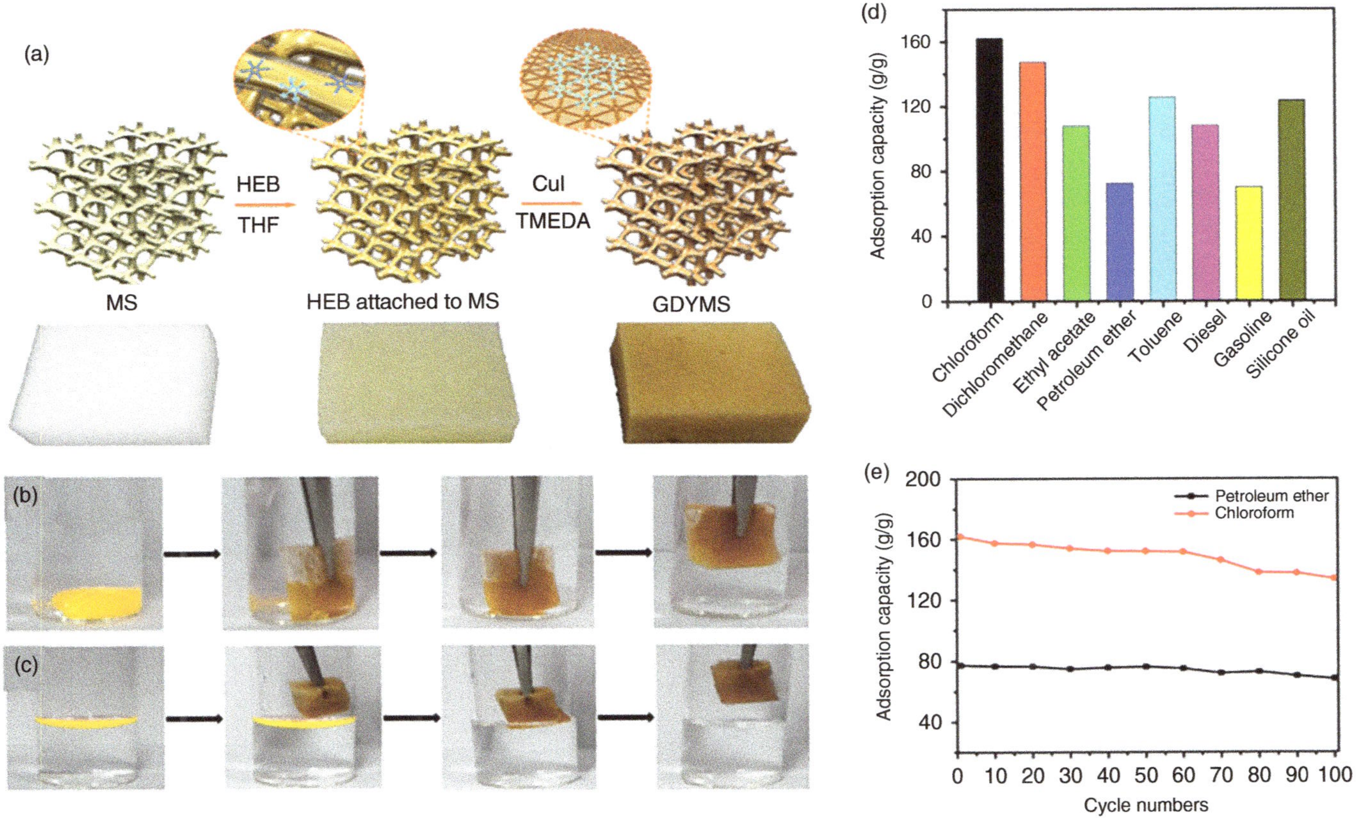

Figure 9.11 (a) Schematic illustration of fabrication procedure of GDY-coated melamine sponge (GDYMS) and the corresponding photograph. Note that the spreading situations of the water on melamine sponge and on GDYMS are quite different. The photographs of the total adsorption process of chloroform (b) and gasoline (c) on the as-prepared GDYMS. (d) Adsorption capacities of GDYMS-15 toward various organic solvents and oils. (e) Adsorption recyclability of GDYMS-15 toward petroleum ether and $CHCl_3$. Source: Li et al. [60]. © 2019, American Chemical Society.

pollutants with different densities due to its excellent hydrophobicity and porous 3D structure (Figure 9.11b,c). Moreover, the adsorption capacity of GDYMS-15 was also quantified. Notably, this as-prepared composite did not adsorb water, which made the adsorbed oil and organic pollutants recyclable. Compared with activated carbon, nanowire membranes, carbon nanotube sponges, and graphene, GDYMS-15 exhibited remarkable and extensive adsorption toward $CHCl_3$, CH_2Cl_2, ethyl acetate, petroleum ether, toluene, diesel, gasoline, and silicone oil (Figure 9.11d). The adsorption capacity remained high after 100 cycles, suggesting the excellent recyclability of GDYMS-15 (Figure 9.11e).

Similarly, Zhang and coworkers designed an ingenious method to fabricate a uniform layer of GDY-based ordered nanostructures via in situ Glaser-Hay coupling using copper foam as substrate [61]. The graphdiyne-grown copper foam (GDCF) was covered with low-surface-energy poly(dimethylsiloxane) (PDMS) coating, yielding a robust superhydrophobic foam (named as PGDCF). The copper foam skeleton was wrapped by an ordered honeycomb composite coating, and GDY was successfully embedded into the PDMS matrix. The as-prepared PGDCF showed outstanding superhydrophobicity in *n*-hexane or dichloromethane. A static contact angle under n-hexane and dichloromethane was 171.0° and 167.2°, respectively. Therefore, when subjected to an oil/water mixture, the multistage structure of PGDCF can trap the oil, forming a composite interface to maintain the superhydrophobicity required for oil-water separation. The dichloromethane (CH_2Cl_2)/water (H_2O) separation process used the as-prepared foam (PGDCF) as separation membrane. It can be directly observed that when exposed to the mixture of oil and water, PGDCF can selectively block water and allow oil to pass through, with the separation efficiency as high as (>98%). The superhydrophobicity and superlipophilicity of PGDCF are the key to the separation of oil and water. In detail, due to the existence of large microporous structure and excellent hydrophobicity, oil can quickly pass through the pores of PGDCF under the action of gravity, while water remains on the surface due to the high-water repulsion. These demonstrated studies will provide new ideas for the design of novel GDY-based superhydrophobic materials for water remediation.

9.3 Conclusion and Perspective

In this chapter, the applications of GDY and GDY-based materials in sensors and separation were discussed. In spite of their very short history, GDY-based materials have already demonstrated broad prospect of applications. Because of the unique structural properties (e.g. large specific surface area), exceptional electrical properties (e.g. extremely high carrier mobility and capacity), electrochemical properties (e.g. high electron transfer rate), optical properties (e.g. excellent ability to quench fluorescence), GDY-based materials sensors have been employed for detecting biological molecules and small molecules such as DNAs, RNA, amino acids, humidity, H_2O_2, glucose, and hazardous gases. These sensors displayed ultrahigh sensitivity, high specificity, fast response, and recovery performance. However, the full potential of GDY-based materials sensors is far from being reached, some developments are merely proof-of-concept demonstrations and need a step forward to the practical or commercialized uses. Exploring new GDY-based materials and structures is still emerging. Besides, one-atom-thick GDY with natural uniform pores and mechanical robustness is a very promising membrane for gases purification or gases separation. The excellent hydrophobicity of GDY and GDY-based materials (the surface is coated with hydrophobic organic compounds) makes them promising application in the field of oil-water separation. Meanwhile, owing to the interaction between acetylenic links in GDY and cations, GDY can effectively remove toxic cations from contaminated water. Considering the unique structure of GDY, the GDY-based filter has great potential to decontaminate water polluted by heavy metal ions such as V^{5+}, Cr^{6+}, Mn^{4+}, Ni^{2+}, Cd^{2+}, Cu^{2+}, Pd^{2+}, and Co^{3+}. Therefore, the development of GDY-based materials in sensors and separation is beneficial to industrial and medical areas.

References

1 Chang, F. (2017). A short review of synthesis of graphdiyne and its potential applications. *International Journal of Electrochemical Science* 12: 10348–10358.

2 Qiu, H., Xue, M., Shen, C. et al. (2019). Graphynes for water desalination and gas separation. *Advanced Materials* 31: 1803772.

3 Chen, Z., Molina-Jirón, C., Klyatskaya, S. et al. (2017). 1D and 2D graphdiynes: recent advances on the synthesis at interfaces and potential nanotechnological applications. *Annalen der Physik* 529: 1700056.

4 Liu, J., Chen, C., and Zhao, Y. (2019). Progress and prospects of graphdiyne-based materials in biomedical applications. *Advanced Materials* 31: 804386.

5 Liu, Y., Dong, X., and Chen, P. (2012). Biological and chemical sensors based on graphene materials. *Chemical Society Reviews* 41: 2283–2307.

6 Jia, Z., Li, Y., Zuo, Z. et al. (2017). Synthesis and properties of 2D carbon-graphdiyne. *Accounts of Chemical Research* 50: 2470–2478.

7 Bitounis, D., Ali-Boucetta, H., Hong, B.H. et al. (2013). Prospects and challenges of graphene in biomedical applications. *Advanced Materials* 25: 2258–2268.

8 Yang, Y., Asiri, A.M., Tang, Z. et al. (2013). Graphene based materials for biomedical applications. *Materials Today* 16: 365–373.

9 Chen, Y., Tan, C., Zhang, H. et al. (2015). Two-dimensional graphene analogues for biomedical applications. *Chemical Society Reviews* 44: 2681–2701.

10 Wang, C., Yu, P., Guo, S. et al. (2016). Graphdiyne oxide as a platform for fluorescence sensing. *Chemical Communications* 52: 5629–5632.

11 Chang, F., Huang, L., Guo, C. et al. (2019). Graphdiyne-based one-step DNA fluorescent sensing platform for the detection of *Mycobacterium tuberculosis* and its drug-resistant genes. *ACS Applied Materials & Interfaces* 11: 35622–35629.

12 Chandra Shekar, S. and Swathi, R.S. (2014). Stability of nucleobases and base pairs adsorbed on graphyne and graphdiyne. *Journal of Physical Chemistry C* 118: 4516–4528.

13 Parvin, N., Jin, Q., Wei, Y. et al. (2017). Few-layer graphdiyne nanosheets applied for multiplexed real-time DNA detection. *Advanced Materials* 29: 1606755.

14 Wang, M., Yin, H., Zhou, Y. et al. (2019). A novel photoelectrochemical biosensor for the sensitive detection of dual microRNAs using molybdenum carbide nanotubes as nanocarriers and energy transfer between CQDs and AuNPs. *Chemical Engineering Journal* 365: 351–357.

15 Li, Y., Li, X., Meng, Y. et al. (2019). Photoelectrochemical platform for MicroRNA let-7a detection based on graphdiyne loaded with AuNPs modified electrode coupled with alkaline phosphatase. *Biosensors and Bioelectronics* 130: 269–275.

16 Li, X., Li, Y., Zhang, J. et al. (2019). Molybdenum disulfide/graphdiyne-based photoactive material derived photoelectrochemical strategy for highly sensitive MicroRNA assay. *Sensors and Actuators B: Chemical* 297: 126808.

17 Wang, H., Deng, K., Xiao, J. et al. (2020). A sandwich-type photoelectrochemical sensor based on tremella-like graphdiyne as photoelectrochemical platform and graphdiyne oxide nanosheets as signal inhibitor. *Sensors and Actuators B: Chemical* 304: 127363.

18 Chen, X., Gao, P., Guo, L. et al. (2015). Graphdiyne as a promising material for detecting amino acids. *Scientific Reports* 5: 16720.

19 Wu, L., Gao, J., Lu, X. et al. (2020). Graphdiyne: a new promising member of 2D all-carbon nanomaterial as robust electrochemical enzyme biosensor platform. *Carbon* 156: 568–575.

20 Liu, X., Cheng, S., Liu, H. et al. (2012). A survey on gas sensing technology. *Sensors* 12: 9635–9665.

21 Varghese, S.S., Lonkar, S., Singh, K.K. et al. (2015). Recent advances in graphene based gas sensors. *Sensors and Actuators B: Chemical* 218: 160–183.

22 Nagarajan, V. and Chandiramouli, R. (2018). Investigation of NH_3 adsorption behavior on graphdiyne nanosheet and nanotubes: a first-principles study. *Journal of Molecular Liquids* 249: 24–32.

23 Ebadi, M. and Reisi-Vanani, A. (2021). Methanol and carbon monoxide sensing and capturing by pristine and Ca-decorated graphdiyne: a DFT-D2 study. *Physica E: Low-dimensional Systems and Nanostructures* 125: 114425.

24 Nagarajan, V., Srimathi, U., and Chandiramouli, R. (2018). First-principles insights on detection of dimethyl amine and trimethyl amine vapors using graphdiyne nanosheets. *Computational and Theoretical Chemistry* 1123: 119–127.

25 Bhuvaneswari, R., Princy Maria, J., Nagarajan, V. et al. (2020). Graphdiyne nanosheets as a sensing medium for formaldehyde and formic acid–a first-principles outlook. *Computational and Theoretical Chemistry* 1176: 112751.

26 Chen, X., Gao, P., Guo, L. et al. (2017). High-efficient physical adsorption and detection of formaldehyde using Sc- and Ti-decorated graphdiyne. *Physics Letters A* 381: 879–885.

27 Bhuvaneswari, R., Nagarajan, V., and Chandiramouli, R. (2019). Explosive vapor detection using novel graphdiyne nanoribbons—a first-principles investigation. *Structural Chemistry* 31: 709–717.

28 Hussain, T., Sajjad, M., Singh, D. et al. (2020). Sensing of volatile organic compounds on two-dimensional nitrogenated holey graphene, graphdiyne, and their heterostructure. *Carbon* 163: 213–223.

29 Dessler, A.E. and Sherwood, S.C. (2009). A matter of humidity. *Science* 323: 1020–1021.

30 Barnett, T.P., Adam, J.C., and Lettenmaier, D.P. (2005). Potential impacts of a warming climate on water availability in snow-dominated regions. *Nature* 438: 303–309.

31 Yan, H., Zhang, L., Yu, P. et al. (2017). Sensitive and fast humidity sensor based on a redox conducting supramolecular ionic material for respiration monitoring. *Analytical Chemistry* 89: 996–1001.

32 Borini, S., White, R., Wei, D. et al. (2013). Ultrafast graphene oxide humidity sensors. *ACS Nano* 7: 11166–11173.

33 Smith, A.D., Elgammal, K., Niklaus, F. et al. (2015). Resistive graphene humidity sensors with rapid and direct electrical readout. *Nanoscale* 7: 19099–19109.

34 Yan, H., Guo, S., Wu, F. et al. (2018). Carbon atom hybridization matters: ultrafast humidity response of graphdiyne oxides. *Angewandte Chemie International Edition* 57: 3922–3926.

35 Chen, S., Yuan, R., Chai, Y. et al. (2012). Electrochemical sensing of hydrogen peroxide using metal nanoparticles: a review. *Microchimica Acta* 180: 15–32.

36 Zhuang, X., Mao, L., and Li, Y. (2017). In situ synthesis of a Prussian blue nanoparticles/graphdiyne oxide nanocomposite with high stability and electrocatalytic activity. *Electrochemistry Communications* 83: 96–101.

37 Liu, J., Shen, X., Baimanov, D. et al. (2018). Immobilized ferrous ion and glucose oxidase on graphdiyne and its application on one-step glucose detection. *ACS Applied Materials & Interfaces* 11: 2647–2654.

38 Chang, Q. and Tang, H. (2014). Optical determination of glucose and hydrogen peroxide using a nanocomposite prepared from glucose oxidase and magnetite nanoparticles immobilized on graphene oxide. *Microchimica Acta* 181: 527–534.

39 Sun, J., Li, C., Qi, Y. et al. (2016). Optimizing colorimetric assay based on V_2O_5 nanozymes for sensitive detection of H_2O_2 and glucose. *Sensors* 16: 584.

40 Park, S., Boo, H., and Chung, T.D. (2006). Electrochemical non-enzymatic glucose sensors. *Analytica Chimica Acta* 556: 46–57.

41 Moschou, E.A., Sharma, B.V., Deo, S.K. et al. (2004). Fluorescence glucose detection: advances toward the ideal in vivo biosensor. *Journal of Fluorescence* 14: 535–547.

42 Wang, S.S., Liu, H.B., Kan, X.N. et al. (2017). Superlyophilicity-facilitated synthesis reaction at the microscale: ordered graphdiyne stripe arrays. *Small* 13: 1602265.

43 Ohs, B., Falkenberg, M., and Wessling, M. (2019). Optimizing hybrid membrane-pressure swing adsorption processes for biogenic hydrogen recovery. *Chemical Engineering Journal* 364: 452–461.

44 Liu, L., Jin, S., Ko, K. et al. (2020). Alkyl-functionalization of (3-aminopropyl) triethoxysilane-grafted zeolite beta for carbon dioxide capture in temperature swing adsorption. *Chemical Engineering Journal* 382: 122834.

45 Chen, S., Dong, X., Xu, J. et al. (2019). Thermodynamic evaluation of the novel distillation column of the air separation unit with integration of liquefied natural gas (LNG) regasification. *Energy* 171: 341–359.

46 Zhang, L., Li, L., Hu, E. et al. (2020). Boosting ethylene/ethane separation within copper(i)-chelated metal-organic frameworks through tailor-made aperture and specific pi-complexation. *Advanced science* 7: 1901918.

47 Qiu, W., Vaughn, J., Liu, G. et al. (2019). Hyperaging tuning of a carbon molecular-sieve hollow fiber membrane with extraordinary gas-separation performance and stability. *Angewandte Chemie International Edition* 58: 11700–11703.

48 Jiang, X., Li, S., Bai, Y. et al. (2019). Ultra-facile aqueous synthesis of nanoporous zeolitic imidazolate framework membranes for hydrogen purification and olefin/paraffin separation. *Journal of Materials Chemistry A* 7: 10898–10904.

49 Wong, K.C., Goh, P.S., Taniguchi, T. et al. (2019). The role of geometrically different carbon-based fillers on the formation and gas separation performance of nanocomposite membranes. *Carbon* 149: 33–44.

50 Bieri, M., Treier, M., Cai, J. et al. (2009). Porous graphenes: two-dimensional polymer synthesis with atomic precision. *Chemical Communications*: 6919–6921.

51 Zhang, H., He, X., Zhao, M. et al. (2012). Tunable hydrogen separation in sp–sp^2 hybridized carbon membranes: a first-principles prediction. *Journal of Physical Chemistry C* 116: 16634–16638.

52 Zhao, L., Sang, P., Guo, S. et al. (2017). Promising monolayer membranes for $CO_2/N_2/CH_4$ separation: graphdiynes modified respectively with hydrogen, fluorine, and oxygen atoms. *Applied Surface Science* 405: 455–464.

53 Jiao, Y., Du, A., Hankel, M. et al. (2011). Graphdiyne: a versatile nanomaterial for electronics and hydrogen purification. *Chemical Communications* 47: 11843–11845.

54 Zhao, W.-h., Yuan, L.-f., and Yang, J.-l. (2012). Graphdiyne as hydrogen purification membrane. *Chinese Journal of Chemical Physics* 25: 434.

55 Meng, Z., Zhang, X., Zhang, Y. et al. (2016). Graphdiyne as a high-efficiency membrane for separating oxygen from harmful gases: a first-principles study. *ACS Applied Materials & Interfaces* 8: 28166–28170.

56 Bartolomei, M., Carmona-Novillo, E., Hernández, M.I. et al. (2014). Graphdiyne pores: "Ad Hoc" ppenings for helium separation applications. *Journal of Physical Chemistry C* 118: 29966–29972.

57 Xu, J., Li, J., Liu, H.J. et al. (2017). Nitrogen-modified graphdiyne as a promising membrane for helium separation: first-principles and molecular dynamics simulations. *Defect and Diffusion Forum* 381: 20–25.

58 Lin, S. and Buehler, M.J. (2013). Mechanics and molecular filtration performance of graphyne nanoweb membranes for selective water purification. *Nanoscale* 5: 11801–11807.

59 Bartolomei, M., Carmona-Novillo, E., Hernandez, M.I. et al. (2014). Penetration barrier of water through graphynes' pores: first-principles predictions and force field optimization. *Journal of Physical Chemistry Letters* 5: 751–755.

60 Li, J., Chen, Y., Gao, J. et al. (2019). Graphdiyne sponge for direct collection of oils from water. *ACS Applied Materials & Interfaces* 11: 2591–2598.

61 Gao, X., Zhou, J., Du, R. et al. (2016). Robust superhydrophobic foam: a graphdiyne-based hierarchical architecture for oil/water separation. *Advanced Materials* 28: 168–173.

10

Perspectives

Yuliang Li

Institute of Chemistry, Chinese Academy of Sciences, Zhongguancun North First Street 2, Beijing 100190, P. R. China

The rapid development of carbon science has deeply affected the progress of science and technology, promoted the development of multidisciplinary and high-tech fields, and extensively affected social civilization and human progress, thus establishing its strategic position in the twenty-first century. Since 2005, the developed countries in the world have published the development plan of carbon nanotechnology, and considered that carbon nanotechnology is one of the largest and most competitive research fields in the world. Carbon nanomaterials science is a new and rapidly developing interdisciplinary science, involving physics, chemistry, materials, information, biology, medicine, environment, energy, and other fields. It is generally believed that the development of carbon science has brought development space for the new upgrading of industrial technology and will become one of the mainstream science and technologies in the next 10–20 years. Major economies in the world have formulated carbon science development strategies and plans, increased investment, and promoted the rapid development of carbon science.

In the past two decades, carbon materials science and technology have made great progress. To realize the sustainable development of carbon nanotechnology, it is urgent to carry out the original fundamental research of carbon nanomaterials in the fields of chemistry, condensed matter physics, materials, biology, information science, and engineering. By deeply understanding the basic processes of the growth, assembly, and evolution of carbon nanomaterials, we can establish appropriate theoretical calculation methods and accurate computer simulation, design and controllable preparation of carbon nanostructures, and discover new functions of the prepared carbon materials. Systematic foundation research results in this major interdisciplinary field can promote the application of carbon nanotechnology in electronics, information, energy, manufacturing, health, and environment.

In recent years, the carbon materials represented by graphene have developed rapidly, especially the discovery of GDY has brought new vitality to the family of carbon materials. As a new material with unique electronic structure and chemical

Graphdiyne: Fundamentals and Applications in Renewable Energy and Electronics,
First Edition. Edited by Yuliang Li.

structure, it has attracted great attention in fundamental and applied research. Although some important fundamental issues have been addressed, GDY has demonstrated excellent properties and performance in the fields of catalysis, energy, intelligent information, optoelectronics, and optics. However, GDY still has some basic problems to be solved and understood in fundamental science. It also needs more knowledge accumulation in application and links with preparation technology, film-forming technology, and so on.

The synthesis and preparation of graphdiyne and its aggregation structure still need to be further developed. Further study and large-scale application of graphdiyne should focus on the controllable synthesis method of graphdiyne and the preparation of aggregated structure. Earlier work focused on the establishment of GDY synthesis methodology and the development of a variety of methods related to the formation of aggregation structure. Further, we focus more on controlling single- or few-layer GDY as well as aggregation-state structures of different dimensions to study their intrinsic properties.

Understanding the fundamental electronic and physical properties of graphdiyne will help to further expand the fundamental and applied research of graphdiyne in the fields of energy, micro-nano-electronics, catalysis, etc., and open up the research of graphdiyne in catalysis, clean energy, superhard materials, aerospace, information technology, and other fields.

We are pleased to see that graphdiyne research has been highly concerned by chemists, materials scientists, physicists, biologists, semiconductors, electronics, microelectronics, and information scientists around the world, and that the rapid development in recent years has made significant progress in fundamental and applied sciences, showing transformative progress in areas conducive to, for example, catalysis, energy and renewable energy, and optoelectronics. However, the research field in graphdiyne appears for a short time, and there are still many scientific issues that need more time to understand. As a new member of a carbon family, the fundamental difference from traditional carbon materials is the new concepts, new properties and new knowledge caused by its sp and sp^2 electronic structure. New ideas from graphdiyne represent the development trend of advanced materials. We can deeply understand GDY great development potential as a transformative material.

After all, the development time of graphdiyne is short. Retrospecting on the present situation in recent years, we think that the main challenges of graphdiyne in the next decade are: (i) controllable preparation of single-layer films with super-large area and accurate measurement of intrinsic properties of single-layer films. (ii) Development of new synthesis methods, in particular the establishment of a new method for controlling the growth of all-crystalline graphdiyne. (iii) Establishment of a high-purity controlled preparation technology for kilogram grade and its application in energy and efficient catalysts. (iv) The foundations and applications of excellent semiconductor properties in electronic, photonic, and optoelectronic devices. (v) The foundation and application of information fields such as efficient intelligence and chips. (vi) Continuous preparation and application of large-area and high-trend graphyne films. (vii) Establishment of standardization of GDY products.

The research and efficient utilization of GDY are still far from the requirements of humans. We believe that graphdiyne materials will usher in a scientific spring under our persistence and efforts to contribute to the progress of human society.

10.1 Chemical Synthesis Methodology and Aggregate Structures of Graphdiyne

Chemical synthesis methodology of graphdiyne: Develop new chemical synthesis methods, growth and self-assembly technology of large area graphdiyne film and high-purity graphdiyne bulk materials, establish high-efficiency, low-cost, large-scale, and repeatable synthesis methods for graphdiyne, and novel controllable preparation methods of large-scale graphdiyne films, so as to form a unique synthesis methodology and realize the controllable synthesis of graphdiyne films with different layers and thickness.

Methodology for the construction and assembly of graphdiyne aggregation: The formation of aggregated structures of graphdiyne by self-assembly technology, especially the one-dimensional linear or tubular structure, is the best state to study the intrinsic properties of graphdiyne. It is very important to understand the relationship between the structure and properties of graphdiyne. Technologies should be developed to assemble the highly ordered and oriented graphdiyne aggregates at large scale, and establish the new technologies and methods for growth, assembly, self-assembly, and self-organization of graphdiyne.

Exploration of graphdiyne materials with novel structures: Through the design and synthesis of graphdiyne precursors, the carbon hybridization, and the connection mode of carbon atoms, even the topological structure of the carbon networks can be tuned to synthesize new two-dimensional carbon graphdiyne and graphdiyne-like materials, so as to realize the regulation of energy band structure and semiconductor properties of graphdiyne. To develop a method to partially substitute the graphdiyne alkyne bonds with heteroatoms, and provide a basis for further investigation of the effect of heteroatom doping on the properties of graphdiyne.

The structural law and structure–activity relationship of graphdiyne and its aggregated state: The aggregation structure of graphdiyne is regulated by adjusting the growth parameters (such as growth temperature, catalyst type, growth substrate, solution polarity, and electric field), and the change rules should be studied; the weak interactions between graphdiyne layers and the mutual arrangement mode are regulated by organic small molecules to control the charge, energy, and photon transfer in the ordered structure of graphdiyne, thus new aggregated graphdiyne adjustable from one dimension to multidimension and related large-area ordered arrays can be obtained. The effects of the chemical structure, electronic structure, and aggregate structures of graphdiyne on its properties are clarified to further investigate the formation mechanism, growth mechanism, self-assembly process, dynamics, and dynamic process of the aggregated state of graphdiyne. Controllable assembly of one-dimensional ordered aggregation state of graphdiyne will help to study the physical and chemical properties induced by size and dimensions, such as the storage and transfer of carrier and energy, photoelectric conversion, charge separation and transportation, sensing and catalysis. The controllable synthesis of graphdiyne and the regulation of the morphology, size and orientation of graphdiyne films and their aggregation states would be beneficial to realize the design and structural regulation of the aggregation and heterostructure of graphdiyne materials, and to understand the structure–activity relationships.

10.2 Controllable Preparation of Highly Ordered Graphdiyne

Preparation of graphdiyne by template confinement method: The quality of graphdiyne can be improved using graphene as template under solvothermal conditions. At the same time, the introduction of graphene is more conducive to the formation of graphdiyne film, which provides a better basis for subsequent characterization. Based on this method, epitaxial growth of graphdiyne thin films less than 5 nm on graphene surface can be realized. For this method, first, the reaction conditions need to be optimized further to obtain single-layer samples; second, more detailed scanning tunneling microscopy (STM), transmission electron microscope (TEM), and other structural characterization of the samples need to be performed; third, the graphdiyne sample needs to be separated from the graphene template for the purpose of studying the intrinsic properties of graphdiyne.

Preparation of graphdiyne by low-temperature chemical vapor deposition: By designing and synthesizing acetylene-rich precursor molecules, selecting metal substrates with different catalytic activities, we can obtain a covalent network structure with acetylene-rich molecules as the structural unit through the coupling reaction between acetylene-rich molecules on the metal surface, that is, to realize the preparation of large-area and highly ordered monolayer graphdiyne film in the chemical vapor deposition system. At the same time, according to the diversity of molecular structure, we can obtain a variety of graphdiyne structures with different symmetries and different properties through designing and synthesizing the corresponding structural unit precursor. Then, by optimizing the vacuum degree, carrier gas type and flow rate, precursor deposition rate, growth temperature, and other conditions, the growth process of graphdiyne can be explored, the growth mechanism can be studied, and finally the growth method of highly ordered graphdiyne will be developed to provide material basis for the fundamental physical properties research of graphdiyne.

Preparation of graphdiyne monolayer by intercalation stripping method: Highly crystalline graphdiyne nanowalls and nanosheets were prepared by controlling reaction conditions, and they can be used as cathode materials of lithium-ion batteries. Lithium ions were gradually inserted into the graphdiyne layers through multiple charge and discharge processes to peel off the graphdiyne nanowalls and nanosheets. By changing the charge and discharge current, charge and discharge times and other conditions, high crystallinity graphdiyne nano-thin film and single layer can be obtained. The morphology of graphdiyne nano-thin film and single layer should be studied by high-resolution spherical aberration electron microscopy combined with electron diffraction image simulation technology, the atomic phase structure of graphdiyne monolayer should be observed.

10.3 Fundamental Physical Properties and Applications of Graphdiyne

The physical properties of graphdiyne: As a new member of the family of carbon materials, graphdiyne has a large π conjugated system, homogeneous atomic pore structure, and adjustable electronic structure. Therefore, the research on its fundamental physical properties such as optical, thermal, mechanical, and electronical is worth looking forward to. The characterization and fundamental physical properties measurement methods of monolayer graphdiyne should be established through the combination of theoretical prediction with the optical, spectroscopy, electrical properties, and so on. It is still important to find a simple and versatile method to quickly judge the quality of graphyne. At present, the main methods used to characterize graphyne are traditional 2D material characterization methods, which lack their own characteristics. Raman spectroscopy was used to demonstrate the existence of acetylenic linkages in GDY, but this method fails to analyze the percentage of carbon–carbon triple bonds, which is important for judging the structure of graphyne (different kinds of graphynes often have different percentages of acetylenic linkages). In addition, different growth methods often yield samples with different qualities. Therefore, it is necessary to develop an effective way to accurately determine the polymerization degree of GDY samples.

High-resolution structure characterization of graphdiyne: The chemical structures of monolayer graphdiyne materials obtained by different preparation methods will be studied by using the atomic resolution of scanning probe microscopy (SPM) – mainly STM and noncontact atomic force microscopy (NC-AFM). Surface-assisted preparation methods in different reaction environments were developed. For example, in solution and in ultra-high vacuum environment, single-layer stable graphdiyne materials will be prepared by selecting precursor molecules suitable for different reaction environments and using different excitation conditions. The intrinsic electronic structure of graphdiyne can be characterized on the basis of the preparation of single-layer graphdiyne. Scanning tunneling spectroscopy (STS) can be used to characterize the electronic structure of single-layer graphdiyne acetylene materials at the same time.

Key research on the application of graphdiyne thin film materials: Research on the controlled preparation of graphdiyne with specific morphology and graphdiyne-based composite materials, such as designing and preparing the multilevel structure of graphdiyne, combining the ordered structure of graphdiyne with the three-dimensional porous substrate to obtain the multilevel composite structures, these materials can be applied in the fields of oil–water separation, energy storage, etc. At the same time, for composite materials applications, graphdiyne can be used as hole transport materials in solar cells, photoelectrocatalytic water splitting and other aspects by taking advantage of the large π conjugation system and high carrier mobility of graphdiyne. To develop the growth method of graphdiyne on arbitrary substrates will open up a new way for the structural characterization and application of graphdiyne. In terms of devices, the utilization of graphdiyne/Cu and graphdiyne/indium tin oxides (ITOs) substrates as anode of battery devices, multistate memory devices based on graphdiyne nanoparticles, and wide-spectrum photodetectors with high-sensitivity based on graphdiyne nanoparticles and thin films is worthwhile.

Index

Graphdiyne: Fundamentals and Applications in Renewable Energy and Electronics, First Edition. Edited by Yuliang Li.

f

g

www.ingramcontent.com/pod-product-compliance
Ingram Content Group UK Ltd.
Pitfield, Milton Keynes, MK11 3LW, UK
UKHW051034091026
15624UKWH00011B/434

9 783527 347872